RING THEORY

RING THEORY

Proceedings of the Biennial
Ohio State–Denison Conference
May 1992

Editors

S. K. Jain
Ohio University

S. Tariq Rizvi
Ohio State University

World Scientific
Singapore • New Jersey • London • Hong Kong

Published by

World Scientific Publishing Co. Pte. Ltd.
5 Toh Tuck Link, Singapore 596224
USA office: 27 Warren Street, Suite 401-402, Hackensack, NJ 07601
UK office: 57 Shelton Street, Covent Garden, London WC2H 9HE

British Library Cataloguing-in-Publication Data
A catalogue record for this book is available from the British Library.

RING THEORY
Proceedings of the Biennial Ohio State-Denison Conference 1992

ISBN-13 978-981-02-1385-5
ISBN-10 981-02-1385-9

Dedicated

to the Memory of

Hans J. Zassenhaus

PREFACE

This volume is dedicated to the memory of Professor Hans J. Zassenhaus, who passed away on November 21, 1991, at the age of 79. Professor Zassenhaus founded the Ohio State-Denison Mathematics Conference more than twenty-five years ago and continued over all these years as their chief organizer and the unifying spirit behind the various fields of algebra represented. The Midwest community of algebraists will fondly remember his many contributions to the subject, his generous guidance to young colleagues, and above all, his kind and warm personality.

The volume consists of a collection of invited refereed papers on the theory of rings and modules mostly by participants at the 21st biennial Ohio State-Denison Mathematics Conference. The Conference was held at the beautiful campus of Denison University, Granville, Ohio, May 14-16, 1992. The topics of the papers represent many modern trends in the theory of rings and modules. The wide variety of methodologies and techniques will provide valuable tools, in particular to young researchers in the area. Covering a broad range of topics, this book should appeal to a wide spectrum of researchers in Algebra and Number Theory. It is our belief that there are strong relations among various branches of ring theory. The present volume emphasizes this interaction.

Our thanks go to all participants, contributors and referees for their participation and assistance in bringing out this volume within tight time limits. The conference was sponsored by the Mathematics Research Institute and Department of Mathematics of the Ohio State University, both at the Columbus and the Lima campuses. Many thanks are due to Dr. Violet I. Meek, Dean and Director, and Dr. Phillip Heath, Associate Dean at the Ohio State University at Lima, for their continued support in terms of financial assistance. Their constant encouragement in bringing this volume to its completion is greatly appreciated. Our thanks also go to a number of our colleagues who spent a great deal of their time on various arrangements and the logistics for the conference – in particular the initiative taken by Professors Thomas Dowling and Surinder Sehgal was truly responsible for the success of the Conference. We are also grateful to the Department of Mathematics at the Denison University for providing us with the venue and all facilities during the conference. The friendly assistance provided by Professor Bruno J. Müller during the final stages of this volume is also greatly appreciated. Finally, Ms. Cindy White is to be commended for an excellent job of typing.

S.K. Jain, Athens, Ohio
S. Tariq Rizvi, Lima, Ohio
May 1993

ACKNOWLEDGEMENT

We would like to thank the following who have kindly helped us in refereeing the papers promptly:

M. Aunslander, G. Azumaya, H.H. Brungs, V. Camillo, K.R. Goodearl, T. Hodges, S.K. Jain, K.A. Kosler, T.Y. Lam, M.S. Li, E. Letzter, S.R. López-Permouth, P. Loustaunau, B.J. Müller, K. Oshiro, B. Osofsky, D.S. Passman, S.T. Rizvi, G. Szeto, M.L. Teply, M.F. Yousif and J.M. Zelmanowitz.

CONTENTS

PRIMENESS CONDITIONS FOR GROUP GRADED RINGS[1]

GENE ABRAMS and JEREMY HAEFNER
Department of Mathematics
University of Colorado, Colorado Spring, CO 80933

Abstract. We describe a general procedure by which a graded property of a group-graded ring may be shown to imply the corresponding ungraded property. The technique, originally used in the context of finite groups, involves extending the smash product construction of Cohen and Montgomery to infinite groups, and utilizing rings with local units. Specifically, we obtain new results relating the primeness and graded-primeness of group-graded rings.

Introduction.

Many authors (e.g. [2], [5], [8], [10], [11], [13], [16], [17], [23]) have considered questions of the following type: if R is a ring graded by the group G, what types of graded conditions are sufficient to yield the corresponding ungraded conditions for R? Examples of such conditions include: semisimple, Noetherian, prime, and Jacobson. In this article we describe an approach which is applicable to many questions of this type, including the four conditions described above. Using this approach, we show how information about rings which are strongly graded by infinite groups may be parlayed to yield information about more general group-gradings. In particular, we obtain via this method the main result of this article (Theorem 3.2):

> *Let R be graded by the torsion-free group G. Then R is prime if and only if R is graded prime.*

This result had previously been known only in the case where R is assumed to be strongly graded by the torsion-free group G.

Our method involves a consideration of the smash product ring of R with G (denoted by $R\#G$), which was first utilized for rings graded by finite groups by Cohen and Montgomery in [8], extended to infinite groups by Quinn in [25], and modified in the infinite case to a nonunital ring with an eye towards categorical properties by Beattie in [3]. A good reference for the smash product is [23].

---

[1] 1991 Mathematics Subject Classification: 16S35, 16W50.

An outline of the method is given here. We transfer information between R and $R\#G$ via two different avenues. The first way requires a two-step procedure, described explicitly in Section 1. Specifically, we pass ungraded properties of R directly to the Morita equivalent ring $R^f\{G\}$; then, as $R^f\{G\}$ is a skew group ring over $R\#G$, we are able to pass certain types of information from $R^f\{G\}$ to $R\#G$. For the second way, we transfer information about graded properties of R from R to $R\#G$ by direct multiplicative arguments; this is described in Section 2. The analysis is completed by showing that the *ungraded* properties of R produce in $R\#G$ via the first avenue the same structural properties as those produced by the *graded* properties of R via the second avenue.

We note that this method has been used by other authors in somewhat different venues. For instance, Cohen and Montgomery [8] utilize this technique to show (among other things) that if R is graded by the finite group G, then the graded Jacobson radical $J_G(R)$ is contained in the Jacobson radical $J(R)$. Quinn [25] uses this method to investigate primeness and semiprimeness conditions for certain types of graded rings. Specifically, when R is graded by the finite group G, he shows that semiprimeness is inherited by certain subrings of R, and that the semiprimeness of R is equivalent to weak semiprimeness and left nondegeneracy. In addition, when G is infinite and R is nicely graded by G ("component regular"), he obtains equivalent conditions for the non-primeness or non-semiprimeness of R in terms of the subgroup structure of G and multiplicative structure of R. Finally, Beattie [3] uses some of these techniques to show that if R is graded by any group G, then $J_G(R)$ is the largest homogeneous ideal J of R having the property that $J \cap R_1 = J(R_1)$.

There are two main distinctions between our investigation and those mentioned in the previous paragraph. First, we obtain a new result about the primeness of rings graded by torsion free groups in terms of graded primeness (Theorem 3.2). This result differs from the aforementioned theorem of Quinn, as it does not involve an analysis of the subgroups of G. (Additionally, we do not require any special type of grading on the underlying ring.)

Second, our approach may be viewed as module-theoretic in nature, in contrast to the ring-theoretic approaches described above. One of the advantages of this approach is that the ideal structures of R and the skew group ring $R^f\{G\}$ are identical (cf. Lemma 1.1 and Corollary 1.2), as these rings are Morita equivalent. In contrast, the skew group ring $M_G(R)$ investigated by Quinn for infinite G is not Morita equivalent to R.

In Section 1 we present the definitions of the main tools used in our discussion. In particular, we describe rings with local units, and give information regarding the ideal structure of two such Morita equivalent rings. We define the rings $R\#G$ and $R^f\{G\}$ mentioned above; when G is infinite these rings are not unital, but do have local units. We show that $R^f\{G\}$ behaves nicely when viewed as a skew group ring over $R\#G$. These discussions are then applied to obtain information relating prime ideals and graded prime ideals in group graded rings. In Section 2 we analyze the G-stable ideals of $R\#G$ inside the skew group ring $R^f\{G\}$; in particular, we show (Theorem 2.3(4)) that all such ideals arise naturally as smash products of graded ideals in R with G. In Section 3 we show how the results of Sections 1 and 2 may be utilized to obtain the aforementioned theorem. In addition, we obtain a strengthened correspondence between prime ideals and graded prime ideals in this torsion-free setting. We conclude the article by mentioning other venues in which these techniques may be applied.

Throughout this article all rings are associative; however, we do not in general assume the existence of a multiplicative identity. Following [1], the ring R is said to have **local units** in case there exists a set E of idempotents in R such that $R = \bigcup_{e \in E} eRe$. If R has multiplicative identity 1_R, then $\{1_R\}$ is a set of local units for R. An ideal P of the ring R is **prime** in case for each pair I and J of two-sided ideals of R, if $IJ \subset P$ then $I \subset P$ or $J \subset P$. An ideal Q of R is **semiprime** in case for any two-sided ideal K of R, if $K^2 \subset Q$ then $K \subset Q$. The ring R is called **prime** (resp., **semiprime**) in case the zero ideal of R is prime (resp., semiprime). For additional information on terms used here, see [1, 3, 8] and [25].

1 Passing ungraded properties from R to $R\#G$ via $R^f\{G\}$.

We show in this section how (ungraded) properties of R may be passed from R to $R\#G$ via the "first avenue" described in the introduction.

Suppose that R is a unital ring and G is any set. We denote by $RFM_G(R)$ the ring of row-finite matrices over R whose rows and columns are indexed by the elements of G; note that $RFM_G(R)$ is unital for R unital. We define the subset $R^f\{G\}$ of $RFM_G(R)$ to be

$$R^f\{G\} = \{\alpha \in RFM(R) \mid \alpha_{x,y} \neq 0 \ \text{ for at most finitely many } \ x, y \in G\}.$$

If we denote by $e(x,y)$ the usual matrix unit, then $e(x,y) \in R^f\{G\}$ as R is unital; it is easily seen that $R^f\{G\}$ is a ring with local units, where the collection of idempotents can be taken as the set

$$E = \{\sum_{x \in S} e(x,x) \mid S \ \text{ is a finite subset of } G\}.$$

We note that the ring $R^f\{G\}$ equals the ideal $M_G^\bullet(R)$ defined in [25]. For any two-sided ideal I of R we define the set

$$I^f\{G\} = \{\alpha \in R^f\{G\} \mid \alpha_{x,y} \in I \ \text{ for all } x, y \in G\}.$$

A straightforward check shows that $I^f\{G\}$ is a two-sided ideal of $R^f\{G\}$.

Lemma 1.1 *There exists a lattice isomorphism Γ between the lattice of two-sided R-ideals and the lattice of two-sided $R^f\{G\}$-ideals. In addition, $\Gamma(IJ) = \Gamma(I)\Gamma(J)$ for any two-sided ideals I and J of R.*

Proof. It is easy to show that the map $\Gamma : I \mapsto I^f\{G\}$ satisfies the stated properties. We note that the existence of a lattice isomorphism of the type described here is ensured by the fact that R and $R^f\{G\}$ are Morita equivalent; see [1, Section 3].□

In particular, we get immediately:

Corollary 1.2 *Let R be any unital ring, and let G be any set. Then*

1. *I is a prime ideal of R if and only if $I^f\{G\}$ is a prime ideal of $R^f\{G\}$.*
2. *Each prime ideal of $R^f\{G\}$ is of the form $P^f\{G\}$ for some prime ideal P of R.*
3. *R is prime (resp., semiprime) if and only if $R^f\{G\}$ is prime (resp., semiprime).*

Now suppose that R is **graded** by the group G. That is, $R = \coprod_{x\in G} R_x$ as abelian groups, and $R_x R_y \subset R_{xy}$ for all x, y in G. In this setting we denote by $R\#G$ the subset of $R^f\{G\}$ defined via

$$R\#G = \{\alpha \in R^f\{G\} \mid \alpha_{x,y} \in R_{xy^{-1}} \quad \text{for all } x, y \in G\}.$$

We call $R\#G$ the **smash product** of R with G. One easily checks that $R\#G$ is a subring of $R^f\{G\}$. If R is unital, then $R\#G$ contains the same set of local units E defined above (as it is easily shown that $1_R \in R_1$ where 1 denotes the identity element of the group G). The ring $R\#G$ is isomorphic to the ring $R\#G^*$ defined in [3]. When G is finite, $R\#G$ is the usual smash product ring defined originally in [8] and [25]. Note that by definition we have $R\#G \subset R^f\{G\} \subset RFM_G(R)$. We embed G inside $RFM_G(R)$ as follows: for each $g \in G$, set $\overline{g} = \sum_{x\in G} e(x, xg)$. Letting $\overline{G}$ denote $\{\overline{g} \mid g \in G\}$, we get $\overline{G} \subset RFM_G(R)$, and $G \cong \overline{G}$.

If G is a group which acts as automorphisms on the ring B, then we use the notation x^g to denote the action of $g \in G$ on $x \in B$. We denote by $B * G$ the **skew group ring** of G over B; formally, $B * G$ is the left B-module $\coprod_{g\in G} Bg$, with multiplication defined to be the linear extension of $(ag) \cdot (bh) = ab^{g^{-1}} gh$.

Lemma 1.3 *(cf. [25, Lemma 1.2 and 2.2] and [4]) Let R be a unital ring graded by the group G. Then*

1. *$R^f\{G\} = \coprod_{g\in G}(R\#G)\overline{g}$ as a direct sum of abelian groups.*

2. $\overline{g}^{-1}(R\#G)\overline{g} = R\#G$ *for all* $g \in G$. *In particular,* G ***acts as automorphisms on*** $R\#G$ *via* $x^g = \overline{g}^{-1}x\overline{g}$.
3. *Using the action of* $\overline{G}$ *on* $R\#G$ *given in (2),* $R^f\{G\}$ *is a skew group ring of the group* $\overline{G}$ *over the ring* $R\#G$*; that is,* $R^f\{G\} = (R\#G) * \overline{G}$. *Specifically,* $R^f\{G\}_1 = R\#G$.

Proof. The proof is virtually identical to that given in [25], so is omitted here.

For the remainder of this section we will describe relationships between $R\#G$ and $R^f\{G\}$ (with an eye towards primeness conditions), using information given in the above lemma. This type of analysis has appeared before in the literature, for instance in [8] and [25]. The difference between these existing arguments and the arguments presented here is that $R\#G$ need not have an identity, but does contain a set of local units. So, for example, part (3) of the above lemma is not a standard type of statement (since skew group rings are often required to contain multiplicative identities), but the demonstration of its validity is identical to the argument given in [25]. Some additional care is required in this more general setting; however, such care typically requires only a modicum of additional work.

If I is any two-sided ideal of the group-graded ring R, then I is called a **graded** ideal in case $I = \coprod_{g\in G}(I \cap R_g)$. More generally, for any two-sided ideal I of the group-graded ring R we denote by I_{gr} the largest graded ideal contained in I; that is,

$$I_{gr} = \coprod_{g\in G}(I \cap R_g) = \coprod_{g\in G} I_g \subset I.$$

In particular, I is graded exactly when $I = I_{gr}$. If I is a graded (two-sided) ideal of R, then we denote by $I\#G$ the smash product of I with G. Formally, $I\#G$ is the following set of elements from $R^f\{G\}$:

$$\begin{aligned} I\#G &= \left\{\alpha \in R^f\{G\} \mid \alpha_{x,y} \in I_{xy^{-1}} \text{ for all } x, y \in G\right\} \\ &= \left\{\sum_{x,y\in G} \alpha_{x,y}e(x,y) \mid \alpha_{x,y} \in I_{xy^{-1}} \text{ for all } x, y \in G\right\}. \end{aligned}$$

Finally, an ideal X of $R\#G$ is said to be G-**stable** if $X^g = X$ for each $g \in G$. It is easy to show that if I is a graded ideal of R, then $X = I\#G$ is a G-stable ideal of $R\#G$. Furthermore, if X is a G-stable ideal of $R\#G$, then $X * \overline{G}$ is an ideal of $(R\#G)*\overline{G} = R^f\{G\}$.

Theorem 1.4 *Let* I *be a (two-sided) ideal of* R. *Then:*
1. $I^f\{G\} \cap R\#G = I_{gr}\#G$. *In particular,* $I^f\{G\} \cap R\#G$ *is* G-*stable.*
2. *If* I *is graded, then* $I^f\{G\} = (I\#G) * \overline{G} = (I^f\{G\} \cap R\#G) * \overline{G}$.

3. $(I^f\{G\})_{gr} = (I_{gr}\#G) * \overline{G}$.
4. *If J is a graded ideal of $R^f\{G\}$, then there exists a graded ideal $\hat{J}$ of R such that*

$$J = (\hat{J}\#G) * \overline{G} = \hat{J}^f\{G\}.$$

Proof. (1) From the definition of $I_{gr}\#G$, we have

$$\begin{aligned} I_{gr}\#G &= \{\alpha \in R^f\{G\} \mid \alpha = \sum_{x,y\in G} \alpha_{x,y} e(x,y) \text{ where } \alpha_{x,y} \in R_{xy^{-1}} \cap I\} \\ &= I^f\{G\} \cap R\#G. \end{aligned}$$

(2) The second equality follows from (1) above. To prove the first equality, let

$$Y = I^f\{G\} = \{\alpha \in R^f\{G\} \mid \alpha_{x,y} \in I\}$$

and let

$$X = (I\#G) * \overline{G} = \coprod_{g\in G} (I\#G)\overline{g}.$$

If $\beta \in R^f\{G\}$ and $g \in G$, we have

$$\begin{aligned} \beta\overline{g} &= \Big(\sum_{x,y\in G} \beta_{x,y} e(x,y)\Big)\Big(\sum_{z\in G} e(z, zg)\Big) \\ &= \sum_{x,y\in G} \beta_{x,y} e(x, yg). \end{aligned}$$

So if $\alpha = \sum_{x,y\in G} \alpha_{x,y} e(x,y) \in I\#G$, where $\alpha_{x,y} \in I_{xy^{-1}}$, then

$$\alpha\overline{g} = \sum_{x,y\in G} \alpha_{x,y} e(x, yg) \in Y.$$

Hence, $X \subset Y$.

For the other containment, it suffices to show that $\alpha e(x,y) \in X$ where $\alpha \in I$. But since I is graded, $\alpha = \sum_{z\in G} \alpha_z$ where $\alpha_z \in I_z$. Clearly, we can write $\alpha = \sum_{z\in G} \alpha_{xzy^{-1}}$ where $\alpha_{xzy^{-1}} \in I_{xzy^{-1}}$. Consequently,

$$\alpha e(x,y) = \sum_{z\in G} \alpha_{xzy^{-1}} e(x, yz^{-1})\overline{z} \in \coprod_{g\in G} (I\#G)\overline{g} = X.$$

(3) By definition and an application of (1),

$$(I^f\{G\})_{gr} = \coprod_{g\in G} I^f\{G\} \cap R^f\{G\}_g$$

$$
\begin{aligned}
&= \coprod_{g\in G} I^f\{G\} \cap (R\#G)\overline{g} \\
&= \coprod_{g\in G} (I^f\{G\} \cap (R\#G))\overline{g} \\
&= \coprod_{g\in G} (I_{gr}\#G)\overline{g}
\end{aligned}
$$

where a straightforward calculation yields $I^f\{G\} \cap (R\#G)\overline{g} = (I^f\{G\} \cap R\#G)\overline{g}$.

(4) This follows from Lemma 1.1 and (1) - (3) above.□

If R is a ring graded by the group G, then a (two-sided) R-ideal P is **graded prime** provided P is graded, and whenever $IJ \subset P$ for graded ideals I and J, then either $I \subset P$ or $J \subset P$. We say that R is **graded prime** if the zero ideal is graded prime.

Analogously, a (two-sided) R-ideal I is **graded semiprime** provided I is graded, and whenever there is a graded ideal J of R such that $J^2 \subset I$, then $J \subset I$. We say that R is **graded semiprime** if the zero ideal is graded semiprime.

For any ring R, the **prime radical** of R is defined to be the intersection of all the prime ideals of R; that is,

$$P(R) = \cap\{P \mid P \text{ is a prime ideal of } R\}.$$

Similarly, for any ring R graded by the group G, the **graded prime radical** of R is defined to be the intersection of all the graded prime ideals of R; that is,

$$P_{gr}(R) = \cap\{P \mid P \text{ is a graded prime of } R\}.$$

We now turn our attention to the connections between the prime ideals and the graded prime ideals of R. We begin by stating a standard result about the "lying over" property of prime ideals for group graded rings. This result appears in [8, Lemma 5.1] for finite groups; the proof given there in fact holds verbatim for arbitrary groups, hence is omitted.

Lemma 1.5 *Let R be graded by the group G. Then*

1. *If P is a prime ideal of R, then P_{gr} is a graded prime ideal of R.*
2. *If Q is a graded prime ideal of R, then there is a prime ideal of R such that $P_{gr} = Q$.*

Proposition 1.6 *Let R be graded by the group G. Then $P_{gr}(R) \subset P(R)$. Moreover, $P_{gr}(R)$ is the graded ideal associated to $P(R)$; that is, $P_{gr}(R) = (P(R))_{gr}$.*

Proof. From the above lemma we have

$$P_{gr}(R) = \bigcap_{P \text{ graded prime}} P = \bigcap_{Q \text{ prime in } R} Q_{gr} \subset \bigcap_{Q \text{ prime in } R} Q = P(R),$$

which yields the first statement. For the second statement, of course we have $P_{gr}(R)$ is graded, so that $P_{gr}(R) \subset (P(R))_{gr}$. For the other containment, observe that for each $g \in G$

$$P(R) \cap R_g \subset \bigcap_{P \text{ prime in } R} (P \cap R_g).$$

Consequently, we have

$$\begin{aligned}
(P(R))_{gr} &= \coprod_{g\in G} P(R) \cap R_g \\
&\subset \coprod_{g\in G} \left[\bigcap_{P \text{ prime in } R} P \cap R_g \right] \\
&\subset \bigcap_{P \text{ prime in } R} \left[\coprod_{g\in G} P \cap R_g \right] \\
&= \bigcap_{P \text{ prime in } R} P_{gr} \\
&= P_{gr}(R).
\end{aligned}$$

This proves the proposition.□

2 Passing graded properties from R to $R\#G$ directly.

In this section we show how information is transferred via the "second avenue" mentioned in the introduction; specifically, we show how certain multiplicative properties of $R\#G$ arise from the graded structure of R. We show in Theorem 2.3 below that there is an ideal lattice bijection between the graded ideals of R and the G-stable ideals of $R\#G$; this bijection enables us to characterize graded prime ideals of R in terms of the G-prime ideals of $R\#G$ (Proposition 2.5). Throughout this section R is a unital ring graded by the group G, and G acts as automorphisms on $R\#G$ as described in Lemma 1.3. The following result was also obtained by Cohen and Montgomery (for finite groups) in [8] as well as by Beattie and Stewart in [6] using their construction of $R\#G$.

Theorem 2.1 *The G-stable (two-sided) ideals of $R\#G$ are precisely the ideals $I\#G$ where I is a graded (two-sided) ideal of R.*

Proof. We first show that every ideal of the form $I\#G$ is G-stable, for I a graded R-ideal. It suffices to show that $(I\#G)^g \subset I\#G$ for every element g in G. Let $\alpha \in I\#G$ so that

$$\alpha = \sum_{x,y\in G} \alpha_{x,y} e(x,y)$$

where $\alpha_{x,y} \in I_{xy^{-1}}$. Thus,

$$\begin{aligned} \alpha^g &= \bar{g}^{-1}\alpha\bar{g} \\ &= \left(\sum_{z\in G} e(z, zg^{-1})\right)\left(\sum_{x,y\in G} \alpha_{x,y}e(x,y)\right)\left(\sum_{u\in G} e(u,ug)\right) \\ &= \sum_{x,y\in G} \alpha_{x,y}e(xg,yg) \\ &\in I\#G, \end{aligned}$$

since $xg(yg)^{-1} = xgg^{-1}y^{-1} = xy^{-1}$. Note also that α^g is again a finite matrix because a row-and-column-finite matrix times a finite matrix yields a finite matrix.

Now we show that every G-stable ideal of $R\#G$ is of the form $I\#G$ where I is a graded R-ideal.

Let H be a two-sided G-stable ideal of B. Because the set $\{e(g,g) \mid g \in G\}$ is contained in $R\#G$ and because the elements of H are matrices with only finitely many nonzero entries, we can write H as a direct sum of abelian groups:

$$H = \coprod_{g,k\in G} e(k,k)He(g,g).$$

Let $H_{k,g}$ be the set of elements from R that appear in the (k,g)-entry of some element of H. Thus, $e(k,k)He(g,g) = H_{k,g}e(k,g)$. We claim that $H_{k,g} = H_{kx,gx}$ for all group elements $g,k,x \in G$.

For any $b = \sum_{x,y\in G} \beta_{x,y}e(x,y) \in R^f\{G\}$ and any $g \in G$, we have

$$b^g = \sum_{x,y\in G} \beta_{x,y}e(xg,yg).$$

Consequently, if $\alpha_{k,g}e(k,g) \in e(k,k)He(g,g)$, then $(\alpha_{k,g}e(k,g))^x = \alpha_{k,g}e(kx,gx)$ must be an element of $e(kx,kx)He(gx,gx)$ since H is G-stable. This shows that

$$H_{k,g} \subset H_{kx,gx}.$$

Of course, if $\alpha_{kx,gx}e(kx,gx) \in e(kx,kx)He(gx,gx)$, then

$$(\alpha_{kx,gx}e(kx,gx))^{x^{-1}} = \alpha_{kx,gx}e(k,g) \in e(k,k)He(g,g)$$

which shows that

$$H_{kx,gx} \subset H_{k,g}.$$

Thus, $H_{k,g} = H_{kx,gx}$ for all $g, x, k \in G$. In particular, setting $x = g^{-1}$, we have

$$H_{k,g} = H_{kg^{-1},1} \tag{$*$}$$

for all k and $g \in G$.

Now set $I = \coprod_{k\in G} H_{k,1}$. This is a graded R-ideal because $H_{k,1} \subset R_k$. We construct

$$H' = I\#G = \left\{ \sum_{x,y\in G} \alpha_{x,y}e(x,y) \mid \alpha_{x,y} \in I_{xy^{-1}} = H_{xy^{-1},1} \right\}$$

and show $H' = H$. Notice that $H'e(1,1) = I$. Clearly, it suffices to show that $H_{k,g} = H'_{k,g}$ for every k and $g \in G$. But $H_{k,g} = H_{kg^{-1},1}$ by $(*)$ above. By the first part of this proof, H' is G-stable, so we also have $H'_{k,g} = H'_{kg^{-1},1}$. Yet it is clear that

$$H'_{kg^{-1},1} = H_{kg^{-1},1}.$$

This finishes the proof.□

The map that sends a graded ideal I of R to the G-stable ideal $I\#G$ of $R\#G$ is an ideal lattice bijection (Theorem 2.3). To prove this, we need the following lemma.

Lemma 2.2 *If I and J are two-sided, graded ideals of R such that $I = \coprod_{g\in G} I_g$ and $J = \coprod_{h\in G} J_h$, then*

$$IJ = \coprod_{k\in G} \left(\sum_{g\in G} I_g J_{g^{-1}k} \right).$$

Proof. Set $T = \coprod_{k\in G} \left(\sum_{g\in G} I_g J_{g^{-1}k} \right)$. To show that $IJ \subset T$, it suffices to let $\alpha \in I_g$ and $\beta \in J_h$ and show that $\alpha\beta \in T$ for any group elements $g, h \in G$. But

$$\alpha\beta \in I_g J_h \subset \sum_{k\in G} I_k J_{k^{-1}gh} \subset T.$$

For the other direction, it suffices to show that, for a fixed $k \in G$,

$$\sum_{g \in G} I_g J_{g^{-1}k} \subset IJ.$$

In fact, we have

$$\sum_{g \in G} I_g J_{g^{-1}k} \subset IJ \cap R_k$$

from the definition of IJ.□

Theorem 2.3 *Let I and J be two-sided, graded ideals of R.*

1. *$(I\#G) \subset (J\#G)$ if and only if $I \subset J$.*
2. *$(I\#G) = (J\#G)$ if and only if $I = J$.*
3. *$(I\#G)(J\#G) = (IJ\#G)$.*
4. *There is an inclusion-preserving, multiplication-preserving ideal lattice bijection between the graded R-ideals, I, and the G-stable $R\#G$-ideals, H, given by*

$$I \longleftrightarrow H = I\#G \qquad \text{and} \qquad H \longleftrightarrow I = He(1,1).$$

Proof. (1) This follows directly from the definition of $I\#G$; that is,

$$I\#G = \left\{ \sum_{x,y \in G} \alpha_{x,y} e(x,y) \mid \alpha_{x,y} \in I_{xy^{-1}} \right\}.$$

(2) This follows from (1).

(3) We first show that $(I\#G)(J\#G) \subset (IJ\#G)$. Let

$$\sum_{x,y \in G} \alpha_{x,y} e(x,y) \in I\#G \text{ and } \sum_{u,v \in G} \beta_{u,v} e(u,v) \in J\#G$$

where $\alpha_{x,y} \in I_{xy^{-1}}$ and $\beta_{u,v} \in J_{uv^{-1}}$. Then

$$\left(\sum_{x,y \in G} \alpha_{x,y} e(x,y) \right) \left(\sum_{u,v \in G} \beta_{u,v} e(u,v) \right) = \sum_{x,y,v \in G} \alpha_{x,y} \beta_{y,v} e(x,v)$$

where

$$\alpha_{x,y} \beta_{y,v} \in I_{xy^{-1}} J_{yv^{-1}} \subset R_{xv^{-1}}.$$

Thus, it suffices to show that

$$\alpha_{x,y} \beta_{y,v} \in (IJ)_{xv^{-1}}.$$

However,

$$I_{xy^{-1}}J_{yv^{-1}} \subset \sum_{g\in G} I_g J_{g^{-1}xv^{-1}} = (IJ)_{xv^{-1}}$$

from the lemma above.

Next we show that $(IJ\#G) \subset (I\#G)(J\#G)$. Here it suffices to show $\alpha_{x,y}e(x,y) \in (I\#G)(J\#G)$ where $\alpha_{x,y} \in (IJ)(xy^{-1})$. Now

$$\alpha_{x,y} \in (IJ)_{xy^{-1}} = \sum_{g\in G} I_g J_{g^{-1}xy^{-1}} = \sum_{s\in G} I_{xs} J_{s^{-1}y^{-1}}.$$

So

$$\begin{aligned}\alpha_{x,y} &= \sum_{s\in G}\left(\sum_{k\in G}(a_{xs})_k(b_{s^{-1}y^{-1}})_k\right)\\ &= \sum_{k\in G}\left(\sum_{s\in G}(a_{xs})_k(b_{s^{-1}y^{-1}})_k\right)\end{aligned}$$

where $(a_{xs})_k \in I_{xs}$ and $(b_{s^{-1}y^{-1}})_k \in J_{s^{-1}y^{-1}}$. Setting

$$\mathcal{A}_k = \sum_{s\in G}(a_{xs})_k e(x,s^{-1}) \in I\#G$$

and

$$\mathcal{B}_k = \sum_{t\in G}(b_{t^{-1}y^{-1}})_k e(t^{-1},y) \in J\#G,$$

we have that

$$\begin{aligned}\mathcal{A}_k\mathcal{B}_k &= \left(\sum_{s\in G}(a_{xs})_k e(x,s^{-1})\right)\left(\sum_{t\in G}(b_{t^{-1}y^{-1}})_k e(t^{-1},y)\right)\\ &= \sum_{s\in G}(a_{xs})_k(b_{s^{-1}y^{-1}})_k e(x,y).\end{aligned}$$

Thus,

$$\alpha_{x,y}e(x,y) = \sum_{k\in G}\mathcal{A}_k\mathcal{B}_k \in (I\#G)(J\#G).$$

This completes the proof of (3).

(4) This now follows from (1), (2), (3) and Theorem 2.1. Notice that the graded ideal I such that $H = I\#G$ is precisely $He(1,1)$ as seen in the proof of Theorem 2.1.□

Definition 2.4 An ideal Q of the smash product $R\#G$ is called **G-prime** if Q is G-stable, and whenever $IJ \subset Q$ for G-stable ideals I and J of $R\#G$, then either $I \subset Q$ or $J \subset Q$. We call $R\#G$ **G-prime** if the zero ideal is G-prime.

Analogously, an ideal J of $R\#G$ is **G-semiprime** if J is G-stable, and whenever there is a G-stable ideal K of $R\#G$ such that $K^2 \subset J$, then $K \subset J$. We call $R\#G$ **G-semiprime** if the zero ideal is G-semiprime.

We now show that Theorem 2.3 allows us to establish a correspondence between the graded prime ideals of R and the G-prime ideals of $R\#G$.

Proposition 2.5 *Let R be a ring graded by the group G. Let P be a G-stable ideal of $R\#G$ and let Q be a graded ideal of R such that $Q\#G = P$. Then*

1. *P is G-prime if and only if Q is graded prime.*
2. *P is G-semiprime if and only if Q is graded semiprime.*

Proof. (1) ($\Rightarrow$) Assume P is G-prime and suppose that $IJ \subset Q$ where I and J are graded ideals of R. But by the above theorem, we have

$$(I\#G)(J\#G) = IJ\#G \subset Q\#G = P$$

where $I\#G$ and $J\#G$ are G-stable ideals of $R\#G$. Since P is G-prime, then either $I\#G \subset P$ or $J\#G \subset P$. So we have either $I \subset Q$ or $J \subset Q$, whence Q is graded prime.

($\Leftarrow$) Assume Q is graded prime and suppose that I and J are G-stable ideals such that $IJ \subset P = Q\#G$. By the lattice isomorphism given in 2.3, there are graded R-ideals $\hat{I}$ and $\hat{J}$ such that $I = \hat{I}\#G$ and $J = \hat{J}\#G$. This gives $\hat{I}\hat{J} \subset Q$; and since Q is graded prime, either $\hat{I} \subset Q$ or $\hat{J} \subset Q$. So we have either $I \subset P$ or $J \subset P$ and hence P is G-prime.

(2) This is an argument similar to that given in (1) above.□

Corollary 2.6 *A ring R graded by the group G is graded prime (respectively, graded semiprime) if and only if the ring $R\#G$ is G-prime (respectively, G-semiprime).*

We note, in comparison to Corollary 2.6, that Cohen and Montgomery prove a stronger result for finite groups. Namely, if G is a finite group and if R is graded by G such that R is graded semiprime, then $R\#G$ is, in fact, semiprime. See [8, Theorem 2.9], [23, Theorem 4.8], and [5, Theorem 3.4] for further explanation. The crucial observation is that if G is finite, then every nilpotent ideal of $R\#G$ is contained in a nilpotent G-stable ideal; specifically, if G is finite and I is a nilpotent ideal of $R\#G$, then the finite sum $J = \sum_{g\in G} I^g$ is a nilpotent

G-stable ideal of $R\#G$ having $I \subset J$. This result is false in the infinite case; see [6, Example 2.4].

The G-prime radical of $R\#G$ is defined next. Corollary 2.8 shows that the G-prime radical of $R\#G$ equals the smash product of the graded prime radical of R with G.

As one might expect, R is graded semiprime if and only if the graded prime radical of R is zero, and $R\#G$ is G-semiprime if and only if the G-prime radical is zero. We mention that these results use a version of Zorn's lemma, which can be readily adapted to rings with local units (such as $R\#G$). We leave the proofs of these results to the reader.

Definition 2.7 If S is a ring on which G acts as automorphisms, then the **G-prime radical of** S is the intersection of all the G-prime ideals of S; that is,

$$P_G(S) = \cap\{P \mid P \text{ is a } G\text{-prime ideal of } S\}.$$

The following corollary is an extension to all groups of the result for finite groups obtained by Cohen and Montgomery in [8, Theorem 5.3].

Corollary 2.8 *If R is a ring graded by an arbitrary group G, then*

1. *$P_G(R\#G)$ is a G-stable ideal of $R\#G$.*
2. *$P_{gr}(R)$ is a graded ideal of R.*
3. *$P_G(R\#G) = P_{gr}(R)\#G$.*

Proof. (1) Any intersection of G-stable ideals is again G-stable.

(2) This follows from the definition of $P_{gr}(R)$.

(3) By (1) we know that $P_G(R\#G)$ is G-stable. So by Theorem 2.1, there is a graded R-ideal I such that $I\#G = P_G(R\#G)$. If P is a graded prime ideal of R, then $P\#G$ is a G-prime ideal of $R\#G$ and so $I\#G \subset P\#G$. Consequently, by Theorem 2.3, $I \subset P$ for all graded prime ideals P of R, so $I \subset P_{gr}(R)$.

On the other hand, since $P_{gr}(R)$ is graded, then $P_{gr}(R)\#G$ is a G-stable ideal of $R\#G$ which by 2.5 is contained in every G-prime ideal. Thus, $P_{gr}(R)\#G \subset P_G(R\#G) = I\#G$ and so we have that $P_{gr}(R) \subset I$. This gives $P_{gr}(R) = I$, so that $P_G(R\#G) = I\#G = P_{gr}(R)\#G$.□

We close this section by briefly considering strongly group-graded rings. Recall that a ring R is **strongly graded** by the group G if $R_\sigma R_\tau = R_{\sigma\tau}$ for all $\sigma, \tau \in G$. In particular, skew group rings are strongly graded. It is well-known (see [12, Theorem 2.8]) that if R is strongly graded by G, then the category R-gr of graded left R-modules is equivalent to the category R_1-mod of (all) left R_1-modules. Further, it is easy to verify that the functor

$$e(1,1)R\#G \otimes_{R\#G} (--) : R\#G\text{-mod} \to R_1\text{-mod}$$

gives a Morita equivalence. Thus in the strongly graded case we have that the three categories R-gr, R_1-mod and $R\#G$-mod are equivalent.

As a result, it is not surprising that, when R is strongly graded, properties of R_1-mod translate to corresponding properties of R-gr as well as corresponding properties of $R\#G$-mod. In particular, we describe how graded primeness of R is equivalent to G-primeness of R_1.

We say that an ideal I of R_1 is **G-invariant** if $R_{g^{-1}}IR_g = I$ for all $g \in G$. By [17, Corollary I.3.29], there exists a bijection between the set of graded ideals of R and the set of G-invariant ideals of R_1. By defining R_1 to be G-prime and G-semiprime in the obvious way (i.e., for G-invariant ideals) and applying Corollary 2.6, we have the following result:

Corollary 2.9 *Let R be a ring which is strongly graded by the group G. Then the following statements are equivalent:*

1. *R_1 is G-prime (resp., G-semiprime).*
2. *$R\#G$ is G-prime (resp., G-semiprime).*
3. *R is graded prime (resp., graded semiprime).*

3 Primeness conditions for rings graded by torsion-free groups.

In this section we will use the machinery developed in the first two sections to obtain the main result of this article (Theorem 3.2).

Theorem 3.1 *(cf. [23, Corollary 8.5]; for S unital) Let S be a ring with local units which is strongly graded by the torsion-free group G. Then S_1 is G-prime if and only if S is prime.*

Proof. A careful but tedious examination of the proof of Corollary 8.5 given in [23, p.53-85]; reveals that the arguments given there hold verbatim, and therefore yield the analogous result, in this more general setting. We note that the "easy" direction is also obtained directly here: if S is prime, then S is graded prime, so that S_1 is G-prime by Corollary 2.9.□

We are now in a position to prove the result mentioned in the introduction.

Theorem 3.2 *Let R be a ring graded by the torsion-free group G. Then R is prime if and only if R is graded prime.*

Proof. By Corollary 1.2, R is prime if and only if $R^f\{G\}$ is prime. Since $R^f\{G\} = (R\#G) * \overline{G}$ is strongly graded with identity component $R\#G$, Theorem 3.1 yields that $R^f\{G\}$ is prime if and only if $R\#G$ is G-prime. Finally, $R\#G$ is G-prime if and only if R is graded prime by Corollary 2.6.□

We now use Theorem 3.2 together with the results of Section 2 to obtain more specific ideal correspondences between (ungraded) prime ideals and graded prime ideals in rings graded by torsion-free groups.

The next result follows easily from the observation that if R is graded by the group G and I is a graded ideal of R, then R/I is a ring graded by G.

Corollary 3.3 *Let R be graded by the torsion-free group G. Then Q is a graded prime ideal of R if and only if Q is a prime ideal of R.*

Corollary 3.4 *Let R be graded by the torsion-free group G. Then $P(R) = P_{gr}(R) = (P(R))_{gr}$.*

Proof. By the above corollary,

$$P(R) = \bigcap_{P \text{ prime in } R} P = \bigcap_{Q \text{ graded prime in } R} Q = P_{gr}(R).$$

In addition, $P_{gr}(R) = (P(R))_{gr}$ by Proposition 1.6.□

Specifically, Corollary 3.4 immediately yields

Corollary 3.5 *Let R be a ring graded by the torsion-free group G. Then R is semiprime if and only if R is graded semiprime.*

We conclude this article with some remarks about the applicability of these techniques to other types of ring-theoretic conditions. First, Theorem 3.1 may be proved for $\mathbf{Z}$-graded rings by using this method directly, without invoking the "delta methods" of Passman. Secondly, an alternate proof of

> [13, Theorem 4.1], *Let R be a $\mathbf{Z}$-graded ring. If R is graded-Jacobson, then R is Jacobson.*

also follows from these techniques; see [15]. Additional results of this type (e.g. the Noetherian result of [10] for polycyclic-by-finite groups) are also amenable to these methods. Finally, we note that in [15] the second author has obtained results analogous to those given in Theorem 2.3 for one-sided ideals.

Acknowledgements. We'd like to thank Don Passman, Declan Quinn and Margaret Beattie for helpful comments offered through private communications.

We include in this list of references not only those articles cited directly herein, but also additional articles which address similar types of questions.

References

[1] P.N. Ánh and L. Márki, *Morita equivalence for rings without identity*, Tsukuba J. Math., **11** (1987), 1-16.

[2] G. Bergman, *On Jacobson radicals of graded rings*, preprint.

[3] M. Beattie, *A generalization of the smash product of a graded ring*, J. Pure Appl. Algebra **52** (1988), 219-226.

[4] M. Beattie, *Duality theorems for rings with actions or coactions*, J. Algebra **115** (1988), 303-312.

[5] M. Beattie, Liu S. X. and P. Stewart, *Comparing graded versions of the prime radical*, Canadian Math. Bull. **34** (1991), 158-164.

[6] M. Beattie and P. Stewart, *Graded radicals of graded rings*, Acta. Math. Hung. **58** (1991), 227-237.

[7] R. Blattner and S. Montgomery, *A duality theorem for Hopf module algebras*, J. Algebra **95** (1985), 153-172.

[8] M. Cohen and S. Montgomery, *Group-graded rings, smash products, and group actions*, Trans. Amer. Math. Soc. **282** (1984), 237-258.

[9] M. Cohen and S. Montgomery, *Addendum: Group-graded rings, smash products, and group actions*, Trans. Amer. Math. Soc. **300** (1987), 810-811.

[10] W. Chin and D. Quinn, *Rings graded by polycyclic-by-finite groups*, Proc. Amer. Math. Soc. **102** (1988), 235-241.

[11] M. Cohen and L. Rowen, *Group graded rings*, Comm. Algebra **11** (1983), 1253-1270.

[12] E.C. Dade, *Group graded rings and modules*, Math. Z., **174** (1980), 241-262.

[13] M. Ferrero, E. Jespers, E. Puczyloski, *Prime ideals of graded rings and related matters*, Comm. in Algebra, **18** (1990), 3819-3834.

[14] J. W. Fisher and S. Montgomery, *Semiprime skew group rings*, J. Algebra **34** (1978), 241-247.

[15] J. Haefner, *The Jacobson property for graded rings*, preprint.

[16] C. Năstăsecu, *Group rings of graded rings, applications*, J. Pure Appl. Algebra **33** (1984), 315-335.

[17] C. Năstăsescu and F. Van Oystaeyen, *Graded Ring Theory*, North-Holland Mathematical Library **28** (North Holland, Amsterdam, 1982).

[18] D.S. Passman, *It's essentially Maschke's theorem*, Rocky Mountain J. Math. **13** (1978), 37-54.

[19] D.S. Passman, *Infinite crossed products and group graded rings*, Trans. Amer. Math. Soc. **284** (1984), 707-727.

[20] D.S. Passman, *Semiprime crossed products*, Houston J. Math. **11** (1985), 257-267.

[21] D.S. Passman, *Algebraic crossed products*, Cont. Math. **43** (1985), 209-225.

[22] D.S. Passman, *Prime ideals in polycyclic crossed products*, Trans. Amer. Math. Soc. **301** (1987), 737-759.

[23] D.S. Passman, *Infinite Crossed Products*, Academic Press, New York, 1989.

[24] K. Pearson and W. Stephenson, *A skew polynomial ring over a Jacobson ring need not be a Jacobson ring*, Comm. Algebra **5** (1977), 783-794.

[25] D. Quinn, *Group-graded rings and duality*, Trans. Amer. Math. Soc. **292** (1985), 155-167.

Electronic Mail Addresses

abrams@vision.uccs.edu
haefner@vision.uccs.edu

SEPARABILITIES AND G-GALOIS ACTIONS

RICARDO ALFARO[1]
Department of Mathematics, University of Michigan - Flint
Flint, Michigan 48502

Abstract

The theory of separable algebras has been well understood for quite some time. The generalization to separable extensions of non-commutative rings started with Hirata and Sugano, although it seems to be a very weak condition. Later work by Hirata himself in H-separability and by McMahon and Newborn in strong separability, seem to give a more concrete approach to separability in non-commutative rings. We study this separability conditions in the case of group actions on rings.

1. Separability and group actions

Let B be a subring of A with 1. The extension $B \subset A$ is called *separable* (or A is *separable* over B) if any of the following equivalent conditions is satisfied:
1) The multiplication map $\mu : A \otimes_B A \to A$ splits as an $(A$-$A)$-bimodule map.
2) There exists an element $e \in A \otimes_B A$ (called a separability element), such that $\alpha e = e\alpha$ for all $\alpha \in A$ and $\mu(e) = 1$.
We now describe group actions.

Let S be a ring with 1, G be a finite group acting faithfully as automorphisms of S and denote ${}^g r$ the image of $r \in S$ under $g \in G$. Let R be the fixed ring and form the skew group ring $S*G$. It is a free left S-module with basis the elements of G, where $gr = {}^g r g$. Consider S as a right R-module and left $S*G$-module with the action of $S*G$ on S given by $(rg)\cdot s = r\,{}^g s$. We say that the action of G on S is *G-Galois* (or that S is *G-Galois* over R) if any of the following equivalent conditions is satisfied:
1) S is finitely generated projective right R-module and the map $j : S*G \to \mathrm{End}(S_R)$ given by the action of $S*G$ on S is a ring isomorphism.
2) There exist elements a_i, b_i (called a *G-Galois* basis) such that $\sum_i a_i\, {}^g b_i = 1$ if $g = 1$ and the sum is 0 if $g \neq 1$.
3) $S\pi S = S*G$ where $\pi = \sum_{g \in G} g$.
Recall that the "trace map" $tr_G : S \to R$ is given by $tr_G(x) = \sum_{g \in G} {}^g x$, which is

[1] Partially supported by a grant from the Faculty Development Fund of the University of Michigan-Flint and fellowship from the C.R.M. Bellaterra, Spain.

an $(R\text{-}R)$-bimodule map; and for every $g \in G$ define $\phi_g = \{r \in S / \ r\,{}^g s = sr \quad \forall s \in S\}$. If $\phi_g \neq 0 \quad g$ is said to be *ω-inner*, and if $\phi_g = 0$ for every $g \neq 1 \quad G$ is said to be *ω-outer*.

The relations between separability and G-Galois actions are as follows:

PROPOSITION 1.1 If S is G-Galois over R then S is separable over R and $\phi_g \cap Z(S) = 0$ for all $g \neq 1$.

PROOF: Let $\{a_i, b_i\}$ be a G-Galois basis for S, then we have $x = 1 \cdot x = \sum a_i \pi b_i \cdot x = \sum a_i tr(b_i x)$ and also $x = x \cdot 1 = x \cdot \sum a_i \pi b_i = \sum tr(x a_i) b_i$ for all $x \in S$.
Let $e = \sum_i a_i \otimes b_i \in S \otimes_R S$. Then $\mu(e) = \sum_i a_i b_i = 1$ and, if $r \in S$

$$\begin{aligned} re &= \sum_i r a_i \otimes b_i = \sum_i \sum_j a_j tr(b_j r a_i) \otimes b_i = \sum_{i,j} a_j \otimes tr(b_j r a_i) b_i \\ &= \sum_j a_j \otimes \sum_i tr(b_j r a_i) b_i = \sum_j a_j \otimes b_j r = er. \end{aligned}$$

Hence S is separable over R. Now let $r \in \phi_g \cap Z(S)$, then $r\,{}^g s = sr = rs$ for all $s \in S$. But $r = \sum_i a_i b_i r = \sum_i a_i r\,{}^g b_i = \sum_i r a_i\,{}^g b_i = 0$ if $g \neq 1$. Hence $\phi_g \cap Z(S) = 0$ if $g \neq 1$.

Conversely we have:

PROPOSITION 1.2 Assume S is separable over R and G is ω-outer, then S is G-Galois over R.

PROOF: Let $e = \sum_i a_i \otimes b_i$ be a separability element for S. Then $\mu(e) = \sum a_i b_i = 1$.

Define the map $1 \otimes g : S \otimes_R S \to S \otimes_R S$ by $(1 \otimes g)(a \otimes b) = a \otimes {}^g b$, is easy to see that $1 \otimes g$ is a left-S-module map, and since μ is an $(S\text{-}S)$-bimodule map we obtain for any $r \in S$:

$$r\left(\sum_i a_i\,{}^g b_i\right) = r\mu\left((1 \otimes g)(e)\right) = \mu\left((1 \otimes g)(re)\right) = \mu\left((1 \otimes g)(er)\right) = \left(\sum_i a_i\,{}^g b_i\right) {}^g r.$$

Hence $\sum_i a_i\,{}^g b_i \in \phi_g$, but G is ω-outer, so $\sum_i a_i\,{}^g b_i = 0$ if $g \neq 1$ and $\{a_i, b_i\}$ form a G-Galois basis for S over R.

COROLLARY 1.3 Let S be a commutative ring. S is G-Galois over R if and only if S is separable over R and G is ω-outer.

PROOF: Immediatly from proposition 1.1 and proposition 1.2.

We now look at the extension $S \subset S * G$ and we obtain:

PROPOSITION 1.4 $S * G$ is separable over S if and only if there exists a central element in S of trace 1.

PROOF: ($\Leftarrow$) Let $e = \sum \alpha_i \otimes \beta_i$ be the separability element and let $\alpha_i = \sum_{g\in G} r_{ig} g$, $\beta_i = \sum_{h\in G} s_{ih} h$. Then if $k \in G$, $ke = \sum_{g,h,i} {}^k r_{ig} {}^{kg} s_{ih} kg \otimes h = \sum_{g,h,i} {}^k r_{ik^{-1}g} {}^g s_{ih} g \otimes h$, and $ek = \sum_{g,h,i} r_{ig} {}^g s_{ih} g \otimes hk = \sum_{g,h,i} r_{ig} {}^g s_{ihk^{-1}} g \otimes h$. But $ke = ek$ since $\{g \otimes h / \ g, h \in G\}$ is a free basis for the left S-module $S*G \otimes_S S*G$, thus we have:

$$\sum_i {}^k r_{ik^{-1}g} {}^g s_{ih} = \sum_i r_{ig} {}^g s_{ihk^{-1}} \qquad \forall g, h, k \in G.$$

Therefore, for $k = g$ and $h = 1$, we obtain:

$$\sum_i {}^g r_{i1} {}^g s_{i1} = \sum_i r_{ig} {}^g s_{ig^{-1}}.$$

If we call $w = \sum_i r_{i1} s_{i1}$, we have:

$$tr(w) = \sum_g {}^g \left(\sum_i r_{i1} s_{i1} \right) = \sum_{g,i} {}^g r_{i1} {}^g s_{i1} = \sum_{g,i} r_{ig} {}^g s_{ig^{-1}}.$$

But the right hand expression is equal to 1 because

$$\sum_i \alpha_i \beta_i = 1 \quad \Rightarrow \quad \sum_{i,g,h} r_{ig} {}^g s_{ih} gh = 1 \quad \Rightarrow \quad \sum_{g,i} r_{ig} {}^g s_{ig^{-1}} = 1.$$

Thus $tr(w) = 1$.
Now we need to show w is central. Let $x \in S$, then $xe = ex$ implies

$$\sum_i x r_{ig} {}^g s_{ih} = \sum_i r_{ig} {}^g s_{ih} {}^{gh} x \qquad \forall g, h \in G.$$

Therefore if $g = h = 1$,

$$x \left(\sum_i r_{i1} s_{i1} \right) = \left(\sum_i r_{i1} s_{i1} \right) x$$

and thus $xw = wx$, $\forall x \in S$.
($\Rightarrow$) Let w be the element of trace 1. Let $e = \sum_{g\in G} gw \otimes g^{-1} \in S*G \otimes_S S*G$, then $\mu(e) = \sum_{g\in G} gwg^{-1} = \sum_{g\in G} {}^g w = tr_G(w) = 1$. Now let $r \in S$, then

$$re = \sum_{g\in G} rgw \otimes g^{-1} = \sum_{g\in G} gw \, {}^{g^{-1}} r \otimes g^{-1} = \sum_{g\in G} gw \otimes {}^{g^{-1}} r g = er;$$

and if $h \in G$ we have:

$$he = \sum_{g\in G} hgw \otimes g^{-1} = \sum_{g\in G} (hg) w \otimes (hg)^{-1} h = eh.$$

Hence $\alpha e = e\alpha$ for all $\alpha \in S*G$ and $S*G$ is separable over S.

2. Strong and H-separability

Let B be a subring of A with 1. A is said to be *strongly separable* over B if as $(A\text{-}A)$-bimodules $A\otimes_B A \cong K \oplus L$ where $\mathrm{Hom}_{A,A}(K, A) = 0$ and $L \oplus H \cong A^n$ for some $(A\text{-}A)$-bimodules K, L, H and some n. In case $K = 0$ we say that A is *H-separable* over B. Strongly separable extensions are separable but the converse is false, see [3]. There is an equivalent definition of this kinds of separability in terms of the natural $(A\text{-}A)$-bimodule map $\varphi : A\otimes_B A \to \mathrm{Hom}(\Delta_c, A_c)$ where $\varphi(a\otimes b)(x) = axb$, C is the center of A and Δ is the centralizer of B in A, i.e. $\Delta = C_A(B) = \{x \in A/\ xb = bx\ \forall b \in B\}$. A is *strongly separable* over B if and only if Δ_c is finitely generated projective C-module and φ is an split epimorphism. Similarly, A is *H-separable* over B if and only if Δ_c is finitely generated projective C-module and φ is an isomorphism. For details see [2] and [3].

Let S be a ring with 1 and G a finite group acting faithfully as automorphisms of S. If X is a subset of S, let $I(X) = \{g \in G/\ {}^g x = x \quad \forall x \in X\}$ be the *"inertia group"* of X, ($I(X)$ is always a subgroup of G), and $Z(S)$ be the center of S.

In [4], McMahon and Shapiro have shown the following

THEOREM 2.1 ([4, theorems 2.2 and 2.3]) Let S be a simple ring and G a finite outer group of automorphisms of S. Let $F = I(Z(S))$. Then

i). $S*G$ is H-separable over $S*F$.

ii). $S*G$ is H-separable over S if and only if F is trivial.

iii). If $|F|^{-1} \in S$, then $S*G$ is strongly separable over S.

But if S is simple and G is outer, $S*G$ is simple and hence the action of G on S is G-Galois. We'll give a general result relating G-Galois actions with strong and H-separability.

Let N be any normal subgroup of G, $D = C_{S*G}(S*N)$ and $C = Z(S*G)$. The action of G on S induces a faithful action of G on $S*G$ via conjugation, ${}^g\alpha = g\alpha g^{-1}$ for $\alpha \in S*G$ and since N is normal, G also acts on D. Let M be the inertia group of D, then N is a subgroup of M and G/M acts faithfully on D by ${}^{\bar{h}}\alpha = {}^g\alpha$ for any $g \in \bar{h}$.

PROPOSITION 2.2 $D^G = D^{G/M} = C$.

PROOF: The first equality is obvious since M is the inertia group of D. Let $\alpha \in D^G$ then $\alpha g = g\alpha \quad \forall g \in G$ and by definition of D, $s\alpha = \alpha s \quad \forall s \in S$ since $s \in S*N$; hence $\alpha \in C$. Conversely, if $\alpha \in C$, $\alpha g = g\alpha \quad \forall g \in G$ and hence $\alpha \in D^G$, (is clear that $C \subseteq D$).

We now give our main theorem for a faithful group action G on a ring S with 1.

THEOREM 2.3 Let N be a normal subgroup of G, let M be the inertia group of $D=C_{S*G}(S*N)$ and let C be the center of $S*G$. Assumme there is a central element w in S with $tr_M(w)=1$, we have:

i). If D is G/M-Galois over C, then $S*G$ is strongly separable over $S*N$.

ii). If D is G/N-Galois over C and $M=N$ then $S*G$ is H-separable over $S*N$.

QUESTION: Are these conditions also neccessary?

PROOF: i) We will use the natural $(S*G\text{-}S*G)$-bimodule map $\varphi : S*G\otimes_{S*N} S*G \to \mathrm{Hom}(D_C, S*G_C)$. Let $\{a_i, b_i\}$ be a G/M-Galois basis for D over C, then $\sum_i a_i\pi_{G/M}b_i=1$. Moreover $f_i(x)=tr_{G/M}(b_ix)\in \mathrm{Hom}(D_C, C_C)$ and $\{a_i, f_i\}$ form a dual projective basis for D over C, i.e. $\alpha=\sum_i a_if_i(\alpha)$ for all $\alpha\in D$.

Claim 1. $\{f_i\}$ is a basis for $\mathrm{Hom}(D_C, S*G_C)$ as $(S*G\text{-}S*G)$-bimodule. For, let $\alpha\in D$, $f\in \mathrm{Hom}(D_C, S*G_C)$, then $f(\alpha)=f\left(\sum_i a_if_i(\alpha)\right)=\sum_i f(a_i)f_i(\alpha)=\sum_i f_i(\alpha)f(a_i)$. Thus $f=\sum f(a_i)f_i=\sum f_if(a_i)$.

Claim 2. φ is an epimorphism.
First, if $g\in G\setminus N$, $\varphi(g\otimes g^{-1})(\alpha)=g\alpha g^{-1}={}^{\bar g}\alpha$ for all $\alpha\in D$ thus $\varphi(g\otimes g^{-1})$ acts as $\bar g\in G/M$ on D and $\varphi(g\otimes g^{-1})=\varphi(h\otimes h^{-1})$ whenever $\bar g=\bar h$ in G/M.(*)
Let $G/M=\{h_1M,\dots,h_pM\}$, i.e. choose $\{h_1,\dots,h_p\}$ a transversal of M in G, then

$$f_j(x)=tr_{G/M}(b_jx)=\sum_{h_i}{}^{h_i}(b_jx)=\sum_{h_i}{}^{h_i}b_j{}^{h_i}x=\sum_i{}^{h_i}b_j\varphi(h_i\otimes h_i^{-1})(x)=\sum_i\varphi(h_ib_j\otimes h_i^{-1})(x).$$

Therefore $f_j\in \mathrm{Im}(\varphi)$ and by claim 1, φ is an epimorphism. Notice that the expression of f_j above is independent of the choice of the transversal of M in G by (*).

Claim 3. φ splits as $(S*G\text{-}S*G)$-bimodule homomorphism.
Let M be given by the set $\{m_1,\dots,m_q\}$ and let $\{h_1,\dots,h_p\}$ be a transversal of M in G. Let $l_k=\sum_{i,j}h_im_jwb_k\otimes(h_im_j)^{-1}\in S*G\otimes_{S*N}S*G$. Then

$$\begin{aligned}\varphi(l_k) &= \sum_{i,j}{}^{h_im_j}w^{\overline{h_im_j}}b_k\varphi(h_im_j\otimes(h_im_j)^{-1}) \quad\text{and by } (*)\\ &= \sum_{i,j}{}^{h_im_j}w^{\overline{h_i}}b_k\varphi(h_i\otimes h_i^{-1})=\sum_i\left(\sum_j{}^{h_im_j}w\right)\varphi\left(h_ib_k\otimes h_i^{-1}\right)=f_k.\end{aligned}$$

Hence we may define the map $\psi : \mathrm{Hom}(D_C, S*G_C)\to S*G\otimes_{S*N}S*G$ by linearity with $\psi(f_k)=l_k$. To show that ψ is an $(S*G\text{-}S*G)$-bimodule map, we need to show $\alpha l_k=l_k\alpha$ for all $\alpha\in S*G$. Let $r\in S$, since $b_k\in D$ and w is central in S we have:

$$\begin{aligned}rl_k &= \sum_{i,j}r(h_im_j)wb_k\otimes(h_im_j)^{-1}=\sum_{i,j}(h_im_j)^{(h_im_j)^{-1}}rwb_k\otimes(h_im_j)^{-1}\\ &= \sum_{i,j}h_im_jwb_k\otimes{}^{(h_im_j)^{-1}}r(h_im_j)^{-1}=\sum_{i,j}h_im_jwb_k\otimes(h_im_j)^{-1}r=l_kr,\end{aligned}$$

and if $g \in G$, we have:

$$gl_k = \sum_{i,j} gh_i m_j w b_k \otimes (h_i m_j)^{-1} = \sum_{i,j} (gh_i) m_j w b_k \otimes ((gh_i)m_j)^{-1} g \, ,$$

but $\{gh_i\}$ is another transversal of M in G, hence by $(*)$ $gl_k = l_k g$ and therefore ψ is an $(S*G\text{-}S*G)$-bimodule map.
But

$$\varphi(\psi(f)) = \varphi\left(\psi\left(\sum_k f(a_k) f_k\right)\right) = \varphi\left(\sum_k f(a_k) l_k\right) = \sum_k f(a_k)\varphi(l_k) = \sum_k f(a_k) f_k = f.$$

Thus ψ splits φ and therefore $S*G$ is strongly separable over $S*N$.

ii) By i) φ is an epimorphism. Assumme $M = N$, then

$$l_k = \sum_{i,j} h_i m_j w b_k \otimes (h_i m_j)^{-1} = \sum_{i,j} h_i{}^{m_j} w b_k \otimes h_i^{-1} = \sum_i h_i b_k \otimes h_i^{-1}$$

for any transversal of M in G. Hence

$$\begin{aligned} \psi\cdot\varphi(1\otimes 1) &= \sum_k \varphi(1\otimes 1)(a_k) l_k = \sum_k a_k l_k = \sum_k a_k \sum_i h_i b_k \otimes h_i^{-1} \\ &= \sum_i \left(\sum_k a_k{}^{h_i} b_k\right) h_i \otimes h_i^{-1} = h_1 \otimes h_1^{-1} \quad \text{where} \quad \overline{h}_1 = \overline{1} \in G/M. \end{aligned}$$

But then $h_1 \in M$ and $h_1 \otimes h_1^{-1} = 1 \otimes h_1 h_1^{-1} = 1 \otimes 1$. Therefore $\psi\cdot\varphi = \mathrm{id}_{S*G\otimes_{S*N}S*G}$ and thus φ is an isomorphism.

NOTE: If G/N is ω-outer in D then part ii) in the theorem is also a neccesary condition because:

Assumme φ is an isomorphism and $m \in M \setminus N$, then $m \otimes m^{-1} \neq 1 \otimes 1$ in $S*G \otimes_{S*N} S*G$; but if $\alpha \in D$, $\varphi(m \otimes m^{-1})(\alpha) = m\alpha m^{-1} = {}^{\overline{m}}\alpha = \alpha = \varphi(1\otimes 1)(\alpha)$, hence $\varphi(m \otimes m^{-1}) = \varphi(1 \otimes 1)(\Rightarrow\Leftarrow)$ thus $M = N$. On the other hand D is separable over C by [5, proposition 1.3] thus D is G/N-Galois over C by propsoition 1.2.

Examples: We analyze some particular cases

A) Let $N = \{1\}$.
Then $D = C_{S*G}(S) = \Delta$ and applying theorem 2.3 we have:

PROPOSITION 2.4

i) If Δ is G/M-Galois over C and $\exists\ w \in Z(S)/\ tr_M(w) = 1$, then $S*G$ is strongly separable over S.

ii) If Δ is G-Galois over C then $S*G$ is H-separable over S. (*Question: Is the converse true?*)

Moreover, let S be a simple ring and G be outer, then $\Delta = Z(S)$ (by [4,Lemma 2.1]) and therefore G is ω-outer. Since $Z(S)$ is a field, G/M is G/M-Galois on $Z(S)$ where $M = I(Z(S))$. Therefore applying proposition 2.4 and the *Note* above, we obtain an improvement to [4, Theorem 2.3] and part ii) of [4, Theorem 2.2] as follows:

COROLLARY 2.5 Let S be a simple ring and G be outer.
i) If $\exists\, w \in Z(S)$ such that $tr_M(w) = 1$, then $S*G$ is strongly separable over S.
ii) $S*G$ is H-separable over S if and only if $M = 1$.

With proposition 2.4 we also obtain the following:

COROLLARY 2.6 If S is a central Galois extension of R then $S*G$ is H-separable over S.

PROOF: $\exists\, a_i, b_i \in Z(S) / \sum a_i \pi_G b_i = 1$, but $Z(S) \subseteq C_{S*G}(S) = \Delta$ and Δ is G-invariant, hence Δ is G-Galois over $\Delta^G = C$ and by proposition 2.4 $S*G$ is H-separable over S.

Hence if S is a commutative ring, Corollary 2.6 implies part of [4, Theorem 3.3]:

COROLLARY 2.7([4, Theorem 3.3]) Let S be a commutative ring, S is G-Galois over $R = S^G$ if and only if $S*G$ is H-separable over S. In particular if S is a field K then $K*G$ is H-separable over K.

B) Let $N = I(Z(S))$ the inertia group of the center of S.
Then we obtain:

PROPOSITION 2.8 ([4, Theorem 2.2]). Let S be a simple ring and G outer. Then $S*G$ is H-separable over $S*N$.

PROOF: By [4, Lemma 2.1], $D = C_{S*G}(S*N) = C_{S*G}(S) = \Delta = Z(S)$, which is a field, and then G/M is G/M-Galois on D with $M = I(D) = I(Z(S)) = N$. Therefore Theorem 2.3 says that $S*G$ is H-separable over $S*N$.

Acknowlegement. I want to thank Professor George Szeto for pointing out a mistake in the earlier version of this work.

REFERENCES

(1) Alfaro R., *"T-Galois Extensions on Rings and a submodule correspondence"*, Comm. in Algebra, **20(6)**, 1992, pp. 1777-1786.

(2) Hirata K., *"Some types of separable extension of rings"*, Nagoya Math. J. **vol. 33** (1968).

(3) MacMahon E. and Mewborn A.C., *"Separable extensions of non-commutative rings"*, Hokkaido Math. J. XIII, **1**, (1984).

(4) MacMahon E. and Shapiro J., *"On strong and H-separability in ordinary and Skew group rings"*, Houston Journal of Math., **15**, No. 3, (1989).

(5) Sugano K., *"On centralizers in separable extensions"*, Osaka J. Math. **7** (1970).

A CHARACTERIZATION OF SEMI-PERFECT RINGS AND MODULES

GORO AZUMAYA
Department of Mathematics, Indiana University
Bloomington, Indiana 47405

ABSTRACT

The notion of generalized projective covers is introduced to give a natural generalization of a theorem of Bass on perfect rings. Moreover, in terms of this notion, some characterizations of semi-perfect rings and semi-perfect modules are established.

1. Semi-Perfect Rings

Throughout R is a ring with identity element 1 whose Jacobson radical is denoted by J, and a semi-simple ring means an Artinian semi-simple ring, i.e., a ring (with identity element) which is a (finite) direct sum of simple left ideals. Let M be a left R-module. Let P be a projective left R-module and $f : P \to M$ an epimorphism. P is called a projective cover of M with respect to f if $Ker(f)$ is small in P, *i.e.*, $P = Ker(f) + U$ with a submodule U of P always implies $P = U$. In his epoch-making paper [2], Bass investigated two important types of rings in connection with the concept of projective covers, one is "semi-perfect ring" and the other is "perfect ring". Namely, R is called semi-perfect if every cyclic left R-module has a projective cover, while R is called left (or right) perfect if every left (or right) R-module has a projective cover. The following were established by him, where the first one indicates that the concept of semi-perfect rings is left-right symmetric:

Theorem A ([2, Th. 2.1]). The following conditions are equivalent:
(1) R is semi-perfect.
(2) The factor ring R/J is semi-simple and every idempotent of R/J can be lifted to an idempotent of R.
(3) Every finitely generated left R-module has a projective cover.

Theorem B ([2, Th. P]). The following conditions are equivalent:
(1) R is left perfect.
(2) Every flat left R-module is projective.
(3) J is left T-nilpotent and the factor ring R/J is semi-simple.
(4) R satisfies the descending chain condition for principal right ideals.

Here, generally a left ideal L of R is called left T-nilpotent if for any sequence $a_1, a_2, a_3, \cdots$ in L there exists an $n > 0$ such that $a_1 a_2 \cdots a_n = 0$. As for left T-nilpotent ideals, the following are actually verified in [2]:

Proposition C. Let L be a left ideal of R. Then the following conditions are equivalent:
(1) L is left T-nilpotent.
(2) For every non-zero left R-module M, $LM \neq M$.
(3) For every left R-module M, LM is small in M.
(4) For a free left R-module F with an infinite-countable free basis, LF is small in F.

For convenience, we give here a proof for this. Assume (1). Suppose there were a non-zero left R-module M such that $LM = M$. It follows then $a_1 M \neq 0$ for some $a_1 \in L$. Since $a_1 M = a_1 LM$, it follows $a_1 a_2 M \neq 0$ for some $a_2 \in L$. Continuing in this way, we can find $a_3, a_4, \cdots$ in L successively so that $a_1 a_2 \cdots a_n M \neq 0$ whence $a_1 a_2 \cdots a_n \neq 0$ for all n, which contradicts to (1). Thus (1) $\Rightarrow$ (2) is proved. Let next M be a left R-module and U a submodule of M such that $LM + U = M$. But this means that $L(M/U) = M/U$. Therefore, if we assume (2) then it follows that $U = M$, which implies that LM is small in M. This shows (2) $\Rightarrow$ (3). (3) $\Rightarrow$ (4) is trivial. Assume (4). Let F be a free left R-module with a free basis $u_1, u_2, u_3, \cdots$. Let $a_1, a_2, a_3, \cdots$ be a sequence in L. Consider then the submodule G of F generated by $u_1 - a_1 u_2, u_2 - a_2 u_3, u_3 - a_3 u_4, \cdots$. Then we have $F = LF + G$, because $u_i = a_i u_{i+1} + (u_i - a_i u_{i+1})$ for every i. Since LF is small in F, it follows that $F = G$. Therefore, u_1 must be in G, or what is the same thing, $u_1 = r_1(u_1 - a_1 u_2) + r_2(u_2 - a_2 u_3) + \cdots + r_n(u_n - a_n u_{n+1})$ for a sufficiently large n and some $r_1, r_2, \cdots, r_n$ in R. The right side is, however, equal to $r_1 u_1 + (r_2 - r_1 a_1)u_2 + \cdots + (r_n - r_{n-1} a_{n-1})u_n - r_n a_n u_{n+1}$. So by comparing the coefficients, we have $r_1 = 1$, $r_2 = r_1 a_1 = a_1$, $r_3 = r_2 a_2 = a_1 a_2, \cdots, r_n = r_{n-1} a_{n-1} = a_1 a_2 \cdots a_{n-1}$, $0 = r_n a_n = a_1 a_2 \cdots a_n$. This shows that L is left T-nilpotent.

Now let M be a left R-module. Let P be a projective left R-module with an epimorphism $f : P \to M$. We call P a *generalized projective cover* of M with respect to f if $Ker(f) \subset JP$. We point out that every projective cover of M is a generalized projective cover. For, suppose $Ker(f)$ is small in P. Since P is projective, there are a family of homomorphisms $h_i : P \to R$ $(i \in I)$ and a family of elements u_i $(i \in I)$ of P (I being an index set) such that, for each $x \in P$, $h_i(x) = 0$ for almost all i and $x = \sum h_i(x)u_i$. The homomorphic image $h_i(Ker(f))$ of the small submodule $Ker(f)$ of P is a small left ideal of R and so is contained in J for each i. This implies that $Ker(f) \subset \sum Ju_i \subset JP$.

Proposition 1. If a left R-module M has a projective cover then every generalized projective cover of M is a projective cover. More precisely, if P is a generalized projective cover of M with respect to an epimorphism $f : P \to M$ and if Q is a projective cover of M with respect to an epimorphism $g : Q \to M$, then there exists an isomorphism $h : P \to Q$ such that $f = g \circ h$.

Proof. Since P is projective and $g : Q \to M$ is an epimorphism, there exists a homomorphism $h : P \to Q$ such that $f = g \circ h$. Then that $f : P \to M$ is an epimorphism implies that $g(h(P)) = M$, *i.e.*, $Ker(g) + h(P) = Q$. Since $Ker(g)$ is small in Q, it follows that $h(P) = Q$, i.e., $h : P \to Q$ is an epimorphism. But since Q is projective, h splits, or equivalently, $Ker(h)$ is a direct summand of P. Thus if we put $K = Ker(h)$ then $P = K \oplus K'$ with a suitable submodule K' of P. From this follows that $JP = JK \oplus JK'$. On the other hand, it is clear that $K \subset Ker(f)$, and that P is a generalized projective cover of M with respect f means that $Ker(f) \subset JP$. It follows then that $K \subset JP$ and, therefore, $K = JK$. But, as a direct summand of P, K is projective and so $K = 0$ by [2, Prop. 2.7]. Thus h is an isomorphism.

Proposition 2. Let M be a left R-module and suppose M is a direct sum of submodules M_i for $i \in I$. If each M_i has a generalized projective cover P_i with respect to an epimorphism $f_i : P_i \to M_i$ then the direct sum $\oplus P_i$ of P_i is a generalized projective cover of M with respect to the canonical epimorphism $\oplus f_i : \oplus P_i \to M$.

Proof. Let $K_i = Ker(f_i)$ for each $i \in I$. Then $K_i \subset JP_i$ for each i by assumption. Regarding as submodules *of* $\oplus P_i$ in the natural manner, we have then that $\oplus K_i \subset \oplus JP_i = J(\oplus P_i)$. Since $\oplus P_i$ is projective and $\oplus K_i$ is clearly the kernel of $\oplus f_i$, we know that $\oplus P_i$ is a generalized projective cover of M with respect *to* $\oplus f_i$.

We now have the following generalization of Miller [5, Lemma 1.2] that every flat module having a projective cover is projective:

Theorem 1. Let M be a flat left R-module having a generalized projective cover. Then M is projective.

Proof. Let P be a generalized projective cover of M with respect to an epimorphism $f : P \to M$, and let $K = Ker(f)$. Then $K \subset JP$. Since M is flat, it follows from Stenström [7, Prop. 11.1] that K is pure in P and so satisfies $JK = K \cap JP$ by [7, Prop. 11.2]. Thus we have that $JK = K$. Since, however, P is projective, its pure submodule K is, by Zimmermann-Huisgen [9, (1) on p. 236], locally projective in the sense that every epimorphism onto K is locally split, or equivalently, for each $x \in K$ there exists a finite number of homomorphisms $h_i : K \to R$ ($i = 1, 2, \cdots, n$) and the same number of elements $u_i \in K$ ($i = 1, 2, \cdots, n$) such that $x = h_1(x)u_1 + h_2(x)u_2 + \cdots + h_n(x)u_n$. It follows now from [9, Prop. 2.2] that $K = 0$, i.e., f is an isomorphism. Thus M is projective.

From this theorem we can now derive the following, which is regarded as a generalization of the implication (1) $\Rightarrow$ (2) in Theorem B due to Proposition C :

Theorem 2. Let R be a semi-perfect ring and M a flat left R-module. Then M is projective whenever JM is small in M.

Proof. The factor module M/JM can be regarded as a left module over the factor ring R/J in the natural manner. But since R is semi-perfect, R/J is a semi-simple ring. Therefore, M/JM is completely reducible, i.e., is a direct sum of simple submodules S_i $(i \in I)$ as a left R/J -whence as a left R-module. Since R is semi-simple and every simple left R-module is cyclic, each S_i has a projective whence a generalized projective cover P_i with respect to an epimorphism $f_i : P_i \to S_i$. If $P = \oplus P_i$, the direct sum of P_i for all $i \in I$, then by Proposition 2, P becomes a generalized projective cover of M/JM with respect to the canonical epimorphism $f = \oplus f_i : P \to M/JM$, which means that $Ker(f) \subset JP$. Let now $g : M \to M/JM$ be the natural epimorphism. Since P is projective, there should exist a homomorphism $h : P \to M$ such that $g \circ h = f$. It follows then $g(h(P)) = f(P) = M/JM$. Since $JM = Ker(g)$, this implies that $JM + h(P) = M$, so that if we assume that JM is small in M then we have $h(P) = M$, i.e., h is an epimorphism. It is clear that $Ker(h) \subset Ker(f)$ and, therefore, we have $Ker(h) \subset JP$, which means that P is a generalized projective cover of M with respect to h. Since M is flat, M is projective by Theorem 1.

Now Sandomierski proved that R is semi-perfect if and only if every simple left R-module has a projective cover ([6, Th. 4]). The following theorem is to extend this to the case of generalized projective covers:

Theorem 3. The following conditions are equivalent:
(1) R is semi-perfect.
(2) Every simple left R-module has a projective cover.
(3) Every simple left R-module has a generalized projective cover.

Proof. (1) $\Rightarrow$ (2) is clear, since every simple left R-module is cyclic, while (3) follows immediately from (2) because every projective cover is a generalized projective cover. So we need only prove (3) $\Rightarrow$ (1). Assume (3). Let $\bar{R}$ denote the factor ring R/J. Let S be a simple left R-module. Then $JS = 0$ and so S can be regarded as a simple left $\bar{R}$-module in the natural manner; conversely every simple left $\bar{R}$-module is regarded as a simple left R-module. Now S has a generalized projective cover P with respect to an epimorphism $f : P \to S$, so that $Ker(f) \subset JP$. On the other hand, since the factor module $P/Ker(f)$ is isomorphic to S, we have $JP \subset Ker(f)$. These imply that $Ker(f) = JP$ and, therefore, P/JP is isomorphic to S. Since, however, P is a projective left R-module, P/JP is and so S is projective as a left $\bar{R}$-module. Thus we know that every simple left $\bar{R}$-module is projective. But, as is well known, this implies that $\bar{R}$ is a semi-simple ring. (For, if the sum C of all simple left ideals of $\bar{R}$ is a proper left ideal of $\bar{R}$ then C is contained in a maximal left ideal D of $\bar{R}$. $\bar{R}/D$ is then a simple left $\bar{R}$-module and, therefore, projective. This implies that $\bar{R} = D \oplus D'$ with a simple left ideal D' of $\bar{R}$ isomorphic to $\bar{R}/D$. D' is, therefore, contained in C and hence in D, which is a contradiction. Thus, $C = \bar{R}$ and so $\bar{R}$ is semi-simple.) Now let ϵ be an idempotent of $\bar{R}$. We want to show that ϵ can be lifted to an idempotent of R. So we may assume that

ϵ is neither of 0 and $\bar{1}$, the identity element of $\bar{R}$, because 0 and 1 are clearly lifted idempotents of 0 and $\bar{1}$, respectively. Then both $\bar{R}\epsilon$ and $\bar{R}(\bar{1}-\epsilon)$ are non-zero left ideals of $\bar{R}$ and we have $\bar{R} = \bar{R}\epsilon \oplus \bar{R}(\bar{1}-\epsilon)$. Since $\bar{R}$ is semi-simple, every non-zero left ideal of $\bar{R}$ is a finite direct sum of simple left ideals. So let $\bar{R}\epsilon = S_1 \oplus S_2 \oplus \cdots \oplus S_k$ and $\bar{R}(\bar{1}-\epsilon) = S_{k+1} \oplus S_{k+2} \oplus \cdots \oplus S_n$ be direct decompositions into simple left ideals. Each S_i has a generalized projective cover P_i with respect to an epimorphism $f_i : P_i \to S_i$. Let $P = P_1 \oplus P_2 \oplus \cdots \oplus P_n$ and $f = f_1 \oplus f_2 \oplus \cdots \oplus f_n$. Then, by Proposition 2, P is a generalized projective cover of (the left R-module) $\bar{R} = S_1 \oplus S_2 \oplus \cdots \oplus S_n$ with respect to the epimorphism $f : P \to \bar{R}$. Let g be the natural epimorphism $R \to \bar{R} = R/J$. Then R is a projective cover of $\bar{R}$ with respect to g, because J is a small left ideal of R. Therefore, by Proposition 1, there is an isomorphism $h : P \to R$ such that $g \circ h = f$. Let $L = h(P_1 \oplus P_2 \oplus \cdots \oplus P_k)$ and $L' = h(P_{k+1} \oplus P_{k+2} \oplus \cdots \oplus P_n)$. Then L and L' are left ideals of R and $R = L \oplus L'$. Let $1 = e + e'$ with $e \in L$ and $e' \in L'$. Then, as is well known, both e and e' are idempotents satisfying $Re = L$ and $Re' = L'$. Let $\bar{e} = g(e) \in \bar{R}$. Then $\bar{R}\bar{e} = g(Re) = g(L) = g(h(P_1 \oplus P_2 \oplus \cdots \oplus P_k)) = f(P_1 \oplus P_2 \oplus \cdots P_k) = f_1(P_1) \oplus f_2(P_2) \oplus \cdots \oplus f_k(P_k) = S_1 \oplus S_2 \oplus \cdots \oplus S_k = \bar{R}\epsilon$. Similarly, if we put $\bar{e}' = g(e') \in \bar{R}$ then we have $\bar{R}\bar{e}' = \bar{R}(\bar{1}-\epsilon)$. Now, $\bar{1} = g(1) = g(e+e') = g(e) + g(e') = \bar{e} + \bar{e}'$ with $\bar{e} \in \bar{R}\bar{e} = \bar{R}\epsilon$ and $\bar{e}' \in \bar{R}\bar{e}' = \bar{R}(\bar{1}-\epsilon)$, while $\bar{1} = \epsilon + (\bar{1}-\epsilon)$ with $\epsilon \in \bar{R}\epsilon$ and $\bar{1} = \epsilon \in \bar{R}(\bar{1}-\epsilon)$. Therefore, by the uniqueness of the decomposition of $\bar{1}$ with respect to the direct decomposition $\bar{R} = \bar{R}\epsilon \oplus \bar{R}(\bar{1}-\epsilon)$, we know that $\bar{e} = \epsilon$, which means that $\epsilon \in \bar{R} = R/J$ is lifted to the idempotent e. Since $\bar{R}$ is semi-simple, this shows that R is semi-perfect by Theorem A.

Remark 1. Let P be a left R-module. Then the following conditions are equivalent, and P is called locally projective if P satisfies any of these conditions:
(a) If A and B are left R-modules, $f : A \to B$ is an epimorphism and $g : P \to B$ is a homomorphism then for any $x \in P$ there exists a homomorphism $h : P \to A$ such that $f(h(x)) = g(x)$; (b) If M is a left R-module and $f : M \to P$ is an epimorphism then for any $x \in P$ there exists a homomorphism $h : P \to M$ such that $f(h(x)) = x$ (i.e., f is locally split); (c) For any $x \in P$ there exist a finite number of homomorphisms $h_i : P \to R$ ($i = 1, 2, \cdots, n$) and the same number of elements $u_i \in P (i = 1, 2, \cdots, n)$ such that $\sum_{i=1}^{n} h_i(x)u_i = x$. (To see their equivalence, (a) $\Rightarrow$ (b) is clear, because (b) is the particular case of (a) where $A = M$, $B = P$ and $g : P \to P$ is the identity map. For (c) $\Rightarrow$ (a), since $g(u_i)$ is in B and f is an epimorphism, there is an $a_i \in A$ such that $f(a_i) = g(u_i)$ for each i. Let us now define a homomorphism $h : P \to A$ by $h(z) = \sum h_i(z)a_i$ for z in P. Then we have $f(h(x)) = \sum h_i(x)f(a_i) = \sum h_i(x)g(u_i) = g(\sum h_i(x)u_i) = g(x)$. Finally, assume ($b$). Let $\{u_i \mid i \in I\}$ be generators of P, i.e., $P = \sum_{i\in I} Ru_i$. Define an epimorphism $f : R^{(I)} \to P$ by $f([r_i]) = \sum_{i\in I} r_i u_i$ for $[r_i]$ in $R^{(I)}$. Then f must be locally split, *i.e.*, if x is in P, there exists a homomorphism $h : P \to R^{(I)}$ such that $f(h(x)) = x$. Now, for any $i \in I$, by associating each $z \in P$ with the i-th entry of $h(z) (\in R^{(I)})$ we have a homomorphism $h_i : P \to R$. It is then obvious that $f(h(x)) = \sum h_i(z)u_i$

for every $z \in P$ and so in particular $x = f(h(x)) = \sum h_i(x)u_i$. But there is a finite subset I_0 of I such that $h_i(x) = 0$ if $i \in I - I_0$, so that we have $x = \sum_{i \in I_0} h_i(x)u_i$. This proves $(b) \Rightarrow (c)$.) We now review some of useful properties of locally projective modules:

(i) Let P be locally projective. Then every pure submodule of P is locally projective too ([9, (1) on p. 236]). For, let Q be a pure submodule of P. Let y be an element of Q. Since P is locally projective, there exist a finite number of homomorphisms $h_i : P \to R$ and elements $u_i \in P$ $(i = 1, 2, \cdots, n)$ such that $y = \sum h_i(y)u_i$. But since Q is pure in P and y is in Q, it follows that we can find suitable $v_1, v_2, \cdots, v_n$ in Q such that $y = \sum h_i(y)v_i$. Thus, by considering the restrictions of h_i to Q, we know that Q is locally projective.

(ii) Let P be locally projective. Then JP is the intersection of all maximal submodules of P, i.e., the Jacobson radical of P. For the proof, denote by $J(P)$ the Jacobson radical of P. Let $x \in P$. Then by associating each $r \in R$ with $rx \in P$ we have a homomorphism $f : R \to P$. Let U be a maximal submodule of P. Then its inverse image $f^{-1}(U)$ is either equal to R or a maximal left ideal of R and so contains J. It follows, therefore, $Jx = f(J) \subset U$. Since this is true for all $x \in P$ and for all maximal submodules U of P, we have $JP \subset J(P)$. Let next x be any element of $J(P)$. Since P is locally projective, there exist a finite number of homomorphisms $h_i : P \to R$ and elements $u_i \in P(i = 1, 2, \cdots, n)$ such that $x = \sum h_i(x)u_i$. Let L be a maximal left ideal of R. Then its inverse image $h_i^{-1}(L)$ is either equal to P or a maximal submodule of P and, therefore, contains $J(P)$. So in particular we know that $x \in h_i^{-1}(L)$, or equivalently, $h_i(x) \in L$. Since this is true for all L and since J is the intersection of all maximal left ideals, it follows that $h_i(x) \in J$. This is true for every $i = 1, 2, \cdots, n$, which implies that $x \in \sum Ju_i \subset JP$ and so $J(P) \subset JP$. Thus it is shown that $J(P) = JP$.

(iii) Let P be locally projective and satisfy $JP = P$. Then $P = 0$([9, Prop. 2.2]). Since every projective module is locally projective, this is a generalization of [2, Prop. 2.7]. We now give here a somewhat simplified proof. Let x be in P. Since P is locally projective, there exist, as above, a finite number of homomorphisms $h_i : P \to R$ and elements $u_i \in P$ $(i = 1, 2, \cdots, n)$ such that $x = \sum h_i(x)u_i$. Let j be any integer such that $1 \leq j \leq n$, and apply h_j on both sides. Then we have $h_j(x) = \sum_i h_i(x)h_j(u_i)$. On the other hand, our assumption that $P = JP$ implies that $h_j(P) = Jh_j(P) \subset J$. Therefore, it follows from the above equality that $h_j(x) \in \sum h_i(x)J$. This means that if we denote by H the right ideal of R generated by $h_1(x), h_2(x), \cdots, h_n(x)$, i.e., $H = \sum h_i(x)R$, then we have $H \subset HJ$ whence $H = HJ$. But since H is a finitely generated right ideal, it follows from so-called Nakayama's Lemma that $H = 0$. This implies that $h_1(x) = h_2(x) = \cdots = h_n(x) = 0$ and hence $x = 0$. Thus we know that $P = 0$.

Remark 2. It is an open question whether the converse of Theorem 2 is true, i.e., whether if every flat left R-module M for which JM is small in M is always projective then R is a semi-perfect ring.

Remark 3. It was proved in [6, Th. 5] that R is left perfect if (and only if) every completely reducible (= semi-simple) left R-module has a projective cover. We would like to point out that this is regarded as a consequence of Proposition 1. For, if we assume that every completely reducible left R-module has a projective cover then in particular every simple left R-module has a projective cover and, therefore, R is semi-perfect by [6, Th. 4] (or by Theorem 3). Let F be a free left R-module with an infinite-countable basis, and let $f : F \to F/JF$ be the natural epimorphism. Then that $Ker(f) = JF$ implies that F is a generalized projective cover of F/JF with respect to f. On the other hand, since R is semi-perfect, R/J is a semi-simple ring and so F/JF is completely reducible (as a left R/J-whence) as a left R-module. Therefore, by assumption, F/JF has a projective cover. Thus it follows from Proposition 1 that F is a projective cover of F/JF with respect to f. This implies that JF is small in F, which is equivalent to that J is left T-nilpotent according to Proposition C. Thus R is left perfect.

In this connection, we note that R is semi-perfect if and only if every completely reducible left R-module has a generalized projective cover, while R is left perfect if and only if every left R-module has a generalized projective cover. Indeed, the first assertion follows from Theorem 3 and Proposition 2, while the second one is a consequence of Theorems 1 and B.

2. Semi-Perfect Modules

Let M be a left R-module. M is called a semi-perfect module if every homomorphic image of M has a projective cover. We now call M a *generalized semi-perfect* module if every homomorphic image of M has a generalized projective cover. It is clear by definition that the left R-module R is semi-perfect if and only if the ring R is semi-perfect. But since every simple left R-module is cyclic, we know by Theorem 3 that the left R-module R is generalized semi-perfect if and only if the ring R is semi-perfect. We shall later see that this phenomenon holds for projective semi-perfect modules.

Remark. The concept of semi-perfect modules was originally defined for projective modules (Mares [4]), but it has been extended to arbitrary modules in Kasch [3].

Proposition 3. Let a non-zero left R-module M have a generalized projective cover. Then M has a maximal submodule.

Proof. Let P be a generalized projective cover of M. Then P is non-zero and so $JP = J(P) \neq P$(*cf.* Remark 1, (ii) and (iii)). Therefore, P has a maximal submodule U. Let $f : P \to M$ be an epimorphism with respect to which P is a generalized projective cover of M, i.e., $Ker(f) \subset JP$. Since $JP \subset U$, it follows that $Ker(f) \subset U$ and, therefore, the image $f(U)$ of U is a maximal submodule of M.

Proposition 4. Let P be a projective left R-module. Then the following conditions are equivalent:
(1) P is a generalized projective cover of a simple left R-module.
(2) P has a unique maximal submodule.
(3) JP is a maximal submodule of P.

And in this case, a simple left R-module S of which P is a generalized projective cover is unique up to isomorphisms and is projective as a left R/J-module.

Proof. Since P is projective, JP is the intersection of all maximal submodules of P (*cf.* Remark 1, (ii)). Moreover, if $P \neq 0$ then $JP \neq P$ by [2, Prop. 2.7] (cf. Remark 1, (iii)) and, therefore, P has a maximal submodule. Also we know that if P has a unique maximal submodule then it is equal to JP, and conversely if JP is maximal in P then JP is a unique maximal submodule of P. This implies that (2) and (3) are equivalent. Suppose next that P is a generalized projective cover of a simple left R-module S with respect to an epimorphism $f : P \to S$. Then $Ker(f) \subset JP$. But since $P/Ker(f) \cong S$ $Ker(f)$ is a maximal submodule of P and, therefore, $JP \subset Ker(f)$ whence $JP = Ker(f)$. Thus JP is a maximal submodule of P, which shows (1) $\Rightarrow$ (3). Moreover, that $P/JP \cong S$ implies that S is unique up to isomorphisms and is projective as a left R/J-module, because P/JP is projective as a left R/J-module. Finally assume (3). Then P/JP is a simple left R-module. If we denote by $f : P \to P/JP$ the natural epimorphism then $Ker(f) = JP$. This shows that P is a generalized projective cover of P/JP with respect to f. Thus (3) $\Rightarrow$ (1) is proved.

As for projective covers, the following is more or less known, but we give a proof for completeness (*cf.* Ware [8, Prop. 4.1]):

Proposition 5. Let P be a projective left R-module. Then the following conditions are equivalent:
(1) P is a projective cover of a simple left R-module.
(2) P has a maximal submodule which contains all proper submodules of P.
(3) $P \neq 0$ and JP contains all proper submodules of P.
(4) $P \neq 0$ and its endomorphism ring is a local ring.
(5) $P \cong Re$ with a local idempotent e of R (i.e., an idempotent $e \neq 0$ of R for which eRe is a local ring).

A projective left R-module P satisfying any of the equivalent conditions in this proposition is called a local (or projective local) module.

Proof. Suppose that P is a projective cover of a simple left R-module S with respect to an epimorphism $f : P \to S$. Then $Ker(f)$ is maximal and small in P. If U is any proper submodule of P then the smallness of $Ker(f)$ implies that $Ker(f) + U$ is proper in P too. But this contains $Ker(f)$ which is maximal in P, so that we have $Ker(f) = Ker(f) + U$, or equivalently, $U \subset Ker(f)$. This shows (1) $\Rightarrow$ (2). Suppose that P has a maximal submodule K which contains all proper submodules of P. Then clearly $P \neq 0$ and K is a unique maximal submodule of P whence coincides with JP, the intersection of all maximal submodules of P. Thus (2) $\Rightarrow$ (3) is proved. Assume (3). Then JP is a proper submodule of P ([2, Prop. 2.7]), and if U is any proper submodule of P then $U \subset JP$ whence $JP + U = JP$, which shows that JP is maximal and small in P. This implies that P is a projective cover of the simple left R-module P/JP with respect to the natural epimorphism $P \to P/JP$, which proves (3) $\Rightarrow$ (1). Next assume (3) again. Let E be the endomorphism ring of P, which we consider a right operator ring of P. Let $h \in E$ be non-invertible in E, or equivalently, a non-automorphism of P. Suppose that $h : P \to P$ were epimorphism. Then $Ker(h) \neq 0$ and, since P is projective, h must split, i.e., P is a direct sum of $Ker(h)$ and a submodule $P' : P = Ker(h) \oplus P'$. That $P \neq 0$ and $Ker(h) \neq 0$ implies that both $Ker(h)$ and P' are proper submodules of P and hence contained in JP. But then $P = Ker(h) \oplus P'$ is also contained in and so coincides with JP, which contradicts to [2, Prop. 2.7]. Thus we know that any non-invertible element h of E is not an epimorphism $P \to P$, i.e., $Im(h)$ is a proper submodule of P and consequently contained in JP. If h' is any other non-invertible element of E then $Im(h') \subset JP$ too. It follows, therefore, that $Im(h + h') \subset Im(h) + Im(h') \subset JP$, which implies that $h + h'$ is non-invertible. Thus it is shown that E is a local ring. Conversely assume (4). Suppose that there were two proper submodules U and V of P such that $U + V = P$. Let $x \in P$. Then it can be expressed as $x = u + v$ with $u \in U$ and $v \in V$. Although such an expression is not necessarily unique, the coset $u + (U \cap V) \in U/(U \cap V)$ is uniquely determined by x, because if $x = u' + v'$ with $u' \in U$ and $v' \in V$ is another expression of x then $u - u' = v' - v$ is in $U \cap V$. Thus, by associating each $x \in P$ with $u + (U \cap V)$, we have a homomorphism $f : P \to U/(U \cap V)$. Let now $g : U \to U/(U \cap V)$ be the natural epimorphism. Then since P is projective, there exists a homomorphism $h : P \to U$ such that $g \circ h = f$. Since U is a proper submodule of P, h can be regarded as a non-invertible endomorphism ($\in E$) of P. Thus, for each $x \in P$, we have that $xh + (U \cap V) = g(xh) = f(x) = u + (U \cap V)$ and, therefore, $x(1 - h) = x - xh = (u - xh) + v$ is in $(U \cap V) + V = V$, where $x = u + v$ $(u \in U,\ v \in V)$ as above. This shows that $P(1 - h) \subset V$, a proper submodule of P, and so $1 - h$ is also a non-invertible endomorphism of P. But this contradicts to our assumption that E is a local ring, because we have $h + (1 - h) = 1$, the identity map of P. Thus we know that the sum of any two proper submodules of P is always a proper submodule of P. Let U be any proper submodule of P and let V be any maximal submodule of P. Then $U + V$ is also a proper submodule of P. But since $V \subset U + V$ and V is maximal in P, it follows that $V = U + V$, or equivalently $U \subset V$. Thus U is contained in every maximal submodules of P and so is contained in JP, which is the intersection of all maximal submodules of P because of the projectivity of P. This proves (4) $\Rightarrow$ (3), so that the four conditions (1), (2), (3) and (4) are equivalent. Assume again (3) whence (4). Let $x \in P$ be such that $x \notin JP$. Then Rx is a submodule of P not contained in JP and so coincides with P, or what is the same, by associating each $r \in R$ with rx we have an epimorphism $R \to P$. Since P is projective, this epimorphism splits, which implies that P can be imbedded isomorphically as a direct summand of R, i.e., $Re \cong P$ with an idempotent $e \neq 0$ of R. As is well known, the subring eRe of R is identified with the endomorphism ring of Re and, therefore, is isomorphic to the endomorphism ring of P, which is a local ring. Thus we have (5). That (5) implies (4) is now clear.

Theorem 4. Let M be a left R-module. Then M is generalized semi-perfect if and only if (a) every proper submodule of M is contained in a maximal submodule of M and (b) every simple homomorphic image of M has a generalized projective cover.

Proof. Let M be generalized semi-perfect. Let U be a proper submodule of M. Then M/U is a non-zero homomorphic image of M and so has a generalized projective cover. Therefore, by Proposition 3, M/U has a maximal submodule. But this means that U is contained in a maximal submodule of M. Thus (a) is satisfied.
(b) is a particular case of the definition of the generalized semi-perfectness. Conversely assume (a). Let U be any proper submodule of M. Then there is a maximal submodule V of M such that $U \subset V$. Then $S = M/V$ is a simple left R-module and so satisfies $JS = 0$, which means that $JM \subset V$. Thus $JM + U \subset V$ and, therefore, $JM + U$ is also a proper submodule of M. This shows that JM is small in M. Assume further (*b*). Denote by $\bar{M}$ the factor module M/JM. Then since $J\bar{M} = 0$, $\bar{M}$ can be regarded as a left R/J-module in the natural manner. $\bar{M}$ is, however, completely reducible, *i.e.*, a (direct) sum of simple submodules as a left R/J-, or equivalently, as a left R-module. This can be proved in the similar way as in the proof of Theorem 3, but we give a proof for completeness: Let C be the socle of $\bar{M}$, i.e., the sum of all simple submodules of $\bar{M}$. Suppose that C is a proper submodule of $\bar{M}$. Then C is contained in a maximal submodule D of $\bar{M}$, because the condition (a) for M is inherited to its homomorphic image $\bar{M}$. $\bar{M}/D$ is then a simple homomorphic image of $\bar{M}$ whence of M and, therefore, has a generalized projective cover. Thus by Proposition 4, $\bar{M}/D$ is projective as a left R/J-module. This implies that $\bar{M} = D \oplus D'$ with a simple submodule D' isomorphic to $\bar{M}/D$ is, therefore, contained in the socle C of $\bar{M}$. But since $C \subset D$, it follows that $D' \subset D$, which is clearly a contradiction. Thus we know that $C = \bar{M}$, i.e., $\bar{M}$ is completely reducible. Now let $\bar{M}$ be a direct sum of simple submodules S_i $(i \in I)$. Then each S_i is a homomorphic image of $\bar{M}$ whence of M and so has a generalized projective cover P_i with respect to an epimorphism $f_i : P_i \to S_i$, whose kernel is JP_i. Let $P = \oplus P_i$. Then by Proposition 2, P is a generalized projective cover of $\bar{M}$ with respect to the canonical epimorphism $f = \oplus f_i : P \to \bar{M}$, and indeed we have $Ker(f) = JP$. Let g be the natural epimorphism $M \to \bar{M} = M/JM$. Since P is projective, there exists then a homomorphism $h : P \to M$ such that $g \circ h = f$. From this follows that $g(h(P)) = f(P) = M$. Since $JM = Ker(g)$, this implies that $JM + h(P) = M$. We have, however, shown above that JM is small in M and, therefore, we know that $h(P) = M$, i.e., h is an epimorphism. It is clear that

$Ker(h) \subset Ker(f)$ and so $Ker(h) \subset JP$, and thus P is a generalized projective cover of M. Thus we have shown that if M satisfies (a) and (b) then M has a generalized projective cover. If, however, we observe the fact that the condition (a) for M is inherited to every homomorphic image of M and moreover every simple homomorphic image of a homomorphic image of M is a simple homomorphic image of M, we know that every homomorphic image of M has a generalized projective cover, i.e., M is a generalized semi-perfect module under the conditions (a) and (b) for M.

Corollary. Let M be a generalized semi-perfect left R-module. Then JM is small in M, the factor module M/JM is completely reducible and M has a generalized projective cover which is a direct sum of projective left R-modules P_i such that JP_i is a (unique) maximal submodule of P_i for each i.

Theorem 5. Every flat generalized semi-perfect module is projective semi-perfect.

Proof. Let M be a flat generalized semi-perfect left R-module. Then M has a generalized projective cover and, therefore, M is projective by Theorem 1. According to the above corollary, a generalized projective cover P of M can be chosen so that $P = \bigoplus_{i \in I} P_i$ with each P_i projective and having JP_i as a (unique) maximal submodule. Since M can be considered trivially a projective cover of M itself, it follows from Proposition 1 that P is isomorphic to M. This implies that we have a direct decomposition $M = \bigoplus_{i \in I} M_i$ of M into projective submodules M_i each of which has a unique maximal submodule JM_i. Let $\bar{M} = M/JM$ and let $f : M \to \bar{M}$ be the natural epimorphism. For each submodule N of M, we denote by $\bar{N}$ the homomorphic image $f(N)(\subset \bar{M})$ of N by f. Thus $\bar{N} = (N+JM)/JM \cong N/(N \cap JM)$. Now from $M = \bigoplus_{i \in I} M_i$ follows that $JM = \bigoplus_{i \in I} JM_i$ and, therefore, $M_i \cap JM = JM_i$ for each $i \in I$. This implies then that $\bar{M} = \bigoplus_{i \in I} \bar{M}_i$ and $\bar{M}_i \cong M_i/JM_i$ is simple for each $i \in I$. Let N be any submodule of M. Then, as is well known, we can choose a suitable subset I_0 of I such that $\bar{M}$ is a direct sum of two submodules $\bar{N}$ and $\bigoplus_{i \in I_0} \bar{M}_i$; $\bar{M} = \bar{N} \oplus (\bigoplus_{i \in I_0} \bar{M}_i)$. For simplicity, we put $U = \bigoplus_{i \in I_0} M_i$. Then since $\bar{U} = \bigoplus_{i \in I_0} \bar{M}_i$, we have $\bar{M} = \bar{N} \oplus \bar{U}$. This means that by the natural epimorphism $M \to \bar{M}$ the submodule $N + U$ of M is mapped onto M, or what is the same, $M = N + U + JM$. Since, however, JM is small in M by the preceding corollary, we have $M = N + U$. Consider next the submodule $N \cap U$ of M. Clearly $N \cap U$ is mapped by f into $\bar{N} \cap \bar{U}$, which is 0 because of $\bar{M} = \bar{N} \oplus \bar{U}$. Thus $N \cap U$ is contained in the kernel JM of f. It follows then $N \cap U \subset JM \cap U$. But since U is a direct summand of M, we have $JM \cap U = JU$ and besides JU is small in U. Let now g be the natural epimorphism $M \to M/N$. Then that $M = N + U$ means that U is mapped onto M/N by g, or equivalently, the restriction $g' : U \to M/N$ of g to U is an epimorphism. Moreover, the kernel of g' is $N \cap U \subset JU$, which is small in U. Since U, as a direct sum of projective submodules, is projective, this shows that U is a projective cover of M/N with respect to g'. Since this is true for every submodule N, it is proved that M is a projective semi-perfect module.

Corollary. Every projective generalized semi-perfect module is semi-perfect.

Remark. The proof of Theorem 5 is, after the projectivity of M is established, virtually the same as the proof to the implication (3) $\Rightarrow$ (1) in Azumaya [1, Th. 6]; for, each M_i is local by Proposition 5 because M_i is projective and JM_i is maximal and small in M_i. As for other characterizations of projective semi-perfect modules, see [4, Th. 5.1], [1, Th. 9] and [3, 11.3].

Theorem 6. Let M be a generalized semi-perfect left R-module. Then M is semi-perfect if and only if M has a projective cover.

Proof. The "only if" part is clear by the definition of the semi-perfectness. In order to prove the "if" part, assume that M has a projective cover P with respect to an epimorphism $f : P \to M$. Let U be a proper submodule of P. Then, since $Ker(f)$ is small in P, the homomorphic image $f(U)$ of U is also a proper submodule of M. Since M is generalized semi-perfect, it follows from Theorem 4 that $f(U)$ is contained in a maximal submodule N of M. The inverse image $f^{-1}(N)$ of N is then a maximal submodule of P and clearly contains U. Let next S be a simple homomorphic image of P. Then there exists a maximal submodule V of P such that $P/V \cong S$. Since $Ker(f)$ is small in P, $Ker(f)$ must be contained in V. Therefore, the image $f(V)$ of V is also a maximal submodule of M and we have $P/V \cong M/f(V)$ whence $S \cong M/f(V)$. This shows that S is a homomorphic image of M and so has a generalized projective cover, because M is generalized semi-perfect. Thus, by applying Theorem 4 to P, we know that P is generalized semi-perfect and hence, by Corollary to Theorem 5, semi-perfect. Now every homomorphic image of M is necessarily a homomorphic image of P and so has a projective cover due to the semi-perfectness of P. Thus it is proved that M is semi-perfect too.

REFERENCES

1. G. Azumaya, *Characterizations of semi-perfect and perfect modules*, Math. Z. **140** (1974), 95-103.
2. H. Bass, *Finitistic dimension and a homological generalization of semi-primary rings*, Trans. Amer. Math. Soc. **95** (1960), 466-488.
3. F. Kasch, *"Moduln und Ringe"*, B.G. Teubner, Stuttgart, 1977.
4. E.A. Mares, *Semi-perfect modules*, Math. Z. **82** (1963), 347-360.
5. R.W. Miller, *Endomorphism rings of finitely generated projective modules*, Pacific J. Math. **47** (1973), 199-220.
6. F.L. Sandomierski, *On semiperfect and perfect rings*, Proc. Amer. Math. Soc. **21** (1969), 205-207.
7. B. Stenström, *"Rings and Quotients"*, Springer-Verlag, Berlin/Heidelberg/New York, 1975.
8. R. Ware, *Endomorphism rings of projective modules*, Trans. Amer. Math. Soc. **155** (1971), 233-256.

9. B. Zimmermann-Huisgen, *Pure submodules of direct products of free modules*, Math. Ann. **224** (1976), 233-245.

ON UNIVERSAL LOCALIZATION AT SEMIPRIME GOLDIE IDEALS

JOHN A. BEACHY
Department of Mathematical Sciences, Northern Illinois University
DeKalb, IL 60115, U.S.A.

Abstract

In this paper we consider an alternative to Ore localization at a semiprime ideal S of a left Noetherian ring R. In [5], P.M.Cohn introduced the universal $\Sigma(S)$-inverting ring, for the set $\Sigma(S)$ of all square matrices over R that remain regular on reduction modulo S. We give an account of this universal localization from an approach that uses general ring theoretic techniques rather than those of the theory of free rings. Together with a review of a number of known results, we present a simplification of Malcolmson's construction ([8]) of the universal Σ-inverting ring that makes use of properties particular to this situation.

We first recall some known results when P is a prime ideal of a commutative Noetherian ring R. We will use $J(R)$ to denote the Jacobson radical of a ring R. Let R_P denote the localization of R with respect to the multiplicative set $R \setminus P$, with canonical homomorphism $\lambda : R \to R_P$. It is well-known that $R_P/J(R_P)$ is isomorphic to the field of quotients of R/P. Furthermore, if $\phi : R \to T$ is any ring homomorphism such that $P = \phi^{-1}(J(T))$ and the induced mapping $\overline{\phi} : R/P \to T/J(T)$ is the embedding of R/P in its field of fractions, then $\phi(c) + J(T)$ is invertible in $T/J(T)$, for all elements $c \in R \setminus P$. Since an element invertible modulo the Jacobson radical is invertible, it follows that $\phi(c)$ is invertible in T, for all $c \in R \setminus P$. By the definition of R_P, there is a unique homomorphism $\theta : R_P \to T$ with $\theta\lambda = \phi$. It is this property that we will use to define the universal localization of a noncommutative ring R at a semiprime ideal S.

1 Properties of the universal localization

Throughout this section, R will denote a left Noetherian ring (with identity), and S will denote a semiprime ideal of R. Consider the following conditions for a ring T and ring homomorphism $\phi : R \to T$.

J_1: The ring $T/J(T)$ is a semisimple Artinian ring.

J_2: $S = \phi^{-1}(J(T))$

J_3: The ring $T/J(T)$ is a classical ring of left quotients of R/S, under the induced embedding $\overline{\phi} : R/S \to T/J(T)$.

J_4: If $\theta : R \to T'$ is a ring homomorphism such that conditions J_1, J_2, and J_3 are satisfied, then there exists a unique ring homomorphism $\theta' : T \to T'$ such that $\theta = \theta'\phi$.

Since condition J_4 states that T is universal with respect to conditions J_1 through J_3, a standard argument shows that if there exists a ring satisfying conditions J_1 through J_4, then it must be unique. Before considering the existence of such a ring, we give the relevant definition.

Definition 1.1 *Let R be a left Noetherian ring, with semiprime ideal S. A ring satisfying the above conditions J_1 through J_4 is called the universal localization of R at S, and will be denoted by R_S, with canonical homomorphism $\lambda : R \to R_S$.*

For any ideal I of R, the set of elements $c \in R$ that are regular modulo I will be denoted by $C(I)$. We need to extend this definition relative to S, as follows. For any positive integer n, let $\Sigma_n(S)$ denote the set of all matrices C such that C belongs to the $n \times n$ matrix ring $M_n(R)$ and the image of C in $M_n(R/S)$ is a regular element. This will be abbreviated by saying that C is regular modulo S. Note that $C \in \Sigma_n(S)$ if and only if the image of C is invertible under the canonical mapping from $M_n(R)$ into the left classical quotient ring $Q_{cl}(M_n(R/S)) \cong M_n(Q_{cl}(R/S))$. The union over all $n > 0$ of $\Sigma_n(S)$ will be denoted by $\Sigma(S)$.

The universal localization $R_{\Sigma(S)}$ of R at $\Sigma(S)$ is defined as the universal $\Sigma(S)$-inverting ring. It can be constructed as follows (see [4] and [5] for details). For each n and each $n \times n$ matrix $[c_{ij}]$ in $\Sigma(S)$, take a set of n^2 symbols $[d_{ij}]$, and take a ring presentation of $R_{\Sigma(S)}$ consisting of all of the elements of R, as well as all of the elements d_{ij} as generators; as defining relations take all of the relations holding in R, together with all of the relations $[c_{ij}][d_{ij}] = I$ and $[d_{ij}][c_{ij}] = I$ which define all of the inverses of the matrices in $\Sigma(S)$.

Theorem 1.2 *Let R be a left Noetherian ring. For any semiprime ideal S of R, the universal localization R_S exists, and is unique up to isomorphism.*

Proof. The uniqueness follows immediately from the definition. If $\Sigma(S)$ is the set of all square matrices that are regular modulo S, then Theorem 4.1 of [4] shows that the universal $\Sigma(S)$-inverting ring $R_{\Sigma(S)}$ satisfies properties J_1 through J_3. If $\phi : R \to T$ is any ring that satisfies conditions J_1 through J_3, then for any matrix $C \in \Sigma_n(S)$ it follows that $\phi(C)$ is invertible modulo $M_n(J(T)) = J(M_n(T))$, and hence $\phi(C)$ is invertible in $M_n(T)$. Since $R_{\Sigma(S)}$ is the universal $\Sigma(S)$-inverting ring, it satisfies condition J_4. □

Proposition 1.3 *Let R be a left Noetherian ring. If S is a localizable semiprime ideal of R, then the universal localization R_S coincides with the Ore localization of R at S.*

Proof. If $C(S)$ satisfies the left Ore condition, it is well-known that the ring of left quotients of R with respect to the multiplicative set $C(S)$ satisfies conditions J_1 through J_3. Since this ring of left quotients is universal with respect to inverting elements in $C(S)$, the argument used in the proof of the previous theorem can be repeated. □

Proposition 1.4 *Let R be a left Noetherian ring, with semiprime ideal S.*

(a) *The canonical mapping $\lambda : R \to R_S$ is an epimorphism in the category of rings.*

(b) *The ring R_S is flat as a right module over R if and only if S is a left localizable ideal.*

Proof. Part (a) follows from the characterization of R_S as the universal $\Sigma(S)$-inverting ring. Part (b) is Corollary 3.2 of [1]. □

Theorem 1.5 *Let R be left Noetherian, let N be the prime radical of R, and let $K = \ker(\lambda)$, for the canonical homomorphism $\lambda : R \to R_S$.*

(a) *The kernel K is the intersection of all ideals $I \subseteq N$ such that $C(N) \subseteq C(I)$.*

(b) *The ring R/K is a left order in a left Artinian ring, and R_N is naturally isomorphic to $Q_{cl}(R/K)$.*

Proof. Parts (a) and (b) are Proposition 1.3 and Theorem 1.4 of [2], respectively. □

It is shown in Example 4 of [1] that the universal localization at a semiprime ideal of a left Noetherian ring need not be left Noetherian. In fact, the ring given as an example is a Noetherian ring finitely generated (as a module) over its center. On the other hand, it is possible to determine conditions under which the universal localization is left Artinian.

Corollary 1.6 *Let R be left Noetherian, let S be a semiprime ideal of R, and let $K = \ker(\lambda)$, for the canonical homomorphism $\lambda : R \to R_S$.*

(a) *The universal localization R_S is left Artinian if and only if $S^n \subseteq K$ for some $n > 0$.*

(b) *If P is a minimal prime ideal of R, then R_P is left Artinian.*

Proof. See Theorem 1.5 and Corollary 1.6 of [1]. □

The symbolic powers of S will be defined as in the commutative situation, by extending S^n to $R_S\lambda(S^n)R_S$ and then contracting back to R.

Definition 1.7 *Let R be a left Noetherian ring, with semiprime ideal S, and let $\lambda : R \to R_S$ be the canonical homomorphism.*

The nth symbolic power of S, denoted by $S^{(n)}$, is defined as

$$S^{(n)} = \lambda^{-1}(R_S\lambda(S^n)R_S) .$$

Proposition 1.8 *Let R be a left Noetherian ring, with semiprime ideal S, and let $\lambda : R \to R_S$ be the canonical homomorphism.*

(a) $S^{(n)} = \lambda^{-1}(J(R_S)^n)$.

(b) *$S^{(n)}$ is the intersection of all ideals I such that $S^n \subseteq I \subseteq S$ and $C(S) \subseteq C(I)$.*

(c) *$C(S)$ is a left Ore set modulo $S^{(n)}$.*

Proof. Parts (a) and (b) follow from Proposition 2.2 of [2]. Since $S/S^{(n)}$ is the prime radical of $R/S^{(n)}$, part (c) follows from part (b) and Small's Theorem. □

A number of additional results can be proved for the symbolic powers of S. For example, for all positive integers n, m we have $S^{(n)}S^{(m)} \subseteq S^{(n+m)}$. For commutative Noetherian rings it is a standard result that $\ker(\lambda) = \cap_{n=1}^{\infty} P^{(n)}$. This fails in the noncommutative setting, as shown by the following example. Let R be the ring of lower triangular 2×2 matrices with entries from the rational numbers, in which the first entry on the diagonal has odd denominator. If S is the Jacobson radical of R, then R/S is semisimple Artinian, and so $R_S = R$ and $S^{(n)} = S^n$ for all n. Thus $\ker(\lambda) = (0) \neq \cap_{n=1}^{\infty} S^{(n)}$. The following proposition gives some positive information along these lines.

Proposition 1.9 *Let R be a left Noetherian ring, with semiprime ideal S, and let $\lambda : R \to R_S$ be the canonical homomorphism. Then* $\ker(\lambda) \subseteq \cap_{n=1}^{\infty} S^{(n)}$.

Proof. This follows from the fact that the symbolic power $S^{(n)}$ is the kernel of the canonical homomorphism from R into $R_S/J(R_S)^n$, and this homomorphism satisfies properties J_1 through J_3 in the definition of R_S. □

Given a prime ideal P of a two-sided Noetherian ring R, and any positive integer n, the left symbolic powers H_n of P are defined by Goldie [6] as follows: $H_1 = P$, and by induction, H_n is defined as the two-sided $C(P)$ closure of PH_{n-1}. Lemma 2.3 of [2] shows that $P^{(n)} = H_n$, for any positive integer n.

Assume that R is Noetherian and let P be a prime ideal of R. For each positive integer n, let Q_n be the Artinian classical ring of quotients of $R/P^{(n)}$. Then there is a canonical epimorphism $Q_{n+1} \to Q_n$, for $n = 1, 2, \ldots$. Let $\hat{Q}$ be the inverse limit of the rings $\{Q_n\}_{n=1}^{\infty}$ under these epimorphisms, and let $\mu : R \to \hat{Q}$ be the induced homomorphism. Goldie's localization Q of R at P is defined as the intersection of all subrings Q' of $\hat{Q}$ such that $Q'/J(Q')$ is simple Artinian, $\cap J(Q')^n = (0)$, and $\mu(P) \subseteq J(Q') \subseteq J(\hat{Q})$. The proof of Theorem 1 of [7] shows that $Q/J(Q) \cong Q_{cl}(R/P)$, $\cap_{n=1}^{\infty} J(Q)^n = (0)$, and $P^{(n)} = \mu^{-1}(J(Q)^n)$.

Theorem 1.10 *Let P be a prime ideal of the Noetherian ring R. Then Goldie's localization of R at P is isomorphic to $R_P / \cap_{n=1}^{\infty} J(R_P)^n$.*

Proof. See Theorem 2.4 of [1]. □

2 Equivalence of quotients

Throughout this section ${}_R X$ will denote a fixed left R-module, and the direct sum of n copies of X will be denoted by X^n. The notation $x \in X^n$ will be used to denote a row vector with entries in X, and the corresponding column vector will be denoted by x^t. The identity of $M_n(R)$ will be denoted by I_n; the subscript will be omitted when the size is clear from the context.

Throughout the remainder of the paper, Σ will denote a set of square matrices over R such that

(i) Σ contains all permutation matrices;

(ii) if $C, D \in \Sigma$, then $\begin{bmatrix} C & A \\ 0 & D \end{bmatrix} \in \Sigma$ for any matrix A of the appropriate size; and

(iii) if $C, D \in \Sigma$ and CD is defined, then $CD \in \Sigma$.

We note that if R is left Noetherian and S is a semiprime ideal of R, then the set $\Sigma(S)$ of all square matrices regular modulo S satisfies the above conditions.

For the given set Σ, an element $x \in X$ is said to be *Σ-torsion* if x is an entry of a column vector v^t with entries in X such that $Cv^t = 0$ for some $C \in \Sigma$. The proof of Proposition 2.1 of [5] shows that the set of all Σ-torsion elements of X is a submodule, which we denote by $\text{rad}_\Sigma(X)$. Then X is said to be *Σ-torsion* if $\text{rad}_\Sigma(X) = X$.

It should be noted that if R is left Noetherian, S is a semiprime ideal of R, and X is finitely generated, then it is possible to give another characterization of $\Sigma(S)$-torsion modules. By Proposition 1.1 of [2], X is $\Sigma(S)$-torsion if and only if X/SX is a torsion module over R/S.

The elements of a module of quotients, denoted by X_Σ, will be constructed as equivalence classes of ordered triples (a, C, x^t), where $a \in R^n$, $C \in \Sigma_n$, and $x \in X^n$ (for any positive n). The ordered triples are modeled on the element $aC^{-1}x^t$, where C is invertible, as would be the case over R_Σ. Let $a \in R^n$, $C \in \Sigma_n$, $x \in X^n$, $b \in R^m$, $D \in \Sigma_m$, $y \in X^m$, and assume that U, V are invertible $n \times n$ matrices. If C and D are invertible, then we have the following identities.

(i) $[a \;\; b] \begin{bmatrix} C & 0 \\ 0 & D \end{bmatrix}^{-1} \begin{bmatrix} x^t \\ y^t \end{bmatrix} = aC^{-1}x^t + bD^{-1}y^t$

(ii) $aU(VCU)^{-1}Vx^t = aC^{-1}x^t$

(iii) $aC^{-1}0 = 0 = 0C^{-1}x^t$

An addition of triples is based on the first of these identities. The second motivates the definition of the initial equivalence relation for triples. We say that $(a, C, x^t) \equiv (b, D, y^t)$ if there exist invertible matrices U, V in Σ such that $b = aU$, $D = VCU$ and $y^t = Vx^t$. It is easily checked that this defines an equivalence relation. (The proof of transitivity uses the fact that Σ is closed under products.) We note that $(a, C, x^t) \equiv (b, D, y^t)$ only if C and D have the same size. Equation (iii) provides the motivation for the definition of the subsemigroup that induces the final equivalence relation.

Definition 2.1 *Let* (a, C, x^t), (b, D, y^t) *be ordered triples with* $a, b \in R^n$, $C, D \in \Sigma_n$ *and* $x, y \in X^n$, *for some positive integer* n. *If there exist invertible* $n \times n$ *matrices* U, V *in* Σ *such that* $b = aU$, $D = VCU$ *and* $y^t = Vx^t$, *then we say that* (a, C, x^t) *is congruent to* (b, D, y^t) *via* U, V, *written* $(a, C, x^t) \equiv (b, D, y^t)$ *via* U, V.

For $a \in R^n$, $C \in \Sigma_n$ *and* $x \in X^n$, *the notation* $(a : C : x^t)$ *will be used for the equivalence class of the ordered triple* (a, C, x^t) *under the equivalence relation* $\equiv$. *The set of all such equivalence classes, for all positive integers* n, *will be denoted by* $\Sigma^{-1}X$.

The subset of $\Sigma^{-1}X$ *consisting of all equivalence classes of elements of the form*

$$(e_1, E_1, 0), \quad or \quad (0, E_2, e_2^t), \quad or \quad \left([e_1 \;\; 0], \begin{bmatrix} E_1 & 0 \\ 0 & E_2 \end{bmatrix}, \begin{bmatrix} 0 \\ e_2^t \end{bmatrix}\right)$$

for some $e_1 \in R^m$, $E_1 \in \Sigma_m$, $E_2 \in \Sigma_n$, *and* $e_2 \in X^n$ *will be denoted by* $\Sigma_0^{-1}X$.

Proposition 2.2 *The sum of elements* $(a : C : x^t)$, $(b : D : y^t) \in \Sigma^{-1}X$ *defined by*

$$(a : C : x^t) + (b : D : y^t) = \left([a \;\; b] : \begin{bmatrix} C & 0 \\ 0 & D \end{bmatrix} : \begin{bmatrix} x^t \\ y^t \end{bmatrix}\right)$$

yields an associative, commutative binary operation on $\Sigma^{-1}X$.

Proof. If $(a_1, C_1, x_1^t) \equiv (a_2, C_2, x_2^t)$ via invertible matrices $U, V \in \Sigma$ and $(b_1, D_1, y_1^t) \equiv (b_2, D_2, y_2^t)$ via invertible matrices $U', V' \in \Sigma$, then it is easy to check that the respective sums are congruent via the matrices $\begin{bmatrix} U & 0 \\ 0 & U' \end{bmatrix}$ and $\begin{bmatrix} V & 0 \\ 0 & V' \end{bmatrix}$, which are invertible and belong to Σ. Thus addition of equivalence classes is well-defined on $\Sigma^{-1}X$, and it is associative by definition. Using permutation matrices, it is straightforward to check that addition of equivalence classes is commutative. □

Proposition 2.3 *For elements* $\overline{x}, \overline{y} \in \Sigma^{-1}X$, *the relation* $\sim$ *defined by*

$$\overline{x} \sim \overline{y} \quad \text{if there exist } \overline{z}_1, \overline{z}_2 \in \Sigma_0^{-1}X \text{ such that} \quad \overline{x} + \overline{z}_1 = \overline{y} + \overline{z}_2$$

is a congruence on the semigroup $\Sigma^{-1}X$. *The set* $\Sigma^{-1}X/\sim$ *of equivalence classes of this congruence is an abelian group.*

Proof. Using permutation matrices, it is easy to check that $\Sigma_0^{-1}X$ is closed under addition. Since addition in $\Sigma^{-1}X$ is associative and commutative, it follows easily that $\sim$ is a congruence. Therefore addition of equivalence classes in $\Sigma^{-1}X/\sim$ is well-defined and satisfies the associative and commutative laws. The equivalence class of $\Sigma_0^{-1}X$ is the zero element, and the following computation shows the existence of additive inverses.

For any element $(a : C : x^t) \in \Sigma^{-1}X$, we have

$$(a, C, x^t) + (a, C, -x^t) = \left([a \quad a], \begin{bmatrix} C & 0 \\ 0 & C \end{bmatrix}, \begin{bmatrix} x^t \\ -x^t \end{bmatrix}\right).$$

Since $[a \quad a] \begin{bmatrix} I & -I \\ 0 & I \end{bmatrix} = [a \quad 0]$, $\begin{bmatrix} I & I \\ 0 & I \end{bmatrix}\begin{bmatrix} C & 0 \\ 0 & C \end{bmatrix}\begin{bmatrix} I & -I \\ 0 & I \end{bmatrix} = \begin{bmatrix} C & 0 \\ 0 & C \end{bmatrix}$,
and $\begin{bmatrix} I & I \\ 0 & I \end{bmatrix}\begin{bmatrix} x^t \\ -x^t \end{bmatrix} = \begin{bmatrix} 0 \\ -x^t \end{bmatrix}$, we have

$$(a, C, x^t) + (a, C, -x^t) \equiv \left([a \quad 0], \begin{bmatrix} C & 0 \\ 0 & C \end{bmatrix}, \begin{bmatrix} 0 \\ -x^t \end{bmatrix}\right)$$

via $\begin{bmatrix} I & -I \\ 0 & I \end{bmatrix}, \begin{bmatrix} I & I \\ 0 & I \end{bmatrix}$, and the last element belongs to $\Sigma_0^{-1}X$. □

If C_1, C_2 are invertible matrices such that $C_2A_1 = A_2C_1$ for matrices A_1, A_2, then $A_1C_1^{-1} = C_2^{-1}A_2$ and so $aA_1C_1^{-1}x^t = aC_2^{-1}A_2x^t$. This motivates the following lemma for triples $(aA_1 : C_1 : x^t)$ and $(a : C_2 : A_2x^t)$ such that $C_2A_1 = A_2C_1$, a situation reminiscent of the left Ore condition. This lemma will prove to be very useful computationally.

Lemma 2.4 *Let $a \in R^m$, $C_1 \in \Sigma_n$, $x \in X^n$, and let A_1 be any $m \times n$ matrix over R. If there exist an $m \times n$ matrix A_2 and a matrix $C_2 \in \Sigma_m$ such that $C_2A_1 = A_2C_1$, then*

$$(aA_1 : C_1 : x^t) \sim (a : C_2 : A_2x^t).$$

Proof. If a, C_1, C_2, A_1, A_2, and x are as stated, then

$$[a \quad aA_1] \begin{bmatrix} I_m & -A_1 \\ 0 & I_n \end{bmatrix} = [a \quad 0] \quad , \quad \begin{bmatrix} I_m & A_2 \\ 0 & I_n \end{bmatrix}\begin{bmatrix} 0 \\ x^t \end{bmatrix} = \begin{bmatrix} A_2x^t \\ x^t \end{bmatrix}$$

and

$$\begin{bmatrix} I_m & A_2 \\ 0 & I_n \end{bmatrix}\begin{bmatrix} C_2 & 0 \\ 0 & C_1 \end{bmatrix}\begin{bmatrix} I_m & -A_1 \\ 0 & I_n \end{bmatrix} = \begin{bmatrix} C_2 & 0 \\ 0 & C_1 \end{bmatrix}.$$

Therefore

$$\left([a \quad aA_1], \begin{bmatrix} C_2 & 0 \\ 0 & C_1 \end{bmatrix}, \begin{bmatrix} 0 \\ x^t \end{bmatrix}\right) \equiv \left([a \quad 0], \begin{bmatrix} C_2 & 0 \\ 0 & C_1 \end{bmatrix}, \begin{bmatrix} A_2x^t \\ x^t \end{bmatrix}\right).$$

We then have

$$(aA_1 : C_1 : x^t) \sim (a : C_2 : 0) + (aA_1 : C_1 : x^t)$$

$$
\begin{aligned}
&= \left([a \quad aA_1] : \begin{bmatrix} C_2 & 0 \\ 0 & C_1 \end{bmatrix} : \begin{bmatrix} 0 \\ x^t \end{bmatrix}\right) \\
&= \left([a \quad 0] : \begin{bmatrix} C_2 & 0 \\ 0 & C_1 \end{bmatrix} : \begin{bmatrix} A_2x^t \\ x^t \end{bmatrix}\right) \\
&= (a : C_2 : A_2x^t) + (0 : C_1 : x^t) \\
&\sim (a : C_2 : A_2x^t) .
\end{aligned}
$$

This completes the proof. □

If S is a semiprime ideal of R for which $C(S)$ is a left denominator set, then for each $(a : C : x^t) \in \Sigma(S)^{-1}X$ there exist elements $y \in X$ and $d \in C(S)$ such that $(a : C : x^t) \sim (1 : d : y)$. To see this, let $\lambda : R \to R_S$ be the classical left localization of R at $C(S)$. We can assume without loss of generality that λ is one-to-one. Let $(a : C : x^t) \in \Sigma(S)^{-1}X$. Then $\lambda(C)$ is invertible over R_S, so it is possible to find a common denominator $d \in C(S)$ for the entries of $\lambda(a)\lambda(C)^{-1}$. Thus we have elements $d \in C(S)$ and $b \in R^n$ such that $da = bC$, and then it follows from Lemma 2.4 that $(1 \cdot a : C : x^t) \sim (1 : d : bx^t)$.

Proposition 2.5 *Let $a \in R^n$, $C \in \Sigma_n$, and $x \in X^n$.*

(a) *If $b \in R^n$ and $y \in X^n$, then*

$$(a : C : x^t) + (a : C : y^t) \sim (a : C : (x+y)^t)$$

and

$$(a : C : x^t) + (b : C : x^t) \sim (a + b : C : x^t) .$$

(b) *For any matrices P, Q such that $PC, CQ \in \Sigma_n$,*

$$(a : C : x^t) \sim (a : PC : Px^t) \quad \text{and} \quad (a : C : x^t) \sim (aQ : CQ : x^t) .$$

(c) *For any $b \in R^m$, $D \in \Sigma_m$, $y \in X^m$ and any matrices A, B of the appropriate size,*

$$(a : C : x^t) \sim \left([a \quad b] : \begin{bmatrix} C & A \\ 0 & D \end{bmatrix} : \begin{bmatrix} x^t \\ 0 \end{bmatrix}\right)$$

and

$$(a : C : x^t) \sim \left([0 \quad a] : \begin{bmatrix} D & B \\ 0 & C \end{bmatrix} : \begin{bmatrix} y^t \\ x^t \end{bmatrix}\right) .$$

Proof. (a) Since $C\,[I \quad I] = [I \quad I] \begin{bmatrix} C & 0 \\ 0 & C \end{bmatrix}$, it follows from Lemma 2.4 that

$$\left(a\,[I \quad I] : \begin{bmatrix} C & 0 \\ 0 & C \end{bmatrix} : \begin{bmatrix} x^t \\ y^t \end{bmatrix}\right) \sim \left(a : C : [I \quad I] \begin{bmatrix} x^t \\ y^t \end{bmatrix}\right)$$

and so $(a : C : x^t) + (a : C : y^t) \sim (a : C : (x+y)^t)$. The second half of condition (a) follows in a similar fashion.

(b) By Lemma 2.4, we have $(aI : C : x^t) \sim (a : PC : Px^t)$ since $(PC)(I) = (P)(C)$. Similarly, $(C)(Q) = (I)(CQ)$ shows that $(aQ : CQ : x^t) \sim (a : C : Ix^t)$.

(c) Since $\begin{bmatrix} C & A \\ 0 & D \end{bmatrix} \begin{bmatrix} I \\ 0 \end{bmatrix} = \begin{bmatrix} I \\ 0 \end{bmatrix} C$, by Lemma 2.4 we have

$$\left([a \quad b] \begin{bmatrix} I \\ 0 \end{bmatrix} : C : x^t \right) \sim \left([a \quad b] : \begin{bmatrix} C & A \\ 0 & D \end{bmatrix} : \begin{bmatrix} I \\ 0 \end{bmatrix} x^t \right) .$$

Finally, since $C\, [0 \quad I] = [0 \quad I] \begin{bmatrix} D & B \\ 0 & C \end{bmatrix}$, it again follows from Lemma 2.4 that

$$\left(a\, [0 \quad I] : \begin{bmatrix} D & B \\ 0 & C \end{bmatrix} : \begin{bmatrix} y^t \\ x^t \end{bmatrix} \right) \sim \left(a : C : [0 \quad I] \begin{bmatrix} y^t \\ x^t \end{bmatrix} \right) .$$

This completes the proof. □

Proposition 2.6 *Let $(a : C : x^t), (b : D : y^t) \in \Sigma^{-1}X$. Then $(a : C : x^t) \sim (b : D : y^t)$ if and only if there exist vectors e_1, u over R and e_2, v over X (of the appropriate size) and matrices $\begin{bmatrix} E_1 & 0 \\ 0 & E_2 \end{bmatrix}$ and $P, Q \in \Sigma$ such that $uv^t = 0$ and*

$$(a, C, x^t) + (b, D, -y^t) + \left([e_1 \quad 0], \begin{bmatrix} E_1 & 0 \\ 0 & E_2 \end{bmatrix}, \begin{bmatrix} 0 \\ e_2^t \end{bmatrix} \right) = (uQ, PQ, Pv^t) .$$

Proof. First assume that $(a : C : x^t) \sim (b : D : y^t)$. Then $(a : C : x^t) + (b : D : -y^t)$ is equivalent to zero, since $\Sigma^{-1}X/\sim$ is an abelian group and $(b : D : -y^t)$ represents the additive inverse of $(b : D : y^t)$. By definition of $\sim$, there exist $\overline{z}_1, \overline{z}_2 \in \Sigma_0^{-1}X$ such that $(a : C : x^t) + (b : D : -y^t) + \overline{z}_1 = \overline{z}_2$. If $\overline{z}_2 = \left([f_1 \quad 0] : \begin{bmatrix} F_1 & 0 \\ 0 & F_2 \end{bmatrix} : \begin{bmatrix} 0 \\ f_2^t \end{bmatrix} \right)$, then using the definition of the equivalence relation $\equiv$, there exist invertible matrices $U, V \in \Sigma$ of the appropriate size such that

$$(a, C, x^t) + (b, D, -y^t) + z_1 = \left([f_1 \quad 0]U, \ V \begin{bmatrix} F_1 & 0 \\ 0 & F_2 \end{bmatrix} U, \ V \begin{bmatrix} 0 \\ f_2^t \end{bmatrix} \right) .$$

Since z_1 already has the desired form, we only need to factor the right hand side. Let $Q = \begin{bmatrix} I & 0 \\ 0 & F_2 \end{bmatrix} U$ and $P = V \begin{bmatrix} F_1 & 0 \\ 0 & I \end{bmatrix}$. This factorization yields (uQ, PQ, Pv^t), for $u = [f_1 \quad 0]$ and $v = [0 \quad f_2]$, and then $uv^t = 0$.

Conversely, suppose that the given condition holds. Then by the definition of $\sim$ we have $(a : C : x^t) + (b : D : -y^t) \sim (uQ : PQ : Pv^t)$. Using the previous proposition and Lemma 2.4, we have

$$(uQ : PQ : Pv^t) \sim (u : I : v^t) \sim (1 : 1 : uv^t) = (1 : 1 : 0) ,$$

and thus $(a : C : x^t) \sim (b : D : y^t)$. □

Recall that an element $x \in X$ is Σ-torsion if it is an entry in a vector $v \in X^n$ such that $Cv^t = 0$ for some $C \in \Sigma$. Since Σ is closed under products (when defined) and contains all permutation matrices, it can be assumed that x is the first entry of v^t.

Theorem 2.7 *Let $x, y \in X$.*

(a) *In $\Sigma^{-1}X$, $(1 : 1 : x) \sim (1 : 1 : y)$ if and only if $x = av^t$ and $y = bw^t$ for some $a, b \in R^n$, $v, w \in X^n$ and some $n > 0$ such that there exist $C, D, P, Q \in \Sigma_n$ satisfying $aD = bQ$, $Cv^t = Pw^t$ and $CD = PQ$.*

(b) *Furthermore, $x - y \in \mathrm{rad}_\Sigma(X)$ if and only if in condition (a) it is possible to take $a = b$, $C = P$ and $D = Q = I$.*

Proof. (a) If $(1 : 1 : x) \sim (1 : 1 : y)$, then there exist $\overline{z}_1, \overline{z}_2 \in \Sigma_0^{-1}X$ such that $(1, 1, x) + z_1 \equiv (1, 1, y) + z_2$. If $z_1 = \left([e_1 \;\; 0], \begin{bmatrix} E_1 & 0 \\ 0 & E_2 \end{bmatrix}, \begin{bmatrix} 0 \\ e_2^t \end{bmatrix}\right)$, then $(1, 1, x) + z_1$ has the form

$$\left([1 \;\; e_1 \;\; 0], \begin{bmatrix} 1 & 0 & 0 \\ 0 & E_1 & 0 \\ 0 & 0 & E_2 \end{bmatrix}, \begin{bmatrix} x \\ 0 \\ e_2^t \end{bmatrix}\right).$$

This can be factored in the form (aD, CD, Cv^t) for

$$a = [1 \;\; e_1 \;\; 0],\; C = \begin{bmatrix} 1 & 0 & 0 \\ 0 & E_1 & 0 \\ 0 & 0 & I \end{bmatrix},\; D = \begin{bmatrix} 1 & 0 & 0 \\ 0 & I & 0 \\ 0 & 0 & E_2 \end{bmatrix},\; v = [x \;\; 0 \;\; e_2],$$

with $av^t = x$. A similar factorization of $(1, 1, y) + z_2$, multiplied by the invertible matrices obtained from the definition of the relation $\equiv$, gives $(aD, CD, Cv^t) = (bQ, PQ, Pw^t)$, where $bw^t = y$.

Conversely, if the stated condition holds, then we have

$$(1 : 1 : x) \sim (a : I : v^t) \sim (aD : CD : Cv^t) =$$
$$(bQ : PQ : Pw^t) \sim (b : I : w^t) \sim (1 : 1 : y).$$

(b) If $x, y \in X$ with $x - y \in \mathrm{rad}_\Sigma(X)$, then $x - y$ is the first entry of a vector u such that $Cu^t = 0$ for some $C \in \Sigma$. It is possible to write $u = v - w$, where x and y are the first entries of v and w, respectively, so that $Cv^t = Cw^t$. If a denotes the vector over R with 1 as its first entry and zeroes elsewhere, then $av^t = x$ and $aw^t = y$, giving the desired result.

Conversely, if the condition is satisfied, then $C(v^t - w^t) = 0$, so all entries of $v^t - w^t$ belong to $\mathrm{rad}_\Sigma(X)$. It follows that $x - y = a(v^t - w^t) \in \mathrm{rad}_\Sigma(X)$, completing the proof. □

Malcolmson [8] has shown that $\lambda : R \to R_\Sigma$ given by $\lambda(r) = (1 : 1 : r)$ is a ring homomorphism, which inverts the matrices in Σ. It follows immediately that any

Σ-torsion element (torsion on either left or right) must be mapped to zero by λ. The following example shows, since the left Σ-torsion ideal differs from the right Σ-torsion ideal, that it is possible to have equivalent elements $(1:1:r)$ and $(1:1:s)$ for which $r-s$ does not belong to the left Σ-torsion ideal.

Let R be the following ring of lower triangular matrices with entries from the ring of integers or the ring of integers modulo 2, as indicated. Let S be the prime radical of R, let Σ be the set $\Sigma(S)$, and consider the ideal I defined below.

$$R = \begin{bmatrix} Z_2 & 0 & 0 \\ Z_2 & Z & 0 \\ Z_2 & Z_2 & Z_2 \end{bmatrix} \qquad S = \begin{bmatrix} 0 & 0 & 0 \\ Z_2 & 0 & 0 \\ Z_2 & Z_2 & 0 \end{bmatrix} \qquad I = \begin{bmatrix} 0 & 0 & 0 \\ Z_2 & 0 & 0 \\ Z_2 & 0 & 0 \end{bmatrix}$$

A matrix belongs to $C(S)$ if and only if its entries on the main diagonal are all nonzero, so $\begin{bmatrix} 1 & 0 & 0 \\ 0 & 2 & 0 \\ 0 & 0 & 1 \end{bmatrix} \in C(S)$. This element annihilates I/SI, which shows by Proposition 1.1 of [2] that I is left $\Sigma(S)$-torsion.

Furthermore, I is the left $\Sigma(S)$-torsion ideal since S/S^2 is not $\Sigma(S)$-torsion. Arguing by symmetry, the right $\Sigma(S)$-torsion ideal is the bottom row of S, so the kernel of λ must be S. Then $(1:1:x)$ is equivalent to $(1:1:0)$ for the element $x = \begin{bmatrix} 0 & 0 & 0 \\ 0 & 0 & 0 \\ 0 & 1 & 0 \end{bmatrix}$, but $x - 0$ is not $\Sigma(S)$-torsion. This establishes a clear distinction between conditions (a) and (b) of Theorem 2.7.

3 Modules of quotients

Throughout this section ${}_RX$ will denote a fixed left R-module. We begin with the definition of a module of quotients.

Definition 3.1 *The set of equivalence classes of $\Sigma^{-1}X/\sim$ will be denoted by X_Σ. The notation $[a:C:x^t]$ will be used for the class of $(a:C:x^t) \in \Sigma^{-1}X$.*

Proposition 2.6 of Section 2 shows that our definition of equivalence for elements of R_Σ is the same as that of Malcolmson [8]. The multiplication about to be defined coincides with that in Malcolmson's construction, so we have in fact defined the universal Σ-inverting ring. Thus properties of R_Σ may be used in constructing modules of quotients. It should be noted that the scalar multiplication defined below can be used to construct the ring R_Σ, and the necessary properties can be verified using only the techniques of this paper.

Let $a, r \in R^n$, $C \in \Sigma_n$, $b \in R^m$, $D \in \Sigma_m$, and $y \in X^m$. If C, D are invertible, then we have the following identity, which motivates the definition of a scalar multiplication.

$$[a \;\; 0] \begin{bmatrix} C & -r^t b \\ 0 & D \end{bmatrix}^{-1} \begin{bmatrix} 0 \\ y^t \end{bmatrix} = aC^{-1}r^t \cdot bD^{-1}y^t$$

Proposition 3.2 *The scalar product of elements $(a : C : r^t) \in \Sigma^{-1}R$ and $(b : D : y^t) \in \Sigma^{-1}X$ defined by*

$$(a : C : r^t) \cdot (b : D : y^t) = \left([a \;\; 0] : \begin{bmatrix} C & -r^t b \\ 0 & D \end{bmatrix} : \begin{bmatrix} 0 \\ y^t \end{bmatrix} \right)$$

yields a well-defined, associative operation.

Proof. If $(a_1, C_1, r_1^t) \equiv (a_2, C_2, r_2^t)$ via invertible matrices $U, V \in \Sigma$ and $(b_1, D_1, y_1^t) \equiv (b_2, D_2, y_2^t)$ via invertible matrices $U', V' \in \Sigma$, then we have

$$\left([a_1 \;\; 0], \begin{bmatrix} C_1 & -r_1^t b_1 \\ 0 & D_1 \end{bmatrix}, \begin{bmatrix} 0 \\ y_1^t \end{bmatrix} \right) \equiv \left([a_2 \;\; 0], \begin{bmatrix} C_2 & -r_2^t b_2 \\ 0 & D_2 \end{bmatrix}, \begin{bmatrix} 0 \\ y_2^t \end{bmatrix} \right)$$

via the matrices $\begin{bmatrix} U & 0 \\ 0 & U' \end{bmatrix}$ and $\begin{bmatrix} V & 0 \\ 0 & V' \end{bmatrix}$.

If $\overline{p}, \overline{q} \in \Sigma^{-1}R$ and $\overline{x} \in \Sigma^{-1}X$, then $\overline{p}(\overline{q}\,\overline{x}) = (\overline{p}\,\overline{q})\overline{x}$ as a consequence of the way in which matrices are combined. □

Lemma 3.3 *The following conditions hold for scalar multiplication.*

(a) *If $(a : C : r^t) \in \Sigma^{-1}R$ and $(1 : 1 : x) \in \Sigma^{-1}X$, then*

$$(a : C : r^t)(1 : 1 : x) \sim (a : C : r^t x) .$$

(b) *If $(1 : 1 : s) \in \Sigma^{-1}R$ and $(a : C : x^t) \in \Sigma^{-1}X$, then*

$$(1 : 1 : s)(a : C : x^t) \sim (sa : C : x^t) .$$

Proof. (a) Since $C\,[I \;\; 0] = [I \;\; r^t] \begin{bmatrix} C & -r^t \\ 0 & 1 \end{bmatrix}$, we have

$$\begin{aligned} (a : C : r^t)(1 : 1 : x) &= \left(a\,[I \;\; 0] : \begin{bmatrix} C & -r^t \\ 0 & 1 \end{bmatrix} : \begin{bmatrix} 0 \\ x \end{bmatrix} \right) \\ &\sim \left(a : C : [I \;\; r^t] \begin{bmatrix} 0 \\ x \end{bmatrix} \right) = (a : C : r^t x) . \end{aligned}$$

(b) Since $\begin{bmatrix} 1 & -sa \\ 0 & C \end{bmatrix}\begin{bmatrix} sa \\ I \end{bmatrix} = \begin{bmatrix} 0 \\ I \end{bmatrix} C$, we have

$$\begin{aligned} (sa : C : x^t) &= \left([1 \;\; 0] \begin{bmatrix} sa \\ I \end{bmatrix} : C : x^t \right) \\ &\sim \left([1 \;\; 0] : \begin{bmatrix} 1 & -sa \\ 0 & C \end{bmatrix} : \begin{bmatrix} 0 \\ I \end{bmatrix} x^t \right) \\ &= (1 : 1 : s)(a : C : x^t) \end{aligned}$$

This completes the proof. □

Lemma 3.4 *Let* $\overline{p} = (u : P : r^t)$, $\overline{q} = (v : Q : s^t) \in \Sigma^{-1}R$, *and let* $\overline{x} = (a : C : x^t)$, $\overline{y} = (b : D : y^t) \in \Sigma^{-1}X$. *Then*

$$(\overline{p} + \overline{q})\overline{x} \sim \overline{p}\,\overline{x} + \overline{q}\,\overline{x} \quad \text{and} \quad \overline{q}(\overline{x} + \overline{y}) \sim \overline{q}\,\overline{x} + \overline{q}\,\overline{y}\,.$$

Proof. We have the following equalities.

$$(\overline{p} + \overline{q})\overline{x} = \left([u \;\; 0 \;\; v \;\; 0] \begin{bmatrix} I & 0 & 0 \\ 0 & 0 & I \\ 0 & I & 0 \\ 0 & 0 & I \end{bmatrix} : \begin{bmatrix} P & 0 & -r^t a \\ 0 & Q & -s^t a \\ 0 & 0 & C \end{bmatrix} : \begin{bmatrix} 0 \\ 0 \\ x^t \end{bmatrix} \right)$$

$$\overline{p}\,\overline{x} + \overline{q}\,\overline{x} = \left([u \;\; 0 \;\; v \;\; 0] : \begin{bmatrix} P & -r^t a & 0 & 0 \\ 0 & C & 0 & 0 \\ 0 & 0 & Q & -s^t a \\ 0 & 0 & 0 & C \end{bmatrix} : \begin{bmatrix} I & 0 & 0 \\ 0 & 0 & I \\ 0 & I & 0 \\ 0 & 0 & I \end{bmatrix} \begin{bmatrix} 0 \\ 0 \\ x^t \end{bmatrix} \right)$$

The two expressions are equal by Lemma 2.4, since

$$\begin{bmatrix} P & -r^t a & 0 & 0 \\ 0 & C & 0 & 0 \\ 0 & 0 & Q & -s^t a \\ 0 & 0 & 0 & C \end{bmatrix} \begin{bmatrix} I & 0 & 0 \\ 0 & 0 & I \\ 0 & I & 0 \\ 0 & 0 & I \end{bmatrix} = \begin{bmatrix} I & 0 & 0 \\ 0 & 0 & I \\ 0 & I & 0 \\ 0 & 0 & I \end{bmatrix} \begin{bmatrix} P & 0 & -r^t a \\ 0 & Q & -s^t a \\ 0 & 0 & C \end{bmatrix}.$$

Finally, by Lemma 2.4 the expressions given below for $\overline{q}(\overline{x} + \overline{y})$ and $\overline{q}\,\overline{x} + \overline{q}\,\overline{y}$ are equal since we have the following identity.

$$\begin{bmatrix} Q & -s^t a & -s^t b \\ 0 & C & 0 \\ 0 & 0 & D \end{bmatrix} \begin{bmatrix} I & 0 & I & 0 \\ 0 & I & 0 & 0 \\ 0 & 0 & 0 & I \end{bmatrix} = \begin{bmatrix} I & 0 & I & 0 \\ 0 & I & 0 & 0 \\ 0 & 0 & 0 & I \end{bmatrix} \begin{bmatrix} Q & -s^t a & 0 & 0 \\ 0 & C & 0 & 0 \\ 0 & 0 & Q & -s^t b \\ 0 & 0 & 0 & D \end{bmatrix}$$

$$\overline{q}\,\overline{x} + \overline{q}\,\overline{y} = \left([v \;\; 0 \;\; 0] \begin{bmatrix} I & 0 & I & 0 \\ 0 & I & 0 & 0 \\ 0 & 0 & 0 & I \end{bmatrix} : \begin{bmatrix} Q & -s^t a & 0 & 0 \\ 0 & C & 0 & 0 \\ 0 & 0 & Q & -s^t b \\ 0 & 0 & 0 & D \end{bmatrix} : \begin{bmatrix} 0 \\ x^t \\ 0 \\ y^t \end{bmatrix} \right)$$

$$\overline{q}(\overline{x}+\overline{y}) = \left([v\ 0\ 0] : \begin{bmatrix} Q & -s^t a & -s^t b \\ 0 & C & 0 \\ 0 & 0 & D \end{bmatrix} : \begin{bmatrix} I & 0 & I & 0 \\ 0 & I & 0 & 0 \\ 0 & 0 & 0 & I \end{bmatrix} \begin{bmatrix} 0 \\ x^t \\ 0 \\ y^t \end{bmatrix} \right)$$

This completes the proof. □

Theorem 3.5 *For any module ${}_RX$, the set X_Σ is a unital left module over R_Σ.*

Proof. We have shown in Proposition 2.3 that X_Σ is an abelian group under addition. To show that the multiplication defined in Proposition 3.2 respects the equivalence relation $\sim$ on $\Sigma^{-1}X$, let $\overline{p}, \overline{q} \in \Sigma^{-1}R$ and $\overline{x}, \overline{y} \in \Sigma^{-1}X$ with $\overline{p} \sim \overline{q}$ and $\overline{x} \sim \overline{y}$. Then there exist $\overline{z}_1, \overline{z}_2 \in \Sigma_0^{-1}R$ and $\overline{z}_3, \overline{z}_4 \in \Sigma_0^{-1}X$ with $\overline{p}+\overline{z}_1 = \overline{q}+\overline{z}_2$ and $\overline{x}+\overline{z}_3 = \overline{y}+\overline{z}_4$. Thus we have

$$(\overline{p}+\overline{z}_1)\overline{x} = (\overline{q}+\overline{z}_2)\overline{x} \qquad \text{and} \qquad \overline{q}(\overline{x}+\overline{z}_3) = \overline{q}(\overline{y}+\overline{z}_4)\ .$$

It follows from Lemma 3.4 that

$$\overline{p}\,\overline{x}+\overline{z}_1\overline{x} \sim \overline{q}\,\overline{x}+\overline{z}_2\overline{x} \qquad \text{and} \qquad \overline{q}\,\overline{x}+\overline{q}\,\overline{z}_3 \sim \overline{q}\,\overline{y}+\overline{q}\,\overline{z}_4\ .$$

By definition of scalar multiplication, we have $\overline{z}_1\overline{x}, \overline{z}_2\overline{x}, \overline{q}\,\overline{z}_3, \overline{q}\,\overline{z}_4 \in \Sigma_0^{-1}X$, and so we obtain $\overline{p}\,\overline{x} \sim \overline{q}\,\overline{y}$.

The distributive laws hold by Lemma 3.4. Finally, X_Σ is a unital left R_Σ-module by Lemma 3.3. □

Proposition 3.6 *For the module ${}_RX$, define the mapping $\eta : X \to X_\Sigma$ by $\eta(x) = [1 : 1 : x]$, for all $x \in X$. Then η is an R-homomorphism.*

Proof. The ring R acts on X_Σ via the homomorphism $\lambda : R \to R_\Sigma$ defined by $\lambda(r) = [1 : 1 : r]$. Thus by Lemma 3.3 (a), for any $r \in R$ and any $x \in X$ we have $\eta(rx) = [1 : 1 : rx] = [1 : 1 : r][1 : 1 : x] = \lambda(r)\eta(x)$. □

Proposition 3.7 *Let $[a : C : x^t] \in X_\Sigma$. Then $[a : C : x^t] = \lambda(a)\lambda(C)^{-1}\eta(x^t)$ for the canonical mappings $\lambda : R \to R_\Sigma$ and $\eta : X \to X_\Sigma$.*

Proof. Let e_i denote the vector with 1 in the ith entry and zero elsewhere. Assume that $C \in \Sigma_n$, and let $C = [c_{ij}]$ for elements $c_{ij} \in R$. It follows from Lemma 3.3 (b) that for a fixed $k > 0$,

$$\begin{aligned} \sum_{i=1}^{n}[1 : 1 : c_{ki}][e_i : C : e_j^t] &= \sum_{i=1}^{n}[c_{ki}e_i : C : e_j^t] \\ &= [e_k C : C : e_j^t] = [e_k : I : e_j^t] \\ &= [1 : 1 : e_k e_j^t] = \delta_{kj}\ . \end{aligned}$$

(Proposition 2.5 (a) and (b) and Lemma 2.4 have also been used.)

A similar argument holds on the other side, showing that the entries of $\lambda(C)^{-1}$ are just the elements $[e_i : C : e_j^t]$ in R_Σ. Having found $\lambda(C)^{-1}$, it follows that

$$\begin{aligned}
\lambda(a)\lambda(C)^{-1}\eta(x^t) &= \sum_{j=1}^{n}\left(\sum_{i=1}^{n}[1:1:a_i][e_i:C:e_j^t]\right)[1:1:x_j] \\
&= \sum_{j=1}^{n}\left(\sum_{i=1}^{n}[a_ie_i:C:e_j^t]\right)[1:1:x_j] \\
&= \sum_{j=1}^{n}[a:C:e_j^t][1:1:x_j] = \sum_{j=1}^{n}[a:C:e_j^tx_j] \\
&= [a:C:x^t]\,.
\end{aligned}$$

This completes the proof. □

We say that the module ${}_RX$ is *Σ-torsionfree* if $Cx^t = 0$ implies $x = 0$, for all $C \in \Sigma_n$ and all $x \in X^n$. We say that X is *Σ-divisible* if for each $x \in X^n$ and each $C \in \Sigma$ there exists $y \in X^n$ such that $Cy^t = x^t$.

Theorem 3.8 *The homomorphism $\eta : X \to X_\Sigma$ is an isomorphism if and only if X is Σ-torsionfree and Σ-divisible.*

Proof. If η is an isomorphism, then X has a natural structure as a left R_Σ-module, and so $Cx^t = 0$ implies $x^t = \lambda(C)^{-1}Cx^t = 0$ for any $x \in X^n$, showing that X is Σ-torsionfree. Similarly, X is Σ-divisible since for any $x \in X^n$ we have $x^t = C(\lambda(C)^{-1}x^t)$.

Conversely, suppose that X is Σ-torsionfree and Σ-divisible. For any element $[a : C : x^t] \in X_\Sigma$, there exists y such that $x^t = Cy^t$, and then

$$[a:C:x^t] = [a:C:Cy^t] = [a:I:y^t] = [1:1:ay^t]$$

by Proposition 2.5 (b) and Lemma 2.4. Thus $[a : C : x^t] = \eta(ay^t)$ and so η is an epimorphism. Now let $x \in \ker(\eta)$. By Theorem 2.7 (a) there exist $a, b \in R^n$, $v, w \in X^n$, and $C, D, P, Q \in \Sigma_n$ such that $x = av^t$, $bw^t = 0$, $aD = bQ$, $Cv^t = Pw^t$, and $CD = PQ$. Since X is Σ-divisible there exist $v_1, w_1 \in X^n$ such that $v^t = Dv_1^t$ and $w^t = Qw_1^t$. Therefore $CDv_1^t = Cv^t = Pw^t = PQw_1^t$, and so $v_1^t = w_1^t$ since $CD = PQ$ and X is Σ-torsionfree. But then $x = av^t = a(Dv_1^t) = (bQ)w_1^t = bw^t = 0$, and so η is a monomorphism. □

As in Theorem 4.9 of [9], the following corollary implies that if R is a left hereditary ring, then so is the universal localization R_Σ. The corollary can be proved using an argument similar to the standard one for the class of torsionfree divisible modules over an integral domain.

Corollary 3.9 *In the category of left R-modules, the class of left R_Σ-modules is closed under extensions.*

Theorem 3.10 *For any module $_RX$, the module of quotients X_Σ is naturally isomorphic to $R_\Sigma \otimes_R X$.*

Proof. Let $\epsilon : X \to R_\Sigma \otimes_R X$ be the natural homomorphism defined by $\epsilon(x) = 1 \otimes x$, for all $x \in X$. Then since X_Σ is an R_Σ-module, we can define $\theta : R_\Sigma \otimes_R X \to X_\Sigma$ with $\theta(q \otimes x) = q\eta(x)$, for all $q \in R_\Sigma$ and $x \in X$, where η is the canonical mapping from X to X_Σ. Then $\theta\epsilon = \eta$, and for $x \in X$ and $\lambda(a)\lambda(C)^{-1}\lambda(r^t) \in R_\Sigma$, we have

$$\begin{aligned} \theta\left(\lambda(a)\lambda(C)^{-1}\lambda(r^t) \otimes x\right) &= \lambda(a)\lambda(C)^{-1}\lambda(r^t)\eta(x) \\ &= \lambda(a)\lambda(C)^{-1}\eta(r^t x) . \end{aligned}$$

It follows that given $\lambda(a)\lambda(C)^{-1}\eta(x^t) \in X_\Sigma$, we have

$$\lambda(a)\lambda(C)^{-1}\eta(x^t) = \theta\left(\sum_{i=1}^{n} \lambda(a)\lambda(C)^{-1}\lambda(e_i^t) \otimes e_i x^t\right) ,$$

which shows that θ is onto.

To show that θ is one-to-one, we will show that it has an inverse. For $(a : C : x^t) \in \Sigma^{-1}X$, define

$$\phi((a : C : x^t)) = \sum_{i=1}^{n} \lambda(a)\lambda(C)^{-1}\lambda(e_i^t) \otimes e_i x^t .$$

It is clear that ϕ is well-defined on $\Sigma^{-1}X$ and additive. To show that it is well-defined on X_Σ, by Proposition 2.6 it suffices to show that $\phi((a : C : x^t)) = 0$ for any element $(a : C : x^t)$ of the form $(uQ : PQ : Pv^t)$ with $P, Q \in \Sigma$ and $uv^t = 0$. We have

$$\begin{aligned} \phi((uQ : PQ : Pv^t)) &= \sum_{i=1}^{n} \lambda(uQ)\lambda(PQ)^{-1}\lambda(e_i^t) \otimes e_i P v^t \\ &= \sum_{i=1}^{n} \lambda(uQ)\lambda(PQ)^{-1}\lambda(e_i^t) \otimes \left(\sum_{j=1}^{n} (e_i P e_j^t) e_j v^t\right) \\ &= \sum_{j=1}^{n} \lambda(u)\lambda(Q)\lambda(Q)^{-1}\lambda(P)^{-1}\lambda(P)\lambda(e_j^t) \otimes e_j v^t \\ &= \sum_{j=1}^{n} \lambda(u)e_j^t \otimes e_j v^t = \sum_{j=1}^{n} 1 \otimes u e_j^t e_j v^t = 1 \otimes uv^t . \end{aligned}$$

This expression is equal to zero whenever $uv^t = 0$. It can be shown easily that $\phi\theta = 1$, and so θ is an R_Σ-isomorphism.

It is clear that θ is a natural transformation. □

References

[1] J. A. Beachy, Inversive localization at semiprime Goldie ideals, *Manuscripta Math.* **34** (1981) 211-239.

[2] J. A. Beachy, On noncommutative localization, *J. Algebra* **87** (1984) 213-221.

[3] J. A. Beachy and W. D. Blair, Examples in noncommutative localization, *J. Algebra* **99** (1986) 108-113.

[4] P. M. Cohn, *Free rings and their relations* (2nd ed., Academic Press, London–New York, 1985).

[5] P. M. Cohn, Inversive localisation in Noetherian rings, *Commun. Pure Appl. Math.* **26** (1973) 679-691.

[6] A. W. Goldie, Localization in non-commutative Noetherian rings, *J. Algebra* **5** (1967) 89-105.

[7] A. W. Goldie, A note on noncommutative localization, *J. Algebra* **8** (1968) 41-44.

[8] P. Malcolmson, Construction of universal matrix localizations, *Lecture Notes in Math* **V 951** (Springer–Verlag, Berlin–Heidelberg–New York, 1982), pp. 117-131.

[9] A. H. Schofield, Representations of rings over skew fields, *London Mathematical Society Lecture Notes* **92** (Cambridge University Press, Cambridge, 1985).

[10] B. Stenström, B. *Rings of Quotients* (Springer–Verlag, Berlin–Heidelberg–New York, 1975).

Strongly and properly semiprime modules and rings

KONSTANTIN BEIDAR
Chair of Algebra, Moscow State University, CIS
and
ROBERT WISBAUER
Mathematical Institute, University of Düsseldorf, Germany

ABSTRACT

Using torsion theoretic techniques we define *strongly* and *properly semiprime* modules. Strongly semiprime modules M are characterized by the fact that in the M-injective hull $\widehat{M}$, every fully invariant submodule is a direct summand. They extend Handelman's left strongly semiprime rings.
For properly semiprime modules M, fully invariant submodules of $\widehat{M}$, which are finitely generated by elements of M, are direct summands.
Any (non-associative) ring A is called *SSP* or *PSP ring*, if A is strongly (resp. properly) semiprime as module over its multiplication ring. A is SSP if and only if its central closure is a direct sum of simple ideals. The structure of the central closure of PSP rings is close to biregular rings.

1 Introduction

Throughout the paper R will be an associative ring with unit and *R-Mod* the category of unital left R-modules. For unexplained notions we refer to [15].

For any left R-module M, the full subcategory of *R-Mod*, whose objects are submodules of M-generated modules, is denoted by $\sigma[M]$.

The M-injective hull of $N \in \sigma[M]$ is written as $\widehat{N}$ or $I_M(N)$.

By $K \subset M$ we usually mean that K is (isomorphic to) a submodule of M, and $K \trianglelefteq M$ indicates that K is an essential submodule of M.

For $X \subset M$ and $b \in M$, we put $(X : b)_R = \{r \in R \mid rb \in X\}$.

$T_K(M)$ is defined in section 2. *Polyform* modules are described in section 3.

Consider the following properties of an R-module M:

(i) $R/An_R(M)$ is cogenerated by every essential submodule of M;

(ii) for every $N \trianglelefteq M$, $M \in \sigma[N]$;

(iii) for every $N \trianglelefteq M$, $M \subset N^{(\Lambda)}$, for some set Λ;

(iv) for every $N \trianglelefteq M$, $R/An_R(M) \subset N^r$, for some $r \in \mathbb{N}$;

(v) M is cogenerated by every essential submodule of M;

(vi) M is polyform;

(vii) for every submodule $K \subset M$, $M/T_K(M) \in \sigma[K]$;

(viii) for every cyclic submodule $K \subset M$, $M/T_K(M) \in \sigma[K]$;

The conditions (i)-(vi) stated for *every* submodule were considered in [13] transferring primeness conditions from rings to modules. In particular, modules satisfying (ii) for *every* submodule $N \subset M$, were called *strongly prime* modules. From the arguments given there it follows that all these conditions are in fact distinct.

Here we have for any module M the implications $(iii) \Rightarrow (v) \Rightarrow (i)$, $(iii) \Rightarrow (ii) \Rightarrow (i)$, $(iv) \Rightarrow (i)$ and $(vii) \Rightarrow (ii)$.

It is easy to prove that for M projective in $\sigma[M]$, (ii) and (iii) are equivalent. With a little more effort we will see that in this case also (vii) is equivalent to M satisfying both (ii) and (vi) (cf. 7.9, 7.10).

For $M = {}_RR$, the conditions (ii), (iii), (iv) and (vii) are equivalent and characterize the *left strongly semiprime rings* introduced by Handelman [6] and further studied by Kutami-Oshiro [9] (cf. 8.2).

For R commutative and $M = R$, any of the properties (i), (v), (vi) and $(viii)$ determine precisely R to be a semiprime ring (hence the complete quotient ring of R is regular). Properties (ii), (iii) (iv) and (vii) imply that the complete quotient ring of R is a finite product of fields.

Basic definitions and technical preparations are provided in Section 2.

General observations on polyform modules are given in Section 3. Hereby their bimodule properties turn out to be of special interest. Using the central idempotents of the self-injective hull, the *idempotent closure* of a polyform module is introduced.

We take (*vii*) to define ***strongly semiprime modules*** in Section 4. They are characterized by the fact that their self-injective hull is semisimple as a bimodule.

Modules satisfying (*viii*) we call ***properly semiprime***. They are treated in Section 5.

Extending the notion of left fully idempotent rings ***pseudo regular modules*** are defined in Section 6.

The structure of modules which are both polyform and SSP (resp. PSP) is investigated in Section 7. For modules M which are projective in $\sigma[M]$ the PSP property in fact implies polyformity. Applying our results to rings we obtain further characterizations of left SSP rings defined in [6] (Section 8).

Section 9 is devoted to the bimodule structure of rings. We consider any ring A as a module over its multiplication ring $M(A)$. As shown in [13], A is a strongly prime $M(A)$-module if and only if the central closure $\widehat{A}$ is a simple ring. Here we observe that A is strongly semiprime as an $M(A)$-module if and only if $\widehat{A}$ is a direct sum of simple ideals. We are also concerned with A being properly semiprime as an $M(A)$-module. In particular we consider the case when the central closure $\widehat{A}$ is an Azumaya ring.

2 Trace and torsion submodules

We begin with some technical preparations. M always denotes an R-module. For our purpose we need an extended version of the trace $Tr(K, L)$ of a module K in L, namely the trace of the category $\sigma[K]$ in L.

2.1 Trace submodules.

For $K, L \in \sigma[M]$, let $\mathcal{T}^K(L)$ denote the *trace of* $\sigma[K]$ *in* L, i.e.

$$\mathcal{T}^K(L) = \sum\{U \subset L \mid U \in \sigma[K]\}.$$

For cyclic submodules $Ra \subset L$, put $\mathcal{T}^a(L) = \mathcal{T}^{Ra}(L)$.

Notice the following properties:

(1) $\mathcal{T}^K(L)$ *is a fully invariant submodule of* L, $\mathcal{T}^K(L)End_R(L) \subset \mathcal{T}^K(L)$.

(2) $\mathcal{T}^K(\widehat{L}) = Tr(K, \widehat{L}) = K\,Hom_R(K, \widehat{L})$ *is a* K*-injective module.*

(3) $\mathcal{T}^K(L) = L \cap \mathcal{T}^K(\widehat{L})$.

(4) If $K = K_1 + K_2$, *then* $\mathcal{T}^K(\widehat{L}) = \mathcal{T}^{K_1}(\widehat{L}) + \mathcal{T}^{K_2}(\widehat{L})$.

(5) If $K \subset L$, $\mathcal{T}^K(\widehat{L}) = K\,End_R(\widehat{L})$ *and* $\mathcal{T}^K(L) = L \cap K\,End_R(\widehat{L})$.

For any injective object $Q \in \sigma[M]$, the modules $X \in \sigma[M]$ with $Hom_R(X, Q) = 0$, form a ***torsion class*** in $\sigma[M]$ (cf. [12]). A module is called ***torsionfree*** with respect to this torsion theory, if its torsion submodule is zero.

With the notation above, the M-injective hull of $\mathcal{T}^K(M)$ may be regarded as submodule of $\widehat{M}$. Denoting $T = End_R(\widehat{M})$, we have $I_M(\mathcal{T}^K(M)) = I_M(\mathcal{T}^K(\widehat{M})) = I_M(KT)$.

We fix the following notation. For the definition of *reject* $Re(-,-)$ see [15], 14.4.

2.2 Torsion submodules.

For $K, L \in \sigma[M]$, let $\mathcal{T}_{K,M}(L)$ denote the *torsion submodule of* L, with respect to the torsion theory (in $\sigma[M]$) determined by $I_M(\mathcal{T}^K(M))$, i.e.

$$\begin{aligned}\mathcal{T}_{K,M}(L) &= \sum\{U \subset L \mid Hom_R(U, I_M(\mathcal{T}^K(M)) = 0\} \\ &= \bigcap\{Ke\, f \mid f \in Hom_R(L, I_M(\mathcal{T}^K(M))\} = Re\,(L, I_M(\mathcal{T}^K(M)).\end{aligned}$$

The index 'M' refers to the category $\sigma[M]$ we are working in. In case this category is fixed, we simply write $\mathcal{T}_K(L)$ instead of $\mathcal{T}_{K,M}(L)$.

In particular, for cyclic modules $K = Ra$, we put $\mathcal{T}_a(L) = \mathcal{T}_{Ra}(L)$.

$\mathcal{T}_K(-)$ is a left exact radical in $\sigma[M]$ and so $\mathcal{T}_K(L)$ is fully invariant in L (see [12]).

Let us list some properties in the form we will mostly use them:

2.3 Properties.

Let M be an R-module, $K \subset M$ a submodule and $T = End_R(\widehat{M})$.

(1) $\mathcal{T}^K(\widehat{M}) = KT$ *is a K-injective module.*

(2) $\mathcal{T}^K(M) = \mathcal{T}^K(\widehat{M}) \cap M$.

(3) $\mathcal{T}^K(M) \cap \mathcal{T}_K(M) = 0$.

(4) For submodules $L, N \subset M$ with $K = L + N$,
$\mathcal{T}^K(\widehat{M}) = \mathcal{T}^L(\widehat{M}) + \mathcal{T}^N(\widehat{M})$ *and* $\mathcal{T}_K(M) = \mathcal{T}_L(M) \cap \mathcal{T}_N(M)$.

(5) If K is generated by $a_1, \ldots, a_n \in K$, $\mathcal{T}_K(M) = \bigcap_{i=1}^n \mathcal{T}_{a_i}(M)$.

Proof. (1),(2) are special cases of the preceding observations.

(3) For $X \subset \mathcal{T}^K(M) \cap \mathcal{T}_K(M)$, $X \subset KT$ and $Hom_R(X, I_M(KT)) = 0$. This implies $X = 0$.

(4) For $K = L + N$, $KT = LT + NT$. Hence without restriction assume $K = KT$, $L = LT$ and $N = NT$. Clearly $\mathcal{T}_K(M) \subset \mathcal{T}_L(M) \cap \mathcal{T}_N(M)$. Consider an additive complement $N_o \subset N$ of L in K. Then $L \oplus N_o$ is an essential submodule of K and

$$\widehat{K} = \widehat{L} \oplus \widehat{N_o} \subset \widehat{L} \oplus \widehat{N}.$$

From this we conclude $\mathcal{T}_K(M) \supset \mathcal{T}_L(M) \cap \mathcal{T}_N(M)$.

(5) This is derived from (4) by induction. □

3 Polyform modules

A module $N \in \sigma[M]$ is called *singular in* $\sigma[M]$ or *M-singular* if $N \simeq L/K$ for some $L \in \sigma[M]$ and $K \trianglelefteq L$ (see [12]). For $M = R$, instead of *R-singular* we just say *singular*.

A submodule $U \subset M$ is called *rational in M* if $Hom(M/U, \widehat{M}) = 0$. Every rational submodule is in particular essential in M.

M is called *polyform* if every essential submodule is rational. The following characterizations of these modules are easy to verify (e.g. [16], [12]).

3.1 Polyform modules. Characterizations.

For a module M with M-injective hull $\widehat{M}$ the following are equivalent:

(a) M is polyform;

(b) for any submodules $K \trianglelefteq L \subset M$, $Hom_R(L/K, \widehat{M}) = 0$;

(c) for every $N \subset M$, the canonical map $Hom_R(\widehat{N}, \widehat{M}) \to Hom_R(N, \widehat{M})$ is an isomorphism;

(d) $\widehat{M}$ is polyform;

(e) $End_R(\widehat{M})$ is regular.

For such modules M, $End_R(M)$ is a subring of $End_R(\widehat{M})$.
Every monic $f \in End_R(M)$ with $Im f \trianglelefteq M$ is invertible in $End_R(\widehat{M})$.

If the M-singular submodule of M is zero, then M is polyform. A ring R is left polyform if and only if it is left non-singular.

3.2 Properties of polyform modules.

Let M be a polyform R-module, $\widehat{M}$ its M-injective hull $\widehat{M}$ and $T = End_R(\widehat{M})$. Then for any submodule $K \subset M$:

(1) $\mathcal{T}_K(\widehat{M})$ is M-injective.

(2) $\mathcal{T}_{\mathcal{T}_K(\widehat{M})}(\widehat{M}) = I_M(KT)$.

(3) $\widehat{M} = \mathcal{T}_K(\widehat{M}) \oplus I_M(KT)$.

(4) $\mathcal{T}_K(M) + \mathcal{T}^K(M) \trianglelefteq M$.

(5) for any $m \in \widehat{M}$, $An_T(m)$ is generated by an idempotent and $\widehat{M}$ is a non-singular right T-module.

Proof. Put $L = \mathcal{T}_K(\widehat{M})$.

(1) Since $\widehat{M}$ is polyform, $Hom_R(I_M(L), I_M(KT)) = 0$ and hence $I_M(L) \subset L$, i.e. $I_M(L) = L$ is M-injective.

(2) Consider $g \in Hom_R(K, L)$. Since $Ke\,g$ is not essential in K, there exists $0 \neq X \subset K$ with $X \cap Keg = 0$ and $X \simeq (X)g \subset L$, a contradiction. Hence $K \subset \mathcal{T}_L(\widehat{M})$ and $KT \subset \mathcal{T}_L(\widehat{M})$ since $\mathcal{T}_L(\widehat{M})$ is fully invariant.

Now we show that $KT \trianglelefteq \mathcal{T}_L(\widehat{M})$. Assume there is a non-zero submodule $U \subset \mathcal{T}_L(\widehat{M})$ with $U \cap KT = 0$. If $Hom_R(U, I_M(KT)) = 0$, then $U \subset L$ and $U \subset \mathcal{T}^L(\widehat{M}) \cap \mathcal{T}_L(\widehat{M}) = 0$, a contradiction.

Hence there is a non-zero $g : U \to I_M(KT)$. Since $Ke\,g$ is not essential in U, there exists a non-zero submodule $V \subset U$ with $V \cap Keg = 0$. Because of $KT \trianglelefteq I_M(KT)$, we may assume $V \simeq (V)g \subset KT$. Now for some $t \in T$, we have $V = (V)gt \subset U \cap KT = 0$, a contradiction.

(3) By (1), L and $\mathcal{T}_L(\widehat{M})$ are M-injective. Since $L \cap \mathcal{T}_L(\widehat{M}) = 0$ by definiton, $\widehat{M} = L \oplus \mathcal{T}_L(\widehat{M}) \oplus W$ for some $W \subset \widehat{M}$.

Assume there is a non-zero $h \in Hom_R(W, L)$ and $Q \cap Keh = 0$ for some non-zero $Q \subset W$. Then $Q \simeq (Q)h \subset L$ and there is some $t \in T$ with $Q = (Q)ht \subset W \cap L = 0$. This implies $Hom_R(W, L) = 0$ and $W \subset \mathcal{T}_L(\widehat{M})$.

Hence $W = 0$ and by (2), $\widehat{M} = L \oplus \mathcal{T}_L(\widehat{M}) = L \oplus I_M(KT)$.

(4) is an immediate consequence of (3).

(5) For $m \in \widehat{M}$ consider $t \in T$ with $(m)t = 0$. Then $(Rm)t = 0$ and - $\widehat{M}$ being polyform - $I_M(Rm)t = 0$. We have $\widehat{M} = I_M(Rm) \oplus U$ for some R-submodule $U \subset \widehat{M}$. For the related projection (idempotent) $g : \widehat{M} \to I_M(Rm)$, $\widehat{M}gt = 0$. This means $gt = 0$ and $t = (1 - g)t \in (1 - g)T$. Therefore $An_T(m) = (1 - g)T$ implying that $\widehat{M}$ is a non-singular T-module. □

As already shown above, the condition on an R-module to be polyform has a strong influence on the structure of its fully invariant submodules. We collect information about this in our next lemma:

3.3 Bimodule properties of polyform modules.

Let M be a polyform R-module, $\widehat{M}$ its M-injective hull and $T = End_R(\widehat{M})$. Denote by C the center of T (i.e., the endomorphism ring of $\widehat{M}$ as an (R,T)-bimodule). Then:

(1) Every essential (R,T)-submodule of $\widehat{M}$ is essential as an R-submodule.

(2) $\widehat{M}$ is self-injective and polyform as an (R,T)-bimodule.
C is a regular self-injective ring.

(3) For every submodule (subset) $K \subset \widehat{M}$, there exists an idempotent $\varepsilon(K) \in C$, such that $An_C(K) = (1 - \varepsilon(K))C$.

(4) If $K \trianglelefteq L \subset \widehat{M}$, then $\varepsilon(K) = \varepsilon(L)$.

(5) Every finitely generated C-submodule of $\widehat{M}$ is C-injective.

(6) If $\widehat{M}$ is a finitely generated (R,T)-module, $\widehat{M}$ is a generator in C-Mod.

Proof. (1) Let $N \subset \widehat{M}$ be an essential (R,T)-submodule. Then $N \cap \mathcal{T}_N(\widehat{M}) = 0$ implies $\mathcal{T}_N(\widehat{M}) = 0$ and $\widehat{M} = I_M(NT) = I_M(N)$ by 3.2.

So $N \trianglelefteq \widehat{M}$ as an R-submodule.

(2) Again let $N \subset \widehat{M}$ be an essential (R,T)-submodule and $h : N \to \widehat{M}$ an (R,T)-morphism. Since $\widehat{M}$ is a self-injective R-module, there is an $f \in T$ which extends h from N to $\widehat{M}$.

For any $t \in T$ and $n \in N$, $(nt)f - (n)ft = (nt)h - (n)ht = 0$. Hence $N \subset Ke(tf - ft)$. By (1), N is an essential R-submodule and since $\widehat{M}$ is polyform, $tf - ft = 0$, implying that f is an (R,T)-morphism and $\widehat{M}$ is a self-injective (R,T)-module.

The endomorphism ring of the self-injective (R,T)-module $\widehat{M}$ is the center of the regular ring T and hence is also regular. So $\widehat{M}$ is a polyform (R,T)-module by 3.1. This in turn implies that C is self-injective.

(3) By 3.2, there is a bimodule decomposition $\widehat{M} = \mathcal{T}_K(\widehat{M}) \oplus I_M(KT)$. Then the projection $\varepsilon(K) : \widehat{M} \to I_M(KT)$ is an idempotent in C and

$An_C(K) = An_C(I_M(KT)) = (1 - \varepsilon(K))C$.

(4) This property is obvious since $\widehat{M}$ is polyform.

(5) As shown in 3.2, every cyclic C-submodule of $\widehat{M}$ is isomorphic to a direct summand of C and hence is C-injective. Since $\widehat{M}$ is a non-singular C-module, any finite sum of C-injective submodules is again C-injective.

(6) Let $\widehat{M}$ be generated as (R,T)-module by $m_1, \ldots, m_k$. Then the map $C \to \widehat{M}^k$, $c \to (m_1, \ldots, m_k)c$, is a monomorphism. Since C is injective, it is a direct summand of $\widehat{M}^k$ and so $\widehat{M}$ is a generator in C-Mod. □

For later use we state some linear dependence properties of elements in M with respect to $End_R(M)$.

3.4 Independence over the endomorphism ring.

Let M be a self-injective R-module, $T = End_R(M)$ and $m_1, \ldots, m_n \in M$.

(1) Assume $m_1T \cap \sum_{i=2}^n m_iT = 0$. Then $An_R(m_2, \ldots, m_n)m_1 \trianglelefteq Rm_1$.

(2) If M is polyform and $An_R(m_2, \ldots, m_n)m_1 \trianglelefteq Rm_1$, then $m_1T \cap \sum_{i=2}^n m_iT = 0$.

Proof. Put $U = An_R(m_2, \ldots, m_n)$.

(1) Assume there exists a non-zero submodule $V \subset Rm_1$ satisfying $V \cap Um_1 = 0$. Consider the canonical projection $\alpha : V \oplus Um_1 \to V$. M being self-injective α extends to an endomorphism t of M. From $Rm_1t \supset (V + Um_1)t = V \neq 0$ we conclude $m_1t \neq 0$.

We also have $U(m_1t) = (Um_1)t = (Um_1)\alpha = 0$. By Corollary 2.2 in [8], this implies $m_1t \in \sum_{i=2}^n m_iT$, a contradiction.

(2) Assume $0 \neq m_1t \in \sum_{i=2}^n m_iT$ for some $t \in T$. Then $Um_1t \subset U\sum_{i=2}^n m_iT = 0$. Since M is polyform and $Um_1 \trianglelefteq Rm_1$ we have $Rm_1t = 0$, a contradiction. □

There is an extension of a module M contained in the M-injective hull $\widehat{M}$ which turns out to be of some interest.

Definition. Let M be an R-module, $T = End_R(\widehat{M})$ and B the Boolean ring of all central idempotents of T. Then we call $\widetilde{M} = MB$ the *idempotent closure of M*.

This notion is closely related to the *π-injective hull* of M defined in Goel-Jain [5], which can be written as MU, with U the subring generated by all idempotents in T. Hence if all idempotents in T are central, $\widetilde{M}$ is just the π-injective hull of M.

3.5 Idempotent closure of polyform modules.

We use the above notation. Let M be an R-module with idempotent closure $\widetilde{M}$. Then for every $a \in \widetilde{M}$, there exist $m_1, \ldots, m_k \in M$ and pairwise orthogonal $c_1, \ldots, c_k \in B$ such that $a = \sum_{i=1}^k m_ic_i$.

If M is polyform module, there exist pairwise orthogonal $e_1, \ldots, e_k \in B$ such that

(1) $a = \sum_{i=1}^k m_ie_i$;

(2) $e_i = \varepsilon(m_i)e_i$ for $i = 1, \ldots, k$;

(3) $\varepsilon(a) = \sum_{i=1}^k e_i$.

Proof. Write $a = \sum_{j=1}^r u_jb_j$, with $u_1, \ldots, u_r \in M$ and $b_1, \ldots, b_r \in B$.

The Boolean subring of B generated by $b_1, \ldots, b_r$ is finite and hence isomorphic to $(\mathbb{Z}_2)^k$ for some $k \in \mathbb{N}$. So it contains a subset of pairwise orthogonal idempotents $c_1, \ldots, c_k$ such that, for all $j = 1, \ldots, r$, $b_j = \sum_{i \in S(j)} c_i$ with $S(j) \subset \{1, \ldots k\}$. Therefore

$$a = \sum_{i=1}^k m_ic_i \quad \text{with } m_i = \sum \{u_l \mid i \in S(l)\}.$$

Assume M is polyform. Put $e_i = \varepsilon(m_i)c_i\varepsilon(a)$. Since $a\varepsilon(a) = a$ and $m_i\varepsilon(m_i) = m_i$,

$$\sum_{i=1}^k m_ie_i = \sum_{i=1}^k m_i\varepsilon(m_i)c_i\varepsilon(a) = \sum_{i=1}^k m_ic_i\varepsilon(a) = (\sum_{i=1}^k m_ic_i)\varepsilon(a) = a.$$

(2) follows from the definition of the idempotents e_i above.

(3) Clearly $\varepsilon(a)e_i = e_i$ for all $i = 1, \dots, k$ and hence $\varepsilon(a)\sum_{i=1}^{k} e_i = \sum_{i=1}^{k} e_i$.

Since $ae_i = m_i e_i$, $a = \sum_{i=1}^{k} m_i e_i = a(\sum_{i=1}^{k} e_i)$. Therefore $id - \sum_{i=1}^{k} e_i \in An_T(a) = (id - \varepsilon(a))T$ (see 3.3). So $\varepsilon(a)\sum_{i=1}^{k} e_i = \varepsilon(a)$ and $\varepsilon(a) = \sum_{i=1}^{k} e_i$. □

4 Strongly semiprime modules

The two following technical lemmas will be crucial for our investigations.

4.1 Lemma.

Let M be an R-module with submodules $K, L \subset M$. Then the following are equivalent:

(a) $M/L \in \sigma[K]$;

(b) for any $b \in M$, there exists a finite subset $X \subset K$, with $An_R(X)b \subset L$.

Proof. (a) $\Rightarrow$ (b) Assume $M/L \in \sigma[K]$ and $b \in M$. Then $Rb + L/L \subset M/L$ is a cyclic module in $\sigma[K]$ and hence a factor module of a cyclic submodule of $K^{(\mathbb{N})}$. So there exist $x_1, \dots, x_k \in K$ and a morphism

$$R(x_1, \dots, x_k) \to M/L,\ (x_1, \dots, x_k) \mapsto (b + L)/L.$$

This implies $An_R(x_1, \dots, x_k)b \subset L$.

(b) $\Rightarrow$ (a) Consider $b \in M$ and chose $x_1, \dots, x_k \in K$ with $An_R(x_1, \dots, x_k)b \subset L$. Then we can define a map as given above and so $(Rb+L)/L \in \sigma[K]$. Hence $M/L \in \sigma[K]$. □

4.2 Lemma.

Let M be a left R-module, $K \subset M$ a submodule and $T = End_R(\widehat{M})$. Then the following conditions are equivalent:

(a) $M/T_K(M) \in \sigma[K]$;

(b) for any $b \in M$, there exists a finite subset $X \subset K$, with $An_R(X)b \subset T_K(M)$.

(c) every K-injective, T_K-torsionfree module in $\sigma[M]$ is M-injective and K-generated;

(d) $I_M(T^K(M)) \in \sigma[K]$ and $T^K(M) + T_K(M) \trianglelefteq M$;

(e) $\widehat{M} = KT \oplus I_M(T_K(\widehat{M}))$.

Notice that the decomposition of $\widehat{M}$ given in (e) is in *R-Mod*. Though KT obviously is a fully invariant submodule, in general this need not be true for $I_M(\mathcal{T}_K(\widehat{M}))$.

Proof. (a) ⇔ (b) follows from 4.1.

(a) ⇒ (c) Assume $Q \in \sigma[M]$ is K-injective and $\mathcal{T}_K$-torsionfree. For any submodule $L \subset M$, consider a morphism $f : L \to Q$. Since $\mathcal{T}_K(Q) = 0$, f factorizes through $f' : L/\mathcal{T}_K(L) \to Q$. We have the commutative diagram (with canonical mappings)

$$\begin{array}{ccccc} 0 & \to & L & \to & M \\ & & \downarrow & & \downarrow \\ 0 & \to & L/\mathcal{T}_K(L) & \to & M/\mathcal{T}_K(M) \\ & & \downarrow_{f'} & & \\ & & Q & & \end{array}$$

Since Q is K-injective and $M/\mathcal{T}_K(M) \in \sigma[K]$, there exists some $M/\mathcal{T}_K(M) \to Q$ yielding a commutative diagram. Hence Q is M-injective and

$$Q = Tr(M, Q) = Tr(M/\mathcal{T}_K(M), Q) = Tr(K, Q).$$

(c) ⇒ (a) Since $I_M(M/\mathcal{T}_K(M))$ is M-injective and $\mathcal{T}_K$-torsionfree, it is K-generated by (c) and so $M/\mathcal{T}_K(M) \in \sigma[K]$.

(a) ⇒ (d) As shown above, $I_M(M/\mathcal{T}_K(M)) \in \sigma[K]$. For a complement U of $\mathcal{T}_K(M)$ in M, $U \oplus \mathcal{T}_K(M) \trianglelefteq M$ and U is isomorphic to a submodule of $M/\mathcal{T}_K(M)$. Hence $U \subset \mathcal{T}^K(M)$ and $\mathcal{T}^K(M) + \mathcal{T}_K(M) \trianglelefteq M$.

(d) ⇒ (a) Since $\mathcal{T}^K(M) + \mathcal{T}_K(M) \trianglelefteq M$ and $\mathcal{T}^K(M) \cap \mathcal{T}_K(M) = 0$,

$$\mathcal{T}^K(M) \simeq [\mathcal{T}^K(M) + \mathcal{T}_K(M)]/\mathcal{T}_K(M) \trianglelefteq M/\mathcal{T}_K(M)\ .$$

Hence $M/\mathcal{T}_K(M)$ is isomorphic to a submodule of $I_M(\mathcal{T}^K(M)) \in \sigma[K]$.

(d) ⇒ (e) The assumptions imply $\widehat{M} = I_M(\mathcal{T}^K(M)) \oplus I_M(\mathcal{T}_K(M))$. As an injective object, $I_M(\mathcal{T}^K(M)) \in \sigma[K]$ is K-generated and hence $I_M(\mathcal{T}^K(M)) = KT$.
Since $\mathcal{T}^K(M) \trianglelefteq \mathcal{T}^K(\widehat{M})$, $I_M(\mathcal{T}^K(M)) = I_M(\mathcal{T}^K(\widehat{M}))$.

(e) ⇒ (d) As a direct summand of $\widehat{M}$, KT is M-injective. Hence

$$\mathcal{T}^K(M) \trianglelefteq I_M(\mathcal{T}^K(M)) = I_M(\mathcal{T}^K(\widehat{M})) = KT.$$

Since $\mathcal{T}_K(M) \trianglelefteq \mathcal{T}_K(\widehat{M}) \trianglelefteq I_M(\mathcal{T}_K(\widehat{M}))$, we conclude $\mathcal{T}^K(M) + \mathcal{T}_K(M) \trianglelefteq M$. □

The above relations are used for our next definition:

4.3 Strongly semiprime modules.

Let M be a left R-module and $T = End_R(\widehat{M})$. We call M strongly semiprime (SSP) *if it satisfies the following equivalent conditions for every submodule $K \subset M$:*

(a) $M/\mathcal{T}_K(M) \in \sigma[K]$;

(b) for any $b \in M$, there exists a finite subset $X \subset K$, with $An_R(X)b \subset \mathcal{T}_K(M)$;

(c) every K-injective $\mathcal{T}_K$-torsionfree module in $\sigma[M]$ is M-injective and K-generated;

(d) $I_M(\mathcal{T}^K(M)) \in \sigma[K]$ *and* $\mathcal{T}^K(M) + \mathcal{T}_K(M) \trianglelefteq M$;

(e) $\widehat{M} = KT \oplus I_M(\mathcal{T}_K(\widehat{M}))$.

For the relationship with strongly prime modules see 5.4. We state some important properties for these modules.

4.4 Basic properties of SSP modules.

Let M be an R-module and $S = End_R(M)$.

(1) Assume M is SSP. Then for every $N \trianglelefteq M$, $M \in \sigma[N]$.

(2) Assume M is self-injective. Then M is SSP if and only if M is semisimple as (R,S)-bimodule.

Proof. (1) For $N \trianglelefteq M$, $\mathcal{T}^N(M) \cap \mathcal{T}_N(M) = 0$ implies $\mathcal{T}_N(M) = 0$ and $NT = \widehat{M}$, in particular $M \in \sigma[N]$.

(2) Let M be self-injective and $U \subset M$ an essential (R,S)-submodule. Then $U = \mathcal{T}^U(M)$, and $\mathcal{T}^U(M) \cap \mathcal{T}_U(M) = 0$ implies $\mathcal{T}_U(M) = 0$. We see from 4.3 that $U = M$. So M has no proper essential (R,S)-submodule. Hence it is a semisimple (R,S)-module.

Now assume M is a semisimple (R,S)-module and $K \subset M$ an R-submodule. Then $M = KS \oplus L$ for some fully invariant $L \subset M$. This implies $Hom_R(L, KS) = 0$ and so $L \subset \mathcal{T}_K(M)$. Hence $M/\mathcal{T}_K(M) \in \sigma[M/L] = \sigma[K]$ showing that M is SSP. □

4.5 More characterizations of SSP.

Let M be an R-module and $T = End_R(\widehat{M})$. Then the following are equivalent:

(a) ${}_RM$ *is SSP;*

(b) for any essential submodule $N \subset {}_RM$, $M \in \sigma[N]$ and for every submodule $K \subset {}_RM$, $\mathcal{T}^K(M) + \mathcal{T}_K(M) \trianglelefteq_R M$;

(c) ${}_R\widehat{M}$ *is SSP;*

(d) $\widehat{M}$ *is a semisimple (R,T)-bimodule.*

Proof. (a) ⇒ (b) This follows from 4.3 and 4.4.

(b) ⇒ (a) Let $K \subset M$ be any submodule. Since $\mathcal{T}_K(M) + \mathcal{T}^K(M) \trianglelefteq M$, $M \in \sigma[\mathcal{T}_K(M) + \mathcal{T}^K(M)] = \sigma[M]$ by assumption. So $\mathcal{T}_K(M) + K$ is a subgenerator in $\sigma[M]$ and hence it generates the M-injective module $I_M(\mathcal{T}^K(M))$. However, since $Hom_R(\mathcal{T}_K(M), I_M(\mathcal{T}^K(M))) = 0$ we conclude that $I_M(\mathcal{T}^K(M))$ is K-generated and M is an SSP module.

(a) ⇒ (c) For any submodule $N \subset \widehat{M}$, put $K = NT \cap M$. Then $KT \trianglelefteq NT$ and $\mathcal{T}_K(\widehat{M}) = \mathcal{T}_N(\widehat{M})$. Consider any N-injective and $\mathcal{T}_N$-torsionfree module $Q \in \sigma[M]$. Then Q is K-injective and $\mathcal{T}_K$-torsionfree, and hence M-injective and K-generated since M is SSP (cf. 4.3). As easily seen, Q is also N-generated and so $\widehat{M}$ is SSP by 4.3.

(c) ⇒ (a) Essential submodules of SSP modules are SSP (cf. 4.4).

(c) ⇔ (d) This is shown in 4.4. □

4.6 SSP and semisimple modules.
Let M be an R-module.

(1) Assume M has essential socle and for every $N \trianglelefteq M$, $M \in \sigma[N]$. Then M is semisimple.

(2) M is semisimple if and only if every module in $\sigma[M]$ is SSP.

Proof. (1) By assumption, $M \in \sigma[Soc(M)]$ and every module in $\sigma[Soc(M)]$ is semisimple.

(2) We see from (1) that every finitely cogenerated module in $\sigma[M]$ is semisimple and hence every simple module in $\sigma[M]$ is M-injective, i.e. M is co-semisimple.

Let N be the sum of all non-isomorphic simple modules in $\sigma[M]$ and consider $L = M \oplus N$. Then $\mathcal{T}_N(L) \subset Rad(L) = 0$ (cf. [15], 23.1). Since L is SSP, this implies $L/\mathcal{T}_N(L) \in \sigma[N]$. Hence L and M are semisimple modules. □

5 Properly semiprime modules

Weakening the conditions for strongly semiprime modules we define:

5.1 Properly semiprime modules.
Let M be a left R-module and $T = End_R(\widehat{M})$. We call M properly semiprime (PSP) *if it satisfies the following equivalent conditions:*

(a) For every element $a \in M$, $M/\mathcal{T}_a(M) \in \sigma[Ra]$;

(b) *for any* $a, b \in M$, *there exist* $r_1, \ldots, r_n \in R$ *such that*

$$An_R(r_1a, r_2a, \ldots, r_na)b \subset \mathcal{T}_a(M);$$

(c) *for every finitely generated submodule* $K \subset M$, $M/\mathcal{T}_K(M) \in \sigma[K]$;

(d) *for any cyclic* $K \subset M$, *every* K*-injective* $\mathcal{T}_K$*-torsionfree module in* $\sigma[M]$ *is* M*-injective and* K*-generated;*

(e) *for any cyclic* $K \subset M$, $I_M(\mathcal{T}^K(M)) \in \sigma[K]$ *and* $\mathcal{T}^K(M) + \mathcal{T}_K(M) \trianglelefteq M$;

(f) *for any cyclic* $K \subset M$, $\widehat{M} = KT \oplus I_M(\mathcal{T}_K(\widehat{M}))$.

Conditions (d)-(f) also hold for finitely generated submodules.

Proof. The equivalence of (a), (c), (d), (e) and (f) follows from 4.2.

(b) $\Leftrightarrow$ (c) One direction is trivial.

Let $a_1, \ldots, a_k$ be a generating subset of K. By 2.3, $\mathcal{T}_K(M) = \bigcap_{i=1}^k \mathcal{T}_{a_i}(M)$. Hence $M/\mathcal{T}_K(M) \subset \bigoplus_{i=1}^k M/\mathcal{T}_{a_i}(M)$. But $M/\mathcal{T}_{a_i}(M) \in \sigma[Ra_i] \subset \sigma[K]$.

Thus $M/\mathcal{T}_K(M) \in \sigma[K]$. □

5.2 Properties of PSP**-modules.**

Let M *be an* R*-module and* $T = End_R(\widehat{M})$.

(1) *Assume* M *is a* PSP*-module and* $U \subset \widehat{M}$ *an* (R,T)*-submodule of finite uniform dimension. Then* U *is a semisimple* (R,T)*-bimodule.*

(2) *Assume* $\widehat{M}$ *has finite uniform dimension as* (R,T)*-bimodule. Then* M *is an* SSP*-module if and only if* M *is a* PSP*-module.*

Proof. (1) First assume U is a uniform (R,T)-bimodule. Let $V \subset U$ be an (R,T)-submodule, and $N \subset M \cap V$ a finitely generated R-submodule. Then NT is an essential (R,T)-submodule of U. $\mathcal{T}^N(U) \cap \mathcal{T}_N(U) = 0$ implies $\mathcal{T}_N(U) = 0$.

From this we deduce $NT \trianglelefteq V \trianglelefteq U$ as R-modules. Since NT is M-injective, we conclude $NT = V = U$ and U is a simple (R,T)-bimodule.

Now assume U has finite uniform dimension as (R,T)-bimodule. Hence there exist uniform submodules $V_i \subset U$, $i = 1, \ldots, n$, such that $\bigoplus_{i=1}^n V_i \trianglelefteq U$ as (R,T)-submodule.

Let N_i be finitely generated submodule of the left R-module $V_i \cap M$ and $N = \bigoplus_{i=1}^n N_i$. As shown above, all $V_i = N_iT$ are simple (R,T)-bimodules. Hence $NT = \sum_{i=1}^n N_iT = \bigoplus_{i=1}^n V_i \trianglelefteq U$ as R-submodule. Therefore $U = NT = \bigoplus_{i=1}^n V_i$ is a finitely generated semisimple (R,T)-bimodule.

(2) If M is SSP then obviously M is PSP.

If M is PSP and $\widehat{M}$ has finite uniform dimension as (R,T)-bimodule, $\widehat{M}$ is a semisimple (R,T)-bimodule by (1) and hence M is SSP by 4.5. □

5.3 Corollary.

Let M be a finitely generated left R-module and $T = End_R(\widehat{M})$. Then the following are equivalent:

(a) M is an SSP-module;

(b) $\widehat{M}$ is (finitely generated and) semisimple as an (R,T)-bimodule;

(c) M is PSP and $\widehat{M}$ has finite uniform dimension as an (R,T)-bimodule.

Proof. Since M is a finitely generated R-module, $\widehat{M} = MT$ is a finitely generated (R,T)-bimodule. Hence the assertions follow from 4.5 and 5.2. □

Recall that an R-module M is called *strongly prime* if for every submodule $N \subset M$, $M \in \sigma[N]$ (see [13]). With similar arguments as used above we can show:

5.4 Strongly prime modules.

For an R-module M with $T = End_R(\widehat{M})$ the following are equivalent:

(a) M is strongly prime;

(b) M is PSP (or SSP) and $\widehat{M}$ is a uniform (R,T)-bimodule.

(c) $\widehat{M}$ is a simple (R,T)-bimodule.

In particular, for a uniform R-module M, the conditions strongly prime, SSP, and PSP are equivalent.

Proof. $(a) \Rightarrow (b)$ For every submodule $K \subset M$, $KT = \widehat{M}$ and hence the assertion is clear.

$(b) \Rightarrow (a)$ Assume M is PSP and $K \subset M$ is a finitely generated submodule. By the uniformity condition, $T^K(\widehat{M}) \cap T_K(\widehat{M}) = 0$ implies $T_K(\widehat{M}) = 0$ and $\widehat{M} = KT$. So M is strongly prime.

$(a) \Leftrightarrow (c)$ This follows with the same arguments applied in the proof of 4.4. It is also shown in [13], Proposition 2.1. □

Both the definitions of SSP and PSP modules M refer to the category $\sigma[M]$. Nevertheless we can show that submodules of such modules are of the same type.

5.5 Submodules of SSP and PSP modules.

Let M be an R-module and $L \subset M$ a submodule. Then:
If M is SSP (resp. PSP), then L is also SSP (resp. PSP).

Proof. Refer to the notation in 2.2. Let $L \subset M$ be a submodule and $K \in \sigma[M]$. Then $\mathcal{T}^K(L) \subset \mathcal{T}^K(M)$ and $I_L(\mathcal{T}^K(L)) \subset I_M(\mathcal{T}^K(L)) \subset I_M(\mathcal{T}^K(M))$. Therefore

$$\mathcal{T}_{K,M}(L) = Re(L, I_M(\mathcal{T}^K(M)) \subset Re(L, I_L(\mathcal{T}^K(L)) = \mathcal{T}_{K,L}(L).$$

Now assume M is PSP and $K \subset L$ is a submodule. By the above relation, $L/\mathcal{T}_{K,L}(L)$ is a homomorphic image of $L/\mathcal{T}_{K,M}(L)$. Since $\mathcal{T}_{K,M}(-)$ is a left exact functor,

$$\mathcal{T}_{K,M}(L) = \mathcal{T}_{K,M}(M) \cap L, \text{ and } L/\mathcal{T}_{K,M}(L) \subset M/\mathcal{T}_{K,M}(M) \in \sigma[K].$$

Therefore $L/\mathcal{T}_{K,L}(L) \in \sigma[K]$ and L is PSP. □

6 Pseudo regular modules

The class of modules we define now is related to regularity properties of modules and includes left fully idempotent rings.

6.1 Pseudo regular modules.

Let M be a left R-module, $S = End_R(M)$, $T = End_R(\widehat{M})$ and $D = End_T(\widehat{M})$. Then the following conditions are equivalent:

(a) for every $m \in M$, there exists $h \in D$ such that $hM \subset RmS$ and $hm = m$;

(b) for any (R, S)-submodule $N \subset M$ and $m_1, \ldots, m_k \in N$, there exists $h \in D$ such that $hM \subset N$ and $hm_i = m_i$ for all $i = 1, \ldots, k$.

A module satisfying these conditions is called pseudo regular.

Proof. $(a) \Rightarrow (b)$ We can follow an induction argument in the proof of [4], Proposition 11.27. Let $N \subset M$ be an (R, S)-submodule of M and $m_1, \ldots, m_k \in N$. For $k = 1$ there is nothing to show.

Assume the assertion holds for any $k-1$ elements in N. Choose $f \in D$ such that $fm_k = m_k$ and $fM \subset Rm_kS \subset N$.

Clearly $m_i - fm_i \in N$ for all $i = 1, \ldots, k-1$, and by hypothesis, there exists $g \in D$ with $gM \subset N$ and $g(m_i - fm_i) = m_i - fm_i$ for $i = 1, \ldots, k-1$. Put $h = 1-(1-g)(1-f)$. Then $hm = gm - gfm + fm \in N$ for all $m \in M$, i.e. $hM \subset N$, and for $i = 1, \ldots, k$,

$$hm_i = m_i - (1-g)(1-f)m_i = m_i - (1-g)(m_i - fm_i) = m_i.$$

$(b) \Rightarrow (a)$ is obvious. $\square$

Let us check what pseudo regularity means for rings:

6.2 Left fully idempotent rings.

A ring R is left fully idempotent if and only if the left R-module R is pseudo regular.

Proof. Recall $End_R(R) = R$. Assume R is left fully idempotent. Let N be an ideal of R and $m \in N$. Since $(Rm)^2 = Rm$, there exist $h \in RmR$ such that $hm = m$. Clearly $hR \subset N$ and $h \in R \subset Biend_R(\widehat{R})$. Hence $_RR$ is pseudo regular.

Now assume $_RR$ to be pseudo regular and $m \in R$. Then there exists $h \in Biend_R(\widehat{R})$ such that $hm = m$ and $hR \subset RmR$. Hence $r = h1 \in RmR$ and $rm = (h1)m = hm = m$. So $m = rm \in (Rm)^2$, $(Rm)^2 = Rm$ and the ring R is left fully idempotent. $\square$

It follows from the above observation that a commutative ring R is von Neumann regular if and only if $_RR$ is pseudo regular.

Of particular interest for our investigations is the following relationship:

6.3 Pseudo regular and PSP modules.

Let M be a left R-module, $S = End_R(M)$ and $T = End_R(\widehat{M})$. The following conditions are equivalent:

(a) for every $m \in M$, there is a central idempotent $e \in T$ with $RmS = Me$;

(b) for every $m \in M$, $M = RmS \oplus \mathcal{T}_m(M)$ and $\widehat{M} = RmT \oplus \mathcal{T}_m(\widehat{M})$;

(c) for every finitely generated submodule $N \subset M$, there is a central idempotent $e \in T$ with $NS = Me$;

(d) for every finitely generated submodule $N \subset M$, $M = NS \oplus \mathcal{T}_N(M)$ and $\widehat{M} = NT \oplus \mathcal{T}_N(\widehat{M})$;

(e) M is a pseudo regular PSP-module over R.

Proof. Denote $D = End_T(\widehat{M})$.

$(a) \Rightarrow (b)$ For $m \in M$ choose a central idempotent $e \in T$ with $RmS = Me$. Since $\widehat{M} = MT$ we have

$$\begin{aligned} \widehat{M} &= MTe \oplus MT(1-e) = MeT \oplus MT(1-e) \\ &= RmST \oplus MT(1-e) = RmT \oplus MT(1-e). \end{aligned}$$

Hence RmT is M-injective and $MTe = RmT = \mathcal{T}_m(\widehat{M}) = I_M(\mathcal{T}_m(\widehat{M}))$ (see 2.3). Since $\widehat{M}$ is M-injective,

$$\begin{aligned} Hom_R(MT(1-e), I_M(\mathcal{T}_m(\widehat{M}))) &= \{t \in T \mid MT(1-e)t \subset I_M(\mathcal{T}_m(\widehat{M}))\} \\ &= \{t \in T \mid MT(1-e)t \subset MTe\} = 0. \end{aligned}$$

So $MT(1-e) \subset \mathcal{T}_m(\widehat{M})$. But $MTe = RmT$, $\widehat{M} = MTe \oplus MT(1-e)$ and

$$MTe \cap \mathcal{T}_m(\widehat{M}) = \mathcal{T}^m(\widehat{M}) \cap \mathcal{T}_m(\widehat{M}) = 0.$$

Hence $\mathcal{T}_m(\widehat{M}) = MT(1-e)$ and $\widehat{M} = RmT \oplus \mathcal{T}_m(\widehat{M})$. Clearly $RmS = Me = M \cap MTe$ and

$$\mathcal{T}_m(M) = M \cap \mathcal{T}_m(\widehat{M}) = M \cap MT(1-e) = M(1-e).$$

So $M = RmT \oplus \mathcal{T}_m(M)$.

(b) $\Rightarrow$ (a) Let e be the projection of $\widehat{M}$ onto RmT along $\mathcal{T}_m(\widehat{M})$. Since both submodules are fully invariant, e is a central idempotent of T. From $RmS \subset RmT$ and $\mathcal{T}_m(M) \subset \mathcal{T}_m(\widehat{M})$ we conclude $Me = RmS$.

(c) $\Leftrightarrow$ (d) can also be shown with the above proof.

(c) $\Rightarrow$ (a) is obvious.

(a) $\Rightarrow$ (e) Since $M/\mathcal{T}_m(M) = RmS \in \sigma[Rm]$, M is PSP.

Let π be the projection of $\widehat{M}$ onto RmT along $\mathcal{T}_m(\widehat{M})$. Obviously, $\pi \in D$. Since $RmS \subset RmT$ and $\mathcal{T}_m(M) \subset \mathcal{T}_m(\widehat{M})$, $\pi M = RmS$. Hence M is pseudo regular.

(e) $\Rightarrow$ (c) Let $N \subset M$ be a finitely generated submodule. Since M is PSP, $\widehat{M} = NT \oplus I_M(\mathcal{T}_N(\widehat{M}))$.

Suppose $I_M(\mathcal{T}_N(\widehat{M})) \neq \mathcal{T}_N(\widehat{M})$ and consider $x \in I_M(\mathcal{T}_N(\widehat{M})) \setminus \mathcal{T}_N(\widehat{M})$. Then $Hom_R(Rx, NT) \neq 0$ and - by M-injectivity of $\widehat{M}$ - there exists $t \in T$ with $0 \neq xt \in NT$. So $xt = \sum_{i=1}^n m_i s_i$ for some $m_i \in N$, $s_i \in T$. M being pseudo regular, there exists $h \in D$ such that $hM \subset NS$ and $hm_i = m_i$ for $i = 1, \dots, n$. Clearly

$$h(xt) = \sum_{i=1}^n h(m_i s_i) = \sum_{i=1}^n (hm_i)s_i = \sum_{i=1}^n m_i s_i = xt.$$

Since $\widehat{M} = MT$, $h\widehat{M} = (hM)T \subset NT$ and $hx \in NT$. Let π be the projection of $\widehat{M}$ onto $I_M(\mathcal{T}_N(\widehat{M}))$ along NT. Then $0 = (hx)\pi = h(x\pi) = hx$ and $xt = h(xt) = (hx)t = 0$, contradicting $xt \neq 0$. Hence $I_M(\mathcal{T}_N(\widehat{M}) = \mathcal{T}_N(\widehat{M})$ and $\widehat{M} = NT \oplus \mathcal{T}_N(\widehat{M})$.

Now consider $\alpha = 1 - \pi : \widehat{M} \to NT$ and $y \in M$. Then $y\alpha = \sum_{i=1}^n n_i t_i$ for some $m_i \in N$, $t_i \in T$. Since M is pseudo regular, there exists $h \in D$ such that $hM \subset NS$ and $hn_i = n_i$ for $i = 1, \dots, n$. Clearly $h(y\alpha) = y\alpha$ and $hy \in NS \subset NT$. So $y\alpha = h(y\alpha) = (hy)\alpha = hy \in NS$ and $y - y\alpha \in \mathcal{T}_N(\widehat{M}) \cap M = \mathcal{T}_N(M)$. Therefore $M = NS \oplus \mathcal{T}_N(M)$.

□

7 Polyform PSP and SSP modules

In this section we are concerned with the interplay between polyform and SSP properties. One of the crucial observations is that projective PSP modules are polyform. We begin with general modules enjoying both properties.

7.1 Polyform SSP modules.
Let M be a polyform R-module and $T = End_R(\widehat{M})$. Then the following are equivalent:

(a) *M is an SSP-module;*

(b) *for every $N \trianglelefteq M$, $M \in \sigma[N]$;*

(c) *$I_M(KT) \in \sigma[K]$;*

(d) *for any submodule $K \subset M$, $\widehat{M} = KT \oplus \mathcal{T}_K(\widehat{M})$.*

Proof. The statements follow from 3.2, 4.3 and 4.5. □

In [12], Theorem 3.7 it is shown, that any polyform module M is SSP if M has finite uniform dimension, $Hom_R(M,N) \neq 0$ for non-zero $N \subset M$, and $End_R(M)$ is semiprime. We can extend this result to a more general class of modules considered in Zelmanowitz [17]. There an R-module M is called *weakly semisimple* if

(i) M is polyform,

(ii) every finitely generated submodule has finite uniform dimension, and

(iii) for every non-zero submodule $N \subset M$, there exists $f \in Hom_R(M,N)$ with $f|_N \neq 0$ (weakly compressible).

7.2 Weakly semisimple modules.
Every weakly semisimple module is SSP.

Proof. We have to show that for every $N \trianglelefteq M$, $M \in \sigma[N]$.
For any $m \in M$, Rm has finite uniform dimension and $N \cap Rm \trianglelefteq Rm$. Hence by Proposition 1.1 in [17], there exists a monomorphism $Rm \to N \cap Rm$ and so $Rm \in \sigma[N]$. This implies $M \in \sigma[N]$. □

We know that a module M is SSP if and only if $\widehat{M}$ is SSP. In general, for a PSP module M, $\widehat{M}$ need not be PSP. For polyform modules the PSP property extends at least to the idempotent closure:

7.3 Polyform PSP modules.
Let M be a polyform R-module, $T = End_R(\widehat{M})$, B the Boolean ring of central idempotents of T and $\widetilde{M}$ the idempotent closure of M. Then the following are equivalent:

(a) *M is a PSP module;*

(b) *for every $m \in M$, $RmT = \widehat{M}\varepsilon(m)$;*

(c) *for every finitely generated submodule $K \subset M$, $KT = \widehat{M}\varepsilon(K)$;*

(d) $\widetilde{M}$ *is a PSP module.*

Under the given conditions, $\mathcal{T}_m(\widehat{M}) = \widehat{M}(id - \varepsilon(m))$.

Proof. (a) ⇒ (b) By 5.1, (f) and 3.2, (1), $\widehat{M} = RmT \oplus \mathcal{T}_m(\widehat{M})$. Now (b) follows from the definition of the idempotent $\varepsilon(m)$.

(b) ⇒ (a) Since $\widehat{M}\varepsilon(m)$ is M-injective,

$$I_M(\mathcal{T}^m(M)) = I_M(\mathcal{T}^m(\widehat{M})) = I_M(RmT) = RmT \in \sigma[Rm].$$

M being polyform, we have $\mathcal{T}_m(M) + \mathcal{T}^m(M) \trianglelefteq M$ and the assertion follows by 5.1.

(b) ⇔ (c) This is obvious by 5.1 and 3.3.

(b) ⇒ (d) Any $a \in \widetilde{M}$ can be written as $a = \sum_{i=1}^k m_i e_i$, with $m_1, \ldots, m_k \in M$ and pairwise orthogonal $e_1, \ldots, e_k \in B$, satisfying

$$\varepsilon(a) = \sum_{i=1}^k e_i \ \text{ and } e_i = \varepsilon(m_i)e_i \text{ for } i = 1, \ldots, k.$$

By (b), $Rm_iT = \widehat{M}\varepsilon(m_i)$ for $i = 1, \ldots, k$, and

$$RaT = R(\sum_{i=1}^k m_i e_i)T = \sum_{i=1}^k (Rm_iT)e_i = \sum_{i=1}^k \widehat{M}\varepsilon(m_i)e_i = \widehat{M}(\sum_{i=1}^k e_i) = \widehat{M}\varepsilon(a).$$

(d) ⇒ (a) By 5.5, submodules of PSP modules are again PSP. □

By 4.5, any self-injective SSP module is semisimple as a bimodule. For self-injective polyform PSP modules we get a weaker structure theorem:

7.4 Self-injective *PSP* modules.

Let M be a self-injective polyform R-module and $T = End_R(\widehat{M})$. Denote $\Lambda = R \otimes_{\mathbb{Z}} T^o$ and $C = End_\Lambda(M)$. Then the following conditions are equivalent:

(a) ${}_RM$ *is a PSP R-module;*

(b) every cyclic Λ*-submodule of* M *is a direct summand;*

(c) every finitely generated Λ*-submodule of* M *is a direct summand;*

(d) as a Λ*-module,* M *is a selfgenerator;*

(e) for any $m \in M$ *and* $f \in End_C(M)$*, there exists* $h \in \Lambda$ *with* $f(m) = hm$.

Proof. Notice that the Λ-submodules of M are just the fully invariant submodules and that C can be identified with the center of T. Hence C is a commutative regular ring.

$(a) \Leftrightarrow (b) \Leftrightarrow (c)$ is clear by 7.3, $(c) \Rightarrow (d)$ is obvious.

$(d) \Rightarrow (e)$ This follows from the proof of the Density Theorem (e.g. [15], 15.7).

$(e) \Rightarrow (b)$ Choose any $m \in M$ and $\varepsilon(m) \in C$ as defined in 3.3. Since $mC \simeq \varepsilon(m)C$ is a direct summand, for any $n \in M$, there exists $f \in End_C(M)$ with $f(m) = n\varepsilon(m)$.

By (e), $f(m) = hm$ for some $h \in \Lambda$ and hence $M\varepsilon(m) \subset \Lambda m$.

On the other hand, $m = m\varepsilon(m)$ and $\Lambda m \subset (\Lambda M)\varepsilon(m) \subset M\varepsilon(m)$. So $\Lambda m = M\varepsilon(m)$ is a direct summand and the assertion is proved. □

Modules M, whose finitely generated submodules are direct summands, are closely related to M being regular in $\sigma[M]$. In fact, if M is finitely presented in $\sigma[M]$, these two notions coincide (e.g. [15], 37.3, 37.4). As a special case we derive from the above theorem:

7.5 Self-injective finitely presented PSP modules.

Let M be a self-injective polyform R-module and $T = End_R(\widehat{M})$. Denote $\Lambda = R \otimes_{\mathbb{Z}} T^o$ and $C = End_\Lambda(M)$. Assume M is finitely generated as a Λ-module. Then the following conditions are equivalent:

(a) M is a PSP R-module and finitely presented in $\sigma[{}_\Lambda M]$;

(b) ${}_\Lambda M$ is regular and projective in $\sigma[{}_\Lambda M]$;

(c) ${}_\Lambda M$ is a (projective) generator in $\sigma[{}_\Lambda M]$;

(d) for any $m_1, \ldots, m_n \in M$ and $f \in End_C(M)$, there exists $h \in \Lambda$ with $f(m_i) = hm_i$ for $i = 1, \ldots, n$ (density property).

Proof. Since ${}_\Lambda M$ is finitely generated, M_C is a generator in C-*Mod* by 3.3. So M_C is a faithfully flat C-module.

$(a) \Leftrightarrow (b)$ By 7.4, every finitely generated Λ-submodule of M is a direct summand. Now the assertion follows from [15], 37.4.

$(b) \Rightarrow (c) \Rightarrow (d)$ are obvious (Density Theorem, [15], 15.7).

$(d) \Rightarrow (a)$ Since M_C is a generator in C-*Mod*, M is (finitely generated and) projective as an $End_C(M)$-module (e.g. [15], 18.8).

By the Density Property, the categories $\sigma[{}_\Lambda M]$ and $\sigma[{}_{End_C(M)}M]$ coincide (see [15], 15.8). So M is projective (hence finitely presented) in $\sigma[{}_\Lambda M]$.

By 7.4, M is a PSP R-module. □

We have seen in 7.4 and 7.5 that self-injective polyform PSP modules have nice structural properties. Hence we may ask for which modules M, the M-injective hull $\widehat{M}$ is a PSP module. For this we need some properties of self-injective polyform modules.

7.6 Self-injective polyform modules.
Let M be a self-injective polyform R-module, $T = End_R(\widehat{M})$ and $\Lambda = R \otimes_{\mathbb{Z}} T^o$. Then:

(1) For any submodule $N \subset M$ and $m \in M$, $(\mathcal{T}_N(M) : m)_R = An_R(m\varepsilon(N))$.

(2) For $m_1, \ldots, m_n, m \in M$ and $U = An_R(m_2, \ldots, m_n)$ the following are equivalent:

(a) There exists $h \in \Lambda$ with $hm_1 = m\varepsilon(Um_1)$ and $hm_i = 0$ for $i = 2, 3, \ldots, n$;

(b) there exist $r_1, \ldots, r_k \in R$ such that for any $s_1, \ldots, s_n \in R$ the relations

$$\sum_{l=1}^{n} s_l r_j m_l = 0 \quad \textit{for } j = 1, \ldots, k,$$

imply $s_1 m \in \mathcal{T}_{Um_1}(M)$.

Proof. (1) By definition of the idempotent $\varepsilon(N)$ (see 3.3), $M\varepsilon(N) = I_M(NT)$ and $M(1 - \varepsilon(N)) = \mathcal{T}_N(M)$. Hence for $r \in R$, $rm \in \mathcal{T}_N(M)$ if and only if $(rm)\varepsilon(N) = 0$. Now our assertion follows from $(rm)\varepsilon(N) = r(m\varepsilon(N))$.

(2) $(a) \Rightarrow (b)$ Put $e = \varepsilon(Um_1)$. By (1), $(\mathcal{T}_{Um_1}(M) : m)_R = An_R(me)$.
Choose an element $h = \sum_{j=1}^{k} r_j \otimes t_j$ in Λ with

$$hm_1 = me \text{ and } hm_i = 0 \text{ for all } i = 2, \ldots, n.$$

Assume for $s_1, \ldots, s_n \in R$, $\sum_{l=1}^{n} s_l r_j m_l = 0$ for $j = 1, \ldots, k$. Then

$$s_1 me = s_1 hm_1 = \sum_{l=1}^{n} s_l hm_l = \sum_{l=1}^{n} s_l \sum_{j=1}^{k} r_j m_l t_j = \sum_{j=1}^{k} (\sum_{l=1}^{n} s_l r_j m_l) t_j = 0.$$

Hence $s_1 \in An_R(me) = (\mathcal{T}_{Um_1}(M) : m)_R$.

$(b) \Rightarrow (a)$ Put $N = \sum_{i=1}^{n} R(r_1 m_i, \ldots, r_k m_i) \subset M^k$ and consider the assignement

$$\psi : N \to M, \quad \sum_{i=1}^{n} s_i(r_1 m_i, \ldots, r_k m_i) \mapsto s_1 me.$$

To show that ψ is well-defined assume $\sum_{i=1}^{n} s_i(r_1 m_i, \ldots, r_k m_i) = 0$.
Then $\sum_{i=1}^{n} s_i r_j m_i = 0$, for all $j = 1, \ldots, k$.
By assumption, $s_1 \in (\mathcal{T}_{Um_1}(M) : m)_R = An_R(me)$. Hence $s_1 me = 0$ proving that ψ is a well-defined morphism.

M being M-injective, ψ can be extended to a morphism $M^k \to M$, also denoted by ψ. Since $Hom_R(M^k, M) = T^k$ there exist $t_1, \ldots, t_k \in T$ such that

$$(x_1, \ldots, x_k)\psi = \sum_{i=1}^{k} x_i t_i \text{ for all } (x_1, \ldots, x_k) \in M^k.$$

Put $h = \sum_{j=1}^{k} r_j \otimes t_j$. Then

$$hm_1 = \textstyle\sum_{j=1}^{k} r_j m_1 t_j = (r_1 m_1, r_2 m_1, \ldots, r_k m_1)\psi = me, \text{ and}$$
$$hm_l = \textstyle\sum_{j=1}^{k} r_j m_l t_j = (r_1 m_l, r_2 m_l, \ldots, r_k m_l)\psi = 0 \text{ for } l = 2, \ldots, n.$$

□

For modules M with $End_R(\widehat{M})$ commutative we are now able to characterize the density property of $\widehat{M}$ as bimodule.

7.7 Density property of the self-injective hull.

Let M be a polyform R-module, assume $T = End_R(\widehat{M})$ to be commutative and put $\Lambda = R \otimes_{\mathbb{Z}} T$. Then the following are equivalent:

(a) *For any $a_1, \ldots, a_n \in \widehat{M}$ and $f \in End_T(\widehat{M})$, there exists $h \in \Lambda$ with $ha_i = fa_i$ for $i = 1, \ldots, n$.*

(b) *For any $m_1, \ldots, m_n, m \in M$ there exist $r_1, \ldots, r_k \in R$ such that for $s_1, \ldots, s_n \in R$ the relations*
$$\sum_{l=1}^{n} s_l r_j m_l = 0 \quad for\ j = 1, \ldots, k,$$
imply $s_1 m \in T_{Um_1}(M)$ for $U = An_R(m_2, \ldots, m_n)$.

(c) *M is a PSP module and for any $m_1, \ldots, m_n \in M$ there exist $r_1, \ldots, r_k \in R$ such that for $s_1, \ldots, s_n \in R$ the relations*
$$\sum_{l=1}^{n} s_l r_j m_l = 0 \quad for\ j = 1, \ldots, k,$$
imply $s_1 m_1 \in T_{Um_1}(M)$ for $U = An_R(m_2, \ldots, m_n)$.

If $\widehat{M}$ is finitely presented in $\sigma[{}_\Lambda\widehat{M}]$ the above are equivalent to:

(d) *$\widehat{M}$ is a PSP R-module.*

Proof. Put $U = An_R(m_2, \ldots, m_n)$.

(a) ⇒ (b) Put $N = \sum_{i=1}^{n} m_i T$. By 3.3, $\widehat{M}$ is a non-singular and N is an injective T-module. Since in a non-singular module the intersection of injective submodules is again injective, $K = m_1T \cap N$ is T-injective. Hence $m_1T = K \oplus L$ for some submodule $L \subset m_1T$. By 3.3, $An_T(m_1) = (1 - \varepsilon(m_1))T$. This means that the map

$$m_1T \to \varepsilon(m_1)T, \quad m_1t \mapsto \varepsilon(m_1)t,$$

is an isomorphism. So there exist idempotents $u, v \in T$ with the properties

$$(*) \quad uv = 0,\ u + v = \varepsilon(m_1),\ K = m_1uT \simeq uT \text{ and } L = m_1vT \simeq vT.$$

Since $m_1uT = K \subset N$ and $UN = 0$, $Um_1u = 0$ and

$$Um_1 = U[m_1\varepsilon(m_1)] = U[m_1(u+v)] = Um_1v.$$

By 3.4, $Um_1v \trianglelefteq Rm_1v$ and by 3.3, $\varepsilon(Um_1v) = \varepsilon(Rm_1v) = \varepsilon(m_1v)$.
The isomorphism $m_1vT \simeq vT$ implies $\varepsilon(m_1v) = v$ and hence we have

$$(**) \qquad \varepsilon(Um_1) = v\ .$$

Since $m \in M$ and $\mathcal{T}_{Um_1}(M) = M \cap \mathcal{T}_{Um_1}(\widehat{M})$, by 7.6,

$$(\mathcal{T}_{Um_1}(M) : m)_R = (\mathcal{T}_{Um_1}(\widehat{M}) : m)_R = An_R(mv).$$

As an injective submodule, $m_1vT \oplus N$ is a direct summand in $\widehat{M}$. Hence there exists a T-endomorphism ψ of $\widehat{M}$ satisfying $\psi N = 0$ and $\psi m_1v = mv$ (recall $m_1vT \simeq vT$). Since $m_1u \in N$, we have $\psi m_1u = 0$ and $\psi m_1 = \psi[m(u+v)] = mv$.

By assumption, there exists $h \in \Lambda$ satisfying

$$hm_1 = \psi m_1 = mv, \text{ and } hm_i = \psi m_i \text{ for } i = 2, \ldots, n\ .$$

Now the assertion follows from 7.6.

$(b) \Rightarrow (c)$ Putting $m_1 = \cdots = m_n = 0$ we see that M is a PSP module. The second part of the conditions in (c) follows from (b) for $m = m_1$.

$(c) \Rightarrow (a)$ Since $\widehat{M} = MT$ it suffices to show that for any $m_1, \ldots, m_n \in M$ and $f \in End_T(\widehat{M})$ there exists $h \in \Lambda$ such that $fm_i = hm_i$ for all $i = 1, \ldots, n$.

We prove this by induction on the cardinality $|I|$ of minimal subsets $I \subset \{1, \ldots, n\}$ satisfying $\sum_{i \in I} m_iT = \sum_{i=1}^n m_iT$.

Consider the case $|I| = 1$, i.e., $I = \{1\}$. By 7.3, $\widehat{M}\varepsilon(m_1) = \Lambda m_1$. Since

$$fm_1 = f[m_1\varepsilon(m_1)] = (fm_1)\varepsilon(m_1) \in \widehat{M}\varepsilon(m_1) = \Lambda m_1\ ,$$

$fm_1 = hm_1$ for some $h \in \Lambda$. By assumption $\sum_{i=1}^n m_iT = m_1T$, implying $fm_i = hm_i$ for all $i = 1, \ldots, n$.

Now assume $|I| = n$ and consider $N = \sum_{i=2}^{n} m_i T$. As shown above there exist idempotents $u, v \in T$ satisfying $(*)$ and $(**)$. By 7.6, there exists $h_1 \in \Lambda$ with

$$h_1 m_1 = m_1 v \text{ and } h_1 m_i = 0 \text{ for } i = 2, \ldots, n .$$

Notice that $h_1 m_1 u = (h_1 m_1) u = (m_1 v) u = m_1 (vu) = 0$.

By induction hypothesis, there exists $h_2 \in \Lambda$ with

$$h_2 m_i = f m_i \text{ for } i = 2, \ldots, n .$$

Obviously, $h_2 x = f x$ for any $x \in N$. From $(*)$ we obtain $h_2 m_1 u = f m_1 u$. Consider the element $m = f m_1 v - h_2 m_1 v$. Clearly $m = mv$ and so $m \in \widehat{M} v$. By $(*)$, $v\varepsilon(m_1) = v$. Now it follows from 7.3 that

$$\Lambda m_1 v = (\Lambda m_1) v = \widehat{M} \varepsilon(m_1) v = \widehat{M} v .$$

Hence there exists $h_3 \in \Lambda$ with

$$h_3 m_1 v = m = f m_1 v - h_2 m_1 v .$$

Putting $h = h_3 h_1 + h_2$ we have

$$\begin{aligned} h m_1 &= h_3(h_1 m_1 u) + h_3(h_1 m_1 v) + h_2 m_1 u + h_2 m_1 v \\ &= h_3 m_1 v + f m_1 u + h_2 m_1 v \\ &= (f m_1 v - h_2 m_1 v) + f m_1 u + h_2 m_1 v \\ &= f m_1 v + f m_1 u = f m_1 , \\ h m_i &= h_3(h_1 m_i) + h_2 m_i = f m_i \text{ for } i = 2, \ldots, n . \end{aligned}$$

$(a) \Leftrightarrow (d)$ This is clear by 7.5. □

Now we turn to the question which additional conditons on an SSP or PSP module M imply that M is polyform. It is interesting to observe that this is achieved by commutativity conditions on the endomorphism rings as well as by projectivity of the modules. Recall that a ring is said to be *(left and right) duo* if all its one-sided ideals are two-sided.

7.8 Duo endomorphism rings.

Let M be an R-module, $T = End_R(\widehat{M})$ and assume for every $N \trianglelefteq M$, $M \in \sigma[N]$.

(1) $Jac(T) \cap center(T) = 0$.

(2) Suppose T is a duo ring. Then M is polyform and SSP.

Proof. (2) Assume for $f \in T$, $N = Ke\,f \trianglelefteq \widehat{M}$. Then $M \in \sigma[N]$ which is equivalent to $NT = \widehat{M}$. This implies $\widehat{M}f = (NT)f = (Nf)T = 0$ and hence $f = 0$. So the Jacobson radical of T is zero, i.e. M is polyform. By 7.1, M is SSP.

(1) A similar argument also implies this assertion. □

Projectivity makes any PSP module polyform. This applies in particular for the left module structure of the ring itself.

7.9 Projective *PSP* modules.

Let M be a PSP module which is projective in $\sigma[M]$. Then

(1) For any submodules $K, N \subset M$, $\mathcal{T}^N(K) + \mathcal{T}_N(K) \trianglelefteq K$.

(2) M is polyform.

Proof. M is projective in $\sigma[M]$ if and only if $M^{(\Lambda)}$ is self-projective for any set Λ. From this it is obvious that M/X is projective in $\sigma[M/X]$ for every fully invariant submodule $X \subset M$.

(1) First we show $\mathcal{T}^N(K) \neq 0$ for any finitely generated submodules $K, N \subset M$ with $Hom(K, N) \neq 0$. For this we may assume that there is an epimorphism $f : K \to N$. Then $N \subset \mathcal{T}^K(M)$ and

$$\mathcal{T}_K(M) \cap N = 0 = \mathcal{T}_K(M) \cap K.$$

Put $L = \mathcal{T}_K(M)$ and $\overline{M} = M/L$. There are canonical inclusions $N \subset \overline{M}$ and $K \subset \overline{M}$. Since M is PSP, $\overline{M} \in \sigma[K]$ and so $\sigma[\overline{M}] = \sigma[K]$.

As outlined above, $\overline{M}$ is projective in $\sigma[K]$ and hence is a direct summand of $K^{(\Lambda)}$, for some set Λ. Therefore the compostion of the inclusion $N \subset \overline{M}$ with a suitable map $\overline{M} \to K$ yields a non-zero morphism $N \to K$. This means $\mathcal{T}^N(K) \neq 0$.

Now we prove $\mathcal{T}^N(K) + \mathcal{T}_N(K) \trianglelefteq K$ for any submodules $K, N \subset M$.

Consider some $x \in K \setminus \mathcal{T}_N(K)$. Then there exists a non-zero $g : Rx \to I_M(N)$ and $0 \neq (y)g \in N$ for some $y \in Rx$. From the above we know $\mathcal{T}_N(Ry) \neq 0$ implying

$$\mathcal{T}_N(Ry) \subset \mathcal{T}^N(K) \cap Rx \neq 0 \text{ and } \mathcal{T}^N(K) + \mathcal{T}_N(K) \trianglelefteq K.$$

(2) Assume M is not polyform. Then there exist a cyclic submodule $K \subset M$ and a non-zero morphism $f : K \to M$ with $Ke\,f \trianglelefteq K$. Put $N = (K)f$ and $\overline{M} = M/\mathcal{T}_N(M)$. Then $\mathcal{T}^N(K) \cap Ke\,f \trianglelefteq \mathcal{T}^N(K)$.

The map $K \xrightarrow{f} M \to \overline{M}$ factorizes through $\bar{f} : K/\mathcal{T}_N(K) \to \overline{M}$. Since

$$\mathcal{T}_N(K) \cap Ke\,f \trianglelefteq \mathcal{T}^N(K) \text{ and } \mathcal{T}^N(K) \trianglelefteq K/\mathcal{T}_N(K),$$

we conclude $Ke\bar{f} \trianglelefteq K/\mathcal{T}_N(K)$. This means that N is an $\overline{M}$-singular module.

By assumption, $\overline{M} \in \sigma[N] = \sigma[\overline{M}]$. So, in particular, $\overline{M}$ is $\overline{M}$-singular.

However, as noted above, $\overline{M}$ is projective in $\sigma[\overline{M}]$ and hence cannot be $\overline{M}$-singular. Therefore M is polyform. □

7.10 Projective SSP modules.
Let M be projective in $\sigma[M]$ and $T = End_R(\widehat{M})$. Then the following are equivalent:

(a) M is an SSP-module;

(b) for every submodule $K \subset M$, $\widehat{M} = KT \oplus \mathcal{T}_K(\widehat{M})$.

(c) M is polyform and for any $N \trianglelefteq M$, $M \in \sigma[N]$.

Proof. $(a) \Rightarrow (b)$ By 7.9, M is polyform and the decomposition was given in 7.1.

$(b) \Rightarrow (a)$ This decomposition implies in particular that every fully invariant submodule is a direct summand in $\widehat{M}$ as (R,T)-submodule. Hence $\widehat{M}$ is a semisimple (R,T)-bimodule and M is SSP by 4.5.

$(c) \Leftrightarrow (c)$ By 7.9, M is polyform and the assertion follows from 7.1. □

8 Left SSP and PSP rings

We call a ring R *left PSP*, if $_RR$ is a PSP module and R is called *left SSP*, if $_RR$ is an SSP module.

Notice that these definitions also apply to rings without units, considering such rings as modules over rings with units in a canonical way.

Before characterizing PSP and SSP rings we want to describe the torsion modules related to semiprime rings:

8.1 Semiprime rings and torsion modules.
Let R be a semiprime ring and $N \subset R$ a left ideal. Then:

(1) $\mathcal{T}_N(R) = An_R(N)$.

(2) $R/\mathcal{T}_N(R) \in \sigma[N]$ if and only if there exists a finite subset $X \subset N$ with $An_R(X) = An_R(N)$.

Proof. (1) The relation $\mathcal{T}_N(R) \subset An_R(N)$ always holds. Clearly $NAn_R(N)$ is a nilpotent left ideal and hence is zero.

Consider $f \in Hom_R(An_R(N), I_R(N))$ and put $K = (N)f^{-1}$. Then $(Kf)^2 \subset N(Kf) = (NK)f \subset (NAn_R(N))f = 0$. Since R is semiprime, $Kf = 0$ and so $Im\, f \cap N = 0$, implying $f = 0$. Hence $\mathcal{T}_N(R) \supset An_R(N)$.

(2) Assume $R/\mathcal{T}_N(R) \in \sigma[N]$. By 4.1, there exists a finite subset $X \subset N$ with

$$An_R(X)1 \subset \mathcal{T}_N(R) = An_R(N) \subset An_R(X).$$

Now assume $An_R(X) = An_R(N)$ for some finite $X \subset N$. Then $An_R(X) = \mathcal{T}_N(R)$ by (1), and for any $b \in R$, $An_R(X)b \subset \mathcal{T}_N(R)$. Now apply 4.1. □

8.2 Left SSP rings.

For a ring R let $Q = Q(R)$ denote the maximal (complete) left ring of quotients. Then the following are equivalent:

(a) R is left SSP;

(b) for every essential left ideal $N \subset R$, $R \in \sigma[N]$;

(c) every essential left ideal $N \subset R$ contains a finite subset X with $An_R(X) = 0$;

(d) for every left ideal $I \subset R$, $Q = IQ \oplus \mathcal{T}_I(Q)$;

(e) R is semiprime and every left ideal $I \subset R$ contains a finite subset $X \subset I$ with $An_R(X) = An_R(I)$;

(f) Q is a semisimple (R,Q)-module.

If R satisfies these conditions, then Q is left self-injective, von Neumann regular, and a finite product of simple rings.

Left ideals with property (c) above are also called *insulated*. So the rings described here are exactly the *left strongly semiprime rings* of Handelman [6], Theorem 1. They generalize *left strongly prime* rings as considered in Handelman-Lawrence [7] and Viola-Prioli [11].

Proof. $(a) \Rightarrow (b)$ is shown in 4.4.

$(b) \Rightarrow (a)$ Assume for every essential left ideal $N \subset R$, $R \in \sigma[N]$. Any such N is a faithful R-module.

First we show that R is semiprime. For this consider an ideal $I \subset R$ with $I^2 = 0$. Let J be the right annihilator of I and $L \subset R$ any non-zero left ideal. Obviously, $I \subset J$. Assume $L \cap J = 0$. Then $IL \neq 0$. However, $IL \subset L \cap J = 0$, a contradiction. This implies that J is an essential left ideal in R. By our assumption, J is a faithful left module and $IJ = 0$ means $I = 0$.

In view of 4.5 and 3.2 it remains to show that R is left non-singular (compare [9], Lemma 2.4 and [6], Proposition 6).

If the left singular ideal $Z(R) \subset R$ is non-zero, $Z(R) \oplus An_R(Z(R))$ is an essential left ideal in R. Hence there are $a_1, \dots, a_k$ with $An_R(a_1, \dots, a_k) = 0$. From this we see that there is a monomorphism $Z(R) \to Z(R)(b_1, \dots, b_r)$ with $b_1, \dots, b_r \in Z(R)$.

The kernel of this map is $Z(R) \cap An_R(b_1) \cap \dots \cap An_R(b_r)$. Since all the $An_R(b_i)$ are essential left ideals in R, this intersection could not be zero, a contradiction. Hence R is left non-singular.

$(b) \Leftrightarrow (c)$ This is obvious by 4.1.

$(a) \Rightarrow (d)$ By 7.9, R is semiprime and left non-singular. Therefore $Q = \widehat{R}$ and $End_R(Q) = Q$ (see Lambek [10], §4.3). Now the assertion follows from 7.1.

$(d) \Rightarrow (a)$ We show that R is left non-singular.

Assume for $a \in R$, $An_R(a)$ is an essential left ideal in R. Since $Q = RaQ \oplus T_a(Q)$, we have $RaQ = eQ$ for some idempotent $e \in Q$. Write $e = \sum_{i=1}^{n} r_i a q_i$, with $r_i \in R, q_i \in Q$.

Obviously, $L = \bigcap_{i=1}^{n}(An_R(a) : r_i)_R$ is an essential left ideal in R and $Le = 0$. Therefore $L \cap Qe = 0$ and also $L \cap (Qe \cap R) = 0$. However, $L \subset R$ is an essential left ideal and $Qe \cap R \neq 0$, a contradiction. Hence R is left non-singular and so $Q = \widehat{R}$ and $End_R(Q) = Q$. Now apply 7.1.

$(a) \Leftrightarrow (e) \Leftrightarrow (f)$ follow from 7.10, 4.3 and 4.5. □

8.3 Semiprime and left PSP rings.

For a ring R the following conditions are equivalent:

(a) R is a left PSP-ring;

(b) R is semiprime and for every finitely generated left ideal $N \subset R$, $An_R(N) = An_R(X)$ for some finite subset $X \subset N$.

In this case R is left non-singular.

Proof. $(a) \Rightarrow (b)$ First we show that R is semiprime. Consider a cyclic left ideal $N \in R$ with $N^2 = 0$. By assumption, $R/T_N(R) \in \sigma[N]$. Since $NK = 0$ for any $K \in \sigma[N]$, in particular $N(R/T_N(R)) = 0$ and hence $NR \subset T_N(R)$. This implies $NR \subset T^N(R) \cap T_N(R) = 0$ and $N = 0$. Now the assertion is clear by 8.1.

$(b) \Rightarrow (a)$ also follows from 8.1.

By 7.9, R is left non-singular. □

As mentioned in the introduction, a commutative ring is PSP if and only if it is semiprime, i.e. if it has no nilpotent elements. It is interesting to observe that the latter property suffices to make non-commutative rings PSP.

Recall that a ring without non-zero nilpotent elements is said to be *reduced*.

8.4 Reduced rings.

Any reduced ring R is left PSP.

Proof. Assume R is reduced. It is shown in Lemma 1 of chapter 4, §2 in Andrunakievic-Rjabuhin [1] that for any subset $U \subset R$, the left annihilator $An_R(U)$ of U coincides with the right annihilator of U in R. This implies $An_R(U) = An_R(RU)$. Now it is obvious from 8.3 that R is PSP. □

9 Bimodule structure of rings

Let A be a not necessarily associative ring. Left and right multiplication by any $a \in A$,

$$L_a : A \to A,\ x \mapsto ax, \quad R_a : A \to A,\ x \mapsto xa \text{ for } x \in A,$$

define $\mathbb{Z}$-endomorphisms of A, i.e. $L_a, R_a \in End(_{\mathbb{Z}}A)$.

The subring of $End(_{\mathbb{Z}}A)$ generated by all left multiplications in A and the identity map id_A is called the *left multiplication ring $L(A)$ of A*. Similarly the *right multiplication ring* $R(A)$ is defined.

The subring of $End(_{\mathbb{Z}}A)$ generated by all left and right multiplications and id_A is called the *multiplication ring* $M(A)$ of A, i.e.

$$M(A) = < \{L_a, R_a \mid a \in A\} \cup \{id_A\} > \quad \subset End(_{\mathbb{Z}}A).$$

Obviously, $L(A)$, $R(A)$ and $M(A)$ are associative rings with units and we consider A as a left module over these rings.

In particular we form the subcategory $\sigma[_{M(A)}A]$ of $M(A)$-*Mod* and denote it by $\sigma[A]$. $c(A) = End_{M(A)}(A)$ is the *centroid* of A. From [2] or [12], Theorem 4.2 we recall for semiprime rings:

9.1 Central closure of semiprime rings.

Let A be a semiprime ring, $\widehat{A}$ the injective hull of A in $\sigma[A]$ and $T = End_{M(A)}(\widehat{A})$. Then

(1) A is a polyform $M(A)$-module.

(2) T is a commutative, regular and self-injective ring.

(3) $\widehat{A} = AT$ is a semiprime ring with respect to the multiplication

$$(\sum a_i s_i)(\sum b_j t_j) = \sum (a_i b_j)(s_i t_j) \ \text{for any finite } a_i, b_j \in A,\ s_i, t_j \in T$$

and its centroid is T.

T is called the *extended centroid* of A, and the ring $\widehat{A}$ the *central closure* of A.

Definition. A ring A is said to be *left PSP* if A is PSP a an $L(A)$-module. A is called a *PSP ring* if it is PSP a an $M(A)$-module.

Similarly we define SSP and *left (right) SSP* rings.

Obviously, every SSP ring is PSP.

Rings A which are strongly prime as $M(A)$-modules are exactly those prime rings whose central closures are simple rings (cf. [13], Theorem 3.2). In particular they are SSP rings.

Now let us see which PSP rings are semiprime. For this recall the definition of the

associator $\quad (a,b,c) = (ab)c - a(bc)$ for $a,b,c \in A$ and the

right nucleus $\quad n(A) = \{c \in A \mid (x,y,c) = 0 \text{ for all } x,y \in A\}$.

Clearly, $R_c \in End_{L(A)}(A)$ for all $c \in n(A)$.

9.2 PSP and semiprime rings.

Let A be any ring.

(1) Assume A is PSP and no non-zero ideal of A annihilates A. Then A is a semiprime ring.

(2) Assume A is left PSP and no non-zero left ideal of A annihilates the right nucleus $n(A)$. Then A is a semiprime ring.

(3) Assume A is (left) strongly prime and no ideal of A annihilates A. Then A is a prime ring.

Proof. (1) Assume $K \subset A$ is a non-zero finitely generated ideal with $K^2 = 0$. Consider $U = \{L_a \mid a \in K\}$. By assumption, $A/\mathcal{T}_K(A) \in \sigma[K]$. Now $UK = 0$ implies $U(A/\mathcal{T}_K(A)) = 0$ and $KA \subset \mathcal{T}_K(A)$. So $KA \subset \mathcal{T}_K(A) \cap K = 0$, contradicting the given condition.

(2) In the proof above, let K be a left ideal. Then $Kn(A) \subset \mathcal{T}_K(A) \cap \mathcal{T}^K(A) = 0$, contradicting the assumption in (2).

(3) Consider non-zero ideals $I, J \subset A$ with $IJ = 0$. Put $U = \{L_a \mid a \in I\}$. By assumption, $A \in \sigma[J]$ and hence $IA = UA = 0$, a contradiction. □

In general, a strongly prime ring need not be prime. For this consider the cyclic group $\mathbb{Z}_p$ of prime order p with the trivial multiplication $ab = 0$, for all $a, b \in \mathbb{Z}_p$. This is a strongly prime (simple) ring which is not prime.

For associative rings we notice a relationship between the one-sided and two-sided versions of the notions above:

9.3 Associative *PSP* rings.

Let A be an associative semiprime ring with unit.

(1) In $\sigma[A]$, $\mathcal{T}_U(A) = An_A(U)$ for any ideal $U \subset A$.

(2) If A is a PSP ring, then A is left PSP.

(3) If A is an SSP ring, then A is left SSP.

(4) If A is a strongly prime ring, then A is left strongly prime.

Proof. (1) $\mathcal{T}_U(A)$ is an ideal in A and $\mathcal{T}_U(A) \cap U = 0$, hence $\mathcal{T}_U(A) \subset An_A(U)$. Consider $f \in Hom_{M(A)}(An_A(U), I_A(U))$ and put $K = (U)f^{-1}$. Then $(Kf)^2 \subset U(Kf) = (UK)f \subset (UAn_A(U))f = 0$. Since A is semiprime, $Kf = 0$ and so $Im\, f \cap U = 0$, implying $f = 0$. Hence $\mathcal{T}_U(A) \supset An_A(U)$ (compare 8.1).

(2) Consider $a, b \in A$. Since A is PSP, there exist $x_1, \ldots, x_k \in M(A)$ such that

$$An_{M(A)}(x_1 a, \ldots, x_k a)b \subset \mathcal{T}_{M(A)a}(A) = An_A(M(A)a) \subset An_A(L(A)a).$$

In particular,

$$An_{L(A)}(x_1 a, \ldots, x_k a)b \subset An_A(L(A)a).$$

Now assume $x_i = \sum_{j=1}^{m_i} L_{y_{ij}} R_{z_{ij}}$. Then

$$An_{L(A)}(\{y_{ij} a \mid 1 \le j \le m_i, 1 \le i \le k\})b \subset An_A(L(A)a).$$

By 8.1, $\mathcal{T}_{L(A)a}(A) = An_A(L(A)a)$. Hence by 5.1, the above relation implies that A is left PSP.

(3) and (4) are shown in a similar way. □

9.4 Strongly semiprime rings.

Let A be a ring which is not annihilated by any non-zero ideal and $T = End_{M(A)}(\widehat{A})$. Then the following conditions are equivalent:

(a) A is an SSP ring;

(b) A is semiprime and for every essential ideal $U \subset A$, $A \in \sigma[U]$;

(c) for every ideal $I \subset A$, $\widehat{A} = IT \oplus \mathcal{T}_I(\widehat{A})$.

(d) A is semiprime and the central closure $\widehat{A}$ is a direct sum of simple ideals.

If A is associative, then (a)-(d) are equivalent to:

(e) A is semiprime and for every ideal $I \subset A$, $A/An_A(I) \in \sigma[I]$;

Proof. By 9.2, A being SSP implies that A is semiprime.

$(a) \Leftrightarrow (b)$ Since A is semiprime, A is a polyform $M(A)$-module by 9.1. Hence the assertion follows from 7.1.

$(a) \Leftrightarrow (c)$ This is also obtained from 7.1.

$(a) \Leftrightarrow (d)$ The central closure $\widehat{A}$ is a direct sum of simple ideals if and only if it is semisimple as an $(M(A), T)$-bimodule. Now apply 4.5.

$(a) \Leftrightarrow (e)$ By 9.2, R is a semiprime ring. Hence by 9.3, $T_U(A) = An_A(U)$. Now the equivalence is clear by 4.3. □

9.5 Idempotent closure of semiprime rings.

Let A be a semiprime ring, $T = End_{M(A)}(\widehat{A})$, B the Boolean ring of idempotents of T and $\widetilde{A} = AB$ the idempotent closure of A as an $M(A)$-module. Then

(1) For every $a \in \widetilde{A}$, there exist $a_1, \ldots, a_k \in A$ and pairwise orthogonal $e_1, \ldots, e_k \in B$ such that
(i) $a = \sum_{i=1}^{k} a_i e_i$,
(ii) $e_i = \varepsilon(a_i)e_i \quad$ for $i = 1, \ldots, k$ and
(iii) $\varepsilon(a) = \sum_{i=1}^{k} e_i$.

(2) For every prime ideal $K \subset \widetilde{A}$, $P = K \cap A$ is a prime ideal in A and

$$\widetilde{A}/K = (A + K)/K \simeq A/P.$$

The set $x = \{e \in B | \widetilde{A}e \subset K\}$ is a maximal ideal in B and $K = PB + \widetilde{A}x$.

(3) For any prime ideal $P \subset A$, there exists a prime ideal $K \subset \widetilde{A}$ with $K \cap A = P$.

Proof. (1) Since a semiprime ring A is polyform as an $M(A)$-module, the assertion follows immediately from 3.5.

(2) Consider two ideals $I, J \subset A$ with $IJ \subset P$. Then IB and JB are ideals in $\widetilde{A}$ and $(IB)(JB) \subset K$. Hence $IB \subset K$ or $JB \subset K$. Assume $IB \subset K$. Then $I \subset IB \subset K \cap A \subset P$. So P is a prime ideal in A.

Consider $a \in A$ and $e \in B$. Since $(\widetilde{A}e)[\widetilde{A}(1-e)] = 0$, $\widetilde{A}e \subset K$ or $\widetilde{A}(1-e) \subset K$. Hence $ae \in K$ or $a(1-e) \in K$ which means $ae + K = K$ or $ae + K = a + K$.

Therefore for any $d \in \widetilde{A}$, there exists $x \in A$, with $d + K = x + K$. This implies $\widetilde{A}/K = (A + K)/K \simeq A/P$.

A straightforward argument shows that x is a maximal ideal in the Boolean ring B.

Put $U = PB + \widetilde{A}x$. Clearly $U \subset K$. For any $a \in K$, choose $a_1, \ldots, a_k \in A$ and orthogonal idempotents $e_1, \ldots, e_k \in B$ satisfying the conditions (*ii*) and (*iii*) of (1). In case all $e_1, \ldots, e_k \in x$, then $a \in \widetilde{A}x \subset U$.

Assume, without restriction, $e_1 \notin x$. Since $e_i e_1 = 0$ for $i \neq 1$, we have $e_2, \ldots, e_k \in K$. Also $1 - e_1 \in x$. Therefore

$$a_1 = a + a_1(1 - e_1) \in K + \tilde{A}x = K$$

and $a_1 \in P$. Consequently, $a_1 e_1 \in PB$ and

$$a = a_1 e_1 + \sum_{i=2}^{k} a_i e_i \in PB + \tilde{A}x.$$

So in any case $a \in U$, implying $K = U$.

(3) Put $S = A \setminus P$. Obviously, for any $a, b \in S$,

$$\emptyset \neq (M(A)a)(M(A)b) \cap S \subset (M(\widehat{A})a)(M(\widehat{A})b) \cap S.$$

Using this relationship we can show (as in the associative case) that an ideal $K \subset \tilde{A}$, which is maximal with respect to $K \cap S = \emptyset$, is a prime ideal.

Clearly $I = K \cap A \subset P$. As shown above, $\tilde{A}/K = (A + K)/K \simeq A/I$. Obviously, P/I is a prime ideal in A/I. Hence there exists a prime ideal $J \supset K$ of $\tilde{A}$, for which $J/K = (P + K)/K$. This means $J = P + K$, $J \cap A = P + K \cap A = P + I = P$ and $J \cap S = \emptyset$. By the choice of K we conclude $K = J$. □

9.6 Properly semiprime rings.

Let A be a semiprime ring, $T = End_{M(A)}(\widehat{A})$, B the Boolean ring of idempotents of T and $\tilde{A} = AB$ the idempotent closure of A. Then the following are equivalent:

(a) A is a PSP ring;

(b) for every $a \in A$, $M(A)aT = \widehat{A}\varepsilon(a)$;

(c) for every finitely generated ideal $K \subset A$, $KT = \widehat{A}\varepsilon(K)$;

(d) $\tilde{A}$ is a PSP ring.

Under the given conditions, for every $a \in A$, $\mathcal{T}_a(\widehat{A}) = \widehat{A}(id - \varepsilon(a))$.

Proof. Since semiprime rings are polyform as bimodules, these equivalences essentially are obtained from 7.3.

We only have to show that (d), i.e. $\tilde{A}$ is PSP as an $M(\tilde{A})$-module, is equivalent to $\tilde{A}$ being PSP as an $M(A)$-module. This follows readily from $M(\tilde{A}) = M(A)B$. □

A ring A is said to be *fully idempotent* if, for any ideal $I \subset A$, $I = M(A)I^2$. Clearly such rings are semiprime. We show a module property of them:

9.7 Fully idempotent rings.
Let A be a fully idempotent ring. Then A is pseudo regular as $M(A)$-module.

Proof. Put $S = End_{M(A)}(A)$ and $T = End_{M(A)}(\hat{A})$. For any subset $U \subset A$, define $L_o(U) = \{L_u \mid u \in U\}$. Consider $a \in A$ and $I = M(A)a$. By assumption, $I = M(A)I^2 = M(A)L_o(I)I$ and hence $a \in M(A)L_o(I)M(A)a$. Therefore there exists $\alpha \in M(A)L_o(I)M(A)$ with $\alpha a = a$. Obviously, $\alpha \in M(A)$ and $\alpha A \subset I \subset M(A)aT$. So A is a pseudo regular $M(A)$-module (see 6.1). □

A ring A is called *biregular* if every principal ideal of A is a direct summand (as an $M(A)$-module).

If A has a unit, then it is biregular if and only if every principal ideal is generated by an idempotent in the center of A. For simplicity here we only consider biregular rings with units, though our methods also apply to the general case.

The next result reveils the connection between biregular and PSP rings.

9.8 Biregular and PSP rings.
Let A be a semiprime ring with unit, $T = End_{M(A)}(\hat{A})$, B the Boolean ring of idempotents of T and $\tilde{A} = AB$ the idempotent closure of A. Then the following are equivalent:

(a) A is biregular;

(b) A is a fully idempotent PSP ring;

(c) A is semiprime and for every $a \in A$, $M(A)a = A\varepsilon(a)$;

(d) A is semiprime and $\tilde{A}$ is biregular.

Proof. $(a) \Rightarrow (b)$ Assume A is biregular. Then A is fully idempotent.

For any $a \in A$, $M(A)a = Ae$ for some idempotent $e \in C$. e extends to a unique idempotent $\tilde{e} \in T$ and $M(A)aT = AT\tilde{e} = \hat{A}\tilde{e}$. As easily checked, $\tilde{e} = \varepsilon(a)$. So A is PSP by 9.6.

$(b) \Rightarrow (a)$ By 9.7, fully idempotent rings are pseudo regular $M(A)$-modules. Hence the assertion follows from 6.3.

$(b) \Leftrightarrow (c)$ Since A is polyform, these assertions follow from 6.3.

$(b) \Rightarrow (d)$ Since A is polyform, $\tilde{A}$ is PSP by 7.3. We show that $\tilde{A}$ is fully idempotent. Any element in $\tilde{A}$ can be written as $a = \sum_{i=1}^{k} a_i e_i$, for some $a_i \in A$ and pairwise orthogonal $e_i \in B$. Put $K = M(\tilde{A})a$ and $K_i = M(A)a_i$. Clearly $K = \sum_{i=1}^{k} K_i e_i B$. Since A is fully idempotent, $K_i = M(A)K_i^2$. Hence

$$M(\tilde{A})K^2 = \sum_{i=1}^{k} M(A)K_i^2 e_i B = \sum_{i=1}^{k} K_i e_i B = K,$$

and $\tilde{A}$ is fully idempotent. So $\tilde{A}$ is biregular by $(a) \Leftrightarrow (b)$.

$(d) \Rightarrow (a)$ Consider $a \in A$. By assumption, $M(\tilde{A})a = \tilde{A}\varepsilon(a)$. Obviously, $M(\tilde{A}) = M(A)B$ and $M(\tilde{A})a = M(A)aB$. Put $c = \varepsilon(a)$. Clearly $\varepsilon(c) = c$ and $c \in M(A)aB$. By 9.5, there exist $a_1, \ldots, a_k \in M(A)a$ and pairwise orthogonal $e_1, \ldots, e_k \in B$ such that

$$c = \sum_{i=1}^{k} a_i e_i \quad \text{and } c = \varepsilon(c) = \sum_{i=1}^{k} e_i.$$

Therefore $ce_i = a_i e_i$, $ce_i = e_i$ and $(1 - a_i)e_i = 0$ for all $i = 1, \ldots, k$. Put

$$b = (\ldots(((1 - a_1)(1 - a_2))(1 - a_3)\ldots)(1 - a_k).$$

Clearly $be_i = 0$ for all $i = 1, \ldots, k$. Hence $bc = 0$. Since $a_i \in M(A)a$, $b = 1 + d$ where $d \in M(A)a$. Further since $ac = a\varepsilon(a) = a$ and $d \in M(A)a$, $dc = d$. So $0 = bc = (1+d)c = c + dc = c + d$ and $\varepsilon(a) = c = -d \in M(A)a$. Hence $M(A)a \subset A\varepsilon(a) \subset M(A)a$ and $M(A)a = A\varepsilon(a)$. □

Combining our results 7.5, 7.7 and 3.2 in [14] we have:

9.9 Central closure as Azumaya ring.

Let A be a semiprime ring with unit, $T = End_{M(A)}(\widehat{A})$ and $\widehat{A}$ the central closure of A. Then the following are equivalent:

(a) $\widehat{A}$ is an Azumaya ring;

(b) $\widehat{A}$ is a PSP ring and the module ${}_{M(\widehat{A})}\widehat{A}$ is finitely presented in $\sigma[\widehat{A}]$;

(c) $\widehat{A}$ is a biregular ring and the module ${}_{M(\widehat{A})}\widehat{A}$ is projective in $\sigma[\widehat{A}]$;

(d) the module ${}_{M(\widehat{A})}\widehat{A}$ is a generator in $\sigma[\widehat{A}]$;

(e) $M(\widehat{A})$ is a dense subring of $End_T(\widehat{A})$;

(f) for any $m_1, \ldots, m_n, m \in A$ there exist $r_1, \ldots, r_k \in M(A)$ such that for $s_1, \ldots, s_n \in M(A)$ the relations

$$\sum_{l=1}^{n} s_l r_j m_l = 0 \quad \textit{for } j = 1, \ldots, k,$$

imply $s_1 m \in \mathcal{T}_{Um_1}(A)$ for $U = An_{M(A)}(m_2, \ldots, m_n)$.

(g) A is a PSP ring and for any $m_1, \ldots, m_n \in A$ there exist $r_1, \ldots, r_k \in M(A)$ such that for $s_1, \ldots, s_n \in M(A)$ the relations

$$\sum_{l=1}^{n} s_l r_j m_l = 0 \quad \textit{for } j = 1, \ldots, k,$$

imply $s_1 m_1 \in \mathcal{T}_{Um_1}(A)$ for $U = An_{M(A)}(m_2, \ldots, m_n)$.

9.10 Example

We construct a biregular PI-ring A with unit, whose central closure is not biregular. So A is PSP as an M(A)-module but $\widehat{A}$ is not PSP.

Let F be field of non-zero characteristic p, $K = F(X)$ the field of the rational functions and $G = F(X^p)$ a subfield of K. Consider the embedding

$$\alpha : K \to M_p(G) = End_G(K), \quad a \mapsto L_a.$$

Denote by $Q = M_p(G)^\omega$ a countable product of copies of $M_p(G)$, and $I = M_p(G)^{(\omega)}$. We have an embedding

$$\psi : K \to M_p(G)^\omega \subset Q, \quad \psi(a) = \{\alpha(a)\}_\omega \in M_p(G)^\omega \text{ for } a \in K.$$

Clearly I is an ideal in Q. Put $A = I + \psi(K)$. Obviously, A is a biregular ring with unit, whose maximal right ring of quotients is Q, and the center T of Q is the extended centroid of A.

Hence $R = AT$ is the central closure of A. I is an ideal in R and R/I is a commutative ring. Let $y \in Q$ denote an element all whose coefficients are equal to X and put $z = \psi(X)$. Then $y^p = z^p$. Therefore $y - z + I$ is a non-zero nilpotent element in the commutative ring R/I. Hence R cannot be biregular. □

Acknowledgements. This paper was prepared during a stay of the first author at Düsseldorf University supported by the German Academic Exchange Service (DAAD). He is grateful to both institutions for making this stimulating visit possible.

References

[1] Andrunakievic, V.A., Rjabuhin, J.M., *Radicals of algebras and structure theory*, NAUKA, Moscow (1979)

[2] Beidar K.I., *Rings with generalized identities I*, Vestnik Mosk. Univ. Ser. Mat.-Mex. 2, 19 - 26 (1977)

[3] Beidar, K.I., *Quotient rings of semiprime rings*, Vestnik Mosk. Univ. Ser. Mat.-Mex. 5, 36 - 43 (1978)

[4] Faith, C., *Rings, Modules and Categories I.*, Springer, Berlin, Heidelberg, New York (1973)

[5] Goel, V.K., Jain, S.K., *π-injective modules and rings whose cyclics are π-injective*, Comm. Algebra 6, 59-73 (1978)

[6] Handelman, D., *Strongly semiprime rings*, Pac. J. Math. 60, 115-122 (1975)

[7] Handelman, D., Lawrence, J., *Strongly prime rings*, Trans. Amer. Math. Soc. 211, 209-223 (1975)

[8] Johnson, R.E, Wong, E.T., *Quasi-injective modules and irreducible rings*, J. London Math. Soc. 36, 260-268 (1961)

[9] Kutami, M., Oshiro, K., *Strongly semiprime rings and nonsingular quasi-injective modules*, Osaka J. Math. 17, 41-50 (1980)

[10] Lambek, J., *Lectures on Rings and Modules*, Blaisdell Publ., Waltham e.a. (1966)

[11] Viola-Prioli, J., *On absolutely torsion-free rings*, Pacif. J. Math. 56, 275-283 (1975)

[12] Wisbauer, R., *Localization of modules and the central closure of rings*, Comm. Algebra 9, 1455-1493 (1981)

[13] Wisbauer, R., *On Prime Modules and Rings*, Comm. Algebra 11, 2249-2265 (1983)

[14] Wisbauer, R., *Local-Global Results for Modules over Algebras and Azumaya Rings*, J.Algebra 135, 440-455 (1990)

[15] Wisbauer, R., *Foundations of Module and Ring Theory*, Gordon and Breach, Reading, New York e.a. (1991)

[16] Zelmanowitz, J., *Representation of rings with faithful polyform modules*, J. Algebra 25, 554-574 (1986)

[17] Zelmanowitz, J., *Weakly semisimple modules and density theorem*, Comm. Algebra, to appear

ENVELOPING ALGEBRAS OF INFINITE DIMENSIONAL LIE ALGEBRAS AND LIE SUPERALGEBRAS

JEFFREY BERGEN
Department of Mathematics, DePaul University
Chicago, IL 60614, USA

ABSTRACT. In this expository paper we examine the enveloping algebras of infinite dimensional Lie algebras and Lie superalgebras. Our goal is to reduce certain natural ring-theoretic questions from the arbitrary infinite dimensional situation to the more familiar finite-dimensional one. This is done by applying techniques analogous to the Δ-methods previously used to solve questions on group algebras.

1. Introduction

In this expository paper we examine the enveloping algebras of infinite dimensional Lie algebras and Lie superalgebras. Our goal is to reduce certain natural ring-theoretic questions from the arbitrary infinite dimensional situation to the more familiar finite-dimensional one. The proofs of the results in this paper appear in a series of fairly technical papers [3], [4], [5], and [6].

Throughout Sections 1, 2, and 3 of this paper L will be either a Lie algebra or a Lie superalgebra over a field K of characteristic 0 and $U(L)$ will denote its enveloping algebra. When L is a Lie algebra $U(L)$ is a Hopf algebra, whereas when L is a Lie superalgebra, there is a group G of order 2 such that the skew group ring $U(L)\#G$ is a Hopf algebra. Group algebras are a well known example of Hopf algebras and many questions on group algebras have been solved used Δ-methods [9]. Therefore, it was reasonable to try to find similar techniques in the Lie context. To this end, we considered

$$\Delta = \Delta(L) = \{l \in L \mid \dim_K[l, L] < \infty\},$$

which was first introduced in [1]. Δ can be viewed as the Lie analog of the finite-conjugate center of a group.

If we let Δ_L denote the join of all the finite-dimensional Lie ideals of L, then Δ_L is a characteristic ideal of L which is appreciably smaller than Δ. In [3] and [4], we reduce questions from L down to Δ and in [5] and [6] we sharpen these results by further reducing from Δ to Δ_L. Unfortunately, there does not appear to be any way to directly reduce from L to Δ_L without using Δ.

The reduction from L to Δ_L takes place by examining <u>linear</u> and <u>derivation identities</u>. Given any ring R, a linear identity for R is an equation of the form

$$\alpha_1 x \beta_1 + \alpha_2 x \beta_2 + \cdots + \alpha_n x \beta_n = 0$$

for all $x \in R$, where α_i, β_i are fixed elements of R. It is easy to see how linear identities arise in ring theory. For example, the identity $\alpha x - x\alpha = 0$ determines if α is central, whereas the identity $\alpha x \beta = 0$ determines whether R is prime. Similarly, a derivation identity for an enveloping algebra $U(L)$ is an equation of the form

$$\alpha_1^\delta \beta_1 + \alpha_2^\delta \beta_2 + \cdots + \alpha_n^\delta \beta_n = 0$$

which holds for all products δ of derivations or superderivations induced, via the ad map, by elements of L (or elements of L_0 and L_1 in the superalgebra case). For example, the identity $\alpha^\delta 1 - 1^\delta \alpha = 0$ determines if α is a constant in $U(L)$ under the adjoint action of L.

For enveloping algebras of Lie algebras it is easy to see the correspondence between linear and derivation identities.

Lemma 1.1. *If L is a Lie algebra, then $U(L)$ satisfies the linear identity*

$$\alpha_1 x \beta_1 + \alpha_2 x \beta_2 + \cdots + \alpha_n x \beta_n = 0$$

if and only if it satisfies the derivation identity

$$\alpha_1^\delta \beta_1 + \alpha_2^\delta \beta_2 + \cdots + \alpha_n^\delta \beta_n = 0.$$

For Lie superalgebras, linear identities no longer yield corresponding derivation identities. Since the reduction from L to Δ_L actually takes place through the manipulation of derivation identities, the superalgebra case is considerably more complicated.

If L is a Lie superalgebra, then $U(L)$ is graded by $\mathbb{Z}_2$ and we have the decomposition $U(L) = U(L)_0 \oplus U(L)_1$. A subset I of $U(L)$ is called <u>graded</u> if $I = (I \cap U(L)_0) \oplus (I \cap U(L)_1)$. When H is a subsuperalgebra of a Lie superalgebra L, let Δ_H be the join of all the graded, finite-dimensional subspaces of Δ stable under the adjoint action of H. Since Δ_H is a Lie ideal of L it follows, by the Poincaré-Birkhoff-Witt theorem, that $U(L)$ is free both as a left and right $U(\Delta_H)$-module. Choosing a fixed complementary basis B for Δ_H in L, the Poincaré-Birkhoff-Witt monomials in B form a basis M for $U(L)$ over $U(\Delta_H)$. Therefore if $\alpha \in U(L)$, we can uniquely write $\alpha = \sum \mu\alpha_\mu$ and $\alpha = \sum \alpha'_\mu \mu$, where $\alpha_\mu, \alpha'_\mu \in U(\Delta_H)$ and $\mu \in M$. In this case, we say that α is written <u>based on</u> Δ_H. Then the main technical result on linear identities appearing in [6] is

Theorem 1.2. *Let H be a subsuperalgebra of finite codimension in the Lie superalgebra L, let $\alpha_i, \beta_i \in U(L)$ for $i = 1, 2, \ldots, v$ and suppose that*

$$\alpha_1 x \beta_1 + \alpha_2 x \beta_2 + \cdots + \alpha_v x \beta_v = 0$$

for all $x \in U(H)$. Choose a complementary basis for Δ_H in L and use it to write $\alpha_i = \sum_\mu \mu\alpha_{\mu,i}$ and $\beta_i = \sum_\mu \beta_{i,\mu}\mu$ based on Δ_H. Then, for all monomials μ,

$$\sum_{i=1}^{v} \alpha_{\mu,i} x \beta_i = 0 = \sum_{i=1}^{v} \alpha_i x \beta_{i,\mu}$$

for all $x \in U(H)$. *In particular, if some* α_i *is not zero, then* $\{\beta_1, \beta_2, \ldots, \beta_v\}$ *is left linearly dependent over* $U(\Delta_H)$. *Similarly, if some* β_i *is not zero, then* $\{\alpha_1, \alpha_2, \ldots, \alpha_v\}$ *is right linearly dependent over* $U(\Delta_H)$.

Note that although Theorem 1.2 is stated in terms of Lie superalgebras, it can be specialized to Lie algebras. Before stating some applications of Theorem 1.2, we consider a simple example to illustrate the difference between Δ and Δ_L. Suppose A is the commutative Lie algebra with countable basis $\{a_1, a_2, \cdots\}$ and let L be the split extension $A \rtimes K\ell$, where $[a_i, \ell] = a_{i+1}$ for all i. Then it is easy to see that $\Delta = A$, whereas $\Delta_L = 0$. Thus in general, Δ_L can be appreciably smaller than Δ.

2. Enveloping Algebras of Lie Algebras

In this section, we assume that L is a Lie algebra. The proofs of all the results in this section can be found in [5]. We begin with a generalization of the main result of [2].

Theorem 2.1. *If* H *is a Lie subalgebra of* L *of finite codimension, then the centralizer of* H *in* $U(L)$ *is contained in* $U(\Delta_H)$. *In particular, the center of* $U(L)$ *is contained in* $U(\Delta_L)$.

A nonzero element $\alpha \in U(L)$ is said to be a semi-invariant if there exists a linear functional $\lambda : L \to K$ such that $[\alpha, \ell] = \lambda(\ell)\alpha$ for all $\ell \in L$. Thus α is an eigenvector for the adjoint representation of L on $U(L)$ and λ is its corresponding eigenfunction. The semi-center $\mathbf{SZ}(L)$ of $U(L)$ is defined as the K-linear span of the semi-invariants. $\mathbf{SZ}(L)$ has been extensively studied when L is finite dimensional. In particular, it is known that $\mathbf{SZ}(L)$ is a commutative unique factorization domain and every nonzero ideal of $U(L)$ contains a semi-invariant. We next have

Theorem 2.2. $\mathbf{SZ}(L)$ *is a commutative integral domain contained in* $U(\Delta_L)$.

Since $U(L)$ is a domain, it is a prime ring and has a symmetric Martindale ring of quotients $Q_S = Q_S(U(L))$. An automorphism or derivation of $U(L)$ is called X-inner if it is induced by an element of Q_S. The center of Q_S, $C = \mathbf{Z}(Q_S)$, is called the extended centroid of $U(L)$.

Theorem 2.3. *An element* $q \in Q_S$ *induces an X-inner automorphism of* $U(L)$ *if and only if* q *is the quotient of two semi-invariants of* $U(L)$. *In particular,* C *consists of all quotients of semi-invariants which have the same eigenfunction.*

Although $U(L)$ need not be Goldie, $U(\Delta_L)$ is always Goldie and has a classical division ring of quotients $Q_{cl} = Q_{cl}(U(\Delta_L))$. The action of L restricts to an action on $U(\Delta_L)$ and then extends uniquely to Q_{cl}. We let Q_{cl}^L denote the algebra of constants of the action of L on Q_{cl}.

Theorem 2.4. *An element $q \in Q_{cl}$ is an eigenvector for the action of L on Q_{cl} if and only if q is a quotient of two semi-invariants of $U(L)$. In particular, the extended centroid C of $U(L)$ and the algebra of constants Q_{cl}^{L} are naturally isomorphic.*

We next have

Theorem 2.5. *If δ is an X-inner derivation of $U(L)$ which stabilizes L then there exists $\ell \in L$ and a finite-dimensional solvable Lie ideal I of L such that $\delta - \operatorname{ad}\ell$ maps L into I. In particular, if L has no nonzero finite-dimensional solvable ideal then δ is induced by an element of L.*

We should note that if L does have a finite-dimensional solvable ideal then $U(L)$ can have an L-stabilizing X-inner derivation which is not inner. More can be said if we consider the case where L has no nonzero finite-dimensional abelian ideal.

Theorem 2.6. *Suppose L has no nonzero finite-dimensional abelian ideal. Then*

i. *Every semi-invariant of $U(L)$ is central.*
ii. *The identity map is the unique X-inner automorphism of $U(L)$.*
iii. *The extended centroid of $U(L)$ is the field of fractions of $\mathbf{Z}(U(L)) = \mathbf{Z}(U(\Delta_L))$.*
iv. *The X-inner derivations of $U(L)$ which stabilize L are induced by elements of L.*
v. *An ideal A of $U(L)$ contains a nonzero central element if and only if $A \cap U(\Delta_L) \neq 0$.*

We also have the following result on the existence of semi-invariants, which is a direct consequence of Lie's theorem:

Theorem 2.7. *Suppose every finite-dimensional homomorphic image of L is solvable and suppose the field K is algebraically closed. If A is subspace of $U(L)$ which is stable under the action of L, then A contains a semi-invariant if and only if $A \cap U(\Delta_L) \neq 0$.*

In light of Theorem 2.6(v), Theorem 2.7 and the known results on semi-invariants in the finite-dimensional case, it is reasonable to wonder if every ideal of $U(L)$ which intersects $U(\Delta_L)$ non-trivially contains a semi-invariant. However, we have an example of the form $L = A \rtimes B$ where A is a 2-dimensional abelian ideal of L, $A = \Delta = \Delta_L$, yet $\mathbf{SZ}(L)$ equals the ground field K.

The adjoint action of L on $U(L)$ extends, by taking sums of compositions of derivations, to an adjoint action of $U(L)$ on $U(L)$. We let

$$U(L)_{fin} = \{x \in U(L) \mid \dim_K \operatorname{ad} U(L) \cdot x < \infty\}.$$

We conclude this section with

Theorem 2.8. $U(L)_{fin} = U(\Delta_L)$.

3. Enveloping Algebras of Lie Superalgebras

In this section, we assume that $L = L_0 \oplus L_1$ is a Lie superalgebra. The structure of $U(L)$ is now quite different, since $U(L)$ is rarely a domain. The proofs of the results in this section can be found in [6]. We begin with some easy consequences of Theorem 1.2.

Theorem 3.1.

i. *If $U(\Delta_L)$ is prime, then so is $U(L)$.*
ii. *The center of $U(L)$ is contained in $U(\Delta_L)$.*

The adjoint action of the elements of L_0 and L_1 on $U(L)$ extend to an adjoint action of $U(L)$ on $U(L)$. We let the algebra of constants of $U(L)$, the semi-invariants of $U(L)$, and $U(L)_{fin}$ be defined as they were for ordinary Lie algebras. Note that the algebra of constants of $U(L)$ is no longer the same as the center of $U(L)$.

Theorem 3.2. $U(L)_{fin} = U(\Delta_L)$. *In particular, the algebra of constants and the semi-invariants of $U(L)$ are contained in $U(\Delta_L)$.*

When $U(L)$ is semiprime, it has a symmetric Martindale ring of quotients $Q_S = Q_S(U(L))$. The action of L and the $\mathbb{Z}_2$ grading extend uniquely to Q_S. The extended centroid C of $U(L)$ is defined as the center of Q_S and $C = C_0 \oplus C_1$ is also $\mathbb{Z}_2$-graded. We let Q_S^L denote the algebra of constants of Q_S under the action of L.

Theorem 3.3. *If $U(L)$ is semiprime then $Q_S^L = C_0$. In particular, all constants of $U(L)$ are central and of even parity.*

In general, the constants of $U(L)$ need not be central or even. Also, even if $U(L)$ is a domain, the central elements of $U(L)$ need not be either constant or even.

We say that $U(L)$ is a graded prime ring if the product of nonzero graded ideals is nonzero and we say that $U(\Delta_L)$ is graded L-prime if the product of nonzero graded ideals stable under the action of L is nonzero. Similarly, $U(\Delta_L)$ is L-semiprime if the square of nonzero graded ideals stable under the action of L is nonzero. Recall that when $U(\Delta_L)$ is prime, then so is $U(L)$. Therefore, when this occurs, it makes sense to compare their extended centroids.

Theorem 3.4.

i. *$U(L)$ is graded prime if and only if $U(\Delta_L)$ is graded L-prime.*
ii. *$U(L)$ is semiprime if and only if $U(\Delta_L)$ is L-semiprime.*
iii. *If $U(\Delta_L)$ is prime then the extended centroid of $U(L)$ embeds naturally into the extended centroid of $U(\Delta_L)$.*

We will now consider the case where $\Delta_L = 0$. In particular, this implies that $U(L)$ must be prime. If σ is the automorphism of L which is the identity on L_0 and negates elements of L_1, then σ extends uniquely to an automorphism of $U(L)$. The final part of the next theorem is the superalgebra analog of [8, Theorem 2.1].

Theorem 3.5. *Let L be a Lie superalgebra with $\Delta_L = 0$.*

i. *The extended centroid of $U(L)$, the semi-invariants of $U(L)$, and the algebra of constants $Q_S{}^L$ are all equal to the ground field K.*
ii. *If $L_1 \neq 0$ then the automorphism σ of $U(L)$ is not X-inner.*
iii. *The adjoint action of $U(L)$ on itself is faithful.*

4. Restricted Lie Algebras and Lie Superalgebras

Many of the results in this paper fail if the ground field K has characteristic $p > 0$. For instance, let us reconsider the example where A is the commutative Lie algebra with countable basis $\{a_1, a_2, \cdots\}$ and L is the split extension $A \rtimes K\ell$, where $[a_i, \ell] = a_{i+1}$ for all i. Then, regardless of the characteristic, $\Delta_L = 0$. However, if char $K = p > 0$ then the center of $U(L)$ is quite large as $a_i{}^p$ is always central. Nevertheless, results can be obtained for both Lie algebras and Lie superalgebras in the char $K = p > 0$ case, by considering restricted Lie algebras and restricted Lie superalgebras. Such algebras have a pth power map $x \longmapsto x^{[p]}$. The restricted enveloping algebra $u(L)$ is defined as $u(L) = U(L)/I$, where I is the ideal of $U(L)$ generated by all $x - x^{[p]}$. See [1, Chapter V] for additional details. In particular, every enveloping algebra $U(L)$ in characteristic $p > 0$ is equal to the restricted enveloping algebra $u(T)$ of a suitable restricted Lie algebra or Lie superalgebra T.

In [3] and [4] enveloping algebras as well as restricted enveloping algebras are studied. In the restricted case, the Lie ideals Δ and Δ_L are restricted and in fact, $\Delta = \Delta_L$. Therefore, in our study of restricted enveloping algebras, it was enough to reduce from L to Δ.

Acknowledgments The author was supported by grant MSP-08491 from the National Security Agency.

References

[1] Yu. A. Bachturin, *Identities in the universal envelopes of Lie algebras*, J. Austral. Math. Soc. **18** (1974), 10-21.

[2] J. Bergen, C. Dean, and D. S. Passman, *Supporting the center of the enveloping algebra*, (to appear).

[3] J. Bergen and D. S. Passman, *Delta methods in enveloping rings*, J. Algebra **133** (1990), 277-312.

[4] J. Bergen and D. S. Passman, *Delta methods in enveloping algebras of Lie superalgebras*, Trans. AMS **334** (1992), 259-280.

[5] J. Bergen and D. S. Passman, *Delta methods in enveloping rings II*, J. Algebra (to appear).

[6] J. Bergen and D. S. Passman, *Delta methods in enveloping algebras of Lie superalgebras II*, (to appear).

[7] N. Jacobson, *Lie algebras*, Interscience, New York, 1962.

[8] D. S. Passman, *The adjoint representation of group algebras and enveloping algebras*, Publicacions Matemàtiques (to appear).

[9] D. S. Passman, *The algebraic structure of group rings*, Wiley-Interscience, New York, 1977 (Krieger, Malabar, 1985).

COMPLETELY PRIME IDEALS AND ASSOCIATED RADICALS

GARY F. BIRKENMEIER

HENRY E. HEATHERLY

ENOCH K. LEE

Department of Mathematics, University of Southwestern Louisiana, Lafayette, Louisiana 70504, U.S.A.

ABSTRACT

This paper concerns relationships between prime ideals and completely prime ideals, with emphasis on connections to various radicals (prime radical, generalized nil radical, and a new radical introduced here, the $\mathfrak{P}_c$ radical). Two classes of rings which are of particular interest are : (1) the class of rings for which the prime radical coincides with the set of nilpotent elements and (2) the class of rings in which each prime ideal is completely prime. Properties of these classes and of the $\mathfrak{P}_c$ radical are developed.

Introduction

This paper concerns relationships between prime ideals and completely prime ideals, with connections to some associated radicals. Two classes of rings are studied in detail:

(i) those rings in which the prime (Baer lower nil) radical is equal to the set of nilpotent elements, which we call "2-primal rings", and

(ii) rings in which each prime ideal is completely prime. (The latter class of rings we call $\mathfrak{N}^2$. Note all rings in $\mathfrak{N}^2$ are 2-primal. The converse is not true).

Numerous examples and methods of constructing examples are given for each class. Some of the rings which have been attracting considerable attention in the recent literature fall in one or both of these classes, e.g. certain universal enveloping algebras and certain differential operator rings.

The flavor of the results in sections 2 and 3 is given by the following results obtained therein:

(i) the class of 2-primal rings is closed under subrings, direct sums, Dorroh extensions, formation of polynomial rings, and formation of upper triangular matrix rings;

(ii) the class $\mathfrak{R}^2$ is closed under ideals, homomorphic images, direct sums, Dorroh extensions, and formation of upper triangular matrix rings, but not under formation of polynomial rings.

The sum of all ideals in a ring which as rings are in $\mathfrak{R}^2$ we call the $\mathfrak{P}_c$-radical of the ring. This is an Amitsur-Kurosh radical which is supernilpotent. Further properties of the $\mathfrak{P}_c$-radical are developed and its relation to the socle is discussed.

Herein a "ring" will mean an associative, not necessarily commutative ring which may or may not have unity. Throughout this paper R will always denote a ring. The sum of all minimal right ideals of R, the right socle, is denoted by $Soc(R)$. (If R has no minimal right ideals, then $Soc(R)$ is zero). If X is a nonempty subset of R, then $\langle X\rangle_R$ is the ideal in R generated by X. The set of all nilpotent elements of X is denoted by $N(X)$. If no ambiguity can arise we use $\langle X\rangle$ for $\langle X\rangle_R$. We say X is *reduced* if $N(X) = 0$.We use $\mathfrak{P}(R)$ for the prime radical of R, i.e. the intersection of all the prime ideals of R, and we use $\mathfrak{P}_2(R)$ for the generalized nil radical of R, i.e. the intersection of all the completely prime ideals of R. (For alternate characterizations of $\mathfrak{P}_2$ see McCoy [21]). The Jacobson radical is denoted by J. For terminology and basic results concerning radical theory we have (except where noted) followed Divinsky [7].

1 Basic results

Under what condition(s) is a prime ideal completely prime? In a wide variety of settings ring theorists have observed that certain prime ideals (and even all prime ideals) are completely prime. (A representative sampling of where this occurs can be found in [6], [13], [18], [20], [27]). Going back to at least 1967 [23], conditions were given which when imposed on a prime ideal it consequently would be completely prime. We are particularly interested in conditions which when coupled with prime are equivalent to completely prime. The 2-primal condition, which we define next, has the virtues of being such a condition which is natural, useful, and which is implied by the other conditions we found in the literature which were used for a similar purpose.

Definition 1.1. An ideal I of R is a *2-primal ideal* if $\mathfrak{P}(R/I) = N(R/I)$. If the zero ideal of R is a 2-primal ideal, then R is a *2-primal ring.*

Any ring which is nilpotent, reduced, commutative, or anti-commutative is 2-primal. If I is an ideal of R such that R/I has any of these properties, then I is a 2-primal ideal. Further examples and methods of obtaining new 2-primal rings from old ones will be given in the sequel.

Proposition 1.2. Let I be an ideal of R. Then I is a completely prime (completely semiprime) if and only if I is prime (semiprime) and is a 2-primal ideal.

Proof. First recall that a ring which is reduced and prime is a domain. If I is a prime ideal and also a 2-primal ideal of R, then R/I is a prime ring and $N(R/I) = \mathfrak{P}(R/I) = 0$; so R/I is reduced and hence is a domain. Thus I is completely prime. The converse is immediate. The semiprime case is similar.

Before developing properties of 2-primal rings and 2-primal ideals we briefly discuss some conditions found in the literature with which prime implies completely prime. Renault [23] showed that an ideal is completely prime if and only if it is prime and completely semiprime. (For the remainder of this section I will always be an ideal in a ring R). Lambek [17] called I "symmetric" if $abc \in I$ implies $acb \in I$, and he observed that a prime ideal which is symmetric must be completely prime. (Lambek did this in the setting of a ring with unity, but no use of the unity was made in establishing the completely prime result). Mason [19] called I "completely reflexive" if $ab \in I$ implies $ba \in I$. He observed that an ideal is completely prime if and only if it is prime and completely reflexive. (In [19] Mason used the term "left symmetric" for what Lambek called "symmetric"). Bell [3] defined the *insertion-of-factors-property,* I.F.P. , for near-rings, and observes that R has I.F.P. if $ab = 0$ implies $aRb = 0$. Immediately one observes that I is completely prime if and only if it is prime and R/I has I.F.P.. Mason showed that if I is either completely reflexive or if I is symmetric, then R/I has I.F.P.. Reddy and Murty [22] introduced "semi-symmetric" ideals for near-rings: I is *semi-symmetric* if $x^n \in I$ implies $\langle x\rangle^n \subseteq I$. They showed that for zero symmetric near-rings (and hence for rings) an ideal is completely prime if and only if it is both prime and semi-symmetric. From the above mentioned papers and some routine arguments we have the following result.

Proposition 1.3. (i) If I is completely semiprime, then I is both symmetric and completely reflexive.

(ii) If R has a left unity and I is symmetric, then I is completely reflexive.

(iii) If R/I has I.F.P., then I is semi-symmetric.

From the definition of 2-primal we immediately have: I is 2-primal if and only if whenever $x^n \in I$, then $x \in \mathfrak{p}(I)$. (Here $\mathfrak{p}(I)$ is the intersection of all prime ideals of R which contain I). So I semi-symmetric implies I is 2-primal. Also observe that if I is a 2-primal ideal, then $N(R) \subseteq \mathfrak{p}(I)$.

The 2-primal condition is strictly weaker than the semi-symmetric condition, as can be seen by the following.

Example 1.4. Let A be the ring of all 4-by-4 upper triangular matrices over a field F. Let $x = \begin{bmatrix} 0 & 1 & 0 & 0 \\ 0 & 0 & 0 & 0 \\ 0 & 0 & 0 & 1 \\ 0 & 0 & 0 & 0 \end{bmatrix}$. Then $x^2 = 0$, but $\langle x\rangle^2 \neq 0$. So the zero ideal of A is 2-primal but not semi-symmetric. (Note $\mathfrak{P}(A) = N(A)$).

The 2-primal condition then satisfies the criterion of being implied by the various conditions we have found in the literature which taken with prime imply completely prime. It seems fair to say the 2-primal condition is a natural one. We will justify its usefulness as we develop its properties and show its interaction with other concepts. We first collect some useful results in the next proposition, the proofs of which are straightforward and will be omitted.

Proposition 1.5. (i) If M is a maximal ideal of R and M is a 2-primal ideal, then either $R^2 \subseteq M$ or M is completely prime.

(ii) Let θ: $R \to S$ be a surjective ring homomorphism. If I is a 2-primal ideal of R and $Ker\theta \subseteq I$, then $\theta(I)$ is a 2-primal ideal of S. If X is a 2-primal ideal of S, then $\theta^{-1}(X)$ is a 2-primal ideal of R.

We say an ideal X of R is the *minimal prime ideal belonging to* I if X is minimal among prime ideals of R which contain I.

Proposition 1.6. The following are equivalent:

(i) I is a 2-primal ideal of R;

(ii) $\mathfrak{p}(I)$ is a completely semiprime ideal of R;

(iii) every minimal prime ideal belonging to I is completely prime.

Proof. Assume(i). If $x^2 \in \mathfrak{p}(I)$, then $x^2+I \in \mathfrak{P}(R/I)$. So $x^2+I \in N(R/I)$ and hence $(x^2)^m \in I$ for some m. From the observation after Proposition 1.3, we have $x \in \mathfrak{p}(I)$. The implication (ii)$\Rightarrow$(i) is immediate from that same observation.

Assume (ii) and let P be a minimal prime ideal of R belonging to I. Then $\overline{P} = P/\mathfrak{p}(I)$ is a minimal primal ideal of the reduced ring $\overline{R} = R/\mathfrak{p}(I)$. Every minimal prime ideal of a reduced ring is a completely prime ideal [24, Theorem 2.6.11]. (Note: the proof there makes no real use of a unity in the ring). So $\overline{P}$ is completely prime and hence so is its pre-image P in R. Thus (ii) $\Rightarrow$ (iii). Finally, assume (iii). Observe that $\mathfrak{p}(I)$ is the intersection of all the minimal prime ideals belonging to I, each of which is completely prime, so $\mathfrak{p}(I)$ is completely semiprime. From the observation following Proposition 1.3 we have I is 2-primal.

An immediate consequence of Proposition 1.6 is that every prime ideal which contains a 2-primal ideal must contain a completely prime ideal, which in turn contains all the nilpotent elements of the ring.

Recall that an ideal A of a ring R is *ideal essential* in an ideal B of R if every nonzero ideal of R having nonzero intersection with B has nonzero intersection with A.

Lemma 1.7. For any ring R there exists an ideal A of R with $A^2 = 0$, $A \subseteq \mathfrak{P}(R)$, and A is ideal essential in $\mathfrak{P}(R)$.

This lemma is an immediate consequence of Theorem 2.4 in [11] and Proposition 1 in [9]. We use this lemma in the proof of the next proposition.

Proposition 1.8. If I is an ideal of a ring R, then there exists an ideal X of R such that:

(i) $I \subseteq X$, $X^2 \subseteq I$, $\mathfrak{p}(X) = \mathfrak{p}(I)$, and X is ideal essential in $\mathfrak{p}(I)$;

(ii) I is 2-primal ideal if and only if X is a 2-primal ideal.

Proof. Apply Lemma 1.7 to R/I to obtain an ideal X of R with $I \subseteq X$ such that $X^2 \subseteq I$ and X/I is ideal essential in $\mathfrak{P}(R/I) = \mathfrak{p}(I)/I$. One immediate consequence is $\mathfrak{p}(I) = \mathfrak{p}(X)$. Consider an ideal Y of R such that $Y \cap \mathfrak{p}(I) \neq 0$. Without loss of generality we can take $Y \subseteq \mathfrak{p}(I)$. Assume for purpose of contradiction that $Y \cap X = 0$. Thus $Y \not\subseteq I$ and hence $(Y + I)/I$ is a nonzero ideal of R/I and $(Y + I)/I \subseteq \mathfrak{p}(I)/I$. Since X/I is ideal essential in $\mathfrak{p}(I)/I$, we have $((Y + I)/I) \cap X/I \neq 0$. Consequently $Y \cap X \neq 0$, a contradiction!

Since $\mathfrak{p}(I) = \mathfrak{p}(X)$, part (ii) is immediate from Proposition 1.6.

Proposition 1.9. Let X be a nonzero 2-primal ideal of a ring R. Either R is a 2-primal ring or none of XN, NX, or $N \cap X$ is zero. (Here $N = N(R)$)

Proof. Consider R not 2-primal. Then there exists a prime ideal P of R such that $N \not\subseteq P$. Suppose $NX = 0$. Then $\langle N \rangle X = 0$ and hence $X \subseteq P$. From the observation immediately following Proposition 1.6 we have $N \subseteq P$, a contradiction. Thus $NX \neq 0$. (Similarly $XN = 0$). So there exists $b \in N$, $x \in X$ such that $b^m x \neq 0$ and $b^{m+1}x = 0$. Since $(b^m x b^m)^2 = 0$, we have $b^m x b^m \in N \cap X$. If $b^m x b^m = 0$, then $(b^m x)^2 = 0$ and hence $b^m x \in N \cap X$. Thus $N \cap X \neq 0$.

2 2-Primal rings

The earliest results known to us concerning 2-primal rings (albeit not so called at the time) are due to G. Shin [26]. He showed that R is a 2-primal ring if and only if every minimal prime ideal of R is completely prime. Shin proved that every prime ideal of R is completely prime if and only if each ideal of R is 2-primal. Of more concern in [26] than 2-primal is what is there called the "PSI" condition: R is 2-primal and R/T is 2-primal for each T which is the right annihilator of aR, for some $a \in R$. So PSI is stronger than 2-primal, but weaker than the condition that every prime ideal of R is completely prime. Several of Shin's results involving PSI can be obtained using 2-primal.

Hirano [14] considered the 2-primal condition in the context of strongly π-regular rings. He used the term "N-ring " for what we call a 2-primal ring. The 2-primal condition was taken up independently by Sun [28], where in the setting of rings with unity he introduced a condition called "weakly symmetric", which is equivalent to the 2-primal condition for rings. Sun showed that if R is weakly symmetric, then each minimal prime ideal of R is a completely prime ideal, and that the ring of n-by-n upper triangular matrices over R inherits the weakly symmetric condition. The authors came upon the 2-primal condition independently, and originally in the context of left near-rings.

Proposition 2.1. The following are equivalent:

(i) R is a 2-primal ring;

(ii)$\mathfrak{P}(R) = \mathfrak{P}_2(R)$;

(iii) each minimal prime ideal of R is completely prime;

(iv) if P is a prime ideal of R and $A_1, \ldots, A_n$ are nonempty subsets of R such that $A_1 \ldots A_n = 0$, then $A_i \subseteq P$ for some i;

(v) $\mathfrak{P}(R)$ contains a 2-primal ideal of R;

Proof. Assume (i). Then $R/\mathfrak{P}(R)$ is reduced. Every reduced ring has zero generalized nil radical [21, p.150]. So $\mathfrak{P}_2(R) \subseteq \mathfrak{P}(R)$ and we have (ii). Note (ii) implies (i) is immediate. The equivalence of (i) and (iii) is given in [26, Proposition 1.11]. Since any prime ideal of R contains a minimal prime ideal of R, we have (iii) implies (iv). Assume (iv). If $x \in R$ with $x^n = 0$, then each prime ideal of R contains x, and hence $x \in \mathfrak{p}(0)$. So (0) is a 2-primal ideal of R. Finally, assume (v). Let I be a 2-primal ideal of R such that $I \subseteq \mathfrak{P}(R)$. Then $N(R) \subseteq \mathfrak{p}(I)$. But $I \subseteq \mathfrak{P}(R)$ also implies $\mathfrak{p}(I) = \mathfrak{P}(R)$, so $N(R) = \mathfrak{P}(R)$. This completes the logical circuit.

We next consider the class of all 2-primal rings and show this class is closed under several important operators or construction procedures.

Proposition 2.2. The class of all 2-primal rings is closed under subrings and direct sums.

Proof. Recall that for any ring R, $\mathfrak{P}(R)$ is equal to the set of all strongly nilpotent elements of R and that each strongly nilpotent element of R which is also contained in a subring S of R is a strongly nilpotent element of S. (See [16, pp.55-56] for details, and note that the relevant passages there made no use of a unity element). So $S \cap \mathfrak{P}(R) \subseteq \mathfrak{P}(S)$. If R is a 2-primal ring, then $\mathfrak{P}(S) \subseteq N(S) \subseteq S \cap N(R) = S \cap \mathfrak{P}(R) \subseteq \mathfrak{P}(S)$; so S is a 2-primal ring.

Let R_λ be a 2-primal ring, for each λ in a nonempty index set Λ, and let B be the direct sum of this family of rings. If $b \in B$ with $b^n = 0$, then $b^n(\lambda) = 0$, for each $\lambda \in \Lambda$. (Here $b(\lambda)$ is the λ-th component of b). Since $\mathfrak{P}(R_\lambda) = N(R_\lambda)$, we have $b(\lambda) \in \mathfrak{P}(R_\lambda)$. There being only a finite number of $b(\lambda)$ which are nonzero, we sum them to get b is in $\bigoplus(R_\lambda)$, $\lambda \subset \Lambda$. Since $\mathfrak{P}(\bigoplus R_\lambda) = \bigoplus \mathfrak{P}(R_\lambda)$, we have established that $N(B) \subseteq \mathfrak{P}(B)$ and hence B is a 2-primal ring.

Corollary 2.3. The intersection of any finite set of 2-primal ideals of R is a 2-primal ideal of R.

Proof. Let Y_i, $i = 1, ..., n$, be 2-primal ideals and let Y be the intersection of this set. Then $\cap Y_i/Y$ is zero. Hence R/Y is a subdirect product of the rings (R/Y) / $(Y_i/Y) \approx R/Y_i$, each of which is a 2-primal ring. Since the index set is finite, this direct product is a direct sum and by Proposition 2.2 it is a 2-primal ring. Thus R/Y is isomorphic to a subring of a 2-primal ring, yielding R/Y is 2-primal and hence Y is a 2-primal ideal.

Can the above corollary be extended to the intersection of arbitrary sets of 2-primal ideals? A moments reflection shows that the following statements are equivalent:

(i) the intersection of any set of 2-primal ideals in a ring is a 2-primal ideal ;

(ii) the direct product of an arbitrary family of 2-primal rings is 2-primal;

(iii) a subdirect product of an arbitrary family of 2-primal rings is 2-primal.

The Moore-Kuratowski minimality principle [25, pp.32-33] establishes that to show (i) it suffices to just do so for chains of 2-primal ideals in a ring. But we have been unable to resolve whether these statements are true or false, and leave this as an open problem.

Proposition 2.4. (Extensions)

(i) If I is an ideal of R such that both R/I and I are 2-primal rings, then R is a 2-primal ring.

(ii) Let R be an algebra over a unital, commutative ring A and let R^1 be the Dorroh extension of R to a ring with unity, using $A \times R = R^1$. If R is a 2-primal ring, then R^1 is a 2-primal ring (and conversely).

Proof. (i) For the purpose of contradiction assume R is not 2-primal. Then there exists $x \in N(R)$ and some prime ideal P of R such that $x \notin P$. Since I is a 2-primal ideal of R, Proposition 1.6 forces $I \not\subseteq P$. Take $x^n = 0$. Then for any $b \in I$ we have $x^{n-1}bx^{n-1} \in N(I)$ and hence $(bx^{n-1})^2 \in N(I)$. So $bx^{n-1} \in N(I)$. Repeat the process to eventually obtain $bx \in N(I)$. So

$Ix \subseteq N(I)$ and $N(I) = \mathfrak{P}(I) \subseteq \mathfrak{P}(R) \subseteq P$. So $I \cdot \langle x \rangle \subseteq P$, which yields the contradictory $I \subseteq P$ or $x \in P$.

(ii) Embed R as an ideal in R^1 via the Dorroh embedding. Then $R^1/R \approx A$. Since R and A are 2-primal rings, part (i) yields R^1 is 2-primal. The converse follows from Proposition 2.2.

Proposition 2.5. (i) R is a 2-primal ring if and only if the ring of all n-by-n upper triangular matrices over R is a 2-primal ring.

(ii) Let F and K be rings and let M be a F,K-bimodule. Then F and K are 2-primal rings if and only if the formal upper triangular matrix ring, $\begin{bmatrix} F & M \\ 0 & K \end{bmatrix}$, is a 2-primal ring.

Proof. (i) A routine calculation shows that the prime ideals of the ring T of all n-by-n upper triangular matrices over R are exactly those matrices of the form (A_{ij}), where $A_{ij} = R$ except for at most one term, A_{kk}, which is a prime ideal of R. From the format of these ideals we see that $\mathfrak{P}(T)$ has the form (B_{ij}), where $B_{ij} = R$, if $i \neq j$, and $B_{ii} = \mathfrak{P}(R)$. Since $\mathfrak{P}(R) = \mathfrak{P}_2(R)$ we have $\mathfrak{P}(T) = \mathfrak{P}_2(T)$ and T is 2-primal. The converse is immediate.

(ii) The proof here is similar to that in part (i), with $n = 2$.

Proposition 2.6. If R is a 2-primal ring, then $R[\Omega]$ is a 2-primal ring, where Ω is any set of commuting indeterminants over R.

Proof. First we show that if P is a completely prime ideal of R, then $P[x]$ is a completely prime ideal of $R[x]$. Recall [1] that if P is a prime ideal of R, then $P[x]$ is a prime ideal of $R[x]$. Suppose for purposes of contradiction that $P[x]$ is not completely prime. Then it is not completely semiprime, and there exists some $f(x) = a_0 + a_1x + ... + a_kx^k \in R[x]$ such that $(f(x))^2 \in P[x]$, but $f(x) \notin P[x]$. So there is some smallest exponent of x, say m, such that $a_m \notin P$. So $a_0, ..., a_{m-1} \in P$. Let $a_j = 0$ if $j > k$. Then each of the terms $a_{2m-j}a_j$, $j = 0, 1, ..., 2m-1$, is in P and the sum $\sum a_{2m-j}a_j$ is in P. Thus $a_m^2 \in P$ and hence $a_m \in P$, a contradiction.

By induction we extend this to obtain: P a completely prime ideal of

R implies $P[\Lambda]$ is a prime ideal of $R[\Lambda]$, for any finite set Λ of commuting indeterminants over R. Let Ω be infinite. If $f \in R[\Omega]$ such that $f^2 \in P[\Omega]$, then there exists a finite subset Λ of Ω such that $f \in R[\Lambda]$. So $f^2 \in P[\Lambda]$ and hence $f \in P[\Lambda]$, and $P[\Lambda] \subseteq P[\Omega]$. We have established: P completely prime in R implies $P[\Omega]$ is completely prime in $R[\Omega]$.

Observe that $\cap\{P[\Omega] : P \text{ is a completely prime ideal of } R\} = \mathfrak{P}_2(R)[\Omega]$. So from our results above we have $\mathfrak{P}_2(R[\Omega]) \subseteq \mathfrak{P}_2(R)[\Omega]$. However, $\mathfrak{P}_2(R)[\Omega] = \mathfrak{P}(R)[\Omega]$, because R is 2-primal, and $\mathfrak{P}(R)[\Omega] = \mathfrak{P}(R[\Omega])$. (The latter holds for any ring R; see [1,Theorem 4]). So $\mathfrak{P}_2(R[\Omega]) \subseteq \mathfrak{P}(R[\Omega])$, and hence $R[\Omega]$ is 2-primal by Proposition 2.1.

It is worth mentioning some constructions or operations under which the class of 2-primal rings is not closed. Since the full ring of n-by-n matrices over a field has zero prime radical, but (for $n > 1$) has nonzero nilpotent elements, we see that a full matrix ring over a 2-primal ring is never 2-primal. With this in mind we see that a 2-primal ring which has maximal ring of quotients need not have this ring of quotients to be 2-primal: use a 2-by-2 upper triangular matrix ring over a field (its maximal right quotient ring is the full ring of 2-by-2 matrices over that field). More troublesome is the fact that a homomorphic image of a 2-primal ring need not be 2-primal, as the following example illustrates.

Example 2.7. Let R be the ring of quaternions with integer coefficients. So R is a domain and hence is 2-primal. However, for any odd prime integer q, the ring R/qR is isomorphic to the ring of 2-by-2 matrices over the field Z_q, [13, p.24]. So R/qR is not a 2-primal ring and qR is not a 2-primal ideal.

This example suggests one ask the question: in a 2-primal ring what can be said about the prime ideals that are not completely prime? The next proposition addresses this question.

Proposition 2.8. Let R be a 2-primal ring. Any prime ideal of R which is not completely prime must be both right and left essential in R.

Proof. Let P be a prime ideal of R which is not completely prime. Proposition 2.1 gives P is not a minimal prime. The desired result then follows from: in any ring a prime ideal which is not right (left) essential must be a minimal prime. (For a proof of this result see: [10, Theorem 3.2]; the proof dose not require that the ring have unity).

Note: In Example 2.7, $2R$ is a 2-primal ideal, but not prime. Also $2R$ is contained in a maximal ideal which is completely prime and is left and right essential, [13, p.24].

Proposition 2.9. (i) Let R be a simple ring. Then R is a 2-primal ring if and only if R is either a domain or $R^2 = 0$.

(ii) Let $R^2 = R$ and let M be an ideal of R which is maximal as a left (right) ideal. Then R is 2-primal if and only if M is a 2-primal ring.

(iii) R is a 2-primal ring if and only if $\mathfrak{P}_2(R)$ is a 2-primal ring.

(iv) R is a 2-primal ring if and only if the commutator ideal of R, $\langle R, R\rangle = \langle\{ab - ba : a, b \in R\}\rangle$, is a 2-primal ring.

(v) R is a 2-primal ring if and only if the medial quasiradical of R is a 2-primal ring.

Proof. (i) If R is a 2-primal, then either $\mathfrak{P}_2(R) = 0$ and hence R is a domain, or $\mathfrak{P}(R) = R$ so that (0) is not a prime ideal and hence $R^2 = 0$. The converse is immediate.

In each of (iii) through (v), if R is 2-primal, then the relevant subring is 2-primal by Proposition 2.2. For the converse in each of these parts we show the relevant factor ring is 2-primal and use Proposition 2.4(i) to get R is 2-primal. This is done by noting that: R/M is a skewfield, $R/P_2(R)$ is reduced; $R/\langle R, R\rangle$ is commutative; and finally, using $\mathfrak{M}(R)$ for the medial quasiradical and recalling $R/\mathfrak{M}(R)$ is medial and all medial rings have every prime ideal is completely prime [4].

The above proposition suggests some particular examples or classes of examples of 2-primal rings. For any local ring L, $L/J(L)$ is a skewfield. So if $J(L)$ is a 2-primal ring, then so is L. One way this can occur is for L to be left (right) Artinian. This is the case for any group algebra KG, where K is a field and G is a finite p-group.

Somewhat similarly one can use Proposition 2.9 to show certain noncommutative Gelfand rings are 2-primal. In a Gelfand ring all maximal ideals are completely prime and are maximal as right (left) ideals. So the generalized nil radical is contained in the Jacobson radical for Gelfand rings. If a Gelfand ring is right (left) Artinian, then it is 2-primal and the Jacobson radical is equal to the prime radical. (For details on Gelfand rings, including examples see [6, Chapter 8]).

Proposition 2.10. A 2-primal ring with unity is directly finite.

Proof. Let R be a 2-primal ring with unity and let $a, b \in R$ such that $ab = 1$. Then ba is an idempotent. Since $R/\mathfrak{P}(R)$ is reduced and reduced rings are directly finite, we have $ba = 1 + x$, where x is nilpotent. Consequently ba is a unit in R and an idempotent; so $ba = 1$.

Proposition 2.11 Let R be a subdirectly irreducible ring with heart H.

(i) If R is 2-primal, then either $H^2 = 0$ or R is a domain and H is a simple domain. If R also is left (right) Artinian and $H^2 \neq 0$, then R is a skewfield.

(ii) If H is a 2-primal ideal, then either:

(a) R is 2-primal, $\mathfrak{p}(H) = \mathfrak{P}(R)$, and $H^2 = 0$; or

(b) R is a prime ring, H is a simple ring with $N(H) \neq 0$.

Proof. (i) The heart of any subdirectly irreducible ring is either square zero or is a simple ring [7, p.135]. So if $H^2 \neq 0$, then it is a simple 2-primal ring and hence is a domain. Thus $N(R) = \mathfrak{P}(R) = 0$ and R is a reduced ring. Subdirectly irreducible reduced rings are domains. If R is left (right) Artinian and $H^2 \neq 0$, then H contains a nonzero idempotent, which in a domain must be the unity for H. This forces $H = R$. Thus R is a simple left (right) Artinian domain, which can only be a skewfield.

(ii) Recall $N(R) \subseteq \mathfrak{p}(R)$. We consider two cases. First, if $\mathfrak{P}(R) \neq 0$, then (0) is not a prime ideal and $H \subseteq \mathfrak{P}(R)$. So $\mathfrak{p}(H) = \mathfrak{P}(R)$ and hence $N(R) \subseteq \mathfrak{P}(R)$, yielding R is 2-primal. If $H^2 \neq 0$ then H is simple and is a

2-primal ring, yielding H is a domain. But $H \subseteq N(R)$, a contradiction.

Next take $\mathfrak{P}(R) = 0$. So (0) is a prime ideal of R (because R is subdirectly irreducible). Thus R is a prime ring and $H^2 \neq 0$. Consequently H must be a simple ring.

Observe that in the $\mathfrak{P}(R) = 0$ case above either R is 2-primal and hence a domain, or R is not 2-primal and $N(H) \neq 0$.

Proposition 2.12 Let R be von Neumann regular. Then the following are equivalent:

(i) R is 2-primal;

(ii) R is strongly regular;

(iii) R satisfies the PSI condition.

Proof. Regular rings are semiprime and a semiprime 2-primal ring is reduced. It is well known that a reduced regular ring is strongly regular. The implication (ii)⇒(i) is immediate and (iii)⇒(ii) is given in [26, Proposition 1.16]

The above yields the following minor improvement on Theorem 1.17 in [26].

Corollary 2.13. The following are equivalent:

(i) R is a von Neumann regular, prime ring which is 2-primal;

(ii) R is a skewfield.

Proof. Note that a prime ring which is 2-primal is a domain.

3 The Class $\mathfrak{R}^2$

Shin [26,Proposition 1.13] showed that the following are equivalent in a ring R:

(i) every prime ideal of R is completely prime;

(ii) every ideal of R is a 2-primal ideal.

We will use $\mathfrak{N}^2$ for the class of all rings which satisfy these conditions. In this section we investigate $\mathfrak{N}^2$ as a class, give a plethora of examples of rings in $\mathfrak{N}^2$, and give several constructions which yield rings in $\mathfrak{N}^2$.

All rings in $\mathfrak{N}^2$ are 2-primal, but not conversely. Example 2.7 gives a domain which is not in $\mathfrak{N}^2$. Some examples from $\mathfrak{N}^2$ which come readily to mind are: all nilpotent rings; all left (right) duo rings; all $\mathfrak{P}$-radical rings; and any local ring R with $J(R)$ in $\mathfrak{N}^2$. There are several important types of domains which are in $\mathfrak{N}^2$, ones which have been the object of intense study. If K is a skewfield of characteristic zero, then the Weyl algebra $A_n(K)$, [20] is a simple domain. All simple domains are in $\mathfrak{N}^2$. As noted by Lorenz [18, p.1231], if K is a field of positive characteristic, then $A_n(K)$ has a maximal (hence prime) ideal which is not completely prime. For K of characteristic zero the subring S of $A_1(K)$ generated by K and $x \cdot A_1(K)$ is a domain, is not simple and is in $\mathfrak{N}^2$. (Here x is the indeterminant for the Weyl Algebra; see [20, p.20] for details on S). A well-known result due to Dixmier [8] connects $\mathfrak{N}^2$ with enveloping algebras: if K is a field of characteristic zero and $\mathfrak{g}$ is a finite dimensional, solvable Lie algebra over K, then the universal enveloping algebra $\mathfrak{U}(\mathfrak{g})$ has every prime ideal completely prime. This has been generalized in several ways, as we note in the following examples.

Example 3.1. Let R be a ring with unity and let θ_1 be an indeterminant over R and δ a derivation on R. Let R_1 be the differential operator ring $R[\theta_1; \delta]$. Recursively, if R_n has been defined let θ_{n+1} be an indeterminant over R_n and δ_{n+1} a derivation on R_n to form $R_{n+1} = R_n[\theta_{n+1}, \delta_{n+1}]$. Sigurdsson [2, Corollary 2.5] showed the following: if R is a right Noetherian Q-algebra and if the ring $R[x]$ is in $\mathfrak{N}^2$, then R_k is in $\mathfrak{N}^2$ for each k. (Here $R[x]$ is the polynomial ring in the commuting indeterminant x).

Example 3.2. Let R be a commutative, unital, Noetherian, Q-algebra. If $\mathfrak{g}$ is a finitely generated (as an R-module), solvable Lie algebra over R, then the universal enveloping algebra $\mathfrak{U}(\mathfrak{g})$ is in $\mathfrak{N}^2$, [27, Corollary 2.7]. This extends Dixmier's result.

Example 3.3. Let R be a unital ring which is a right Noetherian algebra over a field K, with char$K = 0$. Lorenz [18] showed that if for each extension field F of K the algebra $R \otimes_K F$ is right Noetherian and is in $\mathfrak{N}^2$, then $R[\theta; \delta]$ is in $\mathfrak{N}^2$, where δ is any K-derivation on R.

Example 3.4. R is a *permutation identity ring* if there is a nontrivial permutation σ on n letters such that $x_1 x_2 \cdots x_n = x_{\sigma(1)} x_{\sigma(2)} \cdots x_{\sigma(n)}$, for each $x_1, \cdots x_n \in R$. It is shown in [5,Proposition 2.8] that every permutation identity ring is in $\mathfrak{N}^2$. Also in [4] and [5] are many diverse examples of permutation identity rings.

We next establish some equivalent conditions to being in class $\mathfrak{N}^2$.

Proposition 3.5. The following are equivalent:

(i) $R \in \mathfrak{N}^2$;

(ii) $R/\mathfrak{P}(R)$ is in $\mathfrak{N}^2$;

(iii) every semiprime ideal of R is completely semiprime.

Proof. Assume (i). Each ideal of $R/\mathfrak{P}(R)$ has the form $I/\mathfrak{P}(R)$, where I is an ideal of R containing $\mathfrak{P}(R)$. Since R/I is a 2-primal ring and $(R/\mathfrak{P}(R))/(I/\mathfrak{P}(R)) \approx R/I$ is a 2-primal ring, we have $I/\mathfrak{P}(R)$ is a 2-primal ideal. So $R/\mathfrak{P}(R)$ is in $\mathfrak{N}^2$.

Assume (ii) and let S be a semiprime ideal of R. Then $\mathfrak{P}(R) \subseteq S$ and $S/\mathfrak{P}(R)$ is a semiprime ideal of $R/\mathfrak{P}(R)$. Every semiprime ideal (in an arbitrary ring) is an intersection of prime ideals. So $S/\mathfrak{P}(R)$ is an intersection of completely prime ideals and hence is completely semiprime. The pre-image of a completely semiprime ideal is completely semiprime.

Assume (iii) and let I be an ideal of R. Then the semiprime ideal $\mathfrak{p}(I)$ is completely semiprime. Proposition 1.6 yields I is a 2-primal ideal and hence $R \in \mathfrak{N}^2$.

Proposition 3.6. The class $\mathfrak{N}^2$ is closed under homomorphic images and ideals.

Proof. Let $R \in \mathfrak{N}^2$. For any homomorphism θ from R, the pre-image of a prime ideal in $\theta(R)$ is a prime ideal in R and hence is completely prime (and contains $Ker\theta$). Applying θ to this completely prime ideal gives a completely prime ideal in $\theta(R)$, equal to the original prime ideal. So $\theta(R)$ is in $\mathfrak{N}^2$.

Let I be an ideal of R and let K be an ideal of I. Then $(\langle K\rangle_R)^3 \subseteq K$. (This is the Andrunakievic Lemma and holds in any ring, [7, p.107]). Take K to be a prime ideal of I. Then K is an ideal of R and hence is a 2-primal ideal of R. So R/K and its subring I/K are 2-primal rings. Since I/K is also a prime ring, it is a domain and hence K is a completely prime ideal of I. So $I \in \mathfrak{N}^2$.

Proposition 3.7. Let $\mathcal{F}$ be a family of ideals of R such that if $X \in \mathcal{F}$, then R/X is in $\mathfrak{N}^2$. If $\mathcal{F}$ is finite, then$R/\cap\mathcal{F}$ is in $\mathfrak{N}^2$.

Proof. Consider $\mathcal{F} = \{A, B\}$. Any prime ideal of $R/A\cap B$ has the form $P/A\cap B$, where P is a prime ideal of R and $A\cap B \subseteq P$. So either $A \subseteq P$ or $B \subseteq P$. Without loss of generality take $A \subseteq P$. Since $R/A\cap B$ maps homomorphically onto R/A with kernel $A/A\cap B$, we have $P/A\cap B$ maps onto the prime ideal P/A in R/A. So P/A is a completely prime ideal and hence its pre-image, $P/A\cap B$, is a completely prime ideal of $R/A\cap B$. So $R/A\cap B$ is in $\mathfrak{N}^2$. Inductively this extends to any finite family.

Corollary 3.8. Let R be a subdirect product of rings $R_1, ..., R_n$. Then $R \in \mathfrak{N}^2$ if and only if each $R_i \in \mathfrak{N}^2$, $i = 1, ..., n$.

Proposition 3.9. (Extensions)

(i) If I is an ideal of R such that I and R/I are in $\mathfrak{N}^2$, then $R \in \mathfrak{N}^2$.

(ii) Let R be an algebra over a unital commutative ring A and let R^1 be the Dorroh extension of R via A to a ring with unity. Then $R^1 \in \mathfrak{N}^2$ if and only if $R \in \mathfrak{N}^2$.

Proof. (i) Let P be a prime ideal of R. Since $(R/I)/((I+P)/I) \approx R/(I+P)$, we have $R/(I+P)$ is in $\mathfrak{N}^2$ and hence a 2-primal ring. Also, $(R/P)/((I+P)/P) \approx R/(I+P)$. From $(I+P)/P \approx I/I\cap P$, and since $I/I\cap P$ (being a homomorphic image of I) is in $\mathfrak{N}^2$ and hence is 2-primal, we have the ring $(I+P)/P$ is 2-primal. Proposition 2.4 (i) then yields R/P is a 2-primal ring. Since R/P is also a prime ring, we have R/P is a domain and hence P is a completely prime ideal of R.

(ii) Identify R as the ideal $(0,R)$ in R^1. Since $R^1/R \approx A$ is in $\mathfrak{R}^2$, if $R \in \mathfrak{R}^2$, then $R^1 \in \mathfrak{R}^2$. Proposition 3.6 yields the converse.

We are now in a position to observe that Proposition 2.9 (i), (ii), (iv), (v) hold with "2-primal ring" being replaced by "a ring in $\mathfrak{R}^2$".

Proposition 3.10. If $\mathcal{A}$ is a nonempty set of ideals of a ring R and $A \in \mathfrak{R}^2$ for each $A \in \mathcal{A}$, then the sum of all the ideals from $\mathcal{A}$ is a ring in $\mathfrak{R}^2$. Consequently, the class $\mathfrak{R}^2$ is closed under direct sums.

Proof. Let A, B be ideals of R with $A, B \in \mathfrak{R}^2$. Then $(A+B)/B \approx A/A \cap B$ is in $\mathfrak{R}^2$ (since it is a homomorphic image of A) and consequently $A+B$ is in $\mathfrak{R}^2$. By induction this extends to any finite sum. Let $\mathcal{A}$ be infinite and let T be the sum of the ideals from $\mathcal{A}$. Let S be a semiprime ideal of T and consider $t \in T$ such that $t^2 \in S$. Then t is in some finite sum of ideals from $\mathcal{A}$, say the sum W, and $W \in \mathfrak{R}^2$. Observe that $S \cap W$ is a semiprime ideal of W. (This follows by a straightforward use of the Andrunakievic Lemma mentioned above). Thus $S \cap W$ is a completely semiprime ideal of W and since $t^2 \in S \cap W$ we have $t \in S \cap W$ and hence $t \in S$. So S is a completely semiprime ideal of T, yielding $T \in \mathfrak{R}^2$.

The closure of $\mathfrak{R}^2$ under (external) direct sums follows immediately from the first statement in the proposition.

Two further ways to make new rings in $\mathfrak{R}^2$ out of old ones are given in the next proposition, the proof of which is strictly analogous to that of Proposition 2.5.

Proposition 3.11. (i) R is in $\mathfrak{R}^2$ if and only if the ring of all n-by n upper triangular matrices over R is in $\mathfrak{R}^2$.

(ii) Let F and K be rings and let M be a F,K-bimodule. Then F and K are in $\mathfrak{R}^2$ if and only if $\begin{bmatrix} F & M \\ 0 & K \end{bmatrix}$ is in $\mathfrak{R}^2$.

The class $\mathfrak{N}^2$ is not closed under subrings. For example, the quaternion ring $H(\mathbb{R})$ is in $\mathfrak{N}^2$, but its subring $H(\mathbb{Z})$ is not. (Here $\mathbb{R}$ is the field of all real number and $\mathbb{Z}$ is the ring of rational integers). A partial answer to when subrings inherit the property of being in $\mathfrak{N}^2$ can be obtained using ideas of Heinecke and Robson, (see [24, pp.291-295]). For the purposes of this discussion all rings have unity and any subring of such a ring shares the unity element. Recall [24, p.292], the extension of the ring R to the ring R' has the "LO " condition if for each prime ideal P of R there is some prime ideal P' of R' such that $P' \cap R = P$.

Proposition 3.12. (i) If R to R' is an extension which satisfies the "LO" condition and if $R' \in \mathfrak{N}^2$, then $R \in \mathfrak{N}^2$.

(ii) Let R' be a finite centralizing extension of R. If $R' \in \mathfrak{N}^2$, then $R \in \mathfrak{N}^2$.

(iii) Let R to R_i be an extension which satisfies the "LO" condition and let R' be the direct limit of the R_i. If $R' \in \mathfrak{N}^2$, then $R \in \mathfrak{N}^2$.

Proof. (i) Let P be a prime ideal of R and let $x, y \in R$ such that $xy \in P$. There exists a prime ideal P' of R' such that $P' \cap R = P$. So $xy \in P'$ and since P' is completely prime we have $x \in P'$ or $y \in P'$. So $x \in P$ or $y \in P$ and hence P is completely prime in R.

(ii) All finite centralizing extension satisfy the "LO " condition, [24, Theorem 2.12.48].

(iii) Under the conditions given the extension R to R' satisfies the "LO" condition, [24, Proposition 2.12.49]

Example 3.13. (A ring in $\mathfrak{N}^2$ whose polynomial ring is not in $\mathfrak{N}^2$). Let $A = H(\mathbb{R})$ be the real quaternions. Since this $\mathbb{R}$-algebra is simple and is finite dimensional over its center, $\mathbb{R}$, the prime ideals of $A[x]$ are exactly those of the form $f(x)A[x]$, where $f(x)$ is a central polynomial which is irreducible over $\mathbb{R}$, [27, 3.1]. Thus $(x^2+1)A[x]$ is a prime ideal of $A[x]$, but since $(x-i)(x+i) = x^2+1$, and neither term in the product is in $(x^2+1)A[x]$, we have it is not completely prime.

This is in sharp contrast to the result for 2-primal rings (Proposition 2.6).

Finally we note that whereas the question of whether an arbitrary subdirect product of 2-primal rings is 2-primal was left open (section 2), the analogous question is easily resolved, in the negative, for the class $\mathfrak{N}^2$. Since $\mathfrak{N}^2$ is not closed under subrings, it is not a variety, and hence is not closed under subdirect product (Recall: a class is a variety if and only if the class is closed under homomorphic images and subdirect products [12, p.96]).

4 The $\mathfrak{P}_c$ radical

We use the class $\mathfrak{N}^2$ to define a radical property. For any ring R, define $\mathfrak{P}_c(R)$ to be the sum of all ideals I of R such that $I \in \mathfrak{N}^2$. Proposition 3.10 yields $\mathfrak{P}_c(R)$ is in $\mathfrak{N}^2$. We say R is a $\mathfrak{P}_c$-ring if $R = \mathfrak{P}_c(R)$. Proposition 3.6 gives every homomorphic image of a $\mathfrak{P}_c$-ring is also a $\mathfrak{P}_c$-ring. All that remains to be done to establish that $\mathfrak{P}_c$ is an Amitsur-Kurosh radical is to show $\mathfrak{P}_c(R/\mathfrak{P}_c(R)) = 0$. Write $\mathfrak{P}_c(R/\mathfrak{P}_c(R)) = X/\mathfrak{P}_c(R)$, where X is an ideal of R such that $\mathfrak{P}_c(R) \subseteq X$. Since $\mathfrak{P}_c(R)$ is in $\mathfrak{N}^2$ and $X/\mathfrak{P}_c(R)$ is in $\mathfrak{N}^2$, Proposition 3.9(i) yields $X \in \mathfrak{N}^2$; hence $X/\mathfrak{P}_c(R) = 0$. From Proposition 3.6 we see that every ideal of a $\mathfrak{P}_c$-ring is itself a $\mathfrak{P}_c$-ring. This, together with the observation that every nilpotent ring is a $\mathfrak{P}_c$-ring, establishes that $\mathfrak{P}_c$ is a supernilpotent radical. Consequently, $\mathfrak{P}_c(R)$ is a semiprime ideal, [29, Proposition 11.2], and hence $\mathfrak{P}(R) \subseteq \mathfrak{P}_c(R)$.

In general $\mathfrak{P}_2$ and $\mathfrak{P}_c$ are incomparable, as the following example shows.

Example 4.1. Let $R = \begin{bmatrix} K & M \\ 0 & M \end{bmatrix}$, where K is a skewfield and M is the ring of n-by-n matrices over K, $n > 1$. Then $\mathfrak{P}_c(R) = \begin{bmatrix} K & M \\ 0 & 0 \end{bmatrix}$, $\mathfrak{P}_2(R) = \begin{bmatrix} 0 & M \\ 0 & M \end{bmatrix}$, $\mathfrak{P}(R) = \begin{bmatrix} 0 & M \\ 0 & 0 \end{bmatrix}$.

Note that for a commutative ring R we have $\mathfrak{P}(R) = \mathfrak{P}_2(R) \subseteq \mathfrak{P}_c(R) = R$. So if R is commutative and non-nil, we have $\mathfrak{P}_2(R) \subset \mathfrak{P}_c(R)$.

Example 4.2. Let $R = \begin{bmatrix} F & F \\ F & F \end{bmatrix}$, F a field, and let $X = \begin{bmatrix} F & F \\ 0 & 0 \end{bmatrix}$.

Then X is an ideal of $\begin{bmatrix} F & F \\ 0 & F \end{bmatrix}$ and hence $X \in \mathfrak{N}^2$. But $\mathfrak{P}_c(R) = 0$. Thus R is a $\mathfrak{P}_c$-semisimple ring which contains a right ideal which is a $\mathfrak{P}_c$-radical ring.

We next make some observations about $\mathfrak{P}_c$-semisimple rings.

Proposition 4.3. If R is a semiprime nil ring, then $\mathfrak{P}_c(R) = 0$.

Proof. First observe that R cannot be in $\mathfrak{N}^2$; so $\mathfrak{P}_c(R) \neq R$. If I is any ideal of R such that $I \in \mathfrak{N}^2$, then $I = N(I) = \mathfrak{P}(I) = I \cap \mathfrak{P}(R) = 0$. So $\mathfrak{P}_c(R) = 0$.

Rings satisfying the hypothesis of Proposition 4.3 exist. Baer [2, pp.540-542] gave an example of a ring U which is semiprime and locally nilpotent. Note that in this ring $\mathfrak{P}(U) = \mathfrak{P}_c(U) = 0$ and $\mathfrak{P}_2(U) = U$. (Also see [7, p.46] for this example). Baer's example also shows not all locally nilpotent rings are in $\mathfrak{N}^2$.

It is possible for domains to be $\mathfrak{P}_c$-semisimple. Referring to Example 2.7, where R is the ring of quaternions with integer coefficients, for any odd prime integer q, R/qR is $\mathfrak{P}_c$-semisimple. So $\mathfrak{P}_c(R) \subseteq qR$ and hence $\mathfrak{P}_c(R) \subseteq \cap qR$, where q runs over all odd primes. Then $\mathfrak{P}_c(R) = 0$. In stark contrast are the several types of domains in $\mathfrak{N}^2$ which we exhibited in section 2.

We will use the notation: $\mathfrak{r}(t) = \{r \in R : tr = 0\}$, the right annihilator in R of t.

Lemma 4.4. Let X be a reduced minimal right ideal of R.

(i) $X = eR = eRe$, where e is an idempotent, and X is a skewfield;

(ii) $\langle X \rangle_R = Re$, and Re is the right ideal direct sum of eR and the ideal $(\mathfrak{r}(e))e$, the latter also being square zero;

(iii) e is central if and only if $(\mathfrak{r}(e))e = 0$.

Proof. (i) A minimal right ideal, which is not nilpotent is generated by an idempotent (Brauer's Lemma [15, 16.22]), so $X = eR$, where e is a nonzero idempotent. For any $r \in R$, $(er - ere)^2 = 0$; so $er = ere$ and hence $eR = eRe$. Since $aX' = X$ for each nonzero $a \in X$, we have there exists $b \in X$ so that $ab = e$ and consequently eRe is a skewfield.

(ii) Since $eR \subseteq ReR = R(eRe) \subseteq Re$, we have Re is an ideal of R which contains eR. So $\langle X\rangle_R = Re$. Since $eR\cdot\mathfrak{r}(e) = eRe\mathfrak{r}(e) = 0$, we have $R\cdot\mathfrak{r}(e) \subseteq \mathfrak{r}(e)$ and hence $\mathfrak{r}(e)$ is an ideal of R. A moment reflection shows this yields $(\mathfrak{r}(e))e$ is an ideal of R. The Peirce decomposition gives: $R = \mathfrak{r}(e) + eR$; from whence we obtain $Re = (\mathfrak{r}(e))e + eRe$. It is immediate that $(\mathfrak{r}(e))e$ is square zero.

(iii) Recall $\mathfrak{r}(e) = \{r - er : r \in R\}$. If $(\mathfrak{r}(e))e = 0$, then for each $r \in R$, $r = (r - er) + er$ and hence $re = 0 + ere = er$. Conversely, if e is central, then $\mathfrak{r}(e)e = e\mathfrak{r}(e) = 0$.

Observe that Lemma 4.4(ii) yields $\langle X\rangle_R \subseteq \mathfrak{P}_c(R)$.

The homogeneous component determined by a minimal right ideal X will be denoted by $C(X)$.

Proposition 4.5. Let X be a minimal right ideal of R.

(i) The following are equivalent:

(a) $X \cap \mathfrak{P}_c(R) = 0$;

(b) $N(X) \neq 0$ and $X = eR$, where e is a nonzero idempotent in X.

(ii) If $X \cap \mathfrak{P}_c(R) = 0$, then $\langle X\rangle_R \cap \mathfrak{P}_c(R)$ is square zero.

(iii) Let $X \cap \mathfrak{P}_c(R) = 0$. Then $\langle X\rangle_R$ is a minimal ideal of R if and only if $\langle X\rangle_R \cap \mathfrak{P}_c(R) = 0$.

Proof. (i) Assume (a). Note that X cannot be nilpotent; so by Brauer's Lemma there exists a nonzero idempotent in X such that $X = eR$. Since X reduced implies $X \subseteq \mathfrak{P}_c(R)$, we have $N(X) \neq 0$.

Now assume (b). For purpose of contradiction suppose $X \cap \mathfrak{P}_c(R) \neq 0$. Then $X \subseteq \mathfrak{P}_c(R)$. Since $\mathfrak{P}_c(R)$ is in $\mathfrak{R}^2$, and hence is a 2-primal ring, we have $N(\mathfrak{P}_c(R))$ is an ideal of $\mathfrak{P}_c(R)$. By supposition there exists a nonzero element $t \in N(X)$. Let T be the ideal of $\mathfrak{P}_c(R)$ generated by t. So T is a nil ideal of $\mathfrak{P}_c(R)$. The Andrunakievic Lemma yields: $(\langle t \rangle_R)^3 \subseteq T$. But minimality of X as a right ideal of R forces $X \subseteq \langle t \rangle_R$. Consequently X is nil, a contradiction!

(ii) Let $I = \langle X \rangle_R \cap \mathfrak{P}_c(R)$. Consider $I \neq 0$. Note that $I \subseteq \mathcal{C}(X)$, so I is equal to a sum of minimal right ideals of R each isomorphic (as a right R-module) to X. Let Y be one such minimal right ideal. For purposes of contradiction assume Y is not nilpotent. Then Brauer's Lemma gives $Y = eR$, where e is a nonzero idempotent in Y. By part (i), Y must be reduced. Use Lemma 4.4 to get: $\langle Y \rangle_R \subseteq \mathfrak{P}_c(R)$. Since $\langle Y \rangle_R = \mathcal{C}(Y) = \mathcal{C}(X) = \langle X \rangle_R$, we have $X \subseteq \mathfrak{P}_c(R)$, a contradiction! Thus Y is nilpotent (and minimal), so $Y^2 = 0$.

Consequently $(\langle Y \rangle_R)^2 = 0$, and since I is a sum of such ideals, $I \subseteq \mathfrak{P}(R)$. Recall that for any ring, $(Soc(R)) \cdot J(R) = 0$; so for the case at hand : $I^2 \subseteq (Soc(R)) \cdot \mathfrak{P}(R) \subseteq (Soc(R)) \cdot J(R) = 0$.

(iii) Assume $\langle X \rangle_R$ is minimal. Either $(\langle X \rangle_R)^2 = 0$ or $\langle X \rangle_R$ is a simple ring, [7, p.135]. The former implies $\langle X \rangle_R \subseteq \mathfrak{P}_c(R)$, which violates $X \cap \mathfrak{P}_c(R) = 0$; the latter yields $\langle X \rangle_R \cap \mathfrak{P}_c(R) = 0$.

Conversely, assume $\langle X \rangle_R \cap \mathfrak{P}_c(R) = 0$. Let B be a nonzero ideal of R which is contained in $\langle X \rangle_R$. Since $B \subseteq \mathcal{C}(X)$, there is a minimal right ideal of R, say M, with $M \subseteq B$. Since $B \not\subseteq \mathfrak{P}_c(R)$, M cannot be nilpotent. So $M = eR$, where e is a nonzero idempotent, and hence $B = \langle M \rangle_R = \mathcal{C}(X) = \langle X \rangle_R$.

Theorem 4.6. $Soc(R) = U \oplus V \oplus W$, as a direct sum of right ideals, where:

(i) $U = Soc(R) \cap \mathfrak{P}_c(R) = D \oplus K$, as a direct sum of right ideals, where D is a direct sum of right ideals which are skewfields and K is an ideal which is square zero;

(ii) $(V + W) \cap \mathfrak{P}_c(R) = 0$;

(iii) V is a direct sum of minimal ideals, V_j, where $V_j = \langle e_j R\rangle_R$, e_j an idempotent, and $N(e_j R) \neq 0$;

(iv) W is a direct sum of the minimal right ideals W_j, where W_j is generated by an idempotent, $N(W_j) \neq 0$, and W_j is isomorphic as a right R-module to a nilpotent right ideal of R.

Proof. The minimal right ideals of R which are in $\mathfrak{P}_c(R)$ are (exactly) those which are either reduced or nilpotent (Proposition 4.5(i)). The former must be skewfields (Lemma 4.4(i)), and the latter are square zero. This yields U as desired. Next consider the minimal right ideals which are not in $\mathfrak{P}_c(R)$, and hence which have zero intersection with $\mathfrak{P}_c(R)$. Their sum we call $V + W$. Here V is the direct sum of the minimal ideals given by Proposition 4.5 (iii). Then W is the direct sum of the remaining minimal right ideals given by Proposition 4.5 and these are exactly those given in (ii). For such a minimal right ideal, say X, $0 \neq \langle X\rangle_R \cap \mathfrak{P}_c(R)$ is square zero, yielding $\mathcal{C}(X)$ contains as one of its summands a square zero minimal right ideal of R.

Observe that if R is a semiprime ring, then in Theorem 4.6 we have $Soc(R) = V \bigoplus D$, as a direct sum of ideals. Similarly, if $N(R)$ is an ideal of R, then $Soc(R) \subseteq \mathfrak{P}_c(R)$.

Proposition 4.7 If R is subdirectly irreducible, then either (exclusively):

(i) R is $\mathfrak{P}_c$-semisimple;

(ii) R is a domain with heart in $\mathfrak{R}^2$; or

(iii) R has zero square heart.

Proof. Recall that for any hereditary Amitsur-Kurosh radical a subdirectly irreducible ring with heart H either has H is a radical ring or H is semisimple, [7, Lemma74]. Consider R to not be $\mathfrak{P}_c$-semisimple. Then $\mathfrak{P}_c(H) = H$. So $H \in \mathfrak{R}^2$. In general $H^2 = 0$ or H is a simple ring, [7, Lemma 75]. Take $H^2 \neq 0$. Then H is a simple domain. Because $\mathfrak{P}(R)$ is always nil we then have $\mathfrak{P}(R) = 0$ and R is a semiprime subdirectly irreducible ring. Consequently R is a prime ring and thus is a domain.

A moments reflection assures one that all three cases do indeed occur.

References

[1] S. A. Amitsur, Radicals of polynomial rings, *Canad. J. Math.* 8 (1956), 355-361.

[2] R. Baer, Radical ideals, *Amer. J. Math.* 65 (1943), 537-568.

[3] H. E. Bell, Near-rings in which each element is a power of itself, *Bull. Austral. Math. Soc.* 2 (1970), 363-368.

[4] G. Birkenmeier and H. Heatherly, Medial rings and an associated radical, *Czech. Math. J.* 40 (1990), 258-283.

[5] G. Birkenmeier and H. Heatherly, Permutation identity rings and the medial radical,in *Non-commutative Ring Theory*, ed. S. K. Jain and S. R. Lopez-Permouth (Springer-Verlag, Berlin, 1990).

[6] F. Borceux and G. Van den Bossche, *Algebra in a Localic Topos with Applications to Ring Theory* (Springer-Verlag, Berlin, 1983).

[7] N. J. Divinsky, *Rings and Radicals* (Univ. Toronto Press, Toronto, 1965).

[8] J. Dixmier, *Enveloping Algebras* (North Holland, Amsterdam, 1977).

[9] E. Eslami and P. Stewart, Two-sided essential nilpotence, *Internat. J. Math. Math. Sci.* 15 (1992), 351-354.

[10] C. Faith and S. Page, *FPF Ring Theory: Faithful Modules and Generators of Mod-R*, (Cambridge Univ. Press, Cambridge, 1984).

[11] J. W. Fisher, On the nilpotency of nil subrings, *Canad. J. Math* 22 (1970), 1211-1216.

[12] B. J. Gardner, Radicals and varieties, in *Radical Theory*, ed. L. Marki and R. Wiegandt (North-Holland, Amsterdam, 1985).

[13] K. R. Goodearl and R. B. Warfield, Jr., *An Introduction to Noncommutative Noetherian Rings* (Cambridge Univ. Press, Cambridge, 1989).

[14] Y. Hirano, Some studies on strongly π-regular rings, *Math. J. Okayama Univ.* 20 (1978), 141-149.

[15] T. Y. Lam, *A First Course in Noncommutative Rings* (Springer-Verlag, New York, 1991).

[16] J. Lambek, *Lectures on Rings and Modules* (Blaisdell, Waltham, Mass., 1966).

[17] J. Lambek, On the representation of modules by sheaves of factor modules, *Canad. Math. Bull.* 14 (1971), 359-368.

[18] M. Lorenz, Completely prime ideals on Ore extensions, *Comm. Alg.* 11 (1981), 1227-1232.

[19] G. Mason, Reflexive ideals, *Comm. Alg.* 9 (1981), 1709-1724.

[20] J. C. McConnell and J. C. Robson, *Noncommutative Noetherian Rings* (Wiley-Interscience, Chichester, 1987).

[21] N. H. McCoy, in *Rings, Modules, and Radicals,* ed. A. Kertesz (North-Holland, Amsterdam, 1973).

[22] Y. V. Reddy and C. V. L. N. Murty, Semi-symmetric ideals in near-rings, *Indian J. Pure Appl. Math.* 16 (1985), 17-21.

[23] G. Renault, Anneaux réduits non commutatifs, *J. Math. Pure Appl.* 46 (1967), 203-214.

[24] L. Rowen, *Ring Theory*, vol I (Academic Press, Boston, 1988).

[25] H. Rubin and J. E. Rubin, *Equivalents of the Axiom of Choice*, II (North-Holland, Amsterdam, 1985).

[26] G. Shin, Prime ideals and sheaf representations of a pseudo symmetric ring, *Trans. Amer. Math. Soc.* 184 (1973), 43-60.

[27] G. Sigurdsson, Differential operator rings whose prime factors have bounded Goldie dimension, *Arch. Math.* 42 (1984), 348-353.

[28] S-H Sun, Noncommutative rings in which every prime ideal is contained in a unique maximal ideal, *J. Pure Appl. Alg.* 76 (1991), 179-192.

[29] F. A. Szasz, *Radicals of Rings* (Wiley-Interscience Chichester, 1981)

PRIMES OF SKEW FIELDS

H. H. BRUNGS
Department of Mathematics, University of Alberta
Edmonton, Alberta, Canada T6G 2G1

ABSTRACT

The partially ordered and labelled set $T(F)$ of primes of a skew field F is discussed, and considered in the special case where F is the skew field of quotients or the rational closure of the group ring of a left ordered group.

1. Let F be any skew field. A subgroup P of $(F,+)$ with $1 \notin P$ is called a *prime* of F if $xy \in P$ and $x \notin P$ implies $y \in P$ for any elements x, y in F. We will consider the set $T(F)$ of all primes of F together with the partial order given by containment. The three types of prime segments will be discussed as well as their cofinality types. These aspects of $T(F)$, a rather intractable object in general, will be illustrated by constructing skew fields F as skew fields of fractions of group rings of left ordered groups or their rational closure if the group ring is not Ore.

If P is a prime of F, then $B(P) = \{x \in F^* \mid x^{-1} \notin P\} \cup \{0\}$ is a valuation subring of F. Conversely, if B is a subring of F such that $x \in F \setminus B$ implies $x^{-1} \in B$, i.e. B is a *valuation subring* of F, and we say that (F,B) is a *valued skew field*, then $\mathcal{J}(B)$ the Jacobson radical of B is a prime of F. It follows from this correspondence that the following properties hold for $T(F)$ (see [4]):

i) The set $\{P' \in T(F) \mid P' \subseteq P\}$ of primes of F contained in a fixed prime P is totally ordered;
ii) The union of any ascending chain of primes and the intersection of any descending chain of primes of F is again a prime of F.
iii) For any set $\{P_i \mid i \in I\}$ of primes of F there exists a greatest lower bound $P_0 \in T(F)$.
iv) For any pair of primes $P \supset P'$ of F ('$\supset$' means proper containment) there exists a pair of primes $P_1 \supset P_2$ of F with $P \supseteq P_1 \supset P_2 \supseteq P'$ such that $P_1 \supset P_2$, but no other prime lies between P_1 and P_2; i.e. $P_1 \supset P_2$ is a *prime segment* of F.

These properties are very similar to the results in case $F = K$ is a commutative field (see [18]). If F' is an extension skew field of F and P' is a prime of F', then $P = F \cap P'$ is a prime of F and we say that P' is an extension of P in F'. Whereas for commutative fields $K' \supset K$ any prime P in K has at least

---

The author acknowledges support of NSERC.
1991 Mathematics Classification: Primary 16W60, 16S35, 16L30; Secondary 16W80, 20F60.

one extension P' in K' this does not remain true any longer for skew fields. It follows that for a commutative field K we have $T(K) = \{(0)\}$ or $|T(K)| = \infty$. The first possibility occurs if K is an algebraic extension of a field $GF(p)$ with p elements, p a prime; and K contains $K_0 = \mathbf{Q}$, the rationals, or $K_0 = GF(p)(t)$, a transcendental extension of $GF(p)$, otherwise. This then leads to infinitely many primes of K obtained by extending the infinitely many primes of K_0. However, $1 < |T(F)| < \infty$ is possible in the non-commutative case, since primes of a subfield may or may not have an extension in F. It is not known which finite tree graphs can be represented by $T(F)$ for some F.

2. We consider prime segments $P_1 \supset P_2$ of F and define a subset

$$A(P_1, P_2) = \{a \in P_1 \mid P_1 a P_1 \subset P_1\}.$$

This set is actually defined for any prime P_1 of F, but it is equal to P_1 in case $P_1 = \bigcup P$, $P \subset P_1$, i.e. P_1 is the union of primes P properly contained in P_1. We have the following result ([4]):

THEOREM 1. *Let $P_1 \supset P_2$ be a prime segment of the skew field F and denote with B the valuation ring $B(P_1)$ and with $\overline{B}$ the factor ring B/P_2. Then exactly one of the following cases occurs:*

a) *$A = P_1$ if and only if $\overline{B}$ is invariant, i.e. all one-sided ideals of $\overline{B}$ are two-sided;*

b) *$A = P_2$ if and only if $\overline{B}$ is nearly simple, i.e. $\overline{B}$ has no other ideals besides $\overline{B}$, $\mathcal{J}(\overline{B})$ and (0);*

c) *$P_1 \supset A \supset P_2$ if and only if the image $\overline{A}$ of A in $\overline{B}$ is a prime ideal of $\overline{B}$ that is not completely prime.*

We will say that the prime segment $P_1 \supset P_2$ is *invariant*, *simple* or *exceptional* depending on whether a), b), or c) holds respectively.

The rings $\overline{B}$ in the above theorem are rank-1 valuation rings and the cases a), b), c) describe the three possible types of such rings, see [5] and [2]. If F is finite dimensional over its center, then F has invariant prime segments only. The first examples of nearly simple valuation domains were given in [7] and [14], see also [6]. The question whether exceptional prime segments exist appeared in various contexts (see [17], [16], [5], [11], [8], [19]) but remained open. Dubrovin in 1992 (see [9]) announced that, by giving a partial answer to Malcev's problem about the embeddability of group rings over left ordered groups in skew fields, he had constructed skew fields with exceptional prime segments. Some details of this construction will be given in the last section.

3. The right cofinality type of a valuation ring B is defined as the cofinality type of the set of right ideals $I \neq (0)$ of B with respect to the order given by $I_1 \geq I_2$ if and only if $I_1 \subseteq I_2$. Using the ring $\overline{B}$ in Theorem 1 we can define the right cofinality type of a prime segment $P_1 \supset P_2$ of F as the right cofinality type of $\overline{B}$ and the left cofinality type of $P_1 \supset P_2$ is defined similarly. It follows that these types are equal to ω_0 in case $P_1 \supset P_2$ is invariant or exceptional: The non-zero principal right ideals of an invariant rank-1 valuation ring form a subsemigroup isomorphic to a subsemigroup of $(\mathbf{R}, +)$. If on the other hand, $B \supset \mathcal{J}(B) \supset Q \supset (0)$ is a rank-1 valuation ring and Q is a prime ideal of B which is not completely prime, then $\bigcap Q^n = (0)$ and again ω_0 is the (right and left) cofinality type of B. On the other hand, we will discuss constructions of valued skew fields that have prime segments of a given cofinality type ω_α in the next section.

Nearly simple valuation rings have the property that $P_0 = \bigcap I \neq (0)$ where the intersection is taken over all non-zero ideals I of B. We say that a valuation ring B is *finally simple* if this condition holds and $P_0 \supset (0)$ is then a simple prime segment. The cofinality types of such a ring agree then with the cofinality types of the prime segment $P_0 \supset (0)$.

If (F, B) is a valued skew field, then the set $\{aBb \mid a,\, b \in B\setminus\{0\}\}$ defines a neighbourhood basis for the coarsest ring topology $\mathcal{T}_B$ defined on F such that B is open (see [20] and [10]) and one can use Cauchy filters to define the completions $(\widehat{F}, \widehat{B})$ of (F, B).

If B is not finally simple, then $\mathcal{T}_B$ can be defined using the set of non-zero ideals of B as neighbourhood basis for (0) and it follows that $\mathcal{T}_B$ is a V-valuation and $(\widehat{F}, \widehat{B})$ is a valued skew field (see [12], [20]). That $\widehat{F}$ is not a skew field in general if B is a finally simple valuation ring follows from the next result due to Liepold (see [13]):

THEOREM 2. *Let (F, B) be a valued skew field with B a finally simple valuation ring with right-cofinality type and left-cofinality type equal to ω_0. Then $\widehat{F}$ is not a skew field.*

In the same paper it is also proved that $\widehat{F}$ can be a skew field even in the case that B is finally simple, provided the right and left cofinality types agree and are equal to $\omega_\alpha > \omega_0$.

4. To construct skew fields F with prime segments of prescribed types we consider left ordered groups (G, P). We denote with P a subset of G such that $P \cap P^{-1} = \{1\}$, $P \cup P^{-1} = G$ and $PP \subseteq P$. Then P is called a generalized cone of G and $a \leq_\ell b$ if and only if $a^{-1}b \in P$. A right order for G is defined similarly by such a set P.

A valued skew field (F, B) will be said to be *parameterized* by (G, P) if $G \subseteq (F^*, \cdot)$ and $B \cap G = P$ and every principal right ideal $I \neq (0)$ of B can be written in the form $I = aB$ for $a \in B$, every principal left ideal $L \neq (0)$ of B can be written in the form $L = Bc$ for $c \in P$. It follows that a right ideal $I \neq (0)$ of B then corresponds to the right ideal $I \cap P$ of P where a subset $N \neq \emptyset$ of P is called a right ideal of P if $NP \subseteq N$. Left ideals, ideals, prime ideals and completely prime ideals of P are defined similarly and analogous to the definition of these terms for rings.

It should be noticed that the above correspondence between right (left) ideals $\neq (0)$ of B and right (left) ideals of P does not necessarily induce a correspondence between two-sided ideals $\neq (0)$ of B and two-sided ideals of P.

The following results (see [4], also [7], [14], [6] and [1]) deal with the case in which the group ring is Ore provided the ring of coefficients is Ore.

Let R be any ring, G a group and $\sigma: G \to \mathrm{Aut}(R)$ a homomorphism from G into the automorphism group of R. Then we denote with $R[G,\sigma]$ the skew group ring with elements $\sum g r_g$ with $g \in G$, r_g in R, almost all $r_g = 0$ and $rg = g\sigma_g(r)$ defining the multiplication.

THEOREM 3. *Let R_0 be a domain and (G,P) be a left ordered group such that the skew group ring $R_0[G,\sigma]$ is an Ore domain for a homomorphism $\sigma: G \to Aut(R_0)$. Then $S = \{\sum g r_g \in R[G,\sigma] \mid g \in P,\ r_e \neq 0\}$ is an Ore set in $R = R_0[P,\sigma]$ and $B = RS^{-1}$ is a valuation domain in F, the skew field of quotients of $R_0[G,\sigma]$ parameterized by (G,P). In addition, there is a correspondence between the set of two-sided (prime, completely prime) ideals $\neq (0)$ of B and the set of two-sided (prime, completely prime) ideals of P.*

In order to apply this theorem one has to construct suitable left ordered groups (G,P) and we consider the following situation: Let K be an ordered field, V an ordered K-vector space and $C = (U,O)$ a Dedekind cut of V; i.e. $V = U \cup O$, $U \cap O = \emptyset$ and every element in U is smaller in V than every element in O. Then there exists an extension V_1 of V as K-ordered vector spaces with $V_1 = K\eta \oplus V$ and $U = \{v \in V \mid v < \eta\}$, $O = \{v \in V \mid v > \eta\}$ for an element $\eta \in V_1 \setminus V$. Then $G = \{(a,v) \mid 0 < a \in K,\ v \in V\}$ with $(a,v)(a',v') = (aa', av'+v)$ and $P = P_\eta = \{(a,v) \in G \mid a\eta + v \geq \eta\}$ is a left ordered group with generalized cone P_η and we can apply Theorem 3 to obtain a valued skew field (F, P_η) with F the skew field of fractions of the Ore domain $\mathbb{Q}[G]$, $\mathbb{Q}$ the field of rational numbers. The fact that $\mathbb{Q}[G]$ is an Ore domain follows since G is the semidirect product of two torsion free abelian groups. The prime segments $P_1 \supset P_2$ of F with $\mathcal{J}(B_\eta) \supseteq P_1$, i.e. the prime segments of B_η are determined by P_η according to Theorem 3. With the notation as above we have the following result:

THEOREM 4. *There exists a completely prime ideal D of B_η such that a prime segment $P_1 \supset P_2$ of B_η is invariant if and only if $P_2 \supseteq D$ and it is simple if and only if $D \supseteq P_1$. The simple prime segments of B_η have the same cofinality type as K. The ideal D corresponds to the ideal $\Delta = \{(a,v) \in P_\eta \mid a\eta + v > s > \eta$ for some $s \in V\}$.*

In the next result, conditions on the cut $C_\eta = (U,O)$ of V are given such that B_η has certain specific properties.

THEOREM 5. *With the notation as above the following holds:*

a) $D = \mathcal{J}(B_\eta)$ *if and only if for every* $r > 0$ *in* V *there exists a* u *in* U *with* $u + r \in O$;

b) $D = (0)$ *if and only if* $U = V$ *or* $O = V$;

c) B_η *is finally simple if and only if* $r < \eta < s$ *for some elements* r *and* s *in* V *and there exists an element* v' *in* V *such that for any* v *in* V *there exists* a *in* K *with* $v < av'$.

It follows from Theorem 4 that condition a) describes the case in which all prime segments of B_η are simple and condition b) describes the case in which all prime segments of B_η are invariant.

Next we consider a different localization of the semigroup ring $R_1 = B_0[P, \sigma] \subseteq B_0[G, \sigma]$ where now (F_0, B_0) is assumed to be a valued skew field parameterized by the left ordered group (G_0, P_0). As before, (G, P) is a left ordered group as well, σ is a homomorphism from G into the automorphism group of B_0 with the additional property that $\sigma_g(\mathcal{J}(B_0)) \subseteq \mathcal{J}(B_0)$ for all g in G and $B_0[G, \sigma]$ is a right and left Ore domain with F as skew field of quotients.

THEOREM 6. *The set* $S = \{\sum g r_g \in R_1 \mid \sum r_g B_0 = B_0\}$ *is a right Ore system of* R_1 *and* $B = R_1 S^{-1}$ *is a valuation subring of* F *parameterized by* (G_0, P_0).

In the situation of Theorem 6 it is no longer true that two-sided ideals of P_0 are in one-to-one correspondence with two-sided ideals $\neq (0)$ of B. In fact, one must consider ideals Ω of P_0 that have the following property: If $p \in \Omega$ and $\sigma(p) = p'u$ for $p' \in P_0$ and $u \in U(B)$, the unit group of B, then $p' \in \Omega$; and if $u_0 p = p''v$ for $u_0 \in U(B_0)$, $v \in U(B)$, $p'' \in P_0$, then $p'' \in \Omega$. We say that Ω is σ-invariant and $U(B_0)$ invariant and it is exactly the set of such ideals in Ω that corresponds to the set of non-zero ideals of B.

With the help of this last theorem it is possible to construct valuation rings with chains of prime segments arbitrarily prescribed either as invariant or as simple. However, it is not clear whether the constructions described in Theorems 3 or 6, which deal with group rings that are Ore, can produce exceptional prime segments if such segments were not already present in B_0.

5. Dubrovin in [9] uses left ordered groups (G, P), for which the group ring $\mathbb{Q}[G]$ is not Ore any longer, to construct skew fields with exceptional prime segments. Let K be any field and let $K\{G\} = L$ be the $K[G]$-left and K-right module of all series $\sum g r_g = a$ with $g \in G$, $r_g \in K$ and $\operatorname{supp}(a) = \{g \mid r_g \neq 0\}$ well-ordered with respect to the left order defined on G by P. If (G, P) is an ordered group, then L is a ring and in fact the skew field of generalized power series as considered by Neumann (see [15]) and others. However, in the case considered here, where P is a generalized cone of G, we set $Q = \operatorname{End}(K\{G\}_K)$ and the group ring $K[G]$ is a subring of this endomorphism ring Q.

Let A be any nonempty subset of Q. Then the rational closure $D(A)$ of A in Q is the subring of Q minimal with the following two properties: (i) $A \subseteq D(A)$; (ii) $d \in D(A) \cap U(Q)$ implies $d^{-1} \in D(A)$, where $U(Q)$ denotes the group of units of Q. Such a ring always exists as can be seen by considering the union of the ascending chain of subrings $R_n = R_n(A)$ defined as follows: R_0 is the subring of Q generated by A; R_{n+1} is the subring of Q generated by R_n and the elements r^{-1} in Q with $r \in R_n \cap U(Q)$.

Since (G, P) is a left ordered group we can consider the set G together with the total order $\leq_\ell$ and denote this by $(G, \leq_\ell)$. The group G then operates on this set by left multiplication as an automorphism group. With $\widehat{G}$ we denote the Dedekind closure of $(G, \leq_\ell)$ consisting of all (special) Dedekind cuts of $(G, \leq_\ell)$ which are defined as in Section 4 with the additional property that $U \neq \emptyset \neq O$ and that O has no smallest element if U has no largest element. In particular, G can be embedded into $\widehat{G}$ by mapping $g \in G$ to the cut $C_g = (U_g, O_g)$ with $U_g = \{h \in G \mid h \leq_\ell g\}$. Let ε be an element in $\widehat{G}$, then we denote with $S_\varepsilon = \{g \in G \mid g\varepsilon = \varepsilon\}$ the subgroup of elements in G that stabilize ε.

Dubrovin defines a complexity notion for the elements of $D(K[G])$ which refines the depth function given by the chain of subrings $R_n(K[G])$ of Q. This then leads to the next result which gives conditions under which $D(K[G])$ is a skew field and hence providing a partial answer to Malcev's problem.

THEOREM 7. *Let (G, P) be a left-ordered group, K a skew field. Assume that $K[S_\varepsilon]$ is an Ore domain for every (special) Dedekind cut ε in $\widehat{G}$ and that $H = \{g \in H \mid gPg^{-1} = P\}$ is cofinal in G. Then $F = D(K[G])$ is a skew field which contains a valuation subring B parameterized by (G, P) such that the mapping which assigns $I \cap P$ to an ideal $I \neq (0)$ of B defines a correspondence between non-zero ideals of B and ideals of P.*

The valuation ring B in this theorem is defined in the following way: For every q in Q we can define a mapping from $(G, \leq) \cup \infty$ to itself by setting $V_q(h) = \min(\mathrm{supp}\,(q[h])$, where $h \in G \subseteq \{G\}$ and $q[h]$ is the image of h under q in $K\{G\}$. We set $V_q(h) = \infty$ if $q[h] = 0$ and $V_q(\infty) = \infty$. It can be proved that

$B = \{d \in D(K[G]) \mid V_d(\ell_G) \geq \ell_G\}$, with ℓ_g the unit in G, is a valuation subring of $D(K[G])$ with the properties stated in the theorem.

In order to construct skew fields with exceptional prime ideals it remains to find left ordered groups (G, P) that on the one hand satisfy the conditions of Theorem 7 and on the other hand contain prime ideals $\Pi \subseteq P$ that are not completely prime. The example considered in [8] is the group G generated by x and y with the relation $xy^2x = y$ and P the submonoid generated by x and y. This group can be considered as a knot group or a braid group. In [3] it is shown that the universal covering group of $SL(2, \mathbf{R})$ again satisfies the conditions of the above theorem for a generalized cone P that contains prime ideals that are not completely prime.

6. References.

1. U. Albrecht, G. Törner, *Group rings and generalized valuations, Comm. Algebra* **12** (1984), 2243–2272.
2. C. Bessenrodt, H. H. Brungs, G. Törner, *Right Chain Rings*, Schriftenreihe F. B. Mathematik **181**, Duisburg (1990).
3. H. H. Brungs, N. I. Dubrovin, *The universal covering group of $SL(2, \mathbb{R})$ as value group,* (1992).
4. H. H. Brungs, M. Schröder, *Prime segments of skew fields*, preprint (1992).
5. H. H. Brungs, G. Törner, *Chain rings and prime ideals*, Arch. Math. **27** (1976), 253–260.
6. H. H. Brungs, G. Törner, *Extensions of chain rings,* Math. Z. **185** (1984), 93–104.
7. N. I. Dubrovin, *Chain domains,* Moscow Univ. Math. Bull. **36** (1980), 56–60.
8. N. I. Dubrovin, *An example of a chain prime ring with nilpotent elements,* Math. USSR Sbornik **48** (1984), 437–444.
9. N. I. Dubrovin, *The rational closure of group rings of left-ordered groups,* preprint (Russian) (1992), Edmonton Notes (1992).
10. P. Hartmann, *Stellen und Topologien von Schiefkörpern and Alternativkörpern,* Arch. Math. **51** (1988), 274–282.
11. S. K. Jain, Problem section, No. 8; in: *Ring Theory*, F. van Oystaeyen (Ed.) Marcel Dekker, New York, 1978.
12. I. Kaplansky, *Topological methods in valuation theory*, Duke Math. J. **14** (1947), 527–541.
13. F. Liepold, *Zur Vervollständigung bewerteter Schiefkörper,* Results Math. *19* (1991), 122–142.
14. K. Mathiak, *Valuation of skew fields and projective Hjelmslev spaces,* Lecture Notes in Math. **1175**, Springer Verlag, Berlin, 1986.
15. B. H. Neumann, *On ordered division rings,* Trans. Amer. Math. Soc. **66** (1949), 202–252.
16. B. L. Osofsky, *Noncommutative rings whose cyclic modules have cyclic injective hulls,* Pac. J. Math. **25** (1968), 331–340.
17. E. C. Posner, *Left valuation rings and simple radical rings,* Trans. Amer. Math. Soc. **107** (1963), 458–465.
18. P. Ribenboim, *Theorie des valuations,* Presses de l'Université de Montréal, Montréal, 1965.
19. D. M. Smirnov, *Right-ordered groups,* Algebra i Logika **5:6** (1966), 41–59.
20. M. Schröder, *Bewertungsringe von Schiefkörpern, Resultate und offene Probleme,* Results Math. **12** (1981), 191–206.
21. M. Schröder, *N. I. Dubrovin's Ansatz zur Konstruktion von nicht vollprimen Primidealen in Kettenringen,* Results Math. **17** (1990), 296–306.

ON MODULES WITH THE ABSOLUTE DIRECT SUMMAND PROPERTY

W.D. BURGESS
Department of Mathematics, University of Ottawa
Ottawa, Canada K1N 6N5
and
R. RAPHAEL[1]
Department of Mathematics, Concordia University
Montreal, Canada H3G 1M8

ABSTRACT

Modules with the absolute direct summand property are studied. These are modules so that each direct sum decomposition is rigid in the following sense: If $M = A \oplus B$ and C is a complement of A then $M = A \oplus C$. This property, more general than quasi-continuity, does impose structure on M, especially when it is rich in direct sum decompostions. Such modules over left noetherian rings are most amenable to study. The particular module $_RR$ is examined when R is left artinian and when it is von Neumann regular.

0. Introduction.

Various generalizations of injectivity for modules have been studied in recent decades. The monograph by Mohamed and Müller ([11]) is a full account of many of these developements. The properties can often be expressed in terms of the existence of direct sum decompostions. Our subject is also a generalization of injectivity but postulates a property of direct sum decompostions rather than existence. The expression "absolute direct summand" was introduced by L. Fuchs ([5, page 73])[2].

Definition 0.1 *Let M be an R-module. Then M is said to have the* **absolute direct summand property** *when for any decomposition $M = A \oplus B$, if C is a complement of A in M then $M = A \oplus C$. (We say that M is ads or that M is an* ads*-module.)*

By [11, Theorem 2.8], every quasi-continuous module is *ads*. Recall that a module M is said to satisfy C_1 if every non-zero submodule is essential in a direct summand (see [11, Proposition 2.1]). Such modules are also called CS-modules (see e.g., [4]). We have then, for any module, the following strict implications:

---

[1]This research was partially supported by grants A7539 and A7752 of the NSERC.

[2]The topic was first brought to our attention many years ago by G. Renault.

1991 *Mathematics Subject Classification*: 16A53, 16A36.

$$\begin{array}{ccccccc} & & & & & & CS \\ & & & & & & \Uparrow \\ \textit{injective} & \Rightarrow & \textit{quasi-injective} & \Rightarrow & \textit{continuous} & \Rightarrow & \textit{quasi-continuous} \\ & & & & & & \Downarrow \\ & & & & & & ads \end{array}$$

Since every indecomposable module is *ads* and an indecomposable module is quasi-continuous iff it is uniform ([11, Proposition 2.5]), our condition is strictly weaker than quasi-continuity. The $\mathbf{Z}$-module $\mathbf{Z} \oplus \mathbf{Z}$ is *CS*, see [9, Corollary 13] (and also [11, page 19]), but not *ads* by (1.1), below. Indeed, a module M is quasi-continuous if it is both *ads* and CS (see the remark after [13, Lemma C]).[3] Nevertheless the *ads* condition has surprisingly strong consequences for modules which are decomposable. Two main characterizations of *ads*-modules are given which show how they sit in their injective hulls (Section 1). When R is left noetherian more precise descriptions of *ads*-modules are possible. There are full characterizations in some cases, modulo, of course, knowledge about the indecomposable modules (Section 2). The module ${}_RR$ is of particular interest. For R left artinian, ${}_RR$ *ads* implies some important properties of left self-injective rings; however injectivity is much stronger (Section 3). When R is (von Neumann) regular, the condition that ${}_RR$ be *ads* is a weakened form of left continuity (Section 4).

Throughout R will be a unital ring and all modules unital left modules. For a ring R, $Q_l(R)$ will be the left complete ring of quotients and, following [11], if $M = \bigoplus_I M_\alpha$ and $J \subseteq I$ then $M(J) = \bigoplus_J M_\alpha$. The injective hull of a module M is denoted $E(M)$. Other conventions of notation and terminology follow [1] and [11].

1. Characterizations and General Results.

The first description of *ads*-modules is in terms of the mutual injectivity of summands. The idea of relative injectivity is found in [2] and in [11, Definition 1.1]. To recall, for modules A and B, B is A*-injective* if for any submodule $X \subseteq A$ and homomorphism $\phi : X \to B$, ϕ extends to some $\psi : A \to B$. Equivalently, for all $\zeta : E(A) \to E(B)$, $\zeta(A) \subseteq B$ ([11, Lemma 1.13]), an observation which will be used several times.

Proposition 1.1 *An R-module M is ads iff for any decompostion $M = A \oplus B$, B is A- injective.*

Proof. Suppose M is *ads* and that $U \subseteq A$ and $\phi : U \to B$. Set $X = \{u + \phi(u) | u \in U\}$. Now $X \cap B = 0$ and so X lies in a complement, say C, of B in M. By hypothesis $M = C \oplus B$. Then each $a \in A$ has the form $c - b$, $c \in C$ and $b \in B$, so that $c = a + b$. The composition of the projection of M onto C along B followed by the projection of

[3] We were directed to this observation by the referee (who also directed us away from several errors, for which we are grateful).

M onto B along A, applied to A, gives a homomorphism from A to B which extends ϕ.

For the converse, let $M = A \oplus B$ and let C be a complement of B. Define $U = A \cap (B \oplus C)$ so that each $u \in U$ can be written as $b + c$, $b \in B$ and $c \in C$. The projection of $B \oplus C$ onto B along C, restricted to U, gives a homomorphism $\phi : U \to B$ which, by assumption, extends to some $\psi : A \to B$. For $a \in A$, consider $D = R(a - \psi(a)) + C$. Suppose $b = ra - r\psi(a) + c \in D \cap B$, for some $c \in C$ and $r \in R$. Then $ra = r\psi(a) + b - c \in U$ and so $\phi(ra) = r\psi(a) + b = \psi(ra) = r\psi(a)$, showing that $b = 0$ and, hence, that $D \cap B = 0$. This implies that $a - \psi(a) \in C$ for all $a \in A$, and thus $a = (a - \psi(a)) + \psi(a) \in C \oplus B$. □

Corollary 1.2 *(i) If M is an ads-module and $M = A \oplus B$ where A contains an element a with $\mathrm{ann}_R(a) = 0$ then B is injective.*

(ii) If every two-generated R-module is ads then R is semi-simple artinian.

(iii) If the ring R has a complete set of orthogonal central idempotents $\{e_1, \dots, e_k\}$ then a module M is ads iff each e_iM is ads, $i = 1, \dots, k$.

Proof. *(i)*. Since $Ra \cong R$, the Baer criterion for injectivity applies. *(ii)* follows by *(i)* and the well-known theorem of Osofsky ([12] or [13]), and *(iii)* is readily verified. □

The assumption that every cyclic R-module be *ads* is not enough to give R semisimple artinian, as a two dimensional local algebra shows.

Examples. i) As already noted every quasi-continuous module is *ads* (and, *a fortiori*, any continuous, quasi-injective or injective one). ii) Every indecomposable module is *ads*, whereas an indecomposable module is quasi-continuous iff it is uniform ([11, Proposition 2.5]). iii) If R is a ring in which every idempotent is central, then ${}_RR$ can easily be seen to be *ads*. This class of rings includes commutative rings and reduced rings. We shall see that the statement is certainly not true for all rings.

Examples of a different nature will appear later.

The second characterization is an analogue of the description of quasi-continuous modules in [11, Theorem 2.8(3)]. It will be useful when dealing with the module ${}_RR$, as will be seen in Sections 3 and 4.

Proposition 1.3 *An R-module M is ads iff given a direct summand A of M, a decomposition $E(M) = E_1 \oplus E_2$, where E_1 is an injective hull of A, and π the projection of $E(M)$ onto E_1 along E_2, then $\pi(M) \subseteq M$.*

Proof. Suppose M is *ads*. The submodule $B = E_2 \cap M$ is a complement of A in M. Indeed, if C is a complement containing B and $c \in C$, we can write $c = e_1 + e_2$, $e_i \in E_i$. If $e_1 = 0$ then $c = e_2 \in B$. If $e_1 \neq 0$, there is $r \in R$ with $0 \neq re_1 \in A$. Then $rc = re_1 + re_2$ and $re_1 \in A \cap C = 0$, a contradiction. Now since M is *ads*, $M = A \oplus B$ and $\pi(M) = A$.

For the converse, suppose $M = A \oplus B$ and C a complement of A in M. Then $E(M)$ can be written $E(M) = E_1 \oplus E_2$, where E_1 is an injective hull of A chosen so that $A = M \cap E_1$ (which is possible since A is a summand of M), and E_2 is an injective hull of C. If π is the projection of $E(M)$ onto E_1 along E_2, $\pi(M) \subseteq M$. Consider $x \in M$, $x = e_1 + e_2$, $e_i \in E_i$. Then $e_1 \in M$ and so $e_2 \in M$ as well. But $A = M \cap E_1$ and $C \subseteq M \cap E_2$. Since C is a complement of A, $C = M \cap E_2$. Thus $M = A \oplus C$. □

The first corollary is not used later and is given without proof (cf. Example (iii), above).

Corollary 1.4 *Suppose that the idempotents of* $\mathrm{End}_R(E(R))$ *commute with each other (or even that idempotents with the same image commute with each other) then* M *is ads.*

Corollary 1.5 *Let* R *be a ring such that its left complete ring of quotients* $Q = Q_l(R)$ *is left self-injective. Then* ${}_RR$ *is ads iff for any idempotents* $f \in Q$ *and* $e \in R$, $Qf = Qe$ *implies* $f \in R$.

Proof. This is a translation of Proposition 1.3. If $Qf = Qe$ then $C = Q(1-f) \cap R$ is a complement of $Rf \cap R = Re$ in R. To say that $Re \oplus C = R$ is the same as saying that for some $g = g^2 \in R$, $Re = Rg$ and $R(1-g) = C$. But then $gf = g$ and $(1-g)(1-f) = 1-g$, and so $f = g$. □

Proposition 1.3 might suggest that any module M has an *ads*-hull in $E(M)$, constructed by adjoining the images of the projections as in the proposition. This does work in the situation of Corollary 1.5 (see Section 4). However an example can be constructed of a finite dimensional module over a finite dimensional algebra which has no such hull.

Before going on to some more special rings, let us look at direct sums of modules. The next observations are analogues of [11, Proposition 1.18 and Theorem 2.13].

Proposition 1.6 *Let* $M = \bigoplus_I M_\alpha$ *be a direct sum of modules.*

(i) If each M_α, $\alpha \in I$, *is quasi-injective, then* M *is ads iff it is quasi-injective.*

(ii) If each M_α, $\alpha \in I$, *is quasi-continuous, then* M *is ads iff it is quasi-continuous.*

Proof. *(i)* If M is *ads* then by Proposition 1.1, for each $\alpha \in I$, $M(I \setminus \{\alpha\})$ is M_α-injective, and so by [11, Proposition 1.18], M is quasi-injective. The second part is similar but uses [11, Theorem 2.13]. □

In the same spirit we consider which direct sums of indecomposables are *ads*. We need a decomposition which satisfies a strong decomposition property, for example when Warfield's form of the Azumaya Theorem ([16, Theorem 1]) can be invoked.

Proposition 1.7 *Let* $M = \bigoplus_I M_\alpha$ *be a direct sum of indecomposable modules with the property that every direct sum decomposition of* M *refines to one equivalent to the given one. Then* M *is ads iff for each* $J \subset I$, $M(J)$ *and* $M(I \setminus J)$ *are mutually injective. The sum satisfies the condition, for example, when each* M_α *is countably generated and* $\mathrm{End}_R(M_\alpha)$ *is local.*

Proof. The necessity of the condition is clear. To verify the sufficiency, suppose $M = A \oplus B$. By assumption we can write $A = \bigoplus_{K_1} N_\alpha$, $B = \bigoplus_{K_2} N_\beta$, giving a decomposition of M equivalent to the original one. Then there is a bijection $\psi : K_1 \cup K_2 \to I$ so that $A \cong M(J)$, where $J = \psi(K_1)$ and $B \cong M(I \setminus J)$. Then A and B are mutually injective. The additional assumptions are those from [16, Theorem 1]. □

2. Ads modules over left noetherian rings.

The hypothesis that ${}_RR$ be left noetherian allows us to say more about *ads*-modules because direct sums of injectives are injective and because each module can be split into a direct sum of an injective submodule and a *reduced* submodule (i.e., one which has no injective submodules). Some necessary and some sufficient conditions for a module to be *ads* are found. If the ring is left hereditary, there is a complete characterization. Some consequences of the left noetherian assumption for *ads*-modules follow readily.

Proposition 2.1 *Let R be a left noetherian ring.*

(i) If M is an ads-module which is not injective, then M can be written $M = U \oplus V$ where V is torsion and injective, $0 \neq U$ is reduced and there are no non-zero monomorphisms from submodules of V to U.

(ii) Let ${}_RR$ be uniform and M an ads-module which is not injective. Suppose there is $a \in M$ with $\mathrm{ann}_R(a) = 0$, then U, as in (i), is indecomposable.

(iii) If R is left hereditary, an ads-module M which is not injective has the form $U \oplus V$, U reduced, V torsion and injective, and there are no non-zero homomorphisms from submodules of V to U.

Proof. *(i).* We write $M = U \oplus V$ where U is reduced and V is injective. If $0 \neq W \subset V$ and $0 \neq \phi : W \to U$ then ϕ extends to a monomorphism on some non-zero direct summand on V, which contradicts the fact that U is reduced.

(ii). If $a \in M$ has $\mathrm{ann}_R(a) = 0$ we use the decomposition above to write $a = u + v$, $u \in U$ and $v \in V$. By assumption, $\mathrm{ann}_R(u) \cap \mathrm{ann}_R(v) = 0$ implies that one of the two annihilators is zero. We cannot have $\mathrm{ann}_R(v) = 0$ by Corollary 1.2, so $\mathrm{ann}(u) = 0$. If U had a non-trivial decomposition $U = U_1 \oplus U_2$ then the same argument would show one of the summands to be injective.

(iii). Since, now, factor modules of injectives are injective, there are no non-zero homomorphisms from submodules of V to U. □

We now look for sufficient conditions. The gap between the sufficient conditions and the necessary conditions of Proposition 2.1*(i)* is the difference between "homomorphism" and "monomorphism".

Proposition 2.2 *Let R be left noetherian and M an R-module written $M = U \oplus V$ where U is reduced and ads, and V is injective. Suppose further that there are no non-zero homomorphisms from submodules of V to U. Then M is ads.*

If R is also left hereditary then the condition is necessary as well as sufficient.

Proof. Suppose $M = A \oplus B$. We write $A = A_1 \oplus A_2$ and $B = B_1 \oplus B_2$ where A_1 and B_1 are reduced while A_2 and B_2 are injective. By [10, §1.1], $A_1 \oplus B_1 \cong U$ and $A_2 \oplus B_2 \cong V$. Consider a homomorphism $\phi : E(A_1) \oplus A_2 \to E(B_1) \oplus B_2$. It can be represented by a matrix of homomorphisms: $\phi = \begin{pmatrix} \phi_{11} & \phi_{12} \\ \phi_{21} & \phi_{22} \end{pmatrix}$. By the assumptions, $\phi_{21} = 0$ and $\phi_{11}(A_1) \subseteq B_1$ (since U is *ads*). It follows that $\phi(A) \subseteq B$, and hence A is B-injective.

When R is also left hereditary, Proposition 2.1(iii) shows the necessity. □

The proposition applies, for example, to $R \times E(R/P)$ where R is a commutative domain and P is a non-zero prime ideal.

Proposition 2.3 *Let R be left noetherian and set $M = \bigoplus_I M_\alpha$ where each M_α is ads. If for all $\beta \neq \alpha$, $\mathrm{Hom}_R(E(M_\alpha), E(M_\beta)) = 0$, then M is ads.*

Proof. Suppose that $M = A \oplus B$. The hypothesis implies that for each $\beta \in I$, $M_\beta = (M_\beta \cap A) \oplus (M_\beta \cap B)$, and hence that $M = \bigoplus_I ((M_\alpha \cap A) \oplus (M_\alpha \cap B))$. We need to show that B is A-injective. Since our ring is left noetherian,

$$E(M) = E(A) \oplus E(B) = \bigoplus_I (E(M_\alpha \cap A) \oplus E(M_\alpha \cap B)).$$

Let $\phi : E(A) \to E(B)$ and ϕ_α be its restriction to $E(M_\alpha \cap A)$. The map ϕ_α followed by the projection onto $E(M_\beta \cap B)$ is zero for $\beta \neq \alpha$. It follows that $\mathrm{Im}(\phi_\alpha) \subseteq E(M_\alpha \cap B)$ and, since M_α is *ads*, that $M_\alpha \cap B$ is $(M_\alpha \cap A)$-injective, and so $\phi_\alpha(M_\alpha \cap A) \subseteq M_\alpha \cap B$. □

The proposition is true for arbitrary rings if I is finite.

Suppose M is a module so that $E(M) = \bigoplus_I E_\alpha$ where the E_α are pairwise isomorphic indecomposable modules, then M is called *homogeneous*.

Lemma 2.4 *Let R be left noetherian and M a homogeneous module which is ads. Then M is indecomposable or quasi-injective.*

Proof. Suppose that M has a non-trivial decomposition $M = A \oplus B$. Then B is A-injective and, by [2, Theorem 2.5] B is $(\bigoplus_I A)$-injective for any index set I. Since M is homogeneous, there exist a set I and a monomorphism

$$\mu : E(B) \to E(\bigoplus_I A) \cong \bigoplus_I E(A).$$

Since B is $\bigoplus_I A$-injective, $\mu(B) \subseteq \bigoplus_I A$ and, therefore, B must be quasi-injective. Similarly, A is quasi-injective and (1.6) gives the result. □

For commutative noetherian domains of (Krull-)dimension 1 there is a complete characterization of *ads*-modules which elaborates on the second part of Proposition 1.3. We need to recall that for such a ring, R, any torsion module M can be decomposed $M = \bigoplus_{P\in\mathcal{M}} M(P)$. where $\mathcal{M} = \mathrm{mSpec}(R)$, the maximal spectrum, and $M(P) = \{x \in M | \mathrm{ann}_R(x)$ *is P-primary* $\}$. Such a component $M(P)$ is homogeneous (by [10, Proposition 3.4]).

Proposition 2.5 *Let R be a commutative noetherian domain of dimension 1. A module M, which is not injective, is ads iff one of the following condition holds.*

(i) M is indecomposable.

(ii) M is torsion and for each $P \in \mathrm{mSpec}(R)$, $M(P)$ is indecomposable or quasi-injective.

(iii) There is $a \in M$ with $\mathrm{ann}_R(a) = 0$ and $M = U \oplus V$ where $U \neq 0$ is indecomposable and reduced, V is injective torsion, and there are no non-zero homomorphisms from submodules of V to U.

Proof. We have already seen that the three types of module are *ads* (by (2.2) and (2.3)).

Now assume that M is decomposable and *ads*. If M is torsion then *(ii)* holds by (2.4). If not, then M splits as in *(iii)* by (2.1). Suppose there is $0 \neq \phi : W \to Z$ where W and Z are indecomposable summands of V and $E(U)$, respectively. Then W and Z are both P-torsion, for some $P \in \mathrm{mSpec}(R)$, and so are both isomorphic to $E(R/P)$. This contradicts Proposition 2.1*(i)*. □

3. Ads modules over left artinian rings.

If R is a left artinian ring, Proposition 1.7 allows a classification of finitely generated R-modules which are *ads*. For left artinian rings in general, the property that $_RR$ is *ads* (we shall say that R is *self-ads*) imposes many restrictions on R and even though such rings need not be self-injective they share many of the characteristics of self-injective left artinian rings. Finite dimensional algebras of simple structure will show that, as expected, being self-*ads* does not pass to matrix rings, homomorphic images or corner rings.

Proposition 3.1 *Let R be a left artinian ring.*

(i) A finitely generated R-module M is ads iff $M = Q \oplus M_1 \oplus \cdots \oplus M_r$ where Q is quasi-injective, the M_i are pairwise non-isomorphic indecomposable modules which are not quasi-injective, and Q and the M_i are all mutually injective.

(ii) If R is artinian and serial then an R-module M is ads iff it is quasi-injective. In particular $_RR$ is ads iff R is self-injective.

Proof. *(i)*. The module M can be split into a direct sum of indecomposables. All the quasi-injective indecomposable summands can be put together to form Q, so we can write $M = Q \oplus M_1 \oplus \cdots \oplus M_r$. By Propositions 1.6 and 1.7, the conditions are sufficient. On the other hand, if M is *ads*, $M_i \not\cong M_j$ for $i \neq j$, and the condition is necessary.

(ii). By Fuller's [6, Theorem 5.3], every indecomposable R-module is quasi-injective and Proposition 1.6 applies. □

Remark 3.2 *The property that R be self-ads is not preserved by taking matrix rings, homomorphic images or corner rings.*

Proof. The diagram presentations of algebras from [7] are convenient here. The algebras are over any field. Since the following examples are monomial algebras, the question of existence of the algebras is decided by [7, Theorem 3.3]. Consider the finite dimensional algebra whose indecomposable projectives, Re_1 and Re_2, are given by:

$$\begin{array}{ccc} & 1 & \\ / & & \backslash \\ 2 & & 2 \end{array} \qquad \begin{array}{c} 2 \\ | \\ 1 \end{array} .$$

Now Re_1 and Re_2 are readily seen to be mutually injective and so ${}_RR$ is *ads*. (Of course, by the Krull-Schmidt Theorem, we need only concern ourselves with the mutual injectivity of Re_1 and Re_2.) On the other hand, Re_1 is not quasi-injective and it will follow that $M_2(R)$ (with two isomorphic summands which are not quasi-injective) is not self-*ads*.

For the other properties, consider the serial QF-algebra given by the following diagrams for Re_1, Re_2 and Re_3:

$$\begin{array}{ccc} 1 & 2 & 3 \\ | & | & | \\ 2 & 3 & 1 \\ | & | & | \\ 3 & 1 & 2 \end{array} .$$

Clearly R is self-*ads* but the rings R/Re_2R and $(e_1 + e_2)R(e_1 + e_2)$ do not enjoy this property. □

We now look further at the module ${}_RR$. In the first example in the proof of Remark 3.2, ${}_RR$ is not quasi-continuous (Re_1 is not uniform). We shall see that to be self-*ads* is nevertheless a strong requirement. Although we already know that we may lose generality, we sometimes suppose our ring R to be basic (i.e., if $e_1, \ldots, e_n$ is a complete orthogonal set of primitive idempotents then for $i \neq j$, $Re_i \not\cong Re_j$).

Proposition 3.3 *Let R be left artinian and chose a basic orthogonal set of primitive idempotents $e_1, \ldots, e_n$ with corresponding simples $S_i = Re_i/Je_i$. If R is self-ads then for each i, $\mathrm{Soc}(Re_i)$ is homogeneous and if $i \neq j$, $\mathrm{Soc}(Re_i)$ and $\mathrm{Soc}(Re_j)$ are made up of non-isomorphic simples (i.e., there is a permutation σ of $\{1, \ldots n\}$ so that $\mathrm{Soc}(Re_i)$ is a sum of simples isomorphic to $S_{\sigma(i)}$).*

Proof. Suppose S is a simple summand of $\mathrm{Soc}(Re_i)$ and T one of $\mathrm{Soc}(Re_j)$ where $i \neq j$ and $S \cong T$. Then the *ads* property yields homomorphisms $\phi : Re_i \to Re_j$ and $\psi : Re_j \to Re_i$ so that $\phi_{|S}$ and $\psi_{|T}$ are inverse isomorphisms. But since $i \neq j$, $\psi \circ \phi(Re_i) \subset Je_i$ and so $\psi \circ \phi$ is nilpotent. This is impossible since $(\psi \circ \phi)^k$ is never zero on S, for any $k \geq 0$. The rest follows since there are n isomorphism classes of simples. □

This permutation was first studied by Nakayama in QF-rings and is sometimes called the *Nakayama permutation*. A classic result of Utumi ([15, Theorem 7.10]) says that if R is artinian and ${}_RR$ and R_R are continuous then R is quasi-Frobenius. Proposition 3.3 says that left artinian and left self-*ads* already give one of the nice properties of quasi-Frobenius rings.

For the first corollary we recall the notion of the (Gabriel) quiver of a basic left artinian ring R. Using the notation of the proposition, the quiver, Γ of R is an oriented graph with vertices $v_1, \ldots, v_n$ and one arrow $j \to i$ for each copy of Re_j in a projective cover of Je_i.

Corollary 3.4 *Let R be a basic left artinian ring and Γ its quiver. If R is self-ads then for any connected component Δ of Γ which contains more than one vertex, every vertex of Δ has an arrow entering it and one leaving it.*

Proof. Since the connected components of Γ correspond to central idempotents, we may assume Γ is connected by Corollary 1.2. Now if v_i is a vertex with no arrows into it then Re_i is simple. This means that for any j, $j \neq i$, S_i could only appear as a composition factor of Re_j if it were in $\mathrm{Soc}(Re_j)$. But Proposition 3.3 rules that out. If v_i had no arrows into it then S_i would not appear in a composition series of any Re_j, $j \neq i$. But S_i must appear in some $\mathrm{Soc}(Re_j)$, again by Proposition 3.3. Thus Re_i is simple, a case already excluded. □

Corollary 3.5 *Let R be a basic left artinian ring so that $J^2 = 0$. Then R is self-ads iff there is a permutation σ of $\{1, \ldots, n\}$ with the property that for each i, $\mathrm{Soc}(Re_i)$ is a sum of simples isomorphic to $S_{\sigma(i)}$.*

Proof. The proposition says that the condition is necessary. In the other direction the only homomorphism from a proper submodule of Re_i to Re_j, $i \neq j$, is from a homogeneous semisimple module to another of different type. Hence $\mathrm{Hom}_R(Je_i, Re_j) = 0$.
□

Corollary 3.6 *Let R be a basic left artinian ring which is left serial. If R is self-ads then the components of the quiver Γ are single vertices or oriented cycles and all the indecomposable projectives corresponding to the vertices of a component have the same composition length. If R is also a split algebra over a field then R is self-ads iff R is left self-injective.*

Proof. We may again assume that the quiver Γ of R is connected. According to [3, page 160] Γ is either a rooted tree or is like one, but with the root replaced by an oriented cycle. By Corollary 3.4 there can be only a single vertex or a single oriented cycle. In particular, if there is just one vertex, v, then the only possible arrow is a loop at v. If Γ is an oriented cycle with more than one vertex, then the fact that the different Re_i have non-isomorphic socles says that no Je_i can be projective (for $Je_i \cong Re_j$, $j \neq i$, would give two indecomposable projectives with isomorphic socles). But this is exactly the situation of [3, page 161] where all the indecomposable projectives have the same length.

For a split algebra, the remaining conclusion follows from [6, Corollary 3.2]. □

4. Remarks and questions.

By Corollary 1.5, if $Q = Q_l(R)$ is left self-injective then R is self-*ads* iff it contains all idempotents $f \in Q$ such that for some $e^2 = e \in R$, $Qf = Qe$. This shows that such rings have what could be called an *ads-hull*. That is, a unique smallest ring $A(R)$, $R \subseteq A(R) \subseteq Q$, which is self-*ads*. It can be constructed by adjoining, transfinitely if necessary, the needed idempotents from Q. It can also be seen that $A(R)$ is the intersection of all rings T, $R \subseteq T \subseteq Q$ so that T is self-*ads*. Indeed, if S is that intersection and $f^2 = f \in Q$ and $e^2 = e \in S$ with $Qf = Qe$ then f is in all the rings T.

This applies in particular to (von Neumann) regular rings (e.g., [14, XII, Proposition 2.1]). Recall that a regular ring R is left continuous iff it contains all the idempotents of $Q_l(R)$ ([8, Theorem 13.13]). Every left continuous regular ring R is self-*ads* but the latter condition is weaker than the former, as *(i)* of the next remark shows. (This is also illustrated by the ring T in Example *(ii)* following the remark.)

Remark 4.1 *(i) If R is an abelian regular ring then R is self-ads.*

(ii) Let R be a ring such that $Q = Q_l(R)$ is left self-injective. If Q contains a set of $n \times n$ matrix units so that the diagonal matrix units are in R then $A(R) = Q$ (cf. [8, Corollary 13.19]).

(iii) There is a commutative regular ring R so that $M_2(R)$ is not self-ads (and hence a unit regular ring which is not self-ads).

Proof. *(i)* is clear by remarks in Section 1.

(ii). Let $\{e_{ij}\}$ be a set of matrix units in Q so that the e_{ii} are in R. Write $Q = M_n(T)$, for some ring T. For any $i \neq j$ and any $t \in T$ set $f = e_{ii} + te_{ji}$. Then $f^2 = f$, $Qf = Qe_{ii}$ and the entire ring Q can be built from R using such f.

(iii). Here [8, Example 13.8] (but using left ideals) serves as an example. We are given fields $K \subset F$, $K \neq F$, and $R \subset \prod_{\mathbb{N}} F$, $R = \{(x_n) | x_n \in K \textit{ for almost all } n \in \mathbb{N}\}$. Now $Q = Q_l(M_2(R)) = \prod_{\mathbb{N}} M_2(F)$ and *(ii)* now applies. □

Examples. (i) Suppose $Q = Q_l(R)$ is left self-injective and ${}_RR \cong R \oplus R$, then the *ads*-hull of R is Q (by *(ii)* of Remark 4.1). A specific example is found in [8, Examples 9.34 and 10.11].

(ii) In [8, Example 13.8] quoted above, the *ads*-hull is Q. This again follows from *(ii)* of Remark 4.1. Notice that the same reasoning applies to the much smaller ring $S = \{(A_n) | A_n \in M_2(F), A_n \textit{ eventually constant in } M_2(K)\}$. However the subring $T = \{(A_n) | A_n \in M_2(F) \textit{ eventually constant and scalar in } K\}$ can be seen to be self-*ads*.

It is interesting to note that although $A(R)$ is built from R by adding idempotents which are not central, new central ones may nevertheless arise, as happened with the ring S in the example above.

We finish with a list of questions.

1. Is the *ads*-hull of a regular ring regular? If R is directly infinite is $A(R) = Q_l(R)$?

2. For a regular ring R, is the *ads*-hull always attained in one step (in the process of

adding idempotents from $Q_l(R)$)?
3. If R is a split finite dimensional algebra over a field, does ${}_R R$ *ads* imply that the right module R_R is *ads*?
4. Can split finite dimensional self-*ads* algebras be characterized? What if they are diagram algebras ([7]) or even monomial algebras?
5. Is there a non-commutative version of Proposition 2.5?

References

[1] F.W. Anderson and K.R. Fuller, "Rings and Categories of Modules, Second Edition", Springer-Verlag, New York, Heidelberg, Berlin, 1992.

[2] G. Azumaya, F. Mbuntum and K. Varadarayan, On M-projective and M-injective modules, *Pacific J. Math.*, **95** (1975), 9-16.

[3] W.D. Burgess, K.R. Fuller, E. Voss and B. Zimmermann-Huisgen, The Cartan matrix as an indicator of finite global dimension for artinian rings, *Proc. Amer. Math. Soc.*, **95** (1985), 157-165.

[4] V. Camillo and M.F. Yousif, CS-modules with ACC or DCC on essential submodules, *Comm. Algebra*, **19** (1991), 655-662.

[5] L. Fuchs, "Abelian Groups", Pergamon Press, Oxford, London, New York, Toronto, 1960.

[6] K.R. Fuller, On indecomposable injectives over artinian rings, *Pacific J. Math.*, **29** (1969), 115-135.

[7] K.R. Fuller, Algebras from diagrams, *J. Pure and Appl. Algebra*, **48** (1987), 23-37.

[8] K.R. Goodearl,"Von Neumann Regular Rings", Pitman, London, San Francisco, Melbourne, 1979.

[9] M.A. Kamal and B.J. Müller, Extending modules over commutative domains, *Osaka J. Math.*, **25** (1988), 531 538.

[10] E. Matlis, Injective modules over Noetherian rings, *Pacific J. Math.*, **8** (1958), 511-528.

[11] S.H. Mohamed and B.J. Müller, "Continuous and Discrete Modules", LMS Lecture Note Series, **147**, Cambridge University Press, Cambridge, 1990.

[12] B. Osofsky, Rings all of whose finitely generated modules are injective, *Pacific J. Math.*, **14** (1964), 645-650.

[13] B. Osofsky and P.F. Smith, Cyclic modules whose quotients have all complement submodules direct summands, *J. Algebra*, **139** (1991), 342-354.

[14] B. Stenström, "Rings of Quotients", Springer-Verlag, New York, Heidelberg, Berlin, 1975.

[15] Y. Utumi, On continuous rings and self injective rings, *Trans. Amer. Math. Soc.*, **118** (1965), 158-173.

[16] R.B. Warfield, A Krull-Schmidt theorem for infinite sums of modules, *Proc. Amer. Math. Soc.*, **22** (1969), 460-465.

FUNCTORS AND σ-PRODUCTS

JOHN DAUNS
Department of Mathematics, Tulane University
New Orleans, Louisiana 70118, U.S.A.

ABSTRACT

Various functors from ring categories to the category of join semi-lattices and semi-lattice homomorphisms are constructed. The elements or point of a semi-lattice is an equivalence class of modules. The relationship between several such functors is explored by use of natural transformations of functors.

INTRODUCTION

This article is partly expository, but also contains some new results. One purpose is to inform ring and module theorists of some recently developed tools – certain functors, equivalence classes of modules, and partial orders on these equivalence classes – for studying modules and rings. The functors described here are a generalization of some previous functors used for nonsingular modules (i.e. Ξ, [6, 8-11]). The objective here is not so much to study particular rings and modules but functors defined by these, or more precisely the relationship between the modules and rings and the functors. What is new in this article that here for the most part our modules are not restricted to be nonsingular any longer, and also that direct sums have been replaced by σ-products. As is appropriate for a proceedings paper, this article does not give a complete solution to all problems, which would be like a dead end street with nowhere to go afterwards. Rather, it presents a theory which is being presently developed; many unsolved problems are implicit here, and some are actually listed.

1. σ-PRODUCTS

First terminology and symbols are introduced all at once in one place. Then certain subdirect products called σ-products are discussed.

1.1. Notation. Right unital modules over a ring R are used. Submodules are denoted by $<$ and $\leq$, and large or essential ones by $<<$. For $K < M$, and $x \in M$, if $L_x : R_R \longrightarrow M$ is the R-map $L_x(r) = xr$, $r \in R$, then $x^{\perp} =$ kernel $L_x =$ $\ker L_x = \ker(L_x) = \{r \in R \mid xr = 0\}$, and $x^{-1}K = L_x^{-1}(K) = L_x^{-1}K = \{r \in R \mid$

$xr \in K\} \leq R$. Subsets $Y \subset M$ define right ideals $Y^{\perp} = \{r \in R \mid \forall\, y \in Y,\ yr = 0\} = \{r \mid Yr = 0\}$. Ideals are denoted by '$\triangleleft$' in R as well as other rings, e.g. $M^{\perp} \triangleleft R$.

For right R-modules M, right R-injective hulls are denoted by both "$\widehat{\ }$" and "E", e.g. $\widehat{M} = E(M) = EM$, where E is used when M is given by a long formula.

The first and second singular submodules are $ZM = \{m \mid m^{\perp} << R\}$, $Z(M) << Z_2M \leq M$, where $Z[M/(ZM)] = (Z_2M)/ZM$, and Z_2M is called the *torsion submodule* of M. Consequently M is *torsion* if $M = Z_2M$. Since $Z_2M = 0 \Longleftrightarrow ZM = 0$; M is *torsion free* iff $ZM = 0$, abbreivated – "t.f.".

Only for those submodules $K \leq M$ which contain $ZM \subseteq K$ – but not others – define the complement closure $\overline{K}$ of K inside M by $K << \overline{K} \leq M$ by $Z(M/K) = \overline{K}/K$. Thus $\overline{K} = \{x \in M \mid K << K + xR\} = \{x \in M \mid x^{-1}K << R\} = \cap\{C \mid K \subseteq C,\ C \leq M \text{ is a complement}\}$.

For an R-module map $f : A \longrightarrow B$ and $C \leq B$, the restriction of f to C is written as $f|C$. The reject and trace are the fully invariant submodules $\mathrm{rej}_A B = \cap\{h^{-1}0 \mid h \in \mathrm{Hom}_B(B, A)\} \leq B$, $\mathrm{tr}_B A = \mathrm{Hom}_R(A, B)A \leq B$. The symbol "$A \hookrightarrow B$" means that there exists some embedding $A \longrightarrow B$, i.e. that it is possible to embed A in B.

1.2. Partial Orders. For any set X, $\mathcal{P}(X) = \langle \mathcal{P}(X),\ \cup = \vee,\ \cap = \wedge$, complement $= \setminus,\ 1 = X,\ 0 = \emptyset\rangle$ is the usual Boolean lattice; also $\mathcal{P}(X)$ equally well denotes the associated Boolean ring. The term 'Boolean algebra' is not used. The cardinality of a set X is $|X|$. The cofinality of an infinite cardinal σ is a regular cardinal denoted by $\mathrm{cof}(\sigma)$ or $\mathrm{cof}\,\sigma$. (See [20; p.193-195]).

A *quasi-order*, denoted by $\preceq$ or $\prec$, is a binary relation that is reflexive ($a \preceq a$), and transitive ($a \prec b,\ b \prec c \longrightarrow a \preceq c$). Any quasi-order defines an equivalence relation $a \sim b$ iff $a \preceq b$ and $b \preceq a$. The class of all equivalence classes $[a] = \{c \mid a \sim c\}$ is a po-class, where $[a] \leq [b]$ iff $\forall\, a' \in [a],\ b' \in [b],\ a' \preceq b'$ iff $\exists\, a' \in [a],\ b' \in [b],\ a' \preceq b'$ iff $a \preceq b$.

Here L will be called a *join semi-lattice* or just a *semi-lattice* if L is a partially ordered class (po-class) in which for every $x, y \in L$ there exists a least upper bound $x \vee y \in L$. It is *complete* if every nonempty subset $S \subset L$ has a supremum $\vee S \in L$. Note that $\wedge S$ need not necessarily exist, and that subclasses of L are not required to have a supremum in L. A function $f : L_1 \longrightarrow L_2$ of po-classes L_1, L_2 is *complete* ([19; p.39]) if (i) f preserves order; (ii) for any subset $S \subset L$, such that $\sup S = \vee S \in L_1$ exists, also $\vee fS \in L_2$ exists; and $f(\vee S) = \vee(fS)$. If L_1 and L_2 are semi-lattices with smallest elements $0 \in L_i$, so is their direct sum $L_1 \oplus L_2 = L_1 \times L_2$, with $0 \leq (x_1, 0) \vee (0, x_2) = x_1 \vee x_2 = (x_1, x_2) \leq (y_1, y_2) = y_1 \vee y_2$ iff both $x_1 \leq y_1 \in L_1$, and $x_2 \leq y_2 \in L_2$. Thus $L_1, L_2 \subset L_1 \oplus L_2$ are convex. By a category here is meant a large category ([23; p.23]), that is the subobjects of an object need not be a set but may form a proper class. Many ordinary category and semi-lattice concepts carry over to classes, and the customary usual terminology is used, e.g.

dense, direct product, semi-lattice homomorphism, convex semi-sublattice (i.e. a sub- (semi-lattice)). However, the term "lattice" always refers to a set. Everything here could be done in the usual set theoretic axiom system ZFC. Thus for example a function f whose domain is a proper class is merely a short hand abbreviation of statements which could all be phrased in terms of sets whose existance is guaranteed by ZFC, such as $f|C$, where C is a subset.

1.3. σ-Products. For an indexed set $\{W_\gamma \mid \gamma \in \Gamma\}$ of modules, for $x = (x_\gamma)_{\gamma \in \Gamma} = (x_\gamma) \in \prod_\Gamma W_\gamma = \prod W_\gamma$ the *support* of x is $\operatorname{supp} x = \{\gamma \in \Gamma \mid x_\gamma \neq 0\} \subseteq \Gamma$. If x is viewed as a function on Γ, then the zero set of x is $x^{-1}0 = \{\gamma \mid x_\gamma = 0\} = \Gamma \backslash \operatorname{supp} x$. For a fixed infinite cardinal σ, the σ*-product* of a family $\{W_\gamma \mid \gamma \in \Gamma\}$ is the submodule $\prod^\sigma \{W_\gamma \mid \gamma \in \Gamma\} = \prod_\Gamma^\sigma W_\gamma = \prod^\sigma W_\gamma \leq \prod W_\gamma$ consisting of all those $x = (x_\gamma)$ with $|\operatorname{supp} x| < \sigma$. E.g. $\prod_\Gamma^{\aleph_0} W_\gamma = \bigoplus_\Gamma W_\gamma$.

Define $\prod_\Gamma^\infty W_\gamma = \prod_\Gamma W_\gamma$. If one interprets "$\infty$" as an element (not a cardinal) that is larger than every cardinal, then this formally fits the above definition of a σ-product.

For a proof of the next theorem, see [5; p.190, 5.1].

1.4. Fubini Theorem. For modules $\{W_i \mid i \in I\}$ suppose that I is partitioned into disjoint subsets $I(\gamma)$, $\gamma \in \Gamma$; $I = \cup\{I(\gamma) \mid \gamma \in \Gamma\}$. Then for an infinite cardinal σ,

$$\prod_{i \in I}^{\sigma} W_i = \prod_{\gamma \in \Gamma}^{\sigma} \Big(\prod_{i \in I(\gamma)}^{\sigma} W_i \Big)$$

provided either (a) or (b) holds.

(a) σ is regular; or

(b) $|\Gamma| < \operatorname{cof}(\sigma)$.

The following is occasionally useful.

1.5. For any $\aleph_0 \leq \sigma \leq \infty$ and modules $\{W_i^j \mid i \in I,\ j \in J\}$ define $A = \bigoplus_{j \in J} (\prod_{i \in I}^{\sigma} W_i^j)$ and $B = \prod_{i \in I}^{\sigma} (\bigoplus_{j \in J} W_i^j)$. Then there is a canonical inclusion $A \hookrightarrow B$. Moreover, $A = B$ iff J is finite.

2. SEMI-LATTICE FUNCTIONS

Some functions going from rings to semi-lattices are defined. Some of these later will be made into functors.

2.1. Module Types. Let σ be a fixed infinite cardinal or let $\sigma = \infty$. The latter could be interpreted that σ is arbitrary. And let $A, B, C, \ldots$ denote arbitrary right R-modules. On the class of right R-modules, first define a quasi-order $\preceq_{\sigma_1}$ by $A \preceq_{\sigma_1} B$ if there exists some integer m and index sets $J(1), J(2), \ldots, J(m)$ (usually all depending on A and B) such there exists an embedding

$$A \hookrightarrow \prod_{J(m)}{}^{\sigma} \Big[\prod_{J(m-1)}{}^{\sigma} \cdots (\prod_{J(2)}{}^{\sigma} (\prod_{J(1)}{}^{\sigma} B)) \cdots \Big] .$$

Although without loss of generality we can take all index sets to be the same $J = \sup\{|J(i)| \; |1 \leq i \leq m\}$, this would not simplify anything. Then there is an equivalence relation $\sim_{\sigma 1}$ on the class of all R-modules where $A \sim_{\sigma 1} B$ iff $A \preceq_{\sigma 1} B$ and $B \preceq_{\sigma 1} A$. An equivalence class $[A]_{\sigma 1} = \{C \mid C \sim_{\sigma 1} A\}$ will be called a *type* because it consists of modules that are in some respects similar, or of the same type. The class $G^{\sigma 1}(R) = \{[A]^{\sigma 1}, [B]^{\sigma 1}, [C]^{\sigma 1}, \ldots\}$ of all types is partially ordered by $[A]^{\sigma 1} \leq [B]^{\sigma 1}$ iff $A \preceq_{\sigma 1} B$. Thus $[A]^{\sigma 1} < [B]^{\sigma 1}$ if $[A]^{\sigma 1} \leq [B]^{\sigma 1}$ but $[A]^{\sigma 1} \neq [B]^{\sigma 1}$. Also $0 = [0]^{\sigma 1} \leq [A]^{\sigma 1}$ for all A.

Secondly, define $\preceq_{\sigma 2}$ by $A \preceq_{\sigma 2} B$ if for some integer m, there exis and index set J, there exists an embedding of $\widehat{A}$ into an m-fold iterated σ-product of $\widehat{B}$, i.e. $\widehat{A} \hookrightarrow \prod_J{}^{\sigma} [\prod_J{}^{\sigma} (\cdots \prod_J{}^{\sigma} \widehat{B} \cdots)]$. Note that for $\sigma = \infty$ or σ a regular cardinal the latter iterated product can be replaced by $\widehat{A} \hookrightarrow \prod_I{}^{\sigma} \widehat{B}$, where $I = J^m = J \times J \cdots \times J$ by the Fubini theorem. Then $\sim_{\sigma 2}$, $[A]^{\sigma 2}$, and $G^{\sigma 2}(R)$ are defined similarly.

Thirdly, define $A \preceq_{\sigma 3} B$ if there exist three things depending on A and B, an integer $1 \leq m$, index sets $J(1), J(2), \ldots, J(m)$, and an embedding $A \hookrightarrow E \prod_{J(m)}{}^{\sigma} E\{ \prod_{J(m-1)}{}^{\sigma} \ldots E[\prod_{J(2)}{}^{\sigma} (E \prod_{J(1)}{}^{\sigma} B)] \ldots\}$. Again, $\sim_{\sigma 3}$, $[A]^{\sigma 3}$, and $G^{\sigma 3}(R)$ are defined as in the preceding cases.

For any $i = 1, 2, 3$, the equivalence classes of torsion and torsion free modules define two subclasses $G_T^{\sigma i}(R) = \{[A]^{\sigma i} \mid A = Z_2 A\}$ and $G_F^{\sigma i}(R) = \{[A]^{\sigma i} \mid Z_2 A = 0\} = \{[A]^{\sigma i} \mid ZA = 0\}$. They will be called the *torsion* and *torsion free types* respectively. Whenever the greatest lower or least upper bound of $[A]^{\sigma i}, [B]^{\sigma i} \in G^{\sigma i}(R)$ exist in $G^{\sigma i}(R)$, then they are denoted by $[A]^{\sigma i} \wedge [B]^{\sigma i}$ and $[A]^{\sigma i} \vee [B]^{\sigma i}$, and infinite imfima and suprema of subclasses (but mostly subsets) are denoted the same way.

The case $\sigma = \aleph_0$ has been considered before ([11]) and for $i = 3$, extensively ([6] and [9]) and so far has been the most useful. It is also refered to as the finite case for reasons which may become clearer from the following discussion.

2.2. The cases $\sigma = \aleph_0$ or ∞. In this case, we abbreviate "$\aleph_0$" by "O" and write $\preceq_{Oi} = \preceq_{\aleph_0 i}$, $[A]^{Oi} = [A]^{\aleph_0 i} \in G^{Oi}(R) = G^{\aleph_0 i}(R)$, and similarly $G_T^{Oi}(R)$, $G_F^{Oi}(R)$. We summarize what $A \preceq_{\sigma i} B$ means depending on i; there exists some index set depending on A and B for which there exist embeddings

$\sigma = \aleph_0 = 0$	$\sigma = \infty$
(1) $A \hookrightarrow \bigoplus_J B$;	(1) $A \hookrightarrow \prod_J B$
(2) $\widehat{A} \hookrightarrow \bigoplus_J \widehat{B}$;	(2) $\widehat{A} \hookrightarrow \prod_J \widehat{B}$
(3) $A \hookrightarrow E(\bigoplus_J B)$	(3) $\widehat{A} \hookrightarrow \prod_J \widehat{B}$

Note that for $\sigma = \aleph_0$, the iterated application of E's and $\prod_J^{\aleph_0}$'s just reduces to (3). Although for $\sigma = \infty$, $G^{\infty 2}(R) = G^{\infty 3}(R)$ collapse, notationally it will be simpler to allow either $\sigma = \infty$, $i = 2$ or $\sigma = \infty$, $i = 3$ even though they mean the same thing.

2.3. Order Preserving Morphisms. For any $\aleph_0 \leq \sigma \leq \infty$ and any i, j with $1 \leq i \leq j \leq 3$ exclude the case $\aleph_1 \leq \sigma < \infty$, and $(i,j) = (1,2)$. Thus σ, i, j are allowed to be

for $\sigma = \aleph_0 = 0$, or ∞, $1 \leq i \leq j \leq 3$; but
for $\aleph_1 \leq \sigma < \infty$, $1 \leq i \leq 3 = j$ or, $2 \leq i \leq j \leq 3$.

For these values we have that

$$A \preceq_{\sigma i} B \Longrightarrow A \preceq_{\sigma j} B;$$
$$[A]^{\sigma i} \leq [B]^{\sigma i} \Longrightarrow [A]^{\sigma j} \leq [B]^{\sigma j}; \text{ hence}$$
$$[A]^{\sigma i} = [B]^{\sigma i} \Longrightarrow [A]^{\sigma j} = [B]^{\sigma j}$$

and consequently any σ, j equivalence class $[B]^{\sigma j}$ is partitioned into a disjoint union of full σ, i equivalence classes, i.e. $[B]^{\sigma j} = \bigcup_A \{[A]^{\sigma i} \mid [A]^{\sigma j} = [B]^{\sigma j}\}$. Note that $\{[A]^{\sigma i} \mid A \text{ is an } R\text{-module such that } [A]^{\sigma j} = [B]^{\sigma j}\}$ need not be a set but may be a proper class. Hence there is an order preserving surjective function $\eta_i^{\sigma j}(R) : G^{\sigma i}(R) \longrightarrow G^{\sigma j}(R)$ defined by $\eta_i^{\sigma j}(R)[A]^{\sigma i} = [A]^{\sigma j}$ where $\eta_i^{\sigma i}(R) = 1$ is the identity. When R is fixed or understood, we abbreviate $\eta_i^{\sigma j}(R) = \eta_i^{\sigma j}$.

One objective of an article such as this is to bring to the readers attention questions which might have useful and/or interesting answers, such as the following. For what R, i, j, and $\sigma \leq \tau$ is it the case that for all A (or A restricted to some class of R-modules, e.g. t.f.) that $[A]^{\sigma i} = [A]^{\tau j}$? Or, for what R, σ, and i, is $G^{\sigma i}(R)$ a set? For the sake of completeness, and also to provide some notation for discussing here the answer in a few special cases, but also to provide a framework for possible future solutions of these problems, we define some fairly obvious maps of po-classes next.

2.4. For any $\aleph_0 \leq \sigma < \tau \leq \infty$, any j, and all R-modules A and B

$$A \preceq_{\sigma j} B \Longrightarrow A \preceq_{\tau j} B;$$
$$[A]^{\sigma j} \leq [B]^{\sigma i} \Longrightarrow [A]^{\tau j} \leq [B]^{\tau j}; \text{ and}$$
$$\mathcal{I}_{\sigma}^{\tau j} : G^{\sigma j}(R) \longrightarrow G^{\tau j}(R),\ \mathcal{I}_{\sigma}^{\tau j}[A]^{\sigma j} = [B]^{\tau i}$$

is a well defined order preserving function of po-classes. First, if $\sigma, i \leq j$ satisfy 2.3 and $\sigma < \tau$, then the composition $\mathcal{I}_{\sigma}^{\tau j}\eta_i^{\sigma j} : G^{\sigma i}(R) \longrightarrow G^{\tau j}(R)$ is well defined and order preserving, or for $\tau < \sigma$, $\eta_i^{\sigma j}\mathcal{I}_{\tau}^{\sigma i}G^{\tau i}(R) \longrightarrow G^{\sigma j}(R)$. It is not clear what are the kernels of these functions $\eta_i^{\sigma j}$, $\mathcal{I}_{\sigma}^{\tau j}$, or their composites. In general, they would be po-class congruences (semi-lattice, or lattice congruences).

The construction below is done only for torsion and for t.f. modules, but could be done for other classes of modules. For t.f. modules with $\sigma = \aleph_0$ and $i = 3$, it has been used frequently ([6], [9], and [10]).

2.5. For $\aleph_0 \leq \sigma \leq \infty$, and $i = 1, 2, 3$ suppose that all the definitions in 2.1-2.4 remain exactly the same, except that only torsion free modules $A, B, C, \ldots$ are allowed. Then define ${}_F[A]^{\sigma i}$ to be the resulting equivalence class and ${}_FG^{\sigma i}(R) = \{{}_F[A]^{\sigma i},\ {}_F[B]^{\sigma i},\ {}_F[C]^{\sigma i},\ \ldots\}$.

An analogous definition can now be made if instead of t.f. now only torsion modules $A, B, C, \ldots$ only are used, resulting in ${}_T[A]^{\sigma i}$ and ${}_TG^{\sigma i}(R)$. However note that if B is t.f., then so is $\prod_J^{\sigma} B$, but that the corresponding statement for torsion modules in general fails. Thus for $\aleph_1 \leq \sigma$, even if $A = Z_2A$, ${}_T[A]^{\sigma i} \subset [A]^{\sigma i}$, where the inclusion in general is proper because $[A]^{\sigma i}$ will contain t.f. modules.

We next discuss the case $R = \mathbf{Z}$ as a possible model which with suitable modifications might be applicable to other rings. There are even unanwered questions about $G^{\sigma 1}(\mathbf{Z})$. E.g. what is the partial order structure of $G^{\sigma 1}(\mathbf{Z})$ or $G_F^{\sigma 1}(\mathbf{Z})$? The complexity of the case of the supposedly known ring $R = \mathbf{Z}$ should impress upon the reader the value of a subsequent theorem (5.4), which will explicitly identify $G_F^{O3}(R) \cong_F G^{O3}(R)$ as a certain lattice of unique ideals of R. Among other things, 2.6 below shows that $G_T^{O1}(\mathbf{Z})$ is not a set.

2.6. For $R = \mathbf{Z} = 0, \pm 1, \pm 2, \ldots$ and a prime p, and any cardinal $\aleph_0 \leq \sigma \leq \infty$, let A and B be abelian p-groups, with B reduced. For any ordinal μ, $Ap^{\mu+1} = (Ap^{\mu})p$, where $Ap^{\omega} = \bigcap\{Ap^n \mid n = 0, 1, \ldots\}$, and where for a limit ordinal μ, $Ap^{\mu} = \cap\{Ap^{\nu} \mid \nu < \mu\}$. The p-length of A is the smallest ordinal μ for which $Ap^{\mu+1} = Ap^{\mu}$. The latter is a divisible group, and $\mu \leq |A|$ exists. For $a \notin Ap^{\mu}$, the generalized ordinal p-height $h(a)$ of a is defined to be $h(a) = \nu$ if $a \in Ap^{\nu} \backslash Ap^{\nu+1}$. Three facts will be needed. (i) For any $\nu < \mu$, necessarily $Ap^{\nu} \backslash Ap^{\nu+1} \neq \emptyset$, and hence there exists an element $a \in A$ with $h(a) = \nu$. (ii) For any set J and any

$0 \neq b = (b_j) \in \prod_J^{\sigma} B$, $h(b) = \sup\{h(b_j) \mid j \in \text{supp } b\} \leq p - \text{length }(B)$. (iii) There exist p-groups of arbitrarily large p-length.

From (i), (ii), and (iii), we conclude that if $p-\text{length }(B) < p-\text{length }(A) = \mu$, that then there does not exist an embedding $A \hookrightarrow \prod_J^{\sigma} B$ no matter how large J is. Thus $[A]^{\sigma 1} \neq [B]^{\sigma 1} \in G_T^{\sigma 1}(\mathbf{Z})$. For any abelian p-group A, $[A]^{\sigma 1} \leq [\mathbf{Z}(p^{\infty})]^{\sigma 1}$. Consequently all three $\{[A]^{\sigma 1} \mid [A]^{\sigma 1} \leq [\mathbf{Z}(p^{\infty})]^{\sigma 1}\} \subset G_T^{\sigma 1}(\mathbf{Z}) \subset G^{\sigma 1}(\mathbf{Z})$ are not sets. Note that $G_T^{\sigma 1}(\mathbf{Z})$, $G_F^{\sigma 1}(\mathbf{Z})$, and $G^{\sigma 1}(\mathbf{Z})$ contain identity elements, i.e. unique largest elements $[\bigoplus_p \mathbf{Z}(p^{\infty})]^{\sigma 1}$, $[\mathbf{Q}]^{\sigma 1}$, and $[\mathbf{Q} \oplus \bigoplus_p \mathbf{Z}(p^{\infty})]^{\sigma 1}$, where $\mathbf{Q}$ are the rationals.

2.7. Now let $\sigma = \aleph_0$. Define $G_{Tp}^{O1}(\mathbf{Z}) = \{[A]^{O1} \mid A \text{ is a } p\text{-group}\}$. Let $\mathcal{O}$ be the class of ordinals and $\mathcal{O}^* = \mathcal{O} \cup \{\infty\}$ where "∞" is an element that is not an ordinal, but is defined to satisfy $\mu < \infty$ for all $\mu \in \mathcal{O}$. Then there is an order isomorphism $G_{Tp}^{O1}(\mathbf{Z}) \cong \mathcal{O}^*$. Let $\prod_p G_{Tp}^{O1}(\mathbf{Z})$ be partially ordered componentwise. Note that arbitrary subsets have sups and infs computed coordinatewise. If A_p denotes the p-th component of a torsion group A, then $[A]^{O1} \longrightarrow ([A_p]^{O1})_{p=2,3,\ldots}$ is an order preserving bijection $G_T^{O1}(\mathbf{Z}) \longrightarrow \prod_p G_{Tp}^{O1}(\mathbf{Z}) \cong \prod_p \mathcal{O}^*$. Note that $[\bigoplus_p \mathbf{Z}(p^{\infty})]^{O1} \longrightarrow ([\mathbf{Z}(p^{\infty})])_p \longleftrightarrow (\infty)_p \in \prod_p \mathcal{O}^*$.

In $G_F^{O1}(\mathbf{Z})$ the unique largest element is $[\mathbf{Q}]^{O1}$ where $\mathbf{Q}$ are the rationals. It is not the case that every element of $G_F^{O1}(\mathbf{Z})$ is of the form $[C]^{O1}$ where $C \leq \mathbf{Q}$ ([15; Vol.II, p.112, Ex.11]). It is not known what the exact order theoretic structure of $G_F^{O1}(\mathbf{Z})$ is.

2.8. Since injectivity is equivalent to divisibility, it is not difficult to prove that $G^{O3}(\mathbf{Z}) = G^{\sigma 3}(\mathbf{Z}) = G^{\infty 3}(\mathbf{Z})$ for all σ, and similarly for G_T^{*3} and G_F^{*3}.

The known theorem of J. Łoś that $\prod_1^{\infty} \mathbf{Z} \not\hookrightarrow \bigoplus_J \mathbf{Z}$ for all J ([DF2]) implies that $[\mathbf{Z}]^{O1} = [\bigoplus_J \mathbf{Z}]^{O1} \neq [\prod_1^{\infty} \mathbf{Z}]^{O1}$. But $[\mathbf{Z}]^{\infty 1} = [\bigoplus_J \mathbf{Z}]^{\infty 1} = [\prod_1^{\infty} \mathbf{Z}]^{\infty 1}$. Thus $G_F^{O1}(\mathbf{Z}) \neq G_F^{\infty 1}(\mathbf{Z})$, i.e. it is not the case that $[A]^{\infty 1} = [A]^{O1}$ for all A.

2.9. A cardinal κ is *ω − measurable* if there is a non-principal ultrafilter on κ, which is closed under countable intersections. (For definitions and properties of various kinds of cardinals, see [14; pp.19-20; pp.25-28].) The existence or nonexistence of an ω-measurable cardinal cannot be proved in ZFC. Let κ be the smallest ω-measurable cardinal if it exists. Then it is a consequence of some recent results in [12] that for any $\aleph_0 \leq \sigma < \tau < \kappa$, $G^{\sigma 1}(\mathbf{Z}) \neq G^{\tau 1}(\mathbf{Z})$. Moreover, in the latter $\mathbf{Z}$ may be replaced by any ring R provided that R_R is a slender right R-module, generally noncommutative with zero divisors ([13]).

2.10. In regard to $G_F^{\sigma 1} \neq G_F^{\tau 1}$ in 2.9, the following question arises. For what rings R and which (if any) $i = 2$ or 3 is it the case that for all sufficient large $\sigma < \tau$, $G_F^{\sigma i}(R) = G_F^{\tau i}(R)$? Here is a fact which lends some plausibility to such a conjecture. For any ring R if κ is the infinite Goldie dimension of R/Z_2R ([DF1]), then for any regular cardinal $\sigma > \kappa$, and any t.f. modules $\{A_\gamma \mid \gamma \in \Gamma\}$, $\prod_{\gamma\in\Gamma}^{\sigma} \widehat{A}_\gamma$ is injective ([4; p.12, 3.3]).

2.11. A related question is for what rings R for all sufficiently large regular cardinals σ, we have $G_F^{\sigma 2}(R) = G_F^{\sigma 3}(R)$? For $\sigma = \aleph_0$, it is a consequence of the results of [27] and [28] that $G_F^{O2}(R) = G_F^{O3}(R)$ if the Goldie dimension of R/Z_2R is finite.

3. SEMI-LATTICES

Here in this section some of the basic questions about $G^{\sigma i}(R)$ are considered. Is it a set? Is it complete; do arbitrary subsets have suprema? infima? Are $G_T^{\sigma i}(R)$, $G_F^{\sigma i}(R) \subset G^{\sigma i}(R)$ convex subclasses or subsets? When is $G_T^{\sigma i}(R) \cap G_F^{\sigma i}(R) = \{(0)\}$ and $G^{\sigma i}(R) = G_T^{\sigma i}(R) \oplus G_F^{\sigma i}(R)$ their direct sum?

So far the higher cardinal cases $\aleph_1 \leq \sigma$ and $\sigma = \aleph_0$, $i = 1$ or 2 have not been sufficiently studied. Most of the useful answers to the above questions are for $\sigma = \aleph_0$ and $i = 3$ because of the next lemma.

3.1. Lemma. Let X be a set of representatives of isomorphy classes of injective hulls of cyclic R-modules, $X \subset \{E(R/L) \mid L < R\}$. Then for any $\aleph_0 \leq \sigma \leq \infty$,

$$|G^{\sigma 3}(R)| \leq |G^{O3}(R)| \leq |\mathcal{P}(X)| \leq 2^{|\mathcal{P}(R)|}.$$

Proof. In view of the surjective function $\mathcal{I}_0^{\sigma 3} : G^{O3}(R) \longrightarrow G^{\sigma 3}(R)$, we have that $|G^{\sigma 3}(R)| \leq |G^{O3}(R)|$. For $[M]^{O3} \in G^{O3}(R)$, define $f([M]^{O3}) = X$ to be the set of representatives of the isomorphy classes of injective hulls of cyclic submodules of M (without repetitions). The rest of the proof that $f : G^{O3}(R) \longrightarrow \mathcal{P}(X)$ is a well defined function which is also one to one is omitted.

In the next lemma we already see one kind of a module theoretic application of the po-classes. The problem is to find a module M which represents the result of applying order theoretic constructions in $G^{\sigma i}(R)$. For example, given any set $\{[A_\gamma]^{O3} \mid \gamma \in \Gamma\} \subset G_F^{O3}(R)$, since the latter is known to be a complete Boolean lattice, $\inf\{[A_\gamma]^{O3} \mid \gamma \in \Gamma\}$ exists and is represented by a module M, $\wedge[A_\gamma]^{O3} = [M]^{O3} \in G_F^{O3}(R)$. If Γ is infinite, it is not known how to construct the module M out of the A_γ's. If $|\Gamma| = 2$, the answer is given by 3.2 (2) (a) in the next lemma.

3.2. Lemma. For any right R-modules A, B

(1) for any $\aleph_0 \le \sigma \le \infty$ and $i = 1, 2, 3 : [A]^{\sigma i} \vee [B]^{\sigma i} = [A \oplus B]^{\sigma i} \in G^{\sigma i}(R)$.

(2) For $\sigma = \aleph_0$, $i = 3$, and $[A]^{O3}, [B]^{O3} \in G_F^{O3}(R)$:

 (a) $[A]^{O3} \wedge [B]^{O3} = [\mathrm{tr}_{EB} A]^{O3} = [\mathrm{tr}_{EA} B]^{O3} \in G_F^{O3}(R)$.

 (b) $\exists\ !\ [A^C]^{O3} \in G_F^{O3}(R)$ subject to $[A^C]^{O3} \vee [A]^{O3} = 1 = [R/Z_2 R]^{O3} \in G_F^{O3}(R)$, and $[A]^{O3} \wedge [A^C]^{O3} = 0$; $A^C = \mathrm{rej}_B E(R/Z_2)$.

 (c) $[A]^{O3} + [B]^{O3} \equiv ([A]^{O3} \vee [B]^{O3}) \backslash ([A]^{O3} \wedge [B]^{O3})$
 $= [\mathrm{rej}_{EB} EA \oplus \mathrm{rej}_{EA} EB]^{O3}$.

Proof. Conclusion (1) is easy; for (2), see [6; p.59, 3.8 (i); p.60, 3.11, 3.12; and p.63, 3.18 (i)]

3.3. Lemma. For any $\aleph_0 \le \sigma \le \infty$, and $i = 1$, or 3, let $\{[A_\gamma]^{\sigma i} \mid \gamma \in \Gamma\} \subset G^{\sigma i}(R)$ be any subset. Then

$$\sup_{\gamma \in \Gamma}\{[A_\gamma]^{\sigma i}\} = \bigvee_{\gamma \in \Gamma} [A_\gamma]^{\sigma i} = [\bigoplus_{\gamma \in \Gamma} A_\gamma]^{\sigma i} \in G^{\sigma i}(R)$$

holds in the following cases: (1) (a), (b), (c), or (2) (a), (b).

(1) For $i = 1$,
 (a) if σ is regular; if
 (b) $\sigma = \infty$; or
 (c) $|\Gamma| < \mathrm{cof}(\sigma)$.

(2) For $i = 3$
 (a) if $\sigma = \aleph_0$; or
 (b) $\sigma = \infty$.

Proof. (1) Since $[\bigoplus_\Gamma A_\gamma]^{\sigma 1}$ is an upper bound of all the $[A_\gamma]^{\sigma 1}$, it suffices to show that it is smaller than any other upper bound $[D]^{\sigma 1}$. So let $[A_\gamma]^{\sigma 1} \le [D]^{\sigma 1} \in G^{\sigma 1}(R)$ for some module D. It suffices to show that then $[\bigoplus_\Gamma A_\gamma]^{\sigma 1} \le [D]^{\sigma 1}$. There exists a set $I(\gamma)$ and an embedding $A_\gamma \hookrightarrow \prod_{i \in I(\gamma)}^{\sigma} D_i$, where $D_i = D$. Set $I = \cup\{I(\gamma) \mid \gamma \in \Gamma\}$. Then

$$\bigoplus_{\Gamma} A_\gamma \subseteq \prod_{\gamma \in \Gamma}^{\sigma} A_\gamma \subseteq \prod_{i \in \Gamma}^{\sigma} (\prod_{i \subset I(\gamma)}^{\sigma} D_i).$$

Conclusion (b) now is easy, while (a) and (c) now follow from the Fubini theorem; (2) is omitted.

3.4. Corollary. If σ, Γ, and i satisfy one of (1) or (2), then

$$[\bigoplus_{\Gamma} A_\gamma]^{\sigma i} = [\prod_{\gamma \in \Gamma}^{\sigma} A_\gamma]^{\sigma i}.$$

Conjecture: $G_F^{O2}(R)$ is sup-complete. If the supremum of $\{[A_\gamma]^{O2}\}_{\gamma\in\Gamma} \subset G_F^{O2}(R)$ exists, then it will not be equal to $[\bigoplus_\Gamma A_\gamma]^{O2}$ in general. The author believes that if this conjecture is true, that then the proof will require some kind of module theoretic construction similar to the construction of finite infimum modules D in $G_F^{O3}(R)$, where $[A^{(1)}]^{O3} \wedge [A^{(2)}]^{O3} \wedge \cdots \wedge [A^{(n)}]^{O3} = [D]$. (See [11, 4.1 - 4.3].) Wether $G_T^{O2}(R)$ and $G^{O2}(R)$ are complete, presently is a totally open question. It seems that the techniques which would be required to settle the next question would have to be quite different from those appropriate for the last conjectures.

Question. For $\aleph_1 \leq \sigma < \infty$, are any of $G_F^{\sigma 3}(R)$, $G^{\sigma 3}(R)$, or $G_T^{\sigma 3}(R)$ complete?

The most manageable way to summarize the facts in Theorem 3.5 seems to be first to allow σ be arbitrary in $\aleph_0 \leq \sigma \leq \infty$, and then for $\sigma = \aleph_0$ to consider the three cases $i = 1, 2, 3$ separately, as corollaries because in the latter case a lot more can be said.

3.5. Theorem. For any ring R $(1 \in R)$, let $\aleph_0 \leq \sigma \leq \infty$, $i = 1, 2, 3$, and $G_T^{\sigma i}(R)$, $G_F^{\sigma i}(R) \subseteq G^{\sigma i}(R) = \{[A]^{\sigma i} \mid A \in \mathrm{Mod}_R\}$; and $\eta_i^{\sigma j} : G_F^{\sigma i}(R) \longrightarrow G_F^{\sigma j}(R)$ as in 2.1-2.4. Then the following hold.

(i) $G^{\sigma i}(R)$ is a join semi-lattice; $G_F^{\sigma i}(R) \subset G^{\sigma i}(R)$ is convex.

(ii) $G^{\sigma i}(R)$ is complete for $(\sigma, i) = (\aleph_0, 3)$, $(\sigma, i) = (\infty, 3)$; and for $i = 1$, if $\sigma = \infty$ or if σ is regular. For $i = 1$ and any $\sigma < \infty$, any subset of cardinality $<$ cof σ has a supremum in $G^{\sigma i}(R)$.

(iii) For any $1 \leq i \leq j \leq 3$, $\eta^{\sigma j} : G^{\sigma i}(R) \longrightarrow G^{\sigma j}(R)$ is a semi-lattice homomorphism and so is its restrictions and corestrictions $\eta_i^{\sigma j} : G_F^{\sigma i}(R) \longrightarrow G_F^{\sigma j}(R)$. In the cases 3.3 (1) or (2), $\eta_i^{\sigma j}$ preserves suprema of arbitrary subsets.

(iv) For any $\aleph_0 \leq \sigma \leq \infty$, $G^{\sigma 3}(R)$ is a set, hence it is a complete lattice if it is complete.

Proof. (i) By 3.2, least upper bounds of finite subsets exist in all $G^{\sigma i}(R)$. For any t.f. A and any $C \in [B]^{\sigma i} \leq [A]^{\sigma i}$, C can be embedded in an iterated injective σ-product of A's, which is torsion free. Hence $G_F^{\sigma i}(R)$ is a convex subclass of $G^{\sigma i}(R)$.

Conclusion (ii) follows by 3.3. Conclusion (iii) by 2.3. (iv) It will be shown more generally that if any $G^{\sigma i}(R)$ is 1) sup-complete, and 2) a locally small category, then it is inf-complete. Locally small means that the subobjects of any object form a set, i.e. for any $[A]^{\sigma i} \in G^{\sigma i}(R)$, $\{[B]^{\sigma i} \in G^{\sigma i}(R) \mid [B]^{\sigma i} \leq [A]^{\sigma i}\}$ is a set. For the subset $T = \{[A_\gamma]^{\sigma i} \mid \gamma \in \Gamma\} \subset G^{\sigma i}(R)$, let $S = \{[B]^{\sigma i} \in G^{\sigma i}(R) \mid \forall\, \gamma \in \Gamma,\ [B]^{\sigma i} \leq [A_\gamma]^{\sigma i}\}$ be the set of all lower bounds of T. Then $\inf T = \sup S = \bigwedge_{\gamma\in\Gamma} [A_\gamma]^{\sigma i} \in G^{\sigma i}(R)$ exists by 1). Since by 3.1 all the $G^{\sigma 3}(R)$ are sets, they are complete lattices whenever they are complete.

The next four corollaries are in [11] and their proofs are omitted.

3.6. Corollary 1. For $\sigma = \aleph_0$ and $i = 1$, the following hold.

(i) $G^{O1}(R)$ is a complete semi-lattice;

(ii) $G_T^{O1}(R)$, $G_F^{O1}(R) \subset G^{O1}(R)$ are convex and complete semi-sublattices.

(iii) $G_T^{O1}(R) \oplus G_F^{O1}(R) \subseteq G^{O1}(R)$ with the componentwise order on the direct sum.

3.7. Corollary 2. For $\sigma = \aleph_0$, $i = 2$:

(i) $G^{O2}(R)$ is a semi-lattice.

(ii) $G^{O2}(R) = G_T^{O2}(R) \oplus G_F^{O2}$ is a direct sum of convex semi-sublattices.

3.8. Main Corollary 3. With the notation and hypotheses of the last theorem for $\sigma = \aleph_0$ and $i = 3$,

(i) $G^{O3}(R)$ is a set.

(ii) $G^{O3}(R)$ is a complete lattice with largest element $1 = \vee G^{O3}(R) \in G^{O3}(R)$.

(iii) $G^{O3}(R) = G_T^{O3}(R) \oplus G^{O3}(R)$ is a lattice direct sum of convex and complete sublattices $G_T^{O3}(R)$, $G_F^{O3}(R) \subset G^{O3}(R)$.

One can go from one corollary forwards to any one after it via the η_i^{Oj} as explained in the next corollary.

3.9. Corollary 4. For any $1 \leq i \leq j \leq 3$, $\eta_i^{Oj} : G^{Oi}(R) \longrightarrow G^{Oj}(R)$ is a semi-lattice homomorphism and so are also its restrictions and corestrictions $\eta_i^{Oj} : G_T^{Oi}(R) \longrightarrow G_T^{Oj}(R)$, $G_F^{Oi}(R) \longrightarrow G_F^{Oj}(R)$. Moreover, η_1^{O3} preverves arbitrary suprema and so also do the restrictions and corestrictions $\eta_1^{O3} \mid G_T^{O3}(R)$ and $\eta_1^{O3} \mid G_F^{O3}(R)$.

The functor ${}_FG^{O3}$ (sometimes written ${}_FG^{O3} = \Xi$) has been treated in [6], [9], and to a lesser extent in [11]. Also functors naturally isomorphic to ${}_FG^{O3}$ have been used in [8] (as ${}_FG^{O3} \cong \mathcal{I}$) and in [10] (as ${}_FG^{O3} \cong \Sigma$). The next corollary says in particular that the present theory of the $G^{\sigma i}$ also applies to Ξ, $\mathcal{I}$, and Σ.

Note that in general for $\aleph_1 \leq \sigma$, ${}_TG^{\sigma i}(R)$ need not be a subclass of $G^{\sigma i}(R)$.

3.10. Corollary 5. For any $\aleph_0 \leq \sigma \leq \infty$ and any $i = 1, 2, 3$ in the notation of 2.5, let ${}_F[A]^{\sigma i}$ be the equivalence class of the t.f. module A (and where the previous definitions remain verbation the same, except that in all equivalence and quasi-order relations only torsion free modules $A, B, \ldots$ only are allowed), and ${}_FG^{\sigma i}(R) = \{{}_F[A]^{\sigma i}, {}_F[B]^{\sigma i}, \ldots\}$. Then

(i) ${}_F[A]^{\sigma i} = [A]^{\sigma i}$ for all t.f. A.

(ii) ${}_FG^{\sigma i}(R) \longrightarrow G_F^{\sigma i}(R)$, ${}_F[A]^{\sigma i} \longrightarrow [A]^{\sigma i}$ is an isomorphism of po-classes.

If next only torsion modules $A, B, \ldots$ are used, then

(iii) ${}_T[A]^{\sigma i} \subseteq [A]^{\sigma i}$;

(iv) ${}_TG^{\sigma i}(R) \longrightarrow G_T^{\sigma i}(R)$, ${}_T[A]^{\sigma i} \longrightarrow [A]^{\sigma i}$ is a semi-lattice homomorphism.

(v) For $\sigma = \aleph_0$ and all $i = 1,2,3$ the inclusion in (iii) is an equality and the map in (iv) an isomorphism. In particular, ${}_TG^{Oi}(R) \subset G^{Oi}(R)$ canonically.

Proof. (i) and (iii). In general, if an equivalence relation on the class of all modules is reduced by considering only modules belonging to a subclass, such as torsion or t.f. modules, then equivalence classes are decreased, e.g. always ${}_T[A]^{\sigma i} \subseteq [A]^{\sigma i}$, and ${}_F[A]^{\sigma i} \subseteq [A]^{\sigma i}$. The proof of 3.5 (i) also can be used to show that the latter is an equality, and hence (ii) follows immediately.

(iv) Let A be torsion and B, C arbitrary (not assumed torsion) modules such that $V \in {}_T[A]^{\sigma i} \cap [B]^{\sigma i} \neq \emptyset$ and ${}_T[A]^{\sigma i} \cap [C]^{\sigma i} \neq \emptyset$. Then ${}_T[V]^{\sigma i} =_T [A]^{\sigma i}$, and hence $A \in {}_T[V]^{\sigma i} \subseteq [V]^{\sigma i} = [B]^{\sigma i}$. Thus $[B]^{\sigma i} = [A]^{\sigma i} = [C]^{\sigma i}$. Consequently the function in (iv) is a well defined semi-lattice homomorphism.

(v) Let $i = 3$, the other cases $i = 1$ or 2 are similar and are omitted. For $A = Z_2A$ torsion, let $B \in [A]^{O3}$ arbitrary. Then $B \hookrightarrow E(\bigoplus_I A)$ for some I. Since $Z_2(\bigoplus_I \ldots) = \bigoplus_I (Z_2 \ldots)$, and $Z_2E = EZ_2$ (see [4; pp.2-3, 1.2 (9)]), we have $B \subset E(\bigoplus_I A) = E(\bigoplus_I Z_2A) = Z_2E(\bigoplus_I A)$. The latter has two consequences; first $B = Z_2B$ is torsion, and secondly $B \in {}_T[A]^{O3}$. Thus $[A]^{O3} \subseteq {}_T[A]^{O3}$ and hence they are equal by 3.10 (iii). Consequently the map in (iv) is the corestriction of the natural inclusion ${}_TG^{O3}(R) \subset G^{O3}(R)$.

4. FUNCTORS

Categories $\mathcal{A}^* \subset \mathcal{A} \subset \mathcal{C}$ and $\mathcal{B} \subset \mathcal{D}$ are defined so that $G^{\sigma i}, G_T^{\sigma i} : \mathcal{A} \longrightarrow \mathcal{D}$, and $G_F^{\sigma i} : \mathcal{A}^* \longrightarrow \mathcal{B}$ are forced to become functors for $i = 1$ or 3. Here any category such as $\mathcal{A}$ is veiwed as a disjoint union $\mathcal{A} = \mathcal{O}b\mathcal{A} \cup \text{Morph } \mathcal{A}$ of its objects and morphisms.

4.1. Categories. Let $\mathcal{D}$ be the category of sup semi-lattices with smallest elements 0; morphisms are zero preserving semi-lattice homomorphisms which are one to one. Then define $\mathcal{B} \subset \mathcal{D}$ to be the subcategory satisfying the following additional properties. The objects of $\mathcal{B}$ are sup-complete semi-lattices, and morphisms are monic zero preserving semi-lattice homomorphisms which are complete, i.e. preserve suprema of arbitrary subsets whenever they exist ([19; p.39]). Next, $\mathcal{B}^* \subset \mathcal{B}$ denotes the category of complete Boolean lattices (equivalently, Boolean rings) and zero preserving lattice homomorphisms (equivalently, ring homomorphisms) with convex images.

The category of all associative rings with identity and all ring homomorphisms is denoted by $\mathcal{C}$. Then let $\mathcal{A}$ be the category whose objects are also all rings with identity but whose morphisms are identity preserving surjective ring homomorphisms. Next, $\mathcal{A}^* \subset \mathcal{A}$ denotes the subcategory having the same objects as $\mathcal{A}$

but with fewer morphisms. Only those morphisms $\varphi \in \mathcal{A}$ whose kernels $\varphi^{-1}0$ are closed right ideals belong to $\mathcal{A}^*$.

4.2. Notation. For ease of exposition, for a typical $\varphi : R \longrightarrow S$ in $\mathcal{A}$, throughout set $\ker\varphi = I \triangleleft R$ and assume that $\varphi : R \longrightarrow R/I = S$ is the natural projection. Denote right singular submodules and injective hulls with respect to S by Z^S, Z_2^S, and E_S. Throughout for $f : R \longrightarrow T$ any ring homomorphism $f \in \mathcal{C}$ and any T-module P, P_f denotes the induced R-module. However, throughout this section, N will be a right S-module (notation $N = N_S$); N_φ the induced R-module. Note that S-modules can be regarded as those R-modules N such that $NI = 0$. We do not use "E_R" but "E" or "$\widehat{\ }$" for R. Note that $(EN)_\varphi$ and $(ZN)_\varphi$ are meaningless, so define $EN_\varphi = E(N_\varphi)$ and $ZN_\varphi = Z(N_\varphi)$, and similarly for Z_2.

For any right S-module $N, \mathcal{L}_S(N)$ denotes the set of large S-submodules of N; and similarly for a right R-module $M = M_R$ and $\mathcal{L}_R(M)$.

For any $\aleph_0 \le \sigma \le \infty$ and $i = 1, 2$, or 3, the elements of $G^{\sigma i}(S)$ will be written as $[N_S]^{\sigma i} = [N]_S^{\sigma i}$; however for R, we do not use $[\quad]_R^{\sigma i}$ but $[N_\varphi]^{\sigma i} \in G^{\sigma i}(R)$ instead.

The essentials required to go back and forth from S to R-modules are summarized below.

4.3. Facts. Any right S-modules, P, Q, and N satisfy the following.

(1) $\mathrm{Hom}_R(P_\varphi, Q_\varphi) = \mathrm{Hom}_S(P, Q)$. In particular, $P \cong Q$ as S-modules $\iff$ $P_\varphi \cong Q_\varphi$.

(2) $\mathcal{L}_R(N_\varphi) = \mathcal{L}_S(N)$. Consequently, the

(3) essential R-submodules coincide with the essential S-submodules of N. Hence

(4) right R-complement submodules of N_φ coincide with the right S- complements of N.

(5) $N_\varphi << (E_S N)_\varphi$ is a consequence of (3). Thus there exists an inclusion $N_\varphi << (E_S N)_\varphi \le EN_\varphi$. Moreover there is a canonical inclusion because

(6) $E_S N = \{x \in EN_\varphi \mid xI = 0\}$; lastly

(7) $(E_S N)\varphi$ is a quasi-injective R-module by (1).

In order to be able to go from S-torsion modules to R-torsion modules, our ring homomorphism has to be restricted to $\mathcal{A}$ whereas to go from S - t.f. to R-t.f. it has to be even further restricted to $\mathcal{A}^*$ in (4) below.

4.4. Proposition. Let $S, \varphi : R \longrightarrow S = R/I$, N, N_φ, Z^S, and Z_2^S be as in 4.2. Then the following hold:

(1) $Z^S N \subseteq Z(N_\varphi)$.

(2) $Z_2^S N \subseteq Z_2(N_\varphi)$.

(3) $\forall\, P = P_S$, $Q = Q_S$, $Z^S P = P \Longrightarrow ZP_\varphi = P$; $Z_2^S Q = Q \Longrightarrow Z_2 Q_\varphi = Q_\varphi$.

(4) When $I < R$ is a right complement (i.e. $\varphi \in \mathcal{A}^*$), then $\forall\, N = N_S$, $Z^S N = ZN_\varphi$, $Z_2^S N = Z_2 N_\varphi$. In particular $Z^S(R/I) = Z(R/I)$. Moreover $\forall\, P = P_S$, P_S is S-t.f. $\Longrightarrow P_\varphi$ is R-t.f.

In the next lemma in (i) below, at this time the author has been unable to prove the converse (i) $\Longleftarrow$ which would have some valuable consequences. However, the converse does hold for $\sigma = \aleph_0$ or ∞; and we will be able to prove much more in these important special cases.

4.5. Lemma. For $\aleph_0 \le \sigma \le \infty$, $i = 1$ or 3, and for $\varphi : R \longrightarrow S$ in $\mathcal{A}$, for any right S-modules P, Q, and N the following hold:

(i) $[P]_S^{\sigma i} \le [Q]_S^{\sigma i} \Longrightarrow [P_\varphi]^{\sigma i} \le [Q_\varphi]^{\sigma i}$;

(ii) the assignment $\varphi^{\sigma i}([N]_S^{\sigma i}) = [N_\varphi]^{\sigma i}$ is a well defined order preserving function $\varphi^{\sigma i} : G^{\sigma i}(S) \longrightarrow G^{\sigma i}(R)$.

Proof. For $i = 1$, for any ring homomorphism φ in $\mathcal{C}$, both (i) and (ii) hold, as well as the converse $\Longleftarrow$ of (i). So let $i = 3$, and $[P]_S^{\sigma i} \le [Q]_S^{\sigma 3}$. Thus for some integer m, and index sets $\Gamma(1), \ldots, \Gamma(m)$, P can be embedded in an iterated S-injective σ-product $P \subset T_S^\sigma$ where

$$P \subset E_S\Big\{ \prod_{\Gamma(m)}^{\sigma} [E_S \prod_{\Gamma(m-1)}^{\sigma} (\cdots \prod_{\Gamma(2)}^{\sigma} (E_S \prod_{\Gamma(1)}^{\sigma} E_S Q) \cdots)] \Big\} \equiv T_S^\sigma .$$

Above in T_S^σ replace "Q" with Q_φ and all the $m+1$ occurences of "E_S" by "E"; call the resulting R-module T^σ. We next prove that $P \subset T_S^\sigma \subset T^\sigma$. Since $E_S Q \subset E Q_\varphi$, above we may first replace the outermost far left "E_S" to the left of $\Gamma(m)$ by "E" thereby obtaining a larger R-module which contains the R-modules $P \subset T_S^\sigma$. Then repeat this procedure for the next "E_S" before $\Gamma(m-1)$, etc., until $P \subset T_S^\sigma \subseteq T^\sigma$. Hence $[P]^{\sigma 3} \le [Q]^{\sigma 3}$.

(ii) If $P \sim_{\sigma 3} Q$ as S-modules, then $[P]_S^{\sigma 3} = [Q]_S^{\sigma 3}$. Then by (i), $[P_\varphi]^{\sigma 3} \le [Q_\varphi]^{\sigma 3}$ and $[Q_\varphi]^{\sigma 3} \le [P_\varphi]^{\sigma 3}$, and hence also $[P_\varphi]^{\sigma 3} = [Q_\varphi]^{\sigma 3}$. Thus $\varphi^{\sigma 3}$ is well defined and order preserving by (i). The above proof also works for $\sigma = \infty$, just erase "σ" everywhere above.

Note that logically (i) below can be satisfied by some particular, P and Q with $[P]_S^{O3} < [Q]_S^{O3}$ but $[P_\varphi]^{O3} = [Q_\varphi]^{O3}$.

4.6. Proposition. For $i = 1, \sigma$ regular or ∞, and for $i = 3$ but $\sigma = \aleph_0$ or ∞, for $\varphi \in \mathcal{A}$, P, Q, and N as above in the last lemma,

(i) $[P]_S^{\sigma i} \le [Q]_S^{i} \iff [P_\varphi]^{\sigma i} \le [Q_\varphi]^{i}$. Then $\varphi^{\sigma i}([N]_S^{i}) = [N_\varphi]^{\sigma i}$ is a well defined order preserving function $\varphi^{i} : G^{\sigma i}(S) \longrightarrow G^{\sigma i}(R)$ which satisfies the following

(ii) $\varphi^{\sigma i}$ is one to one;

(iii) $\varphi^{\sigma i} G^{\sigma i}(S) \subset G^{\sigma i}(R)$ is convex.

Proof. The case $i = 1$ is clear. Let $i = 3$. (i) $\Longleftarrow$. If $[P_\varphi]^{O3} \le [Q_\varphi]^{O3}$, then $P_\varphi \subset E(\bigoplus_\Gamma Q_\varphi)$ for some Γ. Since $PI = 0$, 4.3 (5) and (6) imply that $P \subset E_S(\bigoplus_\Gamma Q) = \{\xi \in E(\bigoplus_\Gamma Q_\varphi) \mid \xi I = 0\} \subseteq E(\bigoplus_\Gamma Q_\varphi)$. Thus $[P]_S^{O3} \le [Q]_S^{03}$.

(ii) If $\varphi^{O3}[P]_S^{O3} = \varphi^{O3}[Q]_S^{O3}$, then $P_\varphi \subset E(\bigoplus_\Gamma Q_\varphi)$ and $Q_\varphi \subset E(\bigoplus_\Delta P_\varphi)$ for some sets Γ and Δ. Set $N = \bigoplus_\Gamma Q$. Thus $N_\varphi = (\bigoplus_\Gamma Q)_\varphi = \bigoplus_\Gamma Q_\varphi$. By 4.3 (5) and (6), there is a canonical embedding $E_S N = \{\xi \in EN_\varphi \mid \xi I = 0\} \subseteq EN_\varphi$. Since $PI = 0$, we have $P \subset E_S(\bigoplus_\Gamma Q) \subseteq E(\bigoplus_\Gamma Q_\varphi)$. Thus $[P]_S^{O3} \leq [Q]_S^{O3}$. By a similar argument, or by symmetry of hypotheses, we conclude also $[Q]_S^{O3} \leq [P]_S^{O3}$ and hence that $[P]_S^{O3} = [Q]_S^{O3}$. Thus φ^{O3} is one to one. The proof for $\sigma = \infty$ is omitted as well as that of (iii). The proof of (iii) for $\sigma = \aleph_0$ can be found in [11; 3.6].

4.7. Theorem. Let $\mathcal{A} \subset \mathcal{C}$ and $\mathcal{B} \subset \mathcal{D}$ be the ring and semi-lattice categories of 4.1. Let $\varphi : R \longrightarrow S$ be any surjective ring homomorphism of rings with identity. For any $\aleph_0 \leq \sigma \leq \infty$ and $i = 1$ or 3 let $G_T^{\sigma i}(R)$, $G_F^{\sigma i}(R) \subset G^{\sigma i}(R)$ and $\eta_i^{\sigma j}(R) : G^{\sigma i}(R) \longrightarrow G^{\sigma i}(R)$ for $i, j = 1$ or 3 only with $1 \leq i \leq j \leq 3$ be as in 2.1, 2.2, and 2.3. Define $G^{\sigma i}(\varphi) = \varphi^{\sigma i}$ as in 4.5. Then

(i) $G^{\sigma i} : \mathcal{A} \longrightarrow \mathcal{D}$ is a contravariant functor; $\varphi^{\sigma i} : G^{\sigma i}(S) \longrightarrow G^{\sigma i}(R)$ is a zero preserving semi-lattice homomorphism.

(ii) $\eta_j^{\sigma i} : G^{\sigma i} \longrightarrow G^{\sigma j}$ is a natural transformation of functors.

(iii) If either $i = 1$ and σ is regular or ∞, or if $i = 3$ but $\sigma = \aleph_0$ or ∞, then $G^{\sigma i} : \mathcal{A} \longrightarrow \mathcal{B}$, $\phi^{\sigma i}$ is monic, $\phi^{\sigma i} G^{\sigma i}(S) \subset G^{\sigma i}(R)$ is convex.

4.8. Corollary 1. For any $\aleph_0 \leq \sigma \leq \infty$ and $i = 1$, let $f : R \longrightarrow T$ in $\mathcal{C}$ be arbitrary, but let all the other definitions and notation be as before (for $f = \varphi$ in $\mathcal{A}$). In particular for $P_T \in \mathrm{Mod}_T$, $f^{\sigma 1}[P]_T^{\sigma 1} = [P_T]^{\sigma 1} \in G^{\sigma 1}(R)$, and $G^{\sigma 1}(f) = f^{\sigma 1}$. Then

(iv) $G^{\sigma 1} : \mathcal{C} \longrightarrow \mathcal{D}$ is a contravariant functor; conclusions 4.7 (ii) (and 4.7 (iii)) hold.

4.9. Corollary 2. Let $\mathcal{A}^* \subset \mathcal{A}$ be the subcategory in 4.1 and assume $\varphi : R \longrightarrow S$ in $\mathcal{A}^*$ is a surjective ring homomorphism where $\varphi^{-1}0 = I < R$ is a right complement. Then $G^{\sigma i} : \mathcal{A}^* \longrightarrow \mathcal{D}$ is a contravariant functor for $\aleph_0 \leq \sigma \leq \infty$ and $i = 1$ or 3. Moreover

(v) $\varphi^{\sigma i} G_T^{\sigma i}(S) \subseteq G_T^{\sigma i}(R)$ and $\varphi^{\sigma i} G_F^{\sigma i}(S) \subseteq G_F^{\sigma i}(R)$.

A detailed proof of the next corollary for the torsion free case ($G_F^{O3} \simeq_F G^{O3} \cong \Xi$) can be found in [6].

4.10. Main Corollary. Let $\sigma = \aleph_0$, $i = 3$, and $\mathcal{A}^*$, $\mathcal{B}^* \subset \mathcal{B}$ be as in 2.1. Let $\varphi : R \longrightarrow S$ in $\mathcal{A}^*$ be a surjective identity preserving ring homomorphism with kernel $\varphi < R$ a right complement. Then

(1) $G^{O3}(R)$ is a complete lattice with largest element $1 = \vee G^{O3}(R) \in G^{O3}(R)$.

(2) $G^{O3}(R) = G_T^{O3}(R) \oplus G_F^{O3}(R)$ is a lattice direct sum of convex (and complete) sublattices $G_T^{O3}(R)$, $G_F^{O3}(R) \subset G^{O3}(R)$.

(3) $\varphi^{O3}G^{O3}(S) \subseteq G^{O3}(R)$ is a convex and complete sublattice; $\varphi^{O3} : G^{O3}(R) \longrightarrow G^{O3}(R)$ is a zero preserving monic lattice homomorphism which preserves arbitrary infima and suprema. Its corestriction $\varphi^{O3} : G^{O3}(S) \longrightarrow \varphi^{O3}G^{O3}(S)$ is a lattice isomorphism; it and its inverse preserve arbitrary infima and suprema.

(4) $\varphi^{O3}G_T^{O3}(S) \subseteq G_T^{O3}(R)$ and $\varphi^{O3}G_F^{O3}(S) \subseteq G_F^{O3}(R)$.

(5) $G^{O3} : \mathcal{A}^* \longrightarrow \mathcal{B}$ and $G_F^{O3} : \mathcal{A}^* \longrightarrow \mathcal{B}^*$ are contravariant functors; $G^{O3} = G_T^{O3} \oplus G_F^{O3}$ is a direct sum of subfunctors of G^{O3}, i.e.

$$G^{O3}(R) = G_T^{O3}(R) \oplus G_F^{O3}(R) \quad \text{and}$$
$$G^{O3}(\varphi) = G_T^{O3}(\varphi) \oplus G_F^{O3}(\varphi)\ .$$

(6) $G_F^{O3}(R)$ is a complete Boolean lattice; $G_F^{O3}(\varphi) : G_F^{O3}(S) \longrightarrow G_F^{O3}(R)$ is a complete monic ring homomorphism of Boolean rings whose image is convex (and hence a direct summand).

5. CONCLUSION

Below we briefly indicate only one way how the functor G_F^{O3} is applied to (i) module structure, and to (ii) ring theory. There are many others (see [6, 7-11]). It would be much beyond the possible length of this article to even to attempt to give a complete survey some of the concrete applications of functors such as G_F^{O3}, or functors similar to it. In this section only G_F^{O3} will be used; for simplicity we relabel it as $\Xi = G_F^{O3}$ as in [6, 9].

5.1. Definition. A torsion free module A is *atomic* if $A \neq 0$, and if for any $x, y \in A\backslash\{0\}$, $\mathrm{Hom}_R(E(xR), E(yR)) \neq 0$. A t.f. uniform module is atomic. A (nonuniform) t.f. compressible module is atomic. A t.f. module P is *molecular* if every nonzero submodule of P contains an atomic module. A torsion free module Q is *bottomless* if Q contains no atomic submodules. A t.f. module D is discrete if D contains an essential direct sum of uniform submodules, while a t.f. module C is continuous if it contains no uniform submodules. Note that bottomless modules are always continuous. Let D, C, A, CA, and CB denote the following five classes of t.f. modules: discrete, continuous, molecular, continuous molecular, and bottomless.

5.2. DEFINITION. First let M be an injective torsion free module, and let N represent any arbitrary nonzero direct summand of M. Then M is said to be *directly finite* if N is not isomorphic to $N \oplus N$; and M is *abelian* if any two direct summands of M which are isomorphic are equal. The module M is of Type I if every N contains a nonzero abelian submodule. The module M is of Type III if every one of its direct summands N satisfies $N \cong N \oplus N$. Next, M is of Type II if for any N, N is not abelian but also every N contains a nonzero directly finite direct summand. Lastly, any injective or noninjective t.f. module M is defined to be directly finite, abelian, of Type I, II, or III if $\widehat{M}$ has the corresponding property. The symbols I, II, III will denote these classes.

Next, we will briefly describe how $\Xi(R)$ gives a generalization and extension of the previous classification of t.f. modules into Types I, II, and III.

5.3 NOTATION. Recall that $G_F^{O3} = \Xi$, and in view of 3.10, we will actually identify $G_F^{O3} =_F G^{O3} \equiv \Xi$. For simplicity, throughout this section write $[A]$ for $[A]^{O3} = [A] \in \Xi(R)$, where the R-modules A will all be torsion free.

For any class of modules, such as D or III, define functions Ξ_D, Ξ_{III} from rings to posets by $\Xi_D(R) = \{[M] \mid M = M_R \in D\}$, $\Xi_{III}(R) = \{[M] \mid M = M_R \in III\}$. Define similarly $\Xi_C, \Xi_A, \Xi_{CA}, \Xi_{CB}, \Xi_I, \Xi_{II}$, etc.

5.4. Recall the categories of rings $\mathcal{A}^*$ and complete Boolean lattices $\mathcal{B}^*$ defined in 4.1. For $\varphi \in \mathcal{A}^*$ without loss of generality we will continue to assume that $\varphi : R \longrightarrow S = R/I$ is the natural projection, where $I = \varphi^{-1}0 < R$ is a right complement. Furthermore for an S torsion free right S-module $N = N_S$ the notation of 4.2 will be continued for equivalence classes in $\Xi(S)$ by $[N]_S = [N_S] \in \Xi(S)$. As before in 4.2, N_φ denotes the induced R-module N. However define $\varphi^* = \varphi^{O3}$ in 4.5(ii), that is $\varphi^*[N]_S = [N_\varphi] \in \Xi(R)$.

5.5. DEFINITION. A nonempty class of Δ of torsion free modules over some ring R is a *saturated class* if Δ is closed under submodules, direct sums, injective hulls, and isomorphic copies. Note that $\Delta = \{(0)\}$ is the smallest and $\Delta =$ t.f.Mod$_R$, the class of all torsion free right R-modules, is the largest saturated class.

For any saturated class Δ and any t.f. R-module M, define a unique submodule $M_\Delta \leq M$ by $M_\Delta = \Sigma\{V \mid V \leq M,\ V \in \Delta\}$. It can be shown that $E(M_\Delta) = (EM)_\Delta$ and we abbreviate $(EM)_\Delta = EM_\Delta = \widehat{M}_\Delta$. It can also be shown that $M_\Delta \in \Delta$.

A *universal saturated class* is a function Δ on rings, where for every ring R with identity, $\Delta(R)$ is a saturated class of torsion free right R-modules such that for any S-torsion free $N = N_S$, if $N \in \Delta(S)$, then $N_\varphi \in \Delta(R)$. Two examples of universal saturated classes are $\{(0)\}$ and t.f.Mod$_R$. If R is fixed or understood we sometimes abbreviate $\Delta(R)$ by omitting "R" as $\Delta(R) = \Delta$, e.g. in $\{(0)\}, C, D, III$. Strictly speaking, we should write $C(R), D(R), III(R)$.

5.6. THEOREM. In the above notation for $\varphi : R \longrightarrow S \in \mathcal{A}^*$, $\Xi(R)$, $\Xi(S) \in \mathcal{B}^*$, define $\Xi(\varphi) = \varphi^*$. Then

(i) $\Xi^* : \mathcal{A}^* \longrightarrow \mathcal{B}^*$ is a contravariant functor; in particular,

(ii) $\varphi^* : \Xi(S) \longrightarrow \Xi(R)$ is a lattice homomorphism (equivalently ring homomorphism of induced Boolean rings) in $\mathcal{B}^*$.

(iii) For any R and any universal saturated class Δ, there exists an element $[\Delta(R)] \in \Xi(R)$ UNIQUE with respect to the property that

$$\Delta(R) = \bigcup \{x \mid x \in \Xi(R),\ x \leq [\Delta(R)]\} .$$

That is,

(iv) $\forall\ M = M_R,\ M \in \Delta(R) \Longleftrightarrow [M] \leq [\Delta(R)]$.

5.7. COROLLARY 1 TO THEOREM 5.6. With the notation of 5.1 and 5.2, the following are universal saturated classes $\{(0)\}$, t.f.Mod$_R, C, D, A, CA, CB$ and I, II, and III. In particular, for any one of these, such as III for example, there exists a unique largest element $[III] \in \Xi(R)$ such that for any right R-module M,

$$M \text{ is Type } III \Longleftrightarrow [M] \leq [III] .$$

Similar statements hold for all the other unique elements $0, 1, [C], [D] \in \Xi(R)$.

It is well known that any torsion free injective module M is a direct sum $M = M_I \oplus M_{II} \oplus M_{III}$ of unique fully invariant submodules $M_I \in I$, $M_{II} \in II$, $M_{III} \in III$. The next theorem shows how this is merely one instance of a general phenomenon, which can fully be understood only in terms of the functor Ξ (or some other functor naturally isomorphic to Ξ, as in [10]).

5.8. THEOREM. Suppose that for some fixed ring R, $\Delta(1), \ldots, \Delta(n)$ are a finite number of saturated classes of right R-modules. For any t.f. module M, let $M_{\Delta(i)} = \Sigma\{V \mid V \leq M,\ V \in \Delta(i)\}$ and $[\Delta(i)] \in \Xi(R)$ be as previously defined in 5.5 and 5.6 (iii), (iv). Then the following four conditions (1) (a), (b), (c), and (d) are all equivalent.

(1) (a) $\forall$ t.f. injective $M = M_R = \widehat{M},\ M = M_{\Delta(1)} \oplus \ldots \oplus M_{\Delta(n)}$.

(b) $\forall$ t.f. $M = M_R,\ M_{\Delta(1)} + \ldots + M_{\Delta(n)} = M_{\Delta(1)} \oplus \ldots \oplus M_{\Delta(n)} \ll M$; each $M_{\Delta(i)} < M$ is a complement submodule.

(c) $[\Delta(1)], \ldots, [\Delta(n)] \in \Xi(R)$ are pairwise orthogonal and their join is $1 = [R/Z_2R] \in \Xi(R)$. I.e.

$$[\Delta(i)] \wedge [\Delta(j)] = 0 \in \Xi(R),\ i \neq j$$
$$[\Delta(1)] \vee \cdots \vee [\Delta(n)] = 1 \in \Xi(R)$$

Alternatively and equivalently in the Boolean ring $\Xi(R)$ $1 = [\Delta(1)] + \cdots + [\Delta(n)]$, $[\Delta(i)]^2 = [\Delta(i)]$, and $[\Delta(i)][\Delta(j)] = 0$, for $i \neq j$.

(d) $\Xi(R)$ is a lattice direct sum of convex (and complete) sublattices $\Xi(R) = \Xi_{\Delta(1)}(R) \oplus \cdots \oplus \Xi_{\Delta(n)}(R)$, where $\Xi_{\Delta(i)}(R) = \{[M] \mid M \in \Delta(i)\} \triangleleft \Xi(R)$ (equivalently a ring direct sum of convex ideals).

Now in addition assume that $\Delta(1), \ldots, \Delta(n)$ satisfy (1) (a)-(d). Then

(2) $M_{\Delta(i)} < M$ is a fully invariant complement submodule, $i = 1, \ldots, n$.

(3) $M_{\Delta(1)}, \ldots, M_{\Delta(n)} \leq M$ are UNIQUE subject to (1)(b), i.e. if $M_i < M$, $i = 1, \ldots, n$, are complement submodules with $M_i \in \Delta(i)$ satisfying (1) (b), then $M_i = M_{\Delta(i)}$ for all i.

Now in the hypotheses of the theorem assume in addition that $\Delta_1, \ldots, \Delta_n$ are universal saturated classes and that for any or all R, $\Delta(i)$ is now defined as $\Delta(i) = \Delta_i(R)$. Then the above conditions (1) (a), (b), (c), (d) now assumed to hold for all rings R with $1 \in R$ are equivalent to (4) (e), where

(4) (e) $\Xi = \Xi_{\Delta_1} \oplus \ldots \oplus \Xi_{\Delta_n}$ is a direct sum subfunctors $\Xi_{\Delta_i} \leq \Xi$, where $\Xi_{\Delta_i}(R) \equiv \{[M] \in \Xi(R) \mid M \in \Delta_i(R)\}$.

5.9. APPLICATIONS OF THEOREM 5.8. We abbreviate $I(R) = I$, $C(R) = C$, etc. if R is understood as in 5.5. The following n, and $\Delta_1, \ldots, \Delta_n$ (with $\Delta(i) = \Delta_i(R)$) satisfy the conditions (1) (a)-(d), and 4 (e) of the previous theorem:

(i) $n = 3$, $\Delta_1 = I$, $\Delta_2 = II$, $\Delta_3 = III$. In particular in the Boolean ring $\Xi(R)$, $1 = [I] + [II] + [III] \in \Xi(R)$ is a direct sum of three mutually orthogonal idempotents; Ξ is a direct sum of subfunctors $\Xi = \Xi_I \oplus \Xi_{II} \oplus \Xi_{III}$ where $\Xi_{III}(R) = [III] \cdot \Xi(R)$; and $\Xi_{III}(R) = \{[M] \mid M = M_R \in III\}$; $\Xi(R) = \Xi_I(R) \oplus \Xi_{II}(R) \oplus \Xi_{III}(R)$ is a direct sum of convex and complete sublattices, (alternatively a ring direct sum of three ideals).

(ii) $n = 2$, $\Delta_1 = C$, $\Delta_2 = D$.

(iii) $n = 2$, $\Delta_1 = A$, $\Delta_2 = CB$.

(iv) $n = 3$, $\Delta_1 = CA$, $\Delta_2 = CB$, $\Delta_3 = D$.

In particular for any t.f. injective R-module M, $M = M_C \oplus M_D = M_{CA} \oplus M_{CB} \oplus M_D = M_A \oplus M_{CB}$, where $M_C = M_{CA} \oplus M_{CB}$ and $M_A = M_{CA} \oplus M_D$ as in 5.8 (1) (a) and 5.8 (2) and (3).

5.10. REMARKS. (i) The above shows that $\Xi(R)$ gives a genuine generalization of the Murray-Von Neumann-Kaplansky-Goodearl-Boyle ([17]) classification of t.f. injective modules into Types I, II, and III in several ways. (1) First of all, there are new classes of modules such as $n = 2, A, CB$ or $n = 3, CA, CB, D$ for which the same type of theory as in [17] holds. The classes A, CA, and CB are new, and, in fact, were discovered recently only with the aid of the functor Ξ in [6]. (2) Secondly, the modules no longer need be injective, and 5.8 (1)(b) gives the unique analogue for a direct sum decomposition which for t.f. noninjective modules simply is not available. Some of the early work on Types I, II, III assumed that the ring was regular, or nonsingular. Here of course R is arbitrary. (3) The actual definition of Types I, II, III presupposes that these are applicable to all R. However, the direct sum decomposition 5.8 (1)(a) and 5.8 (1)(b), (c), (d) can be applied more generally if R is fixed and we have classes Δ_i of modules which need only be defined for the one single ring R. (4) The last theorem shows that Types I, II, III have a functorial significance as in 5.8 (4)(e) and 5.9 (i). Indeed, the functor Ξ is needed in order to explain why Types I, II, III work so well. To the best of the author's knowledge, previous treatments of Types I, II, III have ignored their functorial aspects. (5) The added generality in 5.8 may possibly allow us to advance the existing theory of Types I, II, III further in new directions. For example, for any t.f. ring R with $ZR = 0$ an application of 5.8 (1)(b) to $M = R_R$ gives for any ring R with identity the existance of unique ideals $R_I, R_{II}, R_{III} \lhd R$ such that as right modules they are complement submodules of R with $R_I \in I$, $R_{II} \in II$, $R_{III} \in III$, and with $R_I \oplus R_{II} \oplus R_{III} << R$. For an extensive treatment of this, see [9] and also [8], or [6].

(ii) The last Theorem 5.8 is not the best possible, (1) (a), (b), and (c) are all equivalent even for an infinite number of pairwise orthogonal saturated classes whose supremum in the complete lattice $\Xi(R)$ is 1. (See [10] and [7]).

The key to applying these functors to ring theory is the fact that every associative ring with identity R contains a unique lattice of ideals $\mathcal{I}(R)$ which is isomorphic to $\Xi(R)$ as explained below (and in [6,8,9]).

5.11. THEOREM. For any ring R with identity, let $\Xi(R)$ be the class of all torsion free saturated classes of right R-modules. Let $\mathcal{I}(R)$ be the set of all right ideals $J \leq R$ such that (a) $Z_2R \subseteq J$, (b) $J \leq R$ is a right complement, and (c) $EJ \leq ER$ is a fully invariant right submodule of ER. Then

(i) $\forall \ J \in \mathcal{I}(R)$, $J \lhd R$, and there is a bijection

$$\Xi(R) \xrightarrow{f} \mathcal{I}(R) \xrightarrow{g} \Xi(R) ,$$
$$\Delta \longrightarrow J \longrightarrow [J] .$$
$$J/Z_2R = (R/Z_2R)_\Delta$$

(ii) In particular for any R with $1 \in R$, the ideal lattice of R contains a UNIQUE sublattice $\mathcal{I}(R)$ which is a complete Boolean lattice under the following operations:

$$\forall \ J, J_1, J_2 \in \mathcal{I}(R) \ \text{ and } \ \forall \ \text{ subset } \ \Gamma = \{J_\alpha\}_\alpha \subseteq \mathcal{I}(R)$$
$$J_1 \wedge J_2 = J_1 \cap J_2, \ J_1 \vee J_2 = (J_1 + J_2)^-; \ R = 1 \in \mathcal{I}(R), \ Z_2R = 0 \in \mathcal{I}(R);$$
$$\exists \,! \ J^c \in \mathcal{I}(R), \ J \wedge J^c = 0, \ J \vee J^c = 1$$
$$\inf \Gamma = \bigwedge_\alpha J_\alpha = \bigcap_\alpha J_\alpha; \ \sup \Gamma = \bigvee_\alpha J_\alpha = (\sum_\alpha J_\alpha)^- .$$

5.12. COROLLARY TO THEOREM 5.11. Under the above hypotheses and in the above notation, for any $J \in \mathcal{I}(R)$, let $C \leq R$ be any (nonunique) right ideal such that $J + C = J \oplus C << R$ the following hold.

(iii) $J^c = (ZR + C)^- \in \mathcal{I}(R)$.

(iv) f, g are lattice isomorphisms (equivalently ring isomorphisms of Boolean rings) which preserve complements, arbitrary infima and suprema, whose inverses g^{-1}, f^{-1} are lattice isomorphisms with the same properties.

(vi) $f^{-1} : \mathcal{I}(R) \longrightarrow \Sigma(R)$ is given by $f^{-1}J = \{M_R$ is t.f. $\mid \forall\ x \in M, \exists$ an embedding $E(xR) \subset E(J/Z_2R)\}$.

(vii) Under the additional hypothesis that $ZR = 0$, $f^{-1} : \Xi(R) \longrightarrow \mathcal{I}(R)$ is given by $f^{-1}([B]) = E(tr_{ER}B) \cap R$ for any $[B] \in \Xi(R)$.

Any Boolean ring R is naturally a poset whose completion Λ by Dedekind cuts ([2]) is isomorphic to its injective hull $\Lambda \cong \widehat{R}$, which in turn is isomorphic to the maximal ring of quotients of R. The following in 5.13 is established in [6; p.76, Theorem V] and [6; p.78, 6.10].

5.13. APPLICATIONS OF THEOREM 5.11. (1) Suppose that R is any Boolean ring and $R \subseteq \Lambda$ is its (unique) completion. Then $\Xi(R) = \Lambda$. In particular, for any complete Boolean ring R, $\Xi(R) = R$.

(2) Suppose that R is any subring $\bigoplus_{x \in X} S_x \subseteq R \subseteq \prod_{x \in X} S_x$ with $1 \in R$, where $\{S_x \mid x \in X\}$ are torsion free rings with identity. Then there is a natural isomorphism $\Xi(R) \cong \prod_{x \in X} \Xi(S_x)$. In particular, if all $S_x = \mathbf{Z}/2\mathbf{Z} = \{\overline{0}, \overline{1}\}$, then $\Xi(R) \cong \prod_{x \in X} \{\overline{0}, \overline{1}\}_x \cong \mathcal{P}(X)$.

5.14. REMARKS. (1) The elements of $\Xi(R)$ could be called unsaturated classes because they are not closed under submodules, whereas the elements of $\Sigma(R)$ are called saturated because they are closed under submodules. In the above treatment in 5.6, 5.8, 5.9 in particular we had to distinguish III and $[III]$, or C and $[C]$, where for any R, $[III] \in \Xi(R)$, while $III = \{M \mid M = M_R,\ [M] \leq [III] \in \Xi(R)\}$.

(2) Although $\Xi \cong \Sigma \cong \mathcal{I}$ are naturally isomorphic functors, they are not interchangeable, because logically or conceptually they are defined very differently, and consequently they each can be used in different ways. Thus for example by replacing "Ξ" by "Σ" a somewhat smoother formulation of 5.6, 5.8, and 5.9 could be given as is done in [10]. However, for the structure of modules such as direct sum decompositions, the functor Ξ and $\Xi(R)$ has definite advantages over Σ or $\mathcal{I}$. The elements of $\Xi(R)$ are not closed under submodules, and hence intuitively speaking, the classes of modules in $\Xi(R)$ are smaller - - and we want to distinguish as different as many submodules as we possibly can.

Given either one of the functors Ξ or Σ, for any concrete class of rings or single ring R, in general it is quite challenging to concretely and explicitly compute $\Xi(R) \cong \Sigma(R)$ as in 5.13. However, $\mathcal{I}(R)$ is more manageable, and 5.11 (a), (b), (c), 5.12 (iii) give criteria for deciding which ideals of R belong to $\mathcal{I}(R)$, and which don't.

Most of this article is not about the torsion free case and even some facts developed in a t.f. context have generalizations and analogues without the t.f. restriction. It is not clear how far and in what form the already developed theory of $G_F^{O3} \cong \Xi$ extends to the other functors G^{O3} and $G^{\sigma i}$. After answering some of the questions concerning the functors $G^{\sigma i}$ with $\aleph_1 \leq \sigma$ raised here, the next step might be to replace the σ-products with more general subdirect products called $\mathcal{F}$-products, of which σ-products are a special case. These $\mathcal{F}$-products have been successfully used in [21] and [22].

REFERENCES

[1] F. Anderson and K. Fuller, *Rings and Categories of Modules*, Springer-Varlag, Berlin-Heidelberg-New York, 1973.

[2] B. Banaschewski, *Hüllensysteme und Erweiterung von Quasi-ordnungen*, Zeitschr. f. math. Logik und Grundlagen d. Math. **2** (1956), 117-130.

[3] G. Birkhoff, *Lattice Theory*, Amer. Math. Soc. Colloquium Pub. XXV, Providence, Rhode Island, 1967.

[4] J. Dauns, *Uniform dimensions and subdirect products*, Pac. J. Math. **136** (1987), 1-19.

[5] J. Dauns, *Subdirect products of injectives*, Comm. in Alg. **17** (1989), 179-196.

[6] J. Dauns, *Torsion free modules*, Annali di Mat. Pura ed Appl., CLIV (1989), 49-81.

[7] J. Dauns, *Semiprime modules and rings, in: Non-Commutative Ring Theory*, Proceedings, Athens OH 1989, Lecture Notes in Math. **1448**, Springer-Verlag, Berlin-New York, 1989, pp.41-62.

[8] J. Dauns, *Classes of Modules*, Forum Math. **3** (1991), 327-338.

[9] J. Dauns, *Torsion free types*, Fundamenta Math. **139** (1991), 25-43.

[10] J. Dauns, *Direct sums and subdirect products, in: Proceedings of the Methods in Module Theory Conference*, University of Colorado, Colorado Springs 1991, Marcel Dekker, New York, 1992.

[11] J. Dauns, *Module classifying functors*, Czech. Math. J., to appear.

[12] J. Dauns and L. Fuchs, *Infinite Goldie dimensions*, J. Alg. **115** (1988), 297-302.

[13] J. Dauns and L. Fuchs, *κ-Products of slender modules*, Acta Scient. Math., to appear.

[14] P. Eklof and A. Mekler, *Almost free modules set-theoretic methods*, North-Holland, New York 1990.

[15] L. Fuchs, *Infinite Abelian Groups*, Academic Press, New York, Vol. **I** (1970), Vol. **II** (1973).

[16] G. Gierz, K. H. Hofmann, K. Keimel, J. D. Lawson, M. Mislove, and D. S. Scott, *A Compendium of Continuous Lattices*, Springer-Verlag, Berlin-Heidelberg-New York, 1980.

[17] K. Goodearl and A. Boyle, *Dimension theory for nonsingular injective modules*, Mem. Amer. Math. Soc. **177** (1976).

[18] G. Grätzer, *General Lattice Theory*, Academic Press, New York, 1978.

[19] P. Halmos, *Lectures on Boolean Algebras*, Math. Studies No. **1**, D. Van Nostrand Co., Princeton, N. J., 1963.

[20] K. Hrbacek and T. Jech, *Introduction to Set Theory*, Marcel Dekker, Second edition, New York, 1984.

[21] P. Loustaunau, *$\mathcal{F}$-products*.

[22] P. Loustaunau, *$\mathcal{F}$-products of injective, flat and projective modules*.

[23] S. MacLane, *Categories for the Working Mathematician*, Springer-Verlag, Berlin-Heidelberg-New York, 1971.

[24] E. Matlis, *Injective modules over noetherian rings*, Pac. J. Math. **8** (1958), 511-528.

[25] R. S. Pierce, *A note on complete Boolean algebras*, Proc. Amer. Math. Soc., **9** (1958), pp. 892-896.

[26] R. Sikorski, *Boolean Algebras*, 2nd. ed., Springer-Verlag, Berlin-Heidelberg-New York, 1964.

[27] M. Teply, *Torsion free injective modules*, Pacific J. Math. **31** (1969), 441-453.

[28] M. Teply, *Some aspects of Goldies' torsion theory*, Pacific J. Math. **29** (1969), 447-459.

PROJECTIVE COVERS OF COMPLEXES

MARK A. GODDARD
Department of Mathematical Sciences
The University of Akron, Akron, OH 44325-4002

ABSTRACT

By generalizing the concept of the projective cover of a module, projective covers of complexes of modules may be defined. The projective cover of a complex shares some but not all of the important properties enjoyed in the module case. The projective cover and minimal free resolution of a complex are compared and contrasted. Necessary and sufficient conditions are given to guarantee that the projective cover and the minimal free resolution are identical.

1. Introduction

In this paper, we shall assume that R always denotes a commutative ring with identity. Suppose that we let M be a finitely generated R-module where R is Noetherian and local as well. A projective module P_0 (which is actually free) can always be chosen so that there exists a map ϕ_0 from P_0 onto M. In the same manner if K_0 is the kernel of ϕ_0, then we can find a map $\phi_1 : P_1 \to P_0$ which maps P_1 onto K_0. Continuing this process, we obtain the projective resolution

$$\dots \to P_n \xrightarrow{\phi_n} P_{n-1} \to \dots \to P_1 \xrightarrow{\phi_1} P_0 \xrightarrow{\phi_0} M \to 0$$

of the module M.

The projective resolution of a module as it is defined above need not be unique, because the free modules P_i in the construction above are not uniquely determined. This problem can be overcome by the use of projective covers of modules.

The projective cover of a module M is a projective module P and a surjective map $P \xrightarrow{\epsilon} M$ satisfying the requirement that $\ker(\epsilon)$ is a superfluous submodule of P. In other words, for any submodule L of P, $L + \ker(\epsilon) = P$ must imply that $L = P$. Every finitely generated module over a Noetherian local ring has a projective cover and it is easily shown that the projective cover is unique. Therefore by use of projective covers at each step, we can obtain a unique projective resolution of a module M.

In order to generalize the concept of a projective cover to complexes, we introduce a more categorical definition. The projective cover of a module can be regarded as a projective module P and a map $P \xrightarrow{\epsilon} M$ satisfying the following two conditions:

1. Given a map $Q \xrightarrow{f} M$ where Q is a projective module, there exists a map $Q \xrightarrow{g} P$ such that $\epsilon \circ g = f$.

2. Any map $P \xrightarrow{h} P$ satisfying $\epsilon \circ h = \epsilon$ must be an automorphism of P.

With this definition in mind, we are almost ready to define the projective cover of a complex of R-modules. First we recall that a complex $C.$ is a sequence of R-module homomorphisms

$$\ldots \to C_n \xrightarrow{\delta_n} C_{n-1} \to \ldots \to C_1 \xrightarrow{\delta_1} C_0 \xrightarrow{\delta_0} \ldots$$

satisfying the condition that $\delta_{i-1} \circ \delta_i = 0$ for all i. A map of complexes $C. \xrightarrow{\phi.} D.$ is a sequence of R-module homomorphisms $C_i \xrightarrow{\phi_i} D_i$ which commute with the boundary maps δ_i and d_i of the complexes $C.$ and $D.$ respectively. In other words, we must have $\phi_{i-1} \circ \delta_i = d_i \circ \phi_i$ for all i. A complex $C.$ will be called bounded above if $C_i = 0$ for all $i < 0$.

We define the projective cover of a complex $C.$ to be a complex $P.$ of projective modules and a map of complexes $P. \xrightarrow{\epsilon.} C.$ satisfying the following two conditions:

1. The diagram

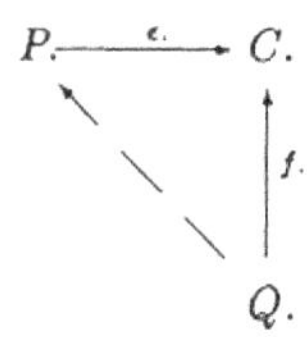

   can be completed for any complex $Q.$ of projective modules and any map of complexes $Q. \xrightarrow{f.} C.$

2. The diagram

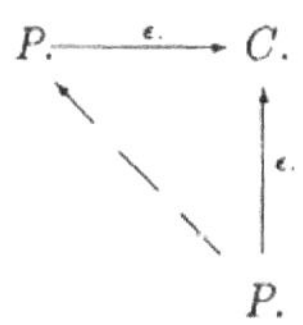

   can only be completed by maps of complexes which are automorphisms of P_i in each degree.

Note: A map of complexes $P. \xrightarrow{\epsilon.} C.$ satisfying only the first condition is called a projective precover of $C.$

2. Properties of Projective Covers of Complexes

We will now consider some of the important properties of the projective cover of a module and see if these properties still hold for complexes. Our first result deals with existence.

Theorem 1 *Let R be a Noetherian local ring and let $C.$ be a bounded above complex of finitely generated R-modules. Then $C.$ has a projective cover.*

The projective cover guaranteed in this theorem is constructed inductively. We allow $P_0 \xrightarrow{\epsilon_0} C_0$ to be the projective cover of C_0. Once we have constructed P_n

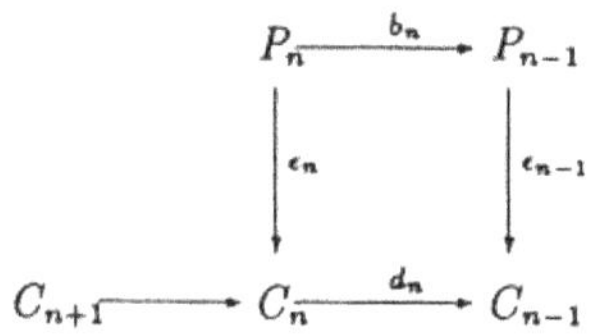

we define the pullback $S = \{(p,c) \in \ker(b_n) \oplus C_{n+1} | d_{n+1}(c) = \epsilon_n(p)\}$. Then we define $P_{n+1} \xrightarrow{\epsilon} S$ to be the projective cover of S. This yields the diagram below

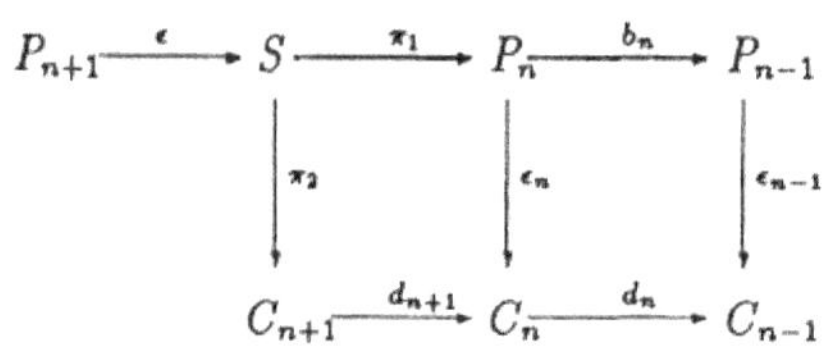

If we let $b_{n+1} = \pi_1 \circ \epsilon$ and $\epsilon_{n+1} = \pi_2 \circ \epsilon$ then our construction is complete. By an extensive diagram chasing argument, it can be shown that $P. \xrightarrow{\epsilon.} C.$ satisfies the two defining conditions of a projective cover.

It should be noted that by replacing "projective" by another type of module in the definition, we can define an injective or flat cover of a complex as well. The theorem above can easily be generalized to provide sufficient conditions for the existence of injective and flat covers.

In the following propositions, we consider three other important properties of the projective cover of a complex which are inherited from the module case.

Proposition 1 *The projective cover of a complex is unique.*

The proof of this proposition is a straightforward generalization of the proof that projective covers of modules are unique.

Proposition 2 *If $C.$ is a bounded above complex of finitely generated modules over a Noetherian local ring with projective cover $P. \xrightarrow{\epsilon.} C.$ then each map $P_i \xrightarrow{\epsilon_i} C_i$ is surjective.*

In order to prove that the projective cover of a module is surjective (using the categorical definition), it is crucial to observe that given any module M, there is a projective module Q which maps onto M. Similarly the proof of proposition 2 depends upon the following lemma. A detailed proof of this lemma appears in Roberts[1].

Lemma 1 *If $C.$ is a bounded above complex of finitely generated modules over a Noetherian local ring then there exists a complex $Q.$ of projective modules and a map of complexes $Q. \xrightarrow{f.} C.$ such that each map $Q_i \xrightarrow{f_i} C_i$ is surjective.*

Like the projective cover, the map of complexes $Q. \xrightarrow{f.} C.$ in the lemma above is constructed inductively. The map $Q_0 \xrightarrow{f_0} C_0$ can be a surjection from any free module onto C_0. After constructing Q_n we have

$$\begin{array}{ccccc} & & Q_n & \xrightarrow{g_n} & Q_{n-1} \\ & & \downarrow{\scriptstyle f_n} & & \downarrow{\scriptstyle f_{n-1}} \\ C_{n+1} & \longrightarrow & C_n & \xrightarrow{d_n} & C_{n-1} \end{array}$$

Define two surjections $Q'_{n+1} \xrightarrow{g'_{n+1}} \ker((\ker g_n) \to H_n(C.))$ and $Q''_{n+1} \xrightarrow{f''_{n+1}} \ker d_{n+1}$ where Q'_{n+1} and Q''_{n+1} are free modules. Let $Q_{n+1} = Q'_{n+1} \oplus Q''_{n+1}$, $g_{n+1} = (g'_{n+1}, 0)$ and $f_{n+1} = (f'_{n+1}, f''_{n+1})$ (where f'_{n+1} is the induced map from Q'_{n+1} to C_{n+1}). This construction yields a map of complexes which is surjective (and also induces an isomorphism on homology) in each degree.

The surjective map constructed by Roberts and shown in the lemma above is called a free resolution of the complex $C.$ The projective cover of the complex $C.$ is in fact itself an example of a surjective free resolution. While the projective cover is unique however, the surjective free resolution will never be unique. In order to obtain a unique free resolution, Roberts defines a minimal free resolution. In the final section of this paper, we will investigate the relationship between the projective cover and the minimal free resolution.

Proposition 3 *Let $P. \xrightarrow{\epsilon.} C.$ and $Q. \xrightarrow{\phi.} D.$ be projective covers of the complexes $C.$ and $D.$ Then $P. \oplus Q. \xrightarrow{(\epsilon,\phi)} C. \oplus D.$ is the projective cover of $C. \oplus D.$*

By generalizing a result by Enochs[2] we can extend the result above to determine necessary and sufficient conditions for a countable direct sum of projective covers of complexes to be a projective cover.

The next proposition indicates an important property of the projective cover which is not inherited from the module case. Recall that a quasi-isomorphism is a map of complexes $C. \xrightarrow{\phi.} D.$ with the property that each induced map on homology $H_i(C.) \to H_i(D.)$ is an isomorphism.

Proposition 4 *If $C.$ is a bounded above complex of finitely generated modules over a Noetherian local ring then the projective cover $P. \xrightarrow{\epsilon.} C.$ of the complex $C.$ is a quasi-isomorphism.*

The last property of the projective cover $P \xrightarrow{\epsilon} M$ of a module M is the fact that P is a projective module. In the next section we consider the idea of a projective complex.

3. Projective Complexes

Generalizing the concept of a projective module, we define a complex $Q.$ to be projective if given maps of complexes $D. \xrightarrow{\phi.} C.$ and $Q. \xrightarrow{\psi.} C.$ where each map $D_i \xrightarrow{\phi_i} C_i$ is surjective, the diagram

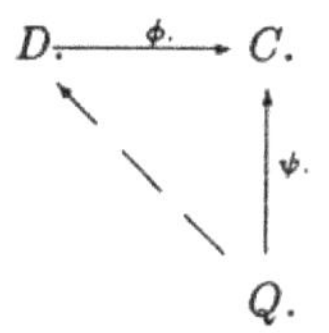

can always be completed by a map of complexes.

With this definition, the following example shows that the projective cover of a complex need not be a projective complex. Let R be the ring of integers and define the complex $C.$ by

$$C. = \ldots \to 0 \to (2) \xrightarrow{i} Z \to 0 \to \ldots$$

which has projective cover $P. \xrightarrow{\epsilon.} C.$ defined by

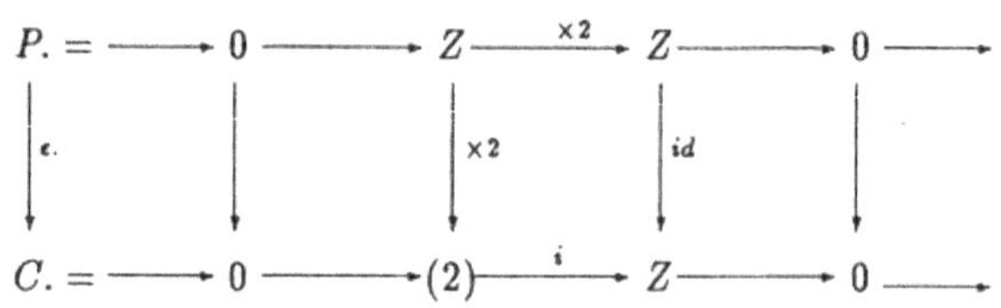

If we define a surjective map of complexes $D. \xrightarrow{\phi.} C.$ by

$$\begin{array}{ccccccccccc}
D. = & \longrightarrow & 0 & \longrightarrow & Z & \xrightarrow{(0,1)} & Z \oplus Z & \longrightarrow & 0 & \longrightarrow \\
\downarrow{\scriptstyle \phi.} & & \downarrow & & \downarrow{\scriptstyle \times 2} & & \downarrow{\scriptstyle \binom{1}{2}} & & \downarrow & \\
C. = & \longrightarrow & 0 & \longrightarrow & (2) & \xrightarrow{i} & Z & \longrightarrow & 0 & \longrightarrow
\end{array}$$

then the diagram

$$\begin{array}{ccc}
D. & \xrightarrow{\phi.} & C. \\
& \nwarrow & \uparrow{\scriptstyle \epsilon.} \\
& & P.
\end{array}$$

cannot be completed by a map of complexes. If there were a map $P. \xrightarrow{h.} D.$ completing the diagram then since $\epsilon_1 = \phi_1 \circ h_1$ we would have $h_1(1) = 1$. Since $h.$ is a map of complexes, $2h_0(1) = h_0 \circ \delta_1(1) = d_1 \circ h_1(1) = (0,1)$ where δ and d are the boundary maps of $P.$ and $D.$ respectively, but $(0,1)$ is not divisible by 2 in $Z \oplus Z$ so we have a contradiction.

Note that this example also shows that every complex of projective modules need not be a projective complex. By placing added constraints upon the complex $D.$ in the definition of a projective complex however, we can force any complex $Q.$ of projective modules to satisfy the projective property.

Proposition 5 *Let $Q.$ be a complex of projective modules and let $C.$ be a bounded above complex. If $Q. \xrightarrow{\psi.} C.$ is a map of complexes and $D. \xrightarrow{\phi.} C.$ is a surjective quasi-isomorphism then the diagram*

$$\begin{array}{ccc}
D. & \xrightarrow{\phi.} & C. \\
& \nwarrow & \uparrow{\scriptstyle \psi.} \\
& & Q.
\end{array}$$

can be completed by a map of complexes.

We will make use of proposition 5 in the next section. We conclude this section by asking when a projective cover is indeed projective. It can be shown that for any projective module P, the complex

$$\ldots \to 0 \to P \xrightarrow{id} P \to 0 \to \ldots$$

is projective. Furthermore, direct sums of projective complexes are projective. It follows that a complex of projective modules $Q.$ is projective if and only if it is an exact sequence. Since the projective cover $P. \xrightarrow{\epsilon} C.$ is a quasi-isomorphism, the homology of the complex $C.$ is preserved in the complex $P.$ and the next theorem follows.

Theorem 2 *A bounded above complex $C.$ of finitely generated R-modules, where R is a Noetherian local ring, has a projective cover which is a projective complex if and only if $C.$ is an exact sequence.*

4. Projective Covers and Minimal Free Resolutions

In order to compare the projective cover and the minimal free resolution, we first formalize several concepts defined by Roberts[1]. In this section we assume that $C.$ denotes a bounded above complex of finitely generated modules over a Noetherian local ring. We define a free resolution of $C.$ to be a bounded above complex $F.$ of finitely generated free modules and a quasi-isomorphism $F. \to C.$ As we observed in lemma 1, free resolutions always exist, if we are working over a Noetherian local ring, but need not be unique. To obtain uniqueness, we must proceed to the minimal free resolution. The minimal free resolution of a complex $C.$ is the unique free resolution with the property that every boundary map $F_i \xrightarrow{\phi_i} F_{i-1}$ is defined by a matrix with entries in the maximal ideal m of R.

As we observed earlier, the projective cover of a complex obtained in theorem 1 is actually an example of a surjective free resolution. Certainly every surjective free resolution cannot be a projective cover but by proposition 5, we do have the following lemma.

Lemma 2 *Any surjective free resolution of a complex $C.$ is a projective precover of $C.$*

Since every projective precover has a direct summand isomorphic to the projective cover, it follows that the projective cover is a direct summand of any surjective free resolution. The minimal free resolution is also a direct summand of the free resolution constructed in lemma 1 so one might suspect that the projective cover and the minimal free resolution are identical. The following example shows that this is not the case. Let $R = Z/8$ which has maximal ideal $m = (2)$. If we define the complex $C.$ by

$$C. = \ldots \to 0 \to Z/8 \xrightarrow{\times 2} Z/8 \xrightarrow{\times 4} (2) \to 0 \to \ldots$$

then the projective cover is given by

$$\begin{array}{ccccccccccccc}
P. = & \rightarrow Z/8 & \xrightarrow{\times 2} & Z/8 & \xrightarrow{\binom{4}{0}} & Z/8 \oplus Z/8 & \xrightarrow{b_2} & Z/8 \oplus Z/8 & \xrightarrow{(4,2)} & Z/8 & \rightarrow 0 \\
\downarrow \epsilon. & \downarrow & & \downarrow & & \downarrow (0,1) & & \downarrow (0,1) & & \downarrow \times 2 & \\
C. = & \rightarrow 0 & \longrightarrow & 0 & \longrightarrow & Z/8 & \xrightarrow{\times 2} & Z/8 & \xrightarrow{\times 4} & (2) & \rightarrow 0
\end{array}$$

where the boundary map b_2 is given by the matrix

$$\begin{pmatrix} 2 & 1 \\ 0 & 2 \end{pmatrix}$$

Since the matrix defining the map b_2 contains an entry not in the maximal ideal (2) of the ring $Z/8$, the projective cover of $C.$ is not equal to the minimal free resolution of $C.$

Although the projective cover and minimal free resolution are not identical, they are closely related. Since the projective cover is a surjective free resolution, we know that the minimal free resolution is a direct summand of the projective cover. It is shown in Roberts[1] that any free resolution is the direct sum of the minimal free resolution and an exact complex of projective modules. Since exact complexes of projective modules are shown in theorem 2 to be projective complexes, we have the following theorem.

Theorem 3 *The projective cover of a complex $C.$ is the direct sum of the minimal free resolution of $C.$ and a projective complex.*

We now see that the projective cover lies somewhere between the surjective free resolution and the minimal free resolution. The surjective free resolution does not always satisfy condition 2 of the definition of the projective cover and hence encompasses too large a class of maps. The minimal free resolution on the other hand is too restrictive, for unlike the projective cover, it may not be surjective in each degree.

Further study of the question of when the projective cover and minimal free resolution may be identical leads to the following three conditions which may be imposed on a complex $C.$ over a local ring with maximal ideal m.

1. $\delta_n(C_n) \subseteq m\ker(\delta_{n-1})$ for all n.

2. The projective cover of $C.$ is the minimal free resolution of $C.$

3. $\delta_n(C_n) \subseteq mC_{n-1}$ for all n.

It can be shown that (1) implies (2) and that (2) implies (3) but none of these are equivalent. In fact, our earlier example of a projective cover which is not a minimal free resolution features a complex which satisfies (3) but not (2). To obtain necessary and sufficient conditions to guarantee (2), we need to define a new submodule D_n of C_n which lies between $m\ker(\delta_{n-1})$ and $(\ker(\delta_{n-1}) \cap mC_{n-1})$. If $P. \overset{\epsilon}{\to} C.$ is the projective cover of $C.$ where b_i and δ_i are the boundary maps of $P.$ and $C.$ respectively and m is the maximal ideal of R then D_i is the set of elements $c \in C_i$ such that there exists $r_1, ..., r_k$ in the maximal ideal m, $c_1, ..., c_k \in C_i$ and $p_1, ..., p_k \in \ker(b_{i-1})$ satisfying the following three conditions:

1. $c = r_1c_1 + ... + r_kc_k$
2. $\epsilon_{i-1}(p_j) = \delta_i(c_j)$ for $1 \leq j \leq k$
3. $r_1p_1 + ... + r_kp_k = 0$

With this definition of D_i we have the following theorem.

Theorem 4 *The projective cover $P.$ of a complex $C.$ is equal to the minimal free resolution of $C.$ if and only if $\delta_n(C_n) \subseteq D_{n-1}$ for all $n \geq 0$.*

Much of the theory which has been developed here for the projective cover of a complex can also be applied to the dual notion of the injective envelope of a complex. The construction of such an injective envelope is possible using an approach very similar to that used in theorem 1 by making use of pushout rather than pullback constructions. It would be of interest to compare this injective envelope with the injective envelope defined by Roberts[1] as a dual to the minimal free resolution.

REFERENCES

1. P. Roberts, *Homological Invarients of Modules over Commutative Rings* (Les Presses de l'Université de Montreal, 1980)
2. E. Enochs, *Injective and flat covers, envelopes and resolvents*, Israel Journal of Mathematics, (3) **39** (1981), 189-209.
3. M. Goddard, *Projective and Exact Complexes*, Unpublished work.
4. J. Rotman, *An Introduction to Homological Algebra* (Academic Press Inc., New York, 1979)
5. L. Silver, *Noncommutative localizations and applications*, Journal of Algebra, **7** (1967), 44-75.
6. D. Lazard, *Autour de la platitude*, Bull. Soc Math, **97** (1968), 82-128.

7. I. Kaplansky, *Projective modules*, Annals of Mathematics, (2) **68** (1958), 372-377.

8. P. Roberts, *Some remarks on the homological algebra of multiple complexes*, Unpublished work.

9. M. Goddard, *Projective Covers and Minimal Free Resolutions*, Unpublished work.

UNIFORM RANKS OF PRIME FACTORS OF SKEW POLYNOMIAL RINGS

K. R. GOODEARL
Department of Mathematics, University of California, Santa Barbara, CA 93106, USA

ABSTRACT

An expository account of some of the work of the author and E. S. Letzter on prime ideals in noetherian skew polynomial rings $R[y;\tau,\delta]$ is presented. The development concentrates on the q-skew case, meaning that $\delta\tau = q\tau\delta$ for some central (τ,δ)-constant q. When this condition holds and q is not a root of unity, the arguments are much simpler than in the general case. For this case, bounds on the uniform ranks of prime factor rings are obtained; namely, the uniform ranks of prime factor rings of $R[y;\tau,\delta]$ are no larger than those of the simpler skew polynomial ring $R[z;\tau]$. In particular, this provides a direct route to the recent result that all prime ideals in the quantum coordinate rings $\mathcal{O}_q(M_n(k))$ are completely prime when the parameter q is not a root of unity.

1. Introduction

The class of algebras known as "quantum groups" provides a rich new source of noetherian rings, whose structure is just beginning to reveal itself. (For a good introduction and a discussion of many of the standard examples, see **[15]**.) Among the various algebras proposed as "quantum coordinate rings", many have the structure of iterated skew polynomial rings. It is tempting to think of these algebras as analogs of enveloping algebras of solvable Lie algebras, although no general construction principle for quantum enveloping algebras in the solvable case has yet been suggested. To support the analogy, one seeks to prove that quantum coordinate rings and solvable enveloping algebras enjoy similar ring-theoretic properties. The structure of both types of algebras as iterated skew polynomial rings is a key tool in this effort.

E. S. Letzter and the author recently proved in **[6]** that all prime ideals are completely prime in $\mathcal{O}_q(M_n(k))$, the coordinate ring of quantum $n \times n$ matrices over a field k, when q is not a root of unity. (In quantum groups, the case where q is not a root of unity is the one analogous to characteristic zero for enveloping algebras.) This result was proved with the aid of a general theorem bounding the uniform ranks of prime factors of iterated skew polynomial rings. This procedure was inspired by Sigurdsson's proof of Dixmier's theorem (the complete primeness

---

1991 *Mathematics Subject Classification.* 16D30, 16P40, 16S36.
This research was partially supported by a National Science Foundation grant.

of prime ideals of solvable enveloping algebras in characteristic zero) by means of a general theorem bounding the ranks of prime factors of iterated differential operator rings [**14**]. The general skew polynomial version, however, requires a considerably sharper analysis of prime ideal behavior, because new phenomena with no analogs in differential operator rings or in skew-Laurent polynomial rings arise. Such an analysis was carried out in [**5**].

For the purpose of proving the complete primeness theorem of [**6**], much of the work in [**5**] is not needed - particularly the portion designed to handle q-skew polynomial rings when q is allowed to be a root of unity. Here we extract an account of that portion of [**5**] on which the complete primeness theorem is based. Our account is expository, with proofs; apart from variations in the arguments and the order of the development, the content is not new. For variety, we apply the main theorem to a version of the coordinate ring of quantum matrices more general than the one discussed in [**6**]. In the same manner, this theorem can be applied to other quantum coordinate rings, as the reader equipped with the requisite generators and relations can easily verify.

All rings and subrings in this paper are assumed to be unital. In particular, when discussing a ring extension $R \subseteq S$ we always assume that R and S share a unit. We write $\operatorname{spec} R$ for the set of prime ideals of a ring R, and $\operatorname{Fract} R$ for the Goldie quotient ring of a semiprime noetherian ring R. Uniform rank (Goldie dimension) will be denoted simply "rank", and classical Krull dimension will be denoted "cl.K.dim".

Recall that a (left) *skew derivation* on a ring R is a pair (τ, δ) where τ is a ring endomorphism of R and δ is a (left) *τ-derivation* on R, that is, an additive map from R to itself such that $\delta(ab) = \tau(a)\delta(b) + \delta(a)b$ for all $a, b \in R$. For any $u \in R$, the rule $\delta_u(a) = ua - \tau(a)u$ defines a τ-derivation δ_u. The τ-derivations of this form are said to be *inner*. An element $q \in R$ is called a *(τ, δ)-constant* if $\tau(q) = q$ and $\delta(q) = 0$. If $\delta\tau = q\tau\delta$ for some fixed central (τ, δ)-constant $q \in R$, then we say that (τ, δ) is a *q-skew derivation* on R.

Given a skew derivation (τ, δ) on R, we can form the corresponding *skew polynomial ring* (or *Ore extension*) $R[y; \tau, \delta]$ (see, e.g., [**8**, pp. 8–11]). Elements of $R[y; \tau, \delta]$ are written as polynomials in the indeterminate y with left-hand coefficients from R; multiplication is induced from that in R using the relations $yr = \tau(r)y + \delta(r)$ for $r \in R$. (Other authors, such as [**12**], use relations $ry = y\tau(r) + \delta'(r)$ based on a right τ-derivation δ'.) In the present paper we only consider the case that τ is an automorphism of R. Then, assuming R noetherian, it follows that $R[y; \tau, \delta]$ is noetherian as well [**8**, Theorem 1.12; **12**, Theorem 1.2.9]. In case (τ, δ) is a q-skew derivation, we refer to $R[y; \tau, \delta]$ as a *q-skew extension* of R, or as a *q-skew polynomial ring* over R. We shall need the standard fact that if δ is an inner τ-derivation, then $R[y; \tau, \delta] \cong R[z; \tau]$ (e.g., [**4**, Lemma 1.5]).

We do not attempt to define the terms "quantum group" and "quantum coordinate ring", nor to give any explicit examples beyond "quantum matrices" (see

Section 5). Very roughly, the use of the term "quantum" in this context is meant to denote the fact that a given algebra is a deformation of a "classical" one such as the enveloping algebra of a Lie algebra or the coordinate ring of an algebraic variety. In the case of "quantum matrices", what is deformed is not the algebra $M_n(k)$ of $n \times n$ matrices over a field k, but the coordinate ring $\mathcal{O}(M_n(k))$, where $M_n(k)$ is viewed as an algebraic variety (isomorphic to n^2-dimensional affine space). Thus a "quantum matrix algebra" is a k-algebra with n^2 generators, say u_{ij} (corresponding to the coordinate function sending each matrix to its (i,j)-entry), which do not commute but satisfy relations involving terms $u_{\ell m}u_{ij} - \alpha_{ij\ell m}u_{ij}u_{\ell m}$ for various scalars $\alpha_{ij\ell m} \in k$. For discussions of the question of which sets of relations might be deemed "meaningful", see [**1**, **16**].

2. General skew polynomial rings

In order to analyze the prime factors of a skew polynomial ring $S = R[y; \tau, \delta]$, we first need to understand how a prime ideal $P \in \operatorname{spec} S$ relates to ideals of R. In the "classical" cases ($\tau = 1$ or $\delta = 0$), there is a well known procedure. For instance, if $\tau = 1$, then the contraction $P \cap R$ is a δ-prime ideal of R; hence, by passing to the ring $S/(P \cap R)S \cong (R/(P \cap R))[y; \delta]$ and then localizing $R/(P \cap R)$, one can reduce to the situation that $P \cap R = 0$ and R is an artinian δ-simple ring. Then the prime ideals of S are generated by their intersections with the center, which is a commutative polynomial ring over a central subfield of R. In case $\delta = 0$, either $y \in P$ or $P \cap R$ is a τ-prime ideal; the first possibility is easily analyzed because $S/yS \cong R$, while the second can be reduced to the case that $P \cap R = 0$ and R is an artinian τ-simple ring. Such nice reductions are not always available in the general case, although we shall be able to use some similar methods in one major subcase of the q-skew situation.

Rather than concentrating on $P \cap R$, we begin by studying the prime ideals minimal over this intersection. With the help of techniques adapted from the theory of noetherian bimodules, we shall show that the prime ideals of R minimal over $P \cap R$ all lie in a single τ-orbit. (The *τ-orbit* of a prime $Q \in \operatorname{spec} R$ refers to the orbit under the action of the cyclic group generated by τ, that is, to the set $\{\tau^n(Q) \mid n \in \mathbb{Z}\}$.) The analysis then splits into two cases, depending on whether this τ-orbit is finite or infinite. These two cases are quite different. For instance, if the prime ideals Q minimal over $P \cap R$ lie in an infinite τ-orbit, we shall be able to recover P from one such Q, while if the τ-orbit of Q is finite, there can be infinitely many prime ideals $P' \in \operatorname{spec} S$ such that Q is minimal over $P' \cap R$.

Up to this point, we do not require any q-skew assumption, only the hypotheses that R is noetherian and τ is an automorphism. In the following section, we shall see that much more control can be obtained in the q-skew case.

To get a grip on the prime ideals minimal over $P \cap R$, we adopt a method from the additivity principle for noetherian ring extensions – namely, we study the annihilators of the composition factors in a bimodule composition series for the Goldie quotient ring $C = \text{Fract}(S/P)$, viewed as an R-C-bimodule. (Note that C, being an artinian ring, also has finite length as an R-C-bimodule.) We shall not need the additivity principle itself, only the following bimodule result excerpted from Warfield's treatment of this principle. In keeping with the expository aims of this paper, we present a full proof.

Proposition 2.1. [17, Lemma 2] *Let $R \subseteq S$ be noetherian rings, $P \in \operatorname{spec} S$, and $C = \text{Fract}(S/P)$. Choose an R-C-bimodule composition series*

$$C_0 = 0 < C_1 < \cdots < C_m = C$$

for C, and let Q_i denote the prime ideal $\text{l.ann}_R(C_i/C_{i-1})$ for $i = 1, \ldots, m$. Then in S there exist R-R-sub-bimodules $L_0 = S > L_1 > \cdots > L_m \geq Q_m S$ such that for $i = 1, \ldots, m$ the left and right R-annihilators of L_{i-1}/L_i are Q_m and Q_i, and L_{i-1}/L_i is torsionfree as a left (R/Q_m)-module and as a right (R/Q_i)-module.

Proof. By, e.g., [8, Proposition 7.19], each Q_i is prime, and C_i/C_{i-1} is torsionfree as a right (R/Q_i)-module.

Without loss of generality, $P = 0$. Recall that since C is simple artinian, there are inverse bijections between the right and left ideals of C given by taking left and right annihilators. These correspondences restrict to inverse bijections between the R-C-sub-bimodules and the C-R-sub-bimodules of C.

Since C is semisimple, there exist orthogonal idempotents $e_1, \ldots, e_m$ in C such that $e_1 + \cdots + e_m = 1$ and $C_i = (e_1 + \cdots + e_i)C$ for all i. Set $D_i = \text{l.ann}_C(C_i) = C(e_{i+1} + \cdots + e_m)$ for each i. Because of the bijections discussed in the previous paragraph, the D_i form a C-R-bimodule composition series

$$D_0 = C > D_1 > \cdots > D_m = 0$$

for C. We claim that $\text{r.ann}(D_{i-1}/D_i)_R = Q_i$ and that D_{i-1}/D_i is a torsionfree right (R/Q_i)-module.

Note first that $C_i = e_i C + C_{i-1}$ and $D_{i-1} = Ce_i + D_i$. Hence, an element $r \in R$ belongs to the right annihilator of D_{i-1}/D_i if and only if $e_i r \in D_i$, if and only if $e_i r C_i = 0$. Since $e_i r C_{i-1} \subseteq e_i C_{i-1} = 0$ and $(1 - e_i)C_i = C_{i-1}$, we have $e_i r C_i = 0$ if and only if $e_i r e_i = 0$, if and only if $r \in \text{l.ann}_R(C_i/C_{i-1})$. Thus $\text{r.ann}(D_{i-1}/D_i)_R = Q_i$. We now conclude from [8, Proposition 7.19] that D_{i-1}/D_i is torsionfree as a right (R/Q_i)-module. This establishes the claim.

Next, set $K_i = C_{m-1} + D_i$ and $L_i = K_i \cap S$ for all i. The K_i are R-R-sub-bimodules of C and the L_i are R-R-sub-bimodules of S, forming descending chains

$$K_0 = C \geq K_1 \geq \cdots \geq K_m = C_{m-1}$$
$$L_0 = S \geq L_1 \geq \cdots \geq L_m.$$

Moreover, $L_m \geq Q_m S$ because $Q_m(C/C_{m-1}) = 0$.

Fix an index $i \in \{1, \ldots, m\}$. As a left R-module, K_{i-1}/K_i is isomorphic to $(C/C_{m-1})e_i$, and so it is a torsionfree left (R/Q_m)-module. As a right R-module, K_{i-1}/K_i is isomorphic to $e_m(D_{i-1}/D_i)$, and so it is a torsionfree right (R/Q_i)-module. Since L_{i-1}/L_i embeds in K_{i-1}/K_i, it too must be a torsionfree left (R/Q_m)-module and a torsionfree right (R/Q_i)-module.

Note that $e_m Ce_i \subseteq K_{i-1}$, and that $e_m Ce_i \cap K_i = 0$ because $e_m K_i e_i = 0$. Hence, there is an embedding $e_m Ce_i \hookrightarrow K_{i-1}/K_i$, which restricts to an embedding $e_m Ce_i \cap S \hookrightarrow L_{i-1}/L_i$. Since S is essential in C as a right or left S-module, we have $e_m C \cap S \neq 0$ and $Ce_i \cap S \neq 0$. The product $(e_m C \cap S)(Ce_i \cap S)$ is nonzero because S is prime, and hence $e_m Ce_i \cap S \neq 0$. Therefore $L_{i-1}/L_i \neq 0$, from which we conclude that $\text{l.ann}_R(L_{i-1}/L_i) = Q_m$ and $\text{r.ann}(L_{i-1}/L_i)_R = Q_i$. □

The list of primes $Q_1, \ldots, Q_m$ appearing in Proposition 2.1 includes all the prime ideals minimal over $P \cap R$ and possibly others which are not minimal. For the case $R = S[y; \tau, \delta]$, we shall see that there are no proper inclusions among the Q_i, whence all the Q_i are minimal over $P \cap R$. The lack of proper inclusions in this case comes from the fact that the rings R/Q_i all have the same classical Krull dimension. In fact, these factor rings are all isomorphic, since, as we shall prove, each $Q_i = \tau^{-j(i)}(Q_m)$ for some $j(i)$. This relationship between Q_i and Q_m will ultimately be derived from Proposition 2.1, together with the special structure of the R-S-bimodule $S/Q_m S$. This bimodule can be represented as a module of polynomials in y with left-hand coefficients from R/Q_m. Note in particular that the set of polynomials in $S/Q_m S$ of degree at most j modulo those of degree at most $j-1$ forms an R-R-bimodule, isomorphic to R/Q_m on the left but to $R/\tau^{-j}(Q_m)$ on the right. It is, however, more convenient to work with the bimodule $\text{Fract}(R/Q_m) \otimes_R S$, and so we first determine the structure of the latter bimodule.

Lemma 2.2. *Let R be a noetherian ring, $Q \in \operatorname{spec} R$, and $A = \text{Fract}(R/Q)$. Let σ be an automorphism of R, and construct an A-R-bimodule A^σ which equals A as a left A-module, but with a right R-module multiplication $*$ given by the rule $a * r = a\sigma(r)$. Then A^σ is a simple A-R-bimodule.*

Proof. Let B be a nonzero A-R-sub-bimodule of A^σ. As a left A-module, B coincides with some nonzero left ideal $C \subseteq A$. Since ${}_A C$ has finite length, so does ${}_A B$. On the right, $AQ = 0$ and A is a torsionfree right (R/Q)-module, whence $A^\sigma * \sigma^{-1}(Q) = 0$ and A^σ is a torsionfree right $(R/\sigma^{-1}(Q))$-module. The same conclusions hold for B_R. Now [**8**, Lemma 7.8] implies that B is divisible as a right $(R/\sigma^{-1}(Q))$-module, and so C is divisible as a right (R/Q)-module. This forces C to be an ideal of A, whence $C = A$ and then $B = A^\sigma$. □

Lemma 2.3. *Let $S = R[y; \tau, \delta]$, where R is noetherian and τ an automorphism,* *let $Q \in \operatorname{spec} R$, and set $A = \operatorname{Fract}(R/Q)$ and $V = A \bigotimes_R S$. Further, set*

$$V_j = \bigoplus_{i=0}^{j-1} (A \otimes y^i)$$

for $j = 0, 1, \ldots$. Then $V_0 = 0 < V_1 < \ldots$ is an exhaustive A-R-bimodule filtration of V. Each of the subfactors V_{j+1}/V_j is a simple A-R-bimodule with ${}_A(V_{j+1}/V_j) \cong {}_AA$ and $\operatorname{r.ann}(V_{j+1}/V_j)_R = \tau^{-j}(Q)$.

Proof. It is clear that the V_j form an ascending chain of A-R-sub-bimodules of V whose union equals V. For any index j, there is a left A-module isomorphism ${}_AA \to {}_A(V_{j+1}/V_j)$ given by the rule $a \mapsto (a \otimes y^j) + V_j$. For $r \in R$, observe that $(1 \otimes y^j)r - \tau^j(r) \otimes y^j \in V_j$, whence $(a \otimes y^j)r + V_j = (a\tau^j(r) \otimes y^j) + V_j$ for all $a \in A$. Thus, using the notation of the previous lemma, $V_{j+1}/V_j \cong A^{\tau^j}$ as A-R-bimodules. The lemma shows that V_{j+1}/V_j is simple as an A-R-bimodule, and it is clear that $\operatorname{r.ann}(V_{j+1}/V_j)_R = \tau^{-j}(Q)$. □

Proposition 2.4. *Let $S = R[y; \tau, \delta]$, where R is noetherian and τ an automorphism, let $P \in \operatorname{spec} S$, and set $C = \operatorname{Fract}(S/P)$. Choose an R-C-bimodule composition series $C_0 = 0 < C_1 < \cdots < C_m = C$ for C, and set $Q_i = \operatorname{l.ann}_R(C_i/C_{i-1})$ for $i = 1, \ldots, m$. Then $Q_1, \ldots, Q_m$ are precisely the primes of R minimal over $P \cap R$, and each $Q_i = \tau^{-j(i)}(Q_m)$ for some $j(i) \geq 0$.*

Proof. Note that $Q_1 Q_2 \cdots Q_m C = 0$ and so $Q_1 Q_2 \cdots Q_m \subseteq P \cap R$. Hence, each prime of R minimal over $P \cap R$ contains one of the Q_i. Since all the Q_i contain $P \cap R$, it follows that each prime of R minimal over $P \cap R$ must equal one of the Q_i.

Fix an index $i \in \{1, \ldots, m\}$. By Proposition 2.1, there exists an R-R-bimodule subfactor B of $S/Q_m S$ such that the left and right R-annihilators of B are Q_m and Q_i, and B is torsionfree as a left (R/Q_m)-module and a right (R/Q_i)-module. Observe that for any $s \in S$, the R-R-bimodule RsR is noetherian as a left or right R-module. Hence, after replacing B by RbR for some nonzero $b \in B$, we may assume that B is noetherian as a left R-module.

Set $A = \operatorname{Fract}(R/Q_m)$, and note that $A \bigotimes_R B$ has finite length as a left A-module. Moreover, because B is a nonzero torsionfree left (R/Q_m)-module, $A \bigotimes_R B$ is a nonzero torsionfree right (R/Q_i)-module. Hence, there exists a simple A-R-sub-bimodule $M \subseteq A \bigotimes_R B$. Then $\operatorname{r.ann}(M_R) = Q_i$ and M is isomorphic to an A-R-bimodule subfactor of $A \bigotimes_R S$. Since M is a simple A-R-bimodule, it follows from Lemma 2.3 that $M \cong V_{j+1}/V_j$ for some $j \geq 0$, and thus $Q_i = \operatorname{r.ann}(M_R) = \tau^{-j}(Q_m)$.

Now the rings R/Q_i are all isomorphic, and so they all have the same classical Krull dimension, say α. Since the primes minimal over $P \cap R$ are included among the Q_i, we find that $\operatorname{cl.K.dim}(R/Q) = \alpha$ for all primes Q minimal over $P \cap R$.

Consequently, no prime minimal over $P \cap R$ can be properly contained in any Q_i, from which we conclude that all the Q_i are minimal over $P \cap R$. □

Under mild additional hypotheses, such as the existence of a symmetric dimension function on R in the sense of [2] (cf. [**8**, Exercise 12J]), the fact that the prime ideals $Q_1, \ldots, Q_m$ in Proposition 2.4 are precisely the primes of R minimal over $P \cap R$ can be obtained from general results on restricted ring extensions, such as [**2**, Corollary 6.6] or [**17**, Corollary 1].

Corollary 2.5. [**5**, Lemma 5.4] *If $S = R[y; \tau, \delta]$ with R noetherian and τ an automorphism, and $P \in \operatorname{spec} S$, then the primes of R minimal over $P \cap R$ all lie in the same τ-orbit.* □

The set of primes of R minimal over $P \cap R$ in Corollary 2.5 actually forms a "segment" of a τ-orbit, namely a set of the form $\{Q, \tau^{-1}(Q), \ldots, \tau^{-r}(Q)\}$ [**5**, Theorem 5.12i].

Our next aim is to show that P can be recovered, in a certain sense, from the primes minimal over $P \cap R$. We shall need to work with one of these primes which, additionally, equals the annihilator of some R-submodule of S/P. The following corollary of Proposition 2.4 guarantees that such a prime actually exists. Recall that an *annihilator prime* of an R-module M is any prime ideal of R which equals the annihilator of some submodule of M. Since R is noetherian, every nonzero R-module has at least one annihilator prime [**8**, Proposition 2.12].

Corollary 2.6. [**5**, Lemma 5.3] *Let $S = R[y; \tau, \delta]$, where R is noetherian and τ an automorphism, let $P \in \operatorname{spec} S$, and let $Q \in \operatorname{spec} R$ be an annihilator prime of $(S/P)_R$. Then Q is minimal over $P \cap R$.*

Proof. Let Q' be an ideal of R that contains Q and is maximal among annihilators of nonzero submodules of $(S/P)_R$. Then Q' is prime (e.g., [**8**, Proposition 2.12]), and it suffices to show that Q' is minimal over $P \cap R$. Hence, we may assume, without loss of generality, that Q is maximal among annihilators of nonzero submodules of $(S/P)_R$.

Set $C = \operatorname{Fract}(S/P)$ and let L be the left annihilator of Q in C. Then L is a C-R-sub-bimodule of C and $\operatorname{r.ann}(L_R) = Q$. Let L' be the right (R/Q)-torsion submodule of L. Then L' is a left ideal of C, whence L' is cyclic as a left C-module, and so $L'z = 0$ for some $z \in \mathcal{C}_R(Q)$. Consequently, $\operatorname{r.ann}(L')_R > Q$, and hence $L' = 0$, by the maximality of Q. Thus L is a torsionfree right (R/Q)-module.

Let K be the right annihilator of L in C. Then K is an R-C-sub-bimodule of C, and, as in the proof of Proposition 2.1, we see that $\operatorname{l.ann}_R(C/K) = Q$. Moreover, since the left annihilator of K is L, we must have $C > K$. Thus we can choose an R-C-bimodule composition series for C which passes through K, and at least one composition factor of this series will have left R-annihilator equal to Q. Therefore by Proposition 2.4, Q must be minimal over $P \cap R$. □

The following lemma provides a bimodule framework relating the prime ideals in a pair of noetherian rings $R \subseteq S$. In particular, it gives a way to express a prime P of S in terms of an annihilator prime Q of $(S/P)_R$. We shall see that P is uniquely determined by Q when $S = R[y; \tau, \delta]$ and the τ-orbit of Q is infinite. (In general, of course, it is not – e.g., when $S = \mathbb{C}[y]$ and Q is the prime 0 in $\mathbb{C}$.)

Lemma 2.7. *Let $R \subseteq S$ be noetherian rings, let $P \in \operatorname{spec} S$, and let $Q \in \operatorname{spec} R$ be an annihilator prime of $(S/P)_R$. Set $A = \operatorname{Fract}(R/Q)$ and $V = A \bigotimes_R S$. Then there exists a proper A-S-sub-bimodule $W < V$ such that $\operatorname{r.ann}(V/W)_S = P$ and V/W is torsionfree as a right (S/P)-module.*

Proof. There is a right R-submodule $X \subseteq S/P$ such that $Q = \operatorname{r.ann}(X_R)$. Now let $C = \operatorname{Fract}(S/P)$; we claim that C/QC is not torsion as a left (R/Q)-module. If it were, then $z.(1 + QC) = 0$ for some $z \in \mathcal{C}_R(Q)$, and so $z \in QC$. However, this would imply that $Xz \subseteq XQC = 0$ and consequently $z \in Q$, which is impossible. This verifies the claim.

Since C/QC is not torsion as a left (R/Q)-module, $A \bigotimes_{R/Q} (C/QC) \neq 0$, and so $A \bigotimes_R C \neq 0$. There is a natural A-S-bimodule homomorphism

$$f : V \to A \bigotimes_R (S/P) \to A \bigotimes_R C$$

such that $f(V)$ generates $A \bigotimes_R C$ as a right C-module. Then f must be nonzero, and so $W = \ker f$ is a proper A-S-sub-bimodule of V. Since V/W is isomorphic to a submodule of the torsionfree right (S/P)-module $A \bigotimes_R C$, we see that $(V/W)P = 0$ and that V/W is a torsionfree right (S/P)-module. Finally, since V/W is nonzero, we conclude that its right S-annihilator must equal P. □

Theorem 2.8. [5, Theorem 7.5.I] *Let $S = R[y; \tau, \delta]$, where R is noetherian and τ an automorphism. Let $P \in \operatorname{spec} S$, let $Q \in \operatorname{spec} R$ be an annihilator prime of $(S/P)_R$, and assume that the τ-orbit of Q is infinite. Set $A = \operatorname{Fract}(R/Q)$ and $V = A \bigotimes_R S$. Then V has a unique maximal proper A-S-sub-bimodule X. Moreover, X is nonzero, V/X has finite length as a left A-module, $\operatorname{r.ann}(V/X)_S = P$, and V/X is a torsionfree right (S/P)-module.*

Proof. By Lemma 2.7, there is a proper A-S-sub-bimodule $W < V$ such that the right S-annihilator of V/W equals P and V/W is a torsionfree right (S/P) module. We first show that $W \neq 0$.

If $W = 0$, then $VP = 0$, and consequently $P \cap R$ annihilates (on the right) all the A-R-bimodule subfactors V_{j+1}/V_j from Lemma 2.3. Thus $P \cap R \subseteq \tau^{-j}(Q)$ for all $j \geq 0$. In view of Corollaries 2.5 and 2.6, Q is minimal over $P \cap R$ and all primes minimal over $P \cap R$ lie in the τ-orbit of Q. Let $\alpha = \operatorname{cl.K.dim}(R/Q)$; then $\operatorname{cl.K.dim}(R/Q') = \alpha$ for all primes Q' minimal over $P \cap R$. For $j \geq 0$, we have $\tau^{-j}(Q) \supseteq P \cap R$ and $\operatorname{cl.K.dim}(R/\tau^{-j}(Q)) = \alpha$, and so $\tau^{-j}(Q)$ must be minimal over $P \cap R$. But these primes are all distinct, because the τ-orbit of Q is infinite.

Since there are only finitely many primes minimal over $P \cap R$, we have reached a contradiction. Therefore $W \neq 0$.

We write elements of V in the form $(v_0 \otimes 1) + \cdots + (v_n \otimes y^n)$ with "coefficients" $v_i \in A$, and we define degrees and leading coefficients in the obvious way. Choose a nonzero element $w \in W$ with degree n and leading coefficient w_n. Since A is simple as an A-R-bimodule (Lemma 2.2), there exist elements $a_j \in A$ and $r_j \in R$ such that $\sum_j a_j w_n r_j = 1$. Now set $w' = \sum_j a_j w \tau^{-n}(r_j)$, and note that w' is a nonzero element of W with degree n and leading coefficient 1. Just as in a polynomial ring, it follows that $V = V_n + W$. Therefore V/W has finite length as a left A-module.

By [**8**, Lemma 7.8], V/W is divisible as a right (S/P)-module. Since V/W has finite length as an A-S-bimodule, V has a maximal proper A-S-sub-bimodule $X \supseteq W$. As a right (S/P)-module, V/X is nonzero and divisible, hence faithful. Thus $\operatorname{r.ann}(V/X)_S = P$. On the other hand, since V/X is noetherian as a left A-module, it cannot be torsion as a right (S/P)-module [**8**, Lemma 7.3]. Since the right (S/P)-torsion submodule of V/X is an A-S-sub-bimodule, we conclude from the bimodule simplicity of V/X that V/X must be torsionfree as a right (S/P)-module.

It remains to show that X is the only maximal proper A-S-sub-bimodule of V. Suppose that Y is another. Then $X + Y = V$ and $V/(X \cap Y) \cong (V/X) \bigoplus (V/Y)$. Moreover, $Y \neq 0$, and it follows as above that V/Y has finite length as a left A-module.

Note that $1 \otimes 1$ generates V as an A-S-bimodule, and so $1 \otimes 1$ cannot lie in either X or Y. Since V_1 is a simple A-R-sub-bimodule of V, it follows that $V_1 \cap X = V_1 \cap Y = 0$. Thus as A-R-bimodules, V/X and V/Y each have a composition factor isomorphic to V_1. Consequently, any A-R-bimodule composition series for $V/(X \cap Y)$ must include at least two composition factors isomorphic to V_1.

Since $V/(X \cap Y)$ has finite length as a left A-module, we see that $V_m + (X \cap Y) = V$ for some m, and so as an A-R-bimodule $V/(X \cap Y)$ is isomorphic to a factor of V_m. However, V_m has an A-R-bimodule composition series $V_0 = 0 < V_1 < \cdots < V_m$ in which the composition factors have distinct right annihilators and so are pairwise nonisomorphic. Thus, it is impossible for $V/(X \cap Y)$ to have an A-R-bimodule composition series with two composition factors isomorphic to V_1.

Therefore X is unique. □

In the situation of Theorem 2.8, the number of primes of R minimal over $P \cap R$ is equal to the length of ${}_A(V/X)$ divided by the length of ${}_AA$ [**5**, Theorem 5.12ii].

3. q-skew polynomial rings

In this section we specialize to skew polynomial rings $S = R[y; \tau, \delta]$ where τ and δ form a q-skew derivation with q not a root of unity. (As before, R will be noetherian and τ an automorphism.) In order to compare ranks of prime factors of S with ranks of prime factors of $R[z;\tau]$, we investigate the corresponding Goldie quotient rings. We shall see that the Goldie quotient ring of a prime factor S/P is in most cases isomorphic to the Goldie quotient ring of a prime factor of $R[z;\tau]$. The possible exception occurs when P is an induced prime, that is, $P = IS$ for some (τ,δ)-invariant prime I of R; however, this is an easy case to analyze directly.

The most immediate benefit of the q-skew assumption is that it allows us to give compact formulas for products in $R[y;\tau,\delta]$, similar to those available in differential operator rings (i.e., when $\tau = 1$). Namely, when (τ,δ) is a q-skew derivation, we have the following *q-Leibniz Rules* noted in [**4**, Lemma 6.2]:

$$\begin{aligned} y^n a &= \sum_{i=0}^{n} \binom{n}{i}_q \tau^{n-i}\delta^i(a) y^{n-i} \\ \delta^n(ab) &= \sum_{i=0}^{n} \binom{n}{i}_q \tau^{n-i}\delta^i(a)\delta^{n-i}(b) \end{aligned}$$

for all $a, b \in R$ and all nonnegative integers n, where the *q-binomial coefficient* $\binom{n}{i}_q$ is the evaluation at $t = q$ of the polynomial

$$\binom{n}{i}_t = \frac{(t^n-1)(t^{n-1}-1)\cdots(t-1)}{(t^i-1)(t^{i-1}-1)\cdots(t-1)(t^{n-i}-1)(t^{n-i-1}-1)\cdots(t-1)} \in \mathbb{Z}[t].$$

We shall need the observation that if $q^i + q^{i-1} + \cdots + 1$ is invertible in R for all positive integers i, then $\binom{n}{i}_q$ is invertible for $n \ge i \ge 0$.

We first consider a prime $P \in \operatorname{spec} S$ such that the primes minimal over $P \cap R$ lie in an infinite τ-orbit. In this case, the bimodule analysis of the previous section together with some minimal degree arguments yields a precise description of $\operatorname{Fract}(S/P)$.

Theorem 3.1. [5, Theorem 8.5.I] *Let $S = R[y;\tau,\delta]$, where R is noetherian and τ an automorphism. Moreover, assume that (τ,δ) is a q-skew derivation for some central unit $q \in R$ such that $q^i + q^{i-1} + \cdots + 1$ is invertible for all $i \ge 0$. Let $P \in \operatorname{spec} S$, let $Q \in \operatorname{spec} R$ be minimal over $P \cap R$, and suppose that the τ-orbit of Q is infinite. Then $P \cap R = Q$ and the embedding $R/Q \hookrightarrow S/P$ extends to an isomorphism of $\operatorname{Fract}(R/Q)$ onto $\operatorname{Fract}(S/P)$. In particular, $\operatorname{rank}(S/P) = \operatorname{rank}(R/Q)$.*

Proof. Let $Q' \in \operatorname{spec} R$ be an annihilator prime of $(S/P)_R$. By Corollaries 2.5 and 2.6, Q' is minimal over $P \cap R$ and Q' lies in the τ-orbit of Q. Hence, the τ-orbit of Q' is infinite. Next, let $A = \operatorname{Fract}(R/Q')$ and $V = A \bigotimes_R S$. By Theorem 2.8, V has a unique nonzero maximal proper A-S-sub-bimodule X, the right S-annihilator of V/X equals P, and V/X is a torsionfree right (S/P)-module.

Choose a nonzero element $x \in X$ with minimal degree, say

$$x = (x_0 \otimes 1) + \cdots + (x_n \otimes y^n)$$

with $x_i \in A$ and $x_n \neq 0$. As in the proof of Theorem 2.8, we can arrange, without loss of generality, that $x_n = 1$. Since X is proper, $n \geq 1$. For any $r \in R$, observe that $xr - \tau^n(r)x$ is an element of X with degree less than n, and so $xr = \tau^n(r)x$. Comparing coefficients of y^{n-1} in this equation, we obtain

$$\binom{n}{1}_q \overline{\tau^{n-1}\delta(r)} + x_{n-1}\tau^{n-1}(r) = \tau^n(r)x_{n-1},$$

where the overbar denotes an image in A. Now $q^{n-1}\tau^{n-1}\delta(r) = \delta\tau^{n-1}(r)$, and our hypothesis on q implies that $\binom{n}{1}_q$ is invertible in R. Set $c = \binom{n}{1}_q^{-1} q^{n-1}x_{n-1} \in A$; then

$$\overline{\delta\tau^{n-1}(r)} = \tau^n(r)c - c\tau^{n-1}(r).$$

Since r is an arbitrary element of R, we may replace it by $\tau^{1-n}(r)$. Therefore we have $\overline{\delta(r)} = \tau(r)c - cr$ for all $r \in R$.

Now set $w = (c \otimes 1) + (1 \otimes y) \in V$, and compute that $wr = \tau(r)w$ for all $r \in R$. Hence, $AwS = \sum_{i=0}^{\infty} Awy^i$, from which it follows that AwS is a proper A-S-sub-bimodule of V. Also, $V = V_1 \bigoplus AwS$ (where $V_1 = A \otimes 1$ as in Lemma 2.3), and so $V/AwS \cong V_1$ as A-R-bimodules. Thus V/AwS is simple as an A-R-bimodule, and hence also as an A-S-bimodule.

Therefore AwS is a maximal proper A-S-sub-bimodule of V. By uniqueness, $X = AwS$, and hence $V/X \cong V_1$ as A-R-bimodules. Consequently, $P \cap R = \operatorname{r.ann}(V_1)_R = Q'$. Thus $P \cap R$ is prime, and so $Q = P \cap R = Q'$.

Finally, let E be the endomorphism ring of the left A-module V/X, and write functions in E on the right of their arguments. Since $V/X \cong V_1 \cong A$ as A-R-bimodules, there is a ring isomorphism $\phi : A \to E$ such that for each $r \in R$, the map ϕ sends the coset $r + Q$ to right multiplication by r. In particular, ϕ sends all regular elements of R/Q to units in E. Since V/X has finite length as a left A-module and is torsionfree as a right (S/P)-module, [8, Lemma 7.8] shows that the right (S/P)-module structure on V/X extends to a right module structure over the ring $B = \operatorname{Fract}(S/P)$ in such a way that V/X becomes an A-B-bimodule. Hence, there is a ring homomorphism $\psi : B \to E$ which sends elements of S/P to right multiplication by those elements. Since B is simple, ψ is injective. Observe that the following diagram is commutative:

$$\begin{array}{ccccc} R/Q & \xrightarrow{\subseteq} & A & \xrightarrow{\phi} & E \\ \downarrow & & & & \| \\ S/P & \xrightarrow{\subseteq} & B & \xrightarrow{\psi} & E \end{array}$$

In particular, the simple artinian ring $\psi(B)$ contains $\phi(R/Q)$. Now for any regular element $z \in R/Q$, we note that $\phi(z)$ is invertible in E, whence $\phi(z)$ is regular in $\psi(B)$, and thus also invertible in $\psi(B)$. Consequently, $E = \phi(A) = \psi(B)$. Therefore $\psi^{-1}\phi$ is the desired isomorphism of A onto B.

The final statement of the theorem is clear. □

We now turn to the case of a prime $P \in \operatorname{spec} S$ such that the primes minimal over $P \cap R$ lie in a finite τ-orbit. This case is closer to the classical situation: although $P \cap R$ need not be invariant under τ or δ, this intersection turns out to contain a "large" (τ, δ)-invariant semiprime ideal I, and we shall of course factor out I and localize. The obvious choice for I is the largest (τ, δ)-invariant ideal contained in $P \cap R$. To show that this ideal has the desired properties, we require a pair of auxiliary results.

Recall that a proper τ-invariant ideal A of R is *τ-prime* provided that whenever B, C are τ-invariant ideals of R such that $BC \subseteq A$, then either $B \subseteq A$ or $C \subseteq A$. The concept of a *(τ, δ)-prime* ideal of R is defined analogously. Because we are assuming that R is noetherian, its τ-prime ideals are precisely the ideals of the form $I = Q \cap \tau(Q) \cap \cdots \cap \tau^{n-1}(Q)$ where $Q \in \operatorname{spec} R$ and $\tau^n(Q) = Q$ [**12**, Lemma 10.6.11, Proposition 10.6.12]. Note in particular that such an ideal I is semiprime, and that the primes minimal over I are $Q, \tau(Q), \dots, \tau^{n-1}(Q)$.

Lemma 3.2. [**4**, Lemma 6.4] *Assume that (τ, δ) is a q-skew derivation on a ring R, for some central unit $q \in R$ such that $q^i + q^{i-1} + \cdots + 1$ is invertible for all $i \geq 0$, and that τ is an automorphism. Then any minimal τ-prime ideal I of R is δ-invariant.*

Proof. Since the proof exactly parallels the proof for invariance of minimal primes under ordinary derivations in characteristic zero given in [**7**, Proposition 1.1], we just give the outline.

First, set $J = \{r \in R \mid \delta^i(r) \in I \text{ for all } i = 0, 1, \dots\}$, and observe that J is a (τ, δ)-ideal contained in I. We claim that J is τ-prime. Once this is verified, $J = I$ by the minimality of I, and the δ-invariance of I follows.

To prove that J is τ-prime, it suffices to show that for any $u, w \in R \setminus J$, there exist $v \in R$ and $t \in \mathbb{Z}$ such that $uv\tau^t(w) \notin J$ (e.g., [**4**, Lemma 2.1(a)]). Choose non-negative integers m and n minimal such that $\delta^m(u), \delta^n(w) \notin I$. By the τ-primeness criterion just cited, there exist $v \in R$ and $s \in \mathbb{Z}$ such that $\delta^m(u)v\tau^s\delta^n(w) \notin I$. Set $a = \tau^{-n}(u)$ and $b = \tau^{-n}(v)$, and set $c = \tau^s(w)$. Then

$$q^{mn}\tau^n\delta^m(a)\tau^n(b)\delta^n(c) = q^{ns}\delta^m(u)v\tau^s\delta^n(w) \notin I,$$

and so $\tau^n\delta^m(a)\tau^n(b)\delta^n(c) \notin I$. On the other hand, $\delta^i(a) = q^{-in}\tau^{-n}\delta^i(u) \in I$ for $0 \leq i < m$, and similarly $\delta^j(c) \in I$ for $0 \leq j < n$.

Now expand $\delta^{m+n}(abc)$ using a double application of the q-Leibniz Rule. All terms but one of the resulting sum lie in I because $\delta^i(a)$ and $\delta^j(c)$ are in I for $i < m$ and $j < n$. The remaining term has the form $\binom{m+n}{n}_q \tau^n\delta^m(a)\tau^n(b)\delta^n(c)$, which is not in I because the q-binomial coefficients are invertible in R. Hence, $\delta^{m+n}(abc) \notin I$, and so $abc \notin J$.

Finally, $uv\tau^{n+s}(w) = \tau^n(abc) \notin J$, and therefore J is τ-prime, as desired. □

Proposition 3.3. [4, Proposition 6.5] *Assume that (τ,δ) is a q-skew derivation on a right noetherian ring R, for some central unit $q \in R$ such that $q^i+q^{i-1}+\cdots+1$ is invertible for all $i \geq 0$, and that τ is an automorphism. Then every (τ,δ)-prime ideal of R is also τ-prime.*

Proof. It suffices to show that if R is a (τ,δ)-prime ring, then it is also a τ-prime ring.

Let $Q_1,\ldots,Q_n$ be the minimal prime ideals of R, and for $i = 1,\ldots,n$ let $J_i = \bigcap_{j\in\mathbb{Z}} \tau^j(Q_i)$. Each J_i is a minimal τ-prime ideal of R, and so is δ-invariant by Lemma 3.2. Some product of the Q_i is zero, and so the corresponding product of the J_i is zero. Since R is (τ,δ)-prime, it follows that some $J_i = 0$, and therefore R is τ-prime. □

Proposition 3.4. [5, Theorem 10.3] *Let $S = R[y;\tau,\delta]$, where R is noetherian, τ an automorphism, and (τ,δ) a q-skew derivation for some central unit $q \in R$ such that $q^i + q^{i-1} + \cdots + 1$ is invertible for all $i \geq 0$. Moreover, let $P \in \operatorname{spec} S$, let $Q \in \operatorname{spec} R$ be minimal over $P \cap R$, and suppose that the τ-orbit of Q is finite. Then there exists a δ-invariant τ-prime ideal $I \subseteq P \cap R$ such that Q is minimal over I.*

Proof. Let $I = \{r \in R \mid \tau^n(r) \in P \text{ for all } n = 0,1,\ldots\}$, and note that I is the largest τ-ideal of R contained in $P \cap R$. Given any $r \in R$, we have

$$\tau^n\delta(r) = q^{-n}\delta\tau^n(r) = q^{-n}[y\tau^n(r) - \tau^{n+1}(r)y] \in P$$

for all $n \geq 0$, whence $\delta(r) \in I$. Thus I is actually a (τ,δ)-ideal.

We next claim that I is (τ,δ)-prime. If A and B are (τ,δ)-ideals of R such that $AB \subseteq I$, then AS and BS are ideals of S such that $(AS)(BS) = ABS \subseteq IS \subseteq P$. Now either $AS \subseteq P$ or $BS \subseteq P$, say the former. Then $A \subseteq P \cap R$, and consequently $A \subseteq I$ because A is a τ-ideal. This verifies the claim. Thus, by Proposition 3.3, I must be τ-prime.

Let $J = \bigcap_{j\in\mathbb{Z}} \tau^j(Q)$, and note that since the τ-orbit of Q is finite, Q is minimal over J. By Corollary 2.5, all the primes minimal over $P \cap R$ lie in the τ-orbit of Q, and so J is contained in the prime radical of $P \cap R$. Consequently, $J^t \subseteq P \cap R$ for some $t \in \mathbb{N}$, whence $J^t \subseteq I$, because J^t is a τ-ideal. Since I is τ-prime, we conclude that $J \subseteq I$, and therefore Q is minimal over I. □

Note that since the ideal I in Proposition 3.4 is τ-prime, it must equal the intersection of the primes in the τ-orbit of Q.

With the help of Proposition 3.4, we shall be able to reduce to the situation that our coefficient ring R is an artinian τ-prime ring. Exempting the case when R and S are simple rings, we shall prove that δ is then inner, whence $S \cong R[z;\tau]$. With the following two lemmas we prepare for this proof.

Lemma 3.5. *Let $S = R[y;\tau,\delta]$, where τ is an automorphism. Assume that (τ,δ) is a q-skew derivation for some central unit $q \in R$ such that $q^i + q^{i-1} + \cdots + 1$ is invertible for all $i \geq 0$. If R is simple but S is not, then δ is an inner τ-derivation.*

Proof. We use the same minimal degree argument as in the proof of Theorem 3.1. Let I be a proper nonzero ideal of S, and let s be a nonzero element of I with minimal degree, say degree n and leading coefficient s_n. Since R is simple, there exist elements $a_i, b_i \in R$ such that $\sum_i a_i s_n b_i = 1$. Then $\sum_i a_i s \tau^{-n}(b_i)$ is an element of I with degree n and leading coefficient 1. Hence, there is no loss of generality in assuming that $s_n = 1$. Since $I \neq S$, we must have $n \geq 1$.

For any $r \in R$, observe that $s\tau^{1-n}(r) - \tau(r)s$ is an element of I with degree less than n, whence $s\tau^{1-n}(r) = \tau(r)s$. If s_{n-1} denotes the coefficient of y^{n-1} in s, then

$$\binom{n}{1}_q \tau^{n-1}\delta\tau^{1-n}(r) + s_{n-1}r = \tau(r)s_{n-1}.$$

Setting $c = \binom{n}{1}_q^{-1} q^{n-1} s_{n-1}$, we conclude that $\delta(r) = \tau(r)c - cr$ for all $r \in R$. Therefore δ is inner. □

Lemma 3.6. [3, Lemma 1.4; **18**, Proposition 5, p. 249] *Let (τ,δ) be a skew derivation on a ring R. Suppose that there exist orthogonal central idempotents $e_1,\dots,e_n \in R$ such that $e_1 + \cdots + e_n = 1$ and $\tau(e_i) = e_{i+1}$ for all i, where indices are interpreted modulo n. If $n \geq 2$, then δ is an inner τ-derivation.*

Proof. For each i, note that $\delta(e_i) = \delta(e_i^2) = e_{i+1}\delta(e_i) + \delta(e_i)e_i$, whence $\delta(e_i) = a_i + b_i$ for some $a_i \in e_iR$ and $b_i \in e_{i+1}R$. Since also

$$0 = \delta(e_ie_{i+1}) = e_{i+1}\delta(e_{i+1}) + \delta(e_i)e_{i+1} = a_{i+1} + b_i,$$

we find that $\delta(e_i) = a_i - a_{i+1}$. For any $r \in R$, we have

$$\tau(r)(a_i - a_{i+1}) + \delta(r)e_i = \delta(re_i) = \delta(e_ir) = e_{i+1}\delta(r) + (a_i - a_{i+1})r.$$

Multiplying this equation by e_i, we obtain $\tau(r)a_i + \delta(r)e_i = a_ir$. Thus

$$\delta(r) = (a_1 + \cdots + a_n)r - \tau(r)(a_1 + \cdots + a_n)$$

for all $r \in R$ □

Theorem 3.7. *Let $S = R[y;\tau,\delta]$, where R is noetherian and τ an automorphism. Assume that (τ,δ) is a q-skew derivation for some central unit $q \in R$ such that $q^i + q^{i-1} + \cdots + 1$ is invertible for all $i \geq 0$. Moreover, let $P \in \operatorname{spec} S$, let $Q \in \operatorname{spec} R$ be minimal over $P \cap R$, and suppose that the τ-orbit of Q is finite. Then either $P = IS$ for some (τ,δ)-invariant prime ideal I of S, or* $\operatorname{Fract}(S/P) \cong \operatorname{Fract}(R[z;\tau]/P')$ *for some prime ideal P' in $R[z;\tau]$.*

Proof. By Proposition 3.4, there exists a δ-invariant τ-prime ideal $I \subseteq P \cap R$ such that Q is minimal over I. After replacing R and S by R/I and S/IS, we may assume that $I = 0$. Hence, R is a τ-prime ring and Q is a minimal prime of R. If R is prime and $P = 0$, we are done. Thus we may assume that either R is not prime or P is nonzero.

Since $P \cap R$ is contained in a minimal prime of R, it is disjoint from $\mathcal{C}_R(0)$ (e.g., [**8**, Proposition 6.5]), and so we may localize with respect to $\mathcal{C}_R(0)$. Thus, without loss of generality, we may assume that R is artinian as well as τ-prime. It follows that R is actually τ-simple. Hence, the primitive central idempotents in R must be transitively permuted by τ, and so we can list them as $e, \tau(e), \ldots, \tau^{n-1}(e)$ where $\tau^n(e) = e$ and $\tau^i(e) \neq \tau^j(e)$ for $0 \leq i < j < n$. If $n = 1$, then R is prime and so $P \neq 0$ by assumption, whence S is not simple. In this case, δ is inner by Lemma 3.5. On the other hand, if $n \geq 2$, then δ is inner by Lemma 3.6. Thus in either case $S \cong R[z; \tau]$, and therefore S/P is isomorphic to a prime factor ring of $R[z; \tau]$. $\square$

4. Bounds on uniform ranks of prime factors

The results of the previous section provide what is needed from [**5**] to prove the q-analog of Sigurdsson's theorem, as follows.

Theorem 4.1. [**6**, Theorem 1.3] *Let $S = R[y; \tau, \delta]$, where R is noetherian and τ an automorphism. Suppose that $\delta\tau = q\tau\delta$ for some central unit $q \in R$ such that $\delta(q) = 0$, $\tau(q) = q$, and $q^i + q^{i-1} + \cdots + 1$ is invertible in R for all positive integers i. Further assume that there is a finite upper bound b for the ranks of all prime factors of $R[z; \tau]$. Then each prime factor of S has rank at most b. In particular, if all prime ideals of $R[z; \tau]$ are completely prime, then all prime ideals of S are completely prime.*

Proof. Given $P \in \operatorname{spec} S$, choose $Q \in \operatorname{spec} R$ minimal over $P \cap R$. If the τ-orbit of Q is infinite, then $\operatorname{rank}(S/P) = \operatorname{rank}(R/Q)$ by Theorem 3.1. Since R/Q is a homomorphic image of $R[z; \tau]$, it is immediate that $\operatorname{rank}(S/P) \leq b$ in this case. Now assume that the τ-orbit of Q is finite. By Theorem 3.7, either $P = IS$ for some (τ, δ)-invariant prime ideal I of S, or $\operatorname{Fract}(S/P) \cong \operatorname{Fract}(R[z; \tau]/P')$ for some prime ideal P' in $R[z; \tau]$. In the latter case, it is immediate that $\operatorname{rank}(S/P) \leq b$. In the former case, $S/P \cong (R/I)[y; \tau, \delta]$, and an easy lemma [**6**, Lemma 1.2] shows that $\operatorname{rank}(S/P) = \operatorname{rank}(R/I)$. Since R/I is a homomorphic image of $R[z; \tau]$, we conclude that $\operatorname{rank}(S/P) \leq b$ in this case also. $\square$

The need to keep track of automorphisms is an obstacle to proving a very general iterated form of Theorem 4.1. The following special version, however, is sufficient to handle many quantum coordinate algebras.

Theorem 4.2. [**6**, Theorem 2.3] *Let $S = k[y_1][y_2; \tau_2, \delta_2] \cdots [y_n; \tau_n, \delta_n]$ be an iterated skew polynomial ring over a field k. Set $S_i = k[y_1][y_2; \tau_2, \delta_2] \cdots [y_i; \tau_i, \delta_i]$ for $1 \leq i \leq n$, and assume that the following conditions hold for $2 \leq i \leq n$:*

(a) τ_i *is a k-algebra automorphism of* S_{i-1} *and* δ_i *is a k-linear* τ_i*-derivation of* S_{i-1}.

(b) *For each* $j = 1, \ldots, i-1$, *there exists* $\lambda_{ij} \in k^\times$ *such that* $\tau_i(y_j) = \lambda_{ij} y_j$.

(c) *There exists* $q_i \in k^\times$ *such that* $\delta_i \tau_i(y_j) = q_i \tau_i \delta_i(y_j)$ *for* $j = 1, \ldots, i-1$.

(d) *Either* q_i *is not a root of unity, or* $q_i = 1$ *and the characteristic of k is zero. Further assume that the subgroup of* $k^\times$ *generated by the* λ_{ij} *is torsionfree. Then all prime ideals of S are completely prime.*

Sketch of proof. Because of assumption (d), the sums $q_i^j + q_i^{j-1} + \cdots + 1$ are invertible for $i = 2, \ldots, n$ and $j = 1, 2, \ldots$.

Observe that $\tau_n^{-1}\delta_n\tau_n$ and $q_n\delta_n$ are τ_n-derivations of S_{n-1} and that they agree on $y_1, \ldots, y_{n-1}$. Since $y_1, \ldots, y_{n-1}$ generate S_{n-1} as a k-algebra, it follows that $\tau_n^{-1}\delta_n\tau_n = q_n\delta_n$, that is, $\delta_n\tau_n = q_n\tau_n\delta_n$. Thus by Theorem 4.1, it suffices to show that all prime ideals of $S_{n-1}[z_n;\tau_n]$ are completely prime. In other words, there is no loss of generality in assuming that $\delta_n = 0$.

Once we have reduced to the case that $\delta_n = 0$, we can rewrite S in the form

$$S_{n-2}[y_n;\tau_n'][y_{n-1};\tau_{n-1}',\delta_{n-1}']$$

where τ_n' denotes the restriction of τ_n to S_{n-2} while τ_{n-1}' and δ_{n-1}' are extensions of τ_{n-1} and δ_{n-1} to $S_{n-2}[y_n;\tau_n']$ such that $\tau_{n-1}'(y_n) = \lambda_{n,n-1}^{-1} y_n$ and $\delta_{n-1}'(y_n) = 0$. As in the preceding paragraph, we see by Theorem 4.1 that it suffices to show that all prime ideals of $S_{n-2}[y_n;\tau_n'][z_{n-1};\tau_{n-1}']$ are completely prime. Hence, there is no loss of generality in assuming that $\delta_{n-1} = 0$.

Continuing in this fashion, we reduce to the case that all the δ_i are zero. A separate analysis is required to handle this case – see [**6**, Theorem 2.1]. □

5. Quantum matrices

The main application of Theorem 4.2 given in [**6**] is to show that the prime ideals in the quantum coordinate rings $\mathcal{O}_q(M_n(k))$ over a field k are completely prime when q is not a root of unity. Special cases had been obtained earlier in work of Hodges, Levasseur, Oh, and Stafford [**9**, **10**, **11**, **13**] – see [**6**, Remarks 3.4] for details. Here we consider the multiparameter version of $\mathcal{O}_q(M_n(k))$ introduced by Artin, Schelter, Tate [**1**] and Sudbery [**16**].

Let λ be an element of k different from 0 and -1, and let $\mathbf{p} = (p_{ij})$ be an $n \times n$ matrix of nonzero elements of k such that $p_{ii} = 1$ and $p_{ji} = p_{ij}^{-1}$ for $i, j = 1, \ldots, n$. We write $\mathcal{O}_{\lambda,\mathbf{p}}(M_n(k))$ for the *multiparameter coordinate ring of quantum* $n \times n$ *matrices* over k with generators and relations as in [**1**, Theorem 1]. The k-algebra $\mathcal{O}_{\lambda,\mathbf{p}}(M_n(k))$ is generated by elements u_{ij} for $i, j = 1, \ldots, n$ subject to the following relations:

$$u_{\ell m}u_{ij} = \begin{cases} p_{\ell i}p_{jm}u_{ij}u_{\ell m} + (\lambda - 1)p_{\ell i}u_{im}u_{\ell j} & \text{when } \ell > i \text{ and } m > j \\ \lambda p_{\ell i}p_{jm}u_{ij}u_{\ell m} & \text{when } \ell > i \text{ and } m \le j \\ p_{jm}u_{ij}u_{\ell m} & \text{when } \ell = i \text{ and } m > j. \end{cases}$$

The one-parameter quantum coordinate ring $\mathcal{O}_q(M_n(k))$ for a nonzero element $q \in k$, as defined in [**15**], then arises as the special case when $\lambda = q^{-4}$ and $p_{ij} = q^2$ for all $i > j$.

If the u_{ij} are adjoined in lexicographic order, $\mathcal{O}_{\lambda,\mathbf{p}}(M_n(k))$ can be presented as an iterated skew polynomial ring

$$k[u_{11}][u_{12};\tau_{12}]\cdots[u_{ij};\tau_{ij},\delta_{ij}]\cdots[u_{nn};\tau_{nn},\delta_{nn}],$$

as shown in [**1**, pp. 890-891]. (This can also be established by direct but tedious calculations.) All the $\tau_{\ell m}$ and $\delta_{\ell m}$ are k-linear, and

$$\tau_{\ell m}(u_{ij}) = \begin{cases} p_{\ell i}p_{jm}u_{ij} & \text{when } \ell \geq i \text{ and } m > j \\ \lambda p_{\ell i}p_{jm}u_{ij} & \text{when } \ell > i \text{ and } m \leq j \end{cases}$$

$$\delta_{\ell m}(u_{ij}) = \begin{cases} (\lambda - 1)p_{\ell i}u_{im}u_{\ell j} & \text{when } \ell > i \text{ and } m > j \\ 0 & \text{when } \ell > i \text{ and } m \leq j \\ 0 & \text{when } \ell = i \text{ and } m > j. \end{cases}$$

Observe that when $\ell > i$ and $m > j$, we have

$$\begin{aligned} \delta_{\ell m}\tau_{\ell m}(u_{ij}) &= p_{\ell i}p_{jm}\delta_{\ell m}(u_{ij}) = p_{\ell i}p_{jm}(\lambda - 1)p_{\ell i}u_{im}u_{\ell j} \\ \tau_{\ell m}\delta_{\ell m}(u_{ij}) &= (\lambda - 1)p_{\ell i}\tau_{\ell m}(u_{im}u_{\ell j}) = (\lambda - 1)p_{\ell i}\lambda p_{\ell i}p_{jm}u_{im}u_{\ell j}. \end{aligned}$$

Hence, $\tau_{\ell m}^{-1}\delta_{\ell m}\tau_{\ell m}(u_{ij}) = \lambda^{-1}\delta_{\ell m}(u_{ij})$ whenever $(\ell, m) > (i, j)$ in the lexicographic order. Therefore $\tau_{\ell m}^{-1}\delta_{\ell m}\tau_{\ell m} = \lambda^{-1}\delta_{\ell m}$, that is, $\delta_{\ell m}\tau_{\ell m} = \lambda^{-1}\tau_{\ell m}\delta_{\ell m}$.

Theorem 5.1. *Let k be a field, λ an element of k different from 0 and -1, and $\mathbf{p} = (p_{ij})$ an $n \times n$ matrix of nonzero elements of k such that $p_{ii} = 1$ and $p_{ji} = p_{ij}^{-1}$ for $i, j = 1, \dots, n$. If the subgroup of $k^\times$ generated by λ together with all the p_{ij} is torsionfree, then all prime ideals of $\mathcal{O}_{\lambda,\mathbf{p}}(M_n(k))$ are completely prime.*

Proof. In case $\lambda = 1$, the result follows from [**6**, Theorem 2.1]. If $\lambda \neq 1$, then because of our torsionfreeness hypothesis, λ^{-1} is not a root of unity. In this case, the result follows from Theorem 4.2, with the scalars $p_{\ell i}p_{jm}$ and $\lambda p_{\ell i}p_{jm}$ taking the role of the λ_{ij} and λ^{-1} taking the role of the q_i. □

The torsionfreeness hypothesis in Theorem 5.1 is essential. For example, let k be a field of characteristic different from 2, choose an arbitrary element $\lambda \neq 0, -1$ in k, and construct the algebra $S = \mathcal{O}_{\lambda,\mathbf{p}}(M_2(k))$ where $\mathbf{p} = \begin{pmatrix} 1 & -1 \\ -1 & 1 \end{pmatrix}$. Observe that $u_{21}S + u_{22}S$ is a two-sided ideal of S, and that $S/(u_{21}S + u_{22}S) \cong k[u_{11}][u_{12};\sigma]$ where σ is the k-algebra automorphism of $k[u_{11}]$ such that $\sigma(u_{11}) = -u_{11}$. Then $u_{11}^2 - 1$ generates a σ-prime ideal of $k[u_{11}]$, and hence $u_{11}^2 - 1$ generates a prime ideal of $k[u_{11}][u_{12};\sigma]$ (see, e.g., [**12**, Proposition 10.6.12]). Thus $(u_{11}^2 - 1)S + u_{21}S + u_{22}S$

is a prime ideal of S. This prime ideal is clearly not completely prime, since it contains the product $(u_{11} + 1)(u_{11} - 1)$ without containing either factor.

6. References

1. M. Artin, W. Schelter, and J. Tate, Quantum deformations of GL_n, *Communic. Pure Appl. Math.* **44** (1991) 879-895.

2. W. Borho, On the Joseph-Small additivity principle for Goldie ranks, *Compositio Math.* **47** (1982) 3-29.

3. G. Cauchon and J. C. Robson, Endomorphisms, derivations, and polynomial rings, *J. Algebra* **53** (1978) 227-238.

4. K. R. Goodearl, Prime ideals in skew polynomial rings and quantized Weyl algebras, *J. Algebra* **150** (1992) 324-377.

5. K. R. Goodearl and E. S. Letzter, Prime ideals in skew and q-skew polynomial rings, *Memoirs Amer. Math. Soc.* (to appear).

6. ———, Prime factor algebras of the coordinate ring of quantum matrices, *Proc. Amer. Math. Soc.* (to appear).

7. K. R. Goodearl and R. B. Warfield, Jr., Primitivity in differential operator rings, *Math. Zeitschr.* **180** (1982) 503-523.

8. ———, *An Introduction to Noncommutative Noetherian Rings*, London Math. Soc. Student Texts 16 (Cambridge Univ. Press., Cambridge, 1989).

9. T. J. Hodges and T. Levasseur, Primitive ideals of $\mathrm{C}_q[\mathrm{SL}(3)]$, *Communic. Math. Phys.* (to appear).

10. T. J. Hodges and S.-Q. Oh, Finite dimensional simple modules over the coordinate ring of quantum matrices, (to appear).

11. T. Levasseur and J. T. Stafford, The quantum coordinate ring of the special linear group, (to appear).

12. J. C. McConnell and J. C. Robson, *Noncommutative Noetherian Rings* (Wiley-Interscience, New York, 1987).

13. S.-Q. Oh, Doctoral dissertation, University of Cincinnati, 1992.

14. G. Sigurdsson, Differential operator rings whose prime factors have bounded Goldie dimension, *Archiv Math.* **42** (1984) 348-353.

15. S. P. Smith, Quantum groups: An introduction and survey for ring theorists, in *Noncommutative Rings*, eds. S. Montgomery and L. Small, M.S.R.I. Publ. 24 (Springer-Verlag, New York, 1992), pp. 131-178.

16. A. Sudbery, Consistent multiparameter quantisation of $GL(n)$, *J. Physics A* **23** (1990) L697-L704.

17. R. B. Warfield, Jr., Prime ideals in ring extensions, *J. London Math. Soc.* (2) **28** (1983) 453-460.

18. E. Wexler-Kreindler, Propriétés de transfert des extensions d'Ore, in *Séminaire d'Algèbre P. Dubreil 1976-77*, Lecture Notes in Math. 641 (Springer-Verlag, Berlin, 1978), pp. 235-251.

WEAKLY PROJECTIVE MODULES

S.K. JAIN[†], S.R. LÓPEZ-PERMOUTH[††] AND MOHAMMAD A. SALEH
Department of Mathematics
Ohio University, Athens, Ohio 45701

ABSTRACT

A module M is said to be weakly projective if and only if it has a projective cover $\pi : P(M) \to M$ and every map from $P(M)$ into a finitely generated (free) module can be factored through M via an epimorphism (not necessarily equal to π). In this paper we investigate the basic properties of weakly projective modules. These properties are dual to those known for weakly injective modules. In particular, we show that over a right perfect ring R there exists a right module K such that for any other module M, the direct sum $M \oplus K$ is weakly projective.

1. Introduction

The purpose of this paper is to study a concept dual to that of weak injectivity as in [1, 5, 6, 9, etc.]. Given a module M with projective cover $\pi : P(M) \to M$ and another module N, we say that M is weakly N-projective if and only if all maps from $P(M)$ into N factor through M via an epimorphism $\sigma : P(M) \to M$ (not necessarily equal to π). We dualize most of the basic results in [6], provide several examples of weakly projective modules which are not projective, and indeed show that, over a perfect ring R there exists a module K such that, for any other R-module M, the direct sum $K \oplus M$ is weakly projective.

We assume all modules are right and unital unless otherwise indicated. Any terminology used but not defined in this paper will be standard unless a specific reference is given. Sources for standard terminology include [2, 3 and 8]. A submodule $N \subset M$ is said to be a small submodule (denoted $N \ll M$) if the only submodule $K \subset M$ such that $K + N = M$ is $K = M$. Given a module M and a submodule $N \subset M$, a supplement of N in M is a submodule $K \subset M$ minimal with respect to the property that $N + K = M$. Equivalently, K is a supplement of N in M if and only if K satisfies both that $K + N = M$ and $K \cap N \ll M$. A module M is said

† Research partially supported by NSF Grant No. INT 9113557.
†† Supported by Ministry of Education and Science, Spain and also by Centre de Recerca Matemàtica, Barcelona.

to be hollow if every proper submodule of M is small in M or, equivalently, if every proper submodule of M has supplement in M equal to M. Unlike the dual concept of the complement of a submodule, not all submodules of a given module need have supplements. A module all of whose submodules have a supplement is said to be a supplemented module. Over a right perfect ring, all right modules are supplemented. A superfluous cover of a module M is a module N together with an epimorphism $p : N \to M$ such that Ker p is small in N. Equivalently, one may think of a superfluous cover for M as being a module N such that $N/K \cong M$ for some small submodule $K \subset N$. A projective superfluous cover will be referred to, as is customary, a projective cover.

For any submodule K of a module N the natural inclusion map will be denoted by $i_K : K \to N$ and the natural projection by $\pi_K : N \to N/K$.

2. Basic Definitions and Results

Given modules M and N we say that M is N-projective if and only if every homomorphism $f : M \to N/K$ into a homomorphic image of N may be lifted to $\hat{f} : M \to N$ through the natural projection $\pi : N \to N/K$. When M has a projective cover one gets the following characterization of relative projectivity.

2.0 Theorem. Let M and N be modules and assume M has a projective cover P via an onto-homomorphism $\pi : P \to M$. Then M is N-projective if and only if for every homomorphism $\varphi : P \to N$ there exists a homomorphism $\hat{\varphi} : M \to N$ such that $\hat{\varphi}\pi = \varphi$. Equivalently, $\varphi(\ker\ \pi) = 0$.

Proof. The proof is basically that of Proposition 2.2 in [10], once you realize that there is neither a need for $M = N$ nor for N to have a projective cover. However, for the sake of completeness (as suggested by the referee) we prove it as below.

Only if direction. Let $\varphi : P \to N$ be a homomorphism. We shall first show that $\varphi(ker\ \pi) = 0$. Let $T = \varphi(ker\ \pi)$ and let $\pi_T : N \to N/T$ be the natural projection. Then φ induces $\hat{\varphi} : M \to N/T$ defined by $\hat{\varphi}(m) = \pi_T\varphi(p)$, where $m = \pi(p)$. Clearly, $\hat{\varphi}\pi = \pi_T\varphi$. Since M is N-projective, there exists a map $\beta : M \to N$ such that $\hat{\varphi} = \pi_T\beta$. Clearly, $(\varphi - \beta\pi)P \subseteq T$. We claim that $\varphi = \beta\pi$.

Let $X = \{p \in P \mid \varphi(p) = \beta\pi(p)\}$. We shall show that $X = P$. Let $x \in P$. Since $(\varphi - \beta\pi)(x) \in T = \varphi(ker\ \pi)$, there exists $k \in ker\ \pi$ such that $(\varphi - \beta\pi)(x) = \varphi(k)$. Therefore, $\varphi(x - k) = \beta\pi(x - k) = 0$, since $\beta\pi(k) = 0$. Thus $x - k \in X$. Therefore, $ker\ \pi + X = P$, which implies $X = P$, since $ker\ \pi$ is small in P. Therefore, $(\varphi - \beta\pi)P = 0$. In particular, $(\varphi - \beta\pi)ker\ \pi = 0$, yielding $\varphi(ker\ \pi) = 0$. Equivalently, there exists $\varphi' : M \to N$ such that $\varphi'\ \pi = \varphi$.

Conversely, let $\psi : M \to N/K$ be a homomorphism. Then by the projectivity of P there exist a homomorphism $\psi' : P \to N$ such that $\psi\pi = \pi_K\psi'$. By our hypothesis there exists $\hat{\psi} : M \to N$ such that $\hat{\psi}\pi = \psi'$. It follows easily that $\pi_K\hat{\psi} = \psi$ as desired.

The above result is dual to a well-known characterization of relative-injectivity and motivates the following definitions which are analogous to those of weak relative-injectivity and weak injectivity [5].

2.1 Definitions. Let M and N be modules and assume M has a projective cover $\pi : P \to M$. We say that M is weakly N-projective if for every map $\varphi : P \to N$ there exists an epimorphism $\sigma : P \to M$ and a homomorphism $\hat{\varphi} : M \to N$ such that $\varphi = \hat{\varphi}\sigma$. If a module M is weakly R^n-projective for all $n \in \mathbf{Z}^+$, we say that M is a *weakly projective* module.

The following characterization proves to be quite useful.

2.2 Theorem. Let M and N be modules and assume M has a projective cover $\pi : P \to M$. Then M is weakly N-projective if and only if for every map $\varphi : P \to N$ there exists a submodule $X \subset Ker\ \varphi$ such that $P/X \cong M$.

Proof. Let $\varphi : P \to N$ be a homomorphism. Assume first that M is weakly N-projective and let the homomorphism $\hat{\varphi} : M \to N$ and the epimorphism $\sigma : P \to M$ be as in the definition of weak relative-projectivity. Since $\varphi = \hat{\varphi}\sigma$, $Ker\ \sigma \subset Ker\ \varphi$. Also, $P/Ker\ \sigma \cong M$. Thus, the implication is proven by choosing $X = Ker\ \sigma$. Conversely, if $X \subset P$ satisfies the condition in the statement of the theorem, then the isomorphism $P/X \cong M$, composed with the natural projection $\pi_X : P \to P/X$ is an epimorphism $\sigma : P \to M$ satisfying that $Ker\ \sigma = X \subset Ker\ \varphi$. It follows that the map $\hat{\varphi} : M \to N$ given by $\hat{\varphi}(m) = \varphi(p)$, whenever $\sigma(p) = m$ is well defined and satisfies $\varphi = \hat{\varphi}\sigma$, proving our claim.

□

Domains of weak projectivity are closed under quotients and submodules, as shown in the next proposition.

2.3 Proposition. Let M and N be modules and assume M has a projective cover $\pi : P \to M$. Then the following statements are equivalent:
(1) M is weakly N-projective,
(2) for every submodule $K \subset N$, M is weakly K-projective, and
(3) for every submodule $K \subset N$, M is weakly N/K-projective.

Proof. Since either condition (2) or (3) trivially implies (1), we need only show that (1) implies both (2) and (3). Assume M is weakly N-projective and let K be a submodule of N and $\varphi : P \to K$ be a homomorphism. Then $\Psi = i_K\varphi : P \to N$ may be expressed as a composition $\Psi = \hat{\Psi}\sigma$, for some homomorphism $\hat{\Psi} : M \to N$ and epimorphism $\sigma : P \to M$. Since σ is onto, the range of $\hat{\Psi}$ equals the range of Ψ and so it is contained in K. Thus, we may define $\hat{\varphi} : M \to K$ via $\hat{\varphi}(m) = \hat{\Psi}(m)$ and then $\varphi = \hat{\varphi}\sigma$, proving that M is weakly K-projective, as claimed. Assume once again that M is weakly N-projective and let $f : P \to N/K$ be a homomorphism. Since P is projective, there exists a map $g : P \to N$ such that $f = \pi_K g$. The weak N-projectivity of M yields an epimorphism $\sigma : P \to M$ and a homomorphism $\hat{g} : M \to N$ such that $g = \hat{g}\sigma$. Let $\hat{f} = \pi_K \hat{g}$. Then $\hat{f}\sigma = \pi_K \hat{g}\sigma = \pi_\kappa g = f$, proving that M is indeed weakly N/K-projective.

□

While checking for weak projectivity one can actually restrict one's attention to epimorphisms, as follows.

2.4 Remark. Let M and N be modules and assume M has a projective cover $\pi : P \to M$. Then M is weakly N-projective if and only if for every submodule $K \subset N$ and for every epimorphism $\varphi : P \to K$ there exist epimorphisms $\sigma : P \to M$ and $\hat{\varphi} : M \to N$ such that $\varphi = \hat{\varphi}\sigma$.

Proof. Clear from the above proposition.

□

One can obtain a dual for [Lemma 1.3, 6] in terms of supplements of submodules.

2.5 Proposition. Let M and N be modules and assume M is supplemented and has a projective cover $\pi : P \to M$. Then M is weakly N-projective if and only if for every submodule $K \subset N$ and for every epimorphism $\varphi : P \to K$ there exist an epimorphism $\hat{\varphi} : M \to K$ such that for every supplement L' of $Ker\ \hat{\varphi}$ in M there exists a submodule $L \subset P$ such that $P/L \cong M/L'$ and $L + Ker\ \varphi = P$.

Proof. Assume M is weakly N-projective and let $\varphi : P \to K$ be an epimorphism onto a submodule $K \subset N$. Then there exists epimorphisms $\sigma : P \to M$ and $\hat{\varphi} : M \to K$ such that $\varphi = \hat{\varphi}\sigma$. Let L' be a supplement of $Ker\ \hat{\varphi}$ in M and let $L = \sigma^{-1}(L')$. For an arbitrary $p \in P$, $\sigma(p)$ may be written as $\sigma(p) = l' + k'$, with $l' \in L'$ and $k' \in Ker\hat{\varphi}$. It follows then that $\varphi(p) = \hat{\varphi}\sigma(p) = \hat{\varphi}(l') + \hat{\varphi}(k') = \hat{\varphi}(l')$. Choose $p_1 \in \sigma^{-1}(l') \subset L$. Then $\sigma(p_1) = l'$. On the other hand, $\varphi(p_1) = \hat{\varphi}\sigma(p_1) = \hat{\varphi}(l') = \varphi(p)$. So $p - p_1 \in Ker\ \varphi$ and so $L + Ker\ \varphi = P$. The fact that $P/L \cong M/L'$ follows since L is the kernel of the onto map $\pi_{L'}\sigma : P \to M/L'$. Conversely, let us assume that for every submodule $K \subset N$ and for every epimorphism $\varphi : P \to K$ there exist an epimorphism $\hat{\varphi} : M \to K$ such that for every supplement L' of $Ker\ \hat{\varphi}$ in M there exists a submodule $L \subset P$ such that $P/L \cong M/L'$ and $L + Ker\ \varphi = P$. Let $\varphi : P \to K$ be an epimorphism and $\hat{\varphi} : M \to K$ be the corresponding epimorphism. All we need is to produce another epimorphism $\sigma : P \to M$ such that $\varphi = \hat{\varphi}\sigma$. Let L' be a supplement for $Ker\ \hat{\varphi}$ and let L be the corresponding submodule of P. Let $\theta : P/L \to M/L'$ be an isomorphism. The Chinese remainder theorem yields that the map $m + Ker\ \hat{\varphi} \cap L' \mapsto (M + Ker\hat{\varphi}, m + L')$ is an isomorphism between $M/(Ker\ \hat{\varphi} \cap L')$ and $M/Ker\ \hat{\varphi} \times M/L'$. Also, $M/Ker\ \hat{\varphi} \cong K$ via $m + Ker\ \hat{\varphi} \mapsto \hat{\varphi}(m)$. So, one gets an isomorphism $\beta : M/Ker\ \hat{\varphi} \cap L' \to K \times M/L'$ such that $\beta(m + Ker\ \hat{\varphi} \cap L') = (\hat{\varphi}(m), \pi_{L'}(m))$. The isomorphism θ induces an onto map $\Psi = \theta\pi_L : P \to M/L'$. Since $Ker\ \varphi + L = P$, the map $\alpha : P \to K \times M/L'$ given by $\alpha(p) = (\varphi(p), \Psi(p))$ is onto. The induced epimorphism $\alpha' = \beta^{-1}\alpha : P \to M/(Ker\hat{\varphi} \cap L')$ may then be lifted to a map $\sigma : P \to M$. Since $Ker\ \hat{\varphi} \cap L' \ll M$, σ is indeed an epimorphism. It only remains to show that $\hat{\varphi}\sigma = \varphi$. Let us refer for the rest of this proof to $\pi_{Ker\ \hat{\varphi} \cap L'}$ simply as π. We do know that $\pi\sigma = \sigma' = \beta'\alpha$, hence $\beta\pi\sigma = \alpha$. Let $p \in P$ be arbitrary. Then $\beta(\sigma(p) + Ker\,\hat{\varphi} \cap L') = \alpha(p) = (\varphi(p), \Psi(p))$. On the other hand, $\beta(\sigma(p) + Ker\hat{\varphi} \cap L') = (\hat{\varphi}(\sigma(p), \sigma(p) + L')$. Comparing the first component in both expressions yields the desired equality. Thus, M is weakly N-projective.

□

2.6 Corollary. Let M be a hollow module with projective cover P and N be an arbitrary module. Then M is weakly N-projective if and only if any submodule K of M, which is a homomorphic image of P, is a homomorphic image of M.

Proof. Straightforward from the above proposition.

□

Modules which are weakly projective relative to a fixed module are closed under finite direct sums and under superfluous covers but not under direct summands. So, in particular, finite direct sums of weakly projective modules are weakly projective and superfluous covers of weakly projective modules are weakly projective.

2.7 Proposition

(1) Let $M_i, i = 1, 2, \cdots n$ be a family of weakly N projective modules. Then the direct sum $\bigoplus_{i=1}^{n} M_i$ is weakly N-projective.

(2) Let M/N be a weakly K-projective module where $N \ll M$. Then M is weakly K-projective.

(3) It is possible to have a direct sum $M \oplus K$ being weakly N-projective while M is not weakly N-projective.

(4) If a module is weakly projective relative to its own projective cover, then the module is indeed projective.

Proof. (1) and (2) are straightforward. To prove (3), let $R = \mathbf{Z}/(4)$. Then $\mathbf{Z}/(2) \times \mathbf{Z}/(4)$ is weakly R-projective (see Proposition 2.11, below) but $\mathbf{Z}/(2)$ is not weakly R-projective, in light of (4). In order to prove (4), consider a module M with projective cover $\pi : P \to M$. If we assume that M is weakly P-projective, then the identity map on P factors through M and this yields that $M \cong P$.

□

2.8 Remark. Over a right perfect ring, any arbitrary direct sum of weakly N-projective modules is weakly N-projective.

Proof. This can be proven in the same way as the proof of Proposition 2.7(1) since, over a right perfect ring, the projective cover of a direct sum of modules is the direct sum of the projective covers of the individual modules.

□

A finitely generated direct summand S of the projective cover of a weakly projective module M yields a direct summand (isomorphic to S) of M.

2.9 Lemma. Let M be a weakly projective module whose projective cover $P(M) = S \oplus K$, where S is finitely generated. Then M has a direct summand isomorphic to S.

Proof. Since S is finitely generated, M is weakly S-projective (Proposition 2.3). Thus the projection map $p : P(M) \to S$ factors through M, yielding an epimorphism $\hat{p} : M \to S$. Since S is projective we get that $M \cong S \times Ker\ \hat{p}$, proving our claim.

□

The next result points out the fact that weakly projective but not projective modules are 'large'.

2.10 Proposition. Every finitely generated projective module is indeed projective. Over a semiperfect ring R, a finite Goldie dimensional weakly projective module is indeed projective.

Proof. If M is finitely generated, then $P(M)$ is also finitely generated and so, by Proposition 2.7(4), M is projective. Suppose $R = \bigoplus_{i=1}^{n} e_i R$ is the representation of the semiperfect ring R as a direct sum of indecomposable projective modules. Let N be a finite Goldie dimensional weakly projective R-module. Write $P(N) = \bigoplus_{i=1}^{n} (e_i R)^{\alpha_i}$ as a direct sum of indecomposable projective modules. If any of the α_i 's were infinite, by Lemma 2.9, N would contain sums of arbitrarily many submodules, contradicting that N has finite Goldie dimension. Therefore, $P(N)$ is finitely generated and hence N is projective.

□

2.11 Proposition. Let R be a right self-injective local ring, but not a division ring, with nonzero right socle. Then for every integer $n > 0$, $R/S \times R^n$ is weakly R^n-projective but not weakly R^{n+1}-projective.

Proof. Since R is local and right self-injective, R is uniform, and so the right socle S is a simple right ideal. However, it is easy to see that S is also a simple left ideal. For, if $s \in S$ then $r \cdot ann(s) \supseteq J$, since $SJ = 0$. But J, being maximal, $r \cdot ann(s) = J$. Then for any $0 \neq x \in S$, the mapping $sR \to xR$ given by $sr \to xr$ is well defined and so by right self-injectivity of R there exists $x' \in R$ such that $x's = x$. This implies $x \in Rs$ and, therefore, $S = Rs$, because S is two-sided. This yields S is also a simple left ideal. Thus $JS = 0$.

Since R is not a division ring, the simple right ideal S is a proper right ideal of R and so is contained in J. This implies that the natural epimorphism $R \to R/S$ gives a projective cover. Furthermore, for every n the canonical epimorphism $R^{n+1} \to R/S \times R^n$ defines a projective cover. We shall prove the result by induction on n. Consider first the case where $n = 1$. Let $\varphi : R \times R \to R$ be a homomorphism. We put $a = \varphi(1,0)$ and $b = \varphi(0,1)$, and we define $u = (1,0)$ or $(-a^{-1}b, 1)$ according as $a \in J$ or $a \notin J$. Then $\varphi(u) = a$ if $a \in J$ and $\varphi(u) = \varphi(-(1,0)a^{-1}b + (0,1)) = -aa^{-1}b + b = 0$ if $a \notin J$. Now clearly the submodule uR of $R \times R$ is isomorphic to R and the submodule uS of uR is isomorphic to S. Moreover, we know that $uS \subset Ker\ \varphi$, because if $a \in J$ then $\varphi(uS) = \varphi(u)S = aS \subset JS = 0$, while if $a \notin J$ then $\varphi(uS) = \varphi(u)S = 0S = 0$. Since $uR \cong R$ is injective, $R \times R = uR \oplus K$ for some $K \subset R \times R$, which is necessarily

$\cong R$. It follows that $(R \times R)/uS = (uR \oplus K)/uS = uR/uS \times K = R/S \times R$. Thus $R/S \times R$ is weakly R-projective by Theorem 2.2.

Let us assume next that $n > 1$ and our proposition holds for $n-1$. Consider a homomorphism $\varphi : R^{n+1} \to R^n$. Let $\pi_i : R^n \to R$ be the projection onto the $i-th$ component. We need to consider two cases depending on whether there is a value of $i (1 \leq i \leq n)$ for which $\pi_i \varphi$ is onto. In the affirmative case, assume without loss of generality that π_1 is onto. Let us refer to the projection onto the complement simply as $\pi : R \times R^{n-1} \to R^{n-1}$. Let $\alpha : R^n \to Ker\ \pi_1 \varphi$ be an isomorphism. Using our inductive hypothesis on $\pi\ \varphi\ \alpha : R^n \to R^{n-1}$, we obtain a $Y \subset Ker\ \pi\ \varphi\alpha$ such that $R^n/Y \cong R/S \times R^{n-1}$. Let $X = \alpha(Y)$. Then clearly $X \subset Ker\ \pi\ \varphi$ and also $X \subset \alpha(R^n) = Ker\ \pi_1\varphi$. But $Ker\ \pi\ \varphi \cap Ker\ \pi_1\varphi = Ker\ \varphi$, and thus we have $X \subset Ker\ \varphi$. Since R^n whence $Ker\ \pi_1\varphi$ is injective, $R^{n+1} = Ker\ \pi_1\varphi\ \oplus L$ for some submodule $L \subset R^{n+1}$ and necessarily $L \cong R$. Then we have $R^{n+1}/X \cong Ker\ \pi_1\varphi\ /X \times L \cong R^n/Y \times R \cong R/S \times R^{n-1} \times R \cong R/S \times R^n$, proving our claim again by Theorem 2.2. To conclude, consider when none of the natural projections $\pi_i : R^n \to R$ satisfies that $\pi_i\varphi$ is onto, or what is the same, $\pi_i(\varphi(R^{n+1})) \subset J$ for $i = 1, 2, \cdots, n$. Denote by S^{n+1} the external product $S \times S \times \cdots \times S$ of $n+1$ copies of S. Then we have $\pi_i(\varphi(S^{n+1})) = \pi_i(\varphi(R^{n+1}S)) = \pi_i(\varphi(R^{n+1}))S \subset JS = 0$ for every i, which means that $\varphi(S^{n+1}) = 0$. Let $X = S \times 0 \times \cdots \times 0$. Then $X \subset S^{n+1} \subset Ker\ \varphi$ and $R^{n+1}/X \cong R/S \times R^n$. This shows that $R/S \times R^{n+1}$ is weakly R^n-projective, as claimed.

Suppose that $R/S \times R^n$ is weakly R^{n+1}-projective. Then $R/S \times R^n$ and hence R/S is projective by Proposition 2.7(4). But this implies that $S = 0$, which is a contradiction. This completes the proof.

□

An important fact in the theory of weakly injective modules is that a quasi-injective weakly injective module is indeed injective [6]. The dual result is

2.12 Proposition. Let N be a module. Then any quasi-projective weakly N-projective module is indeed N-projective.

Proof. Let M be a quasi-projective module M with projective cover $\pi : P \to M$ and assume that M is weakly N-projective. Consider a map $\varphi : M \to N/K$, for some submodule $K \subset N$. The projectivity of P guarantees the existence of a map $\hat{\varphi} : P \to N$ such that $\varphi\pi = \pi_K\,\hat{\varphi}$. Now, since M is weakly N-projective there exist an epimorphism $\sigma : P \to M$ and a map $\hat{\psi} : M \to N$ such that $\hat{\psi}\sigma = \hat{\varphi}$. Since M is quasi-projective there exists $\sigma' : M \to M$ such that $\sigma'\pi = \sigma$, (Theorem 2.0). One easily checks that the map $\hat{\psi}\sigma' : M \to N$ lifts φ, proving our claim.

□

3. Every Module is a Direct Summand of a Weakly Projective Module

It is shown in [9] that over arbitrary rings every semisimple module is a direct summand of a weakly injective module and every module (without restriction) is a direct summand of a tight module. This implies that if the ring is right q.f.d.

(i.e., if all cyclic right modules have finite uniform dimension) then every module is a summand of a weakly injective one. For right q.f.d. rings, one can actually get a stronger result '*with the quantifiers commuted*'. Namely, one can show that not only is it true that for every right module M there exists a right module K such that $M \oplus K$ is weakly injective, but actually that a fixed module K may be chosen such that, for every right module M, the sum $M \oplus K$ is weakly injective [7]. However, the construction of an ad hoc K for a given M, as in [9], has an interest of its own right and can be used in applications where the second construction would not apply. In the dual case, we will show that for right perfect rings, it is also true that every right module is a summand of a weakly projective module.

3.1 Theorem. Over a right perfect ring R, there exists a module K such that the direct sum of K plus any other module yields a weakly projective module.

Proof. Since R is right perfect, we may write $R = \bigoplus_{i=1}^{k}(e_i R)^{n_i}$, where $\{e_1 R, \cdots, e_k R\}$ is a complete set of representatives of indecomposable projective right R-modules. Let $L = \bigoplus I, where I \subset R^n$ for all $n \in Z^+$ be the external sum of all submodules of finitely generated free right R-modules. Let $\aleph$ be an infinite cardinal such that $\aleph >| R |$. Define $K = L \oplus [P(L)]^{(\aleph)}$, where $P(L)$ is the projective cover of L. Consider an arbitrary right R-module M and an integer $n \in \mathbf{Z}^+$. Our aim is to show that the direct sum $N = M \oplus K$ is weakly R^n-projective. Consider an epimorphism $\varphi : P(N) \to I$, where $I \subset R^n$. Let $\pi : P(I) \to I$ be the projective cover map. The projectivity of $P(N)$ yields a map $\hat{\varphi} : P(N) \to P(I)$ such that $\pi\hat{\varphi} = \varphi$. Furthermore, since $Ker\ \pi \ll P(I)$, one gets that $\hat{\varphi}$ is an epimorphism. Since $P(I)$ is projective, $\hat{\varphi}$ splits and, therefore, we may write $P(N) = P \oplus Ker\ \hat{\varphi}$, for some submodule $P \subset P(N)$ isomorphic to $P(I)$. Over a semiperfect ring all projective modules are decomposable as direct sums of indecomposable projective ones. So let us write $P(I) \cong \bigoplus_{i=1}^{k}(e_i R)^{(\alpha_i)} \cong P$, and $Ker\ \hat{\varphi} \cong \bigoplus_{i=1}^{k}(e_i R)^{(\beta_i)}$. Suppose further that $P(L) \cong \bigoplus_{i=1}^{k}(e_i R)^{(C_i)}$. Then $P(K) \cong \bigoplus_{i=1}^{k}(e_i R)^{(C_i \aleph)} \cong \bigoplus_{i=1}^{k}(e_i R)^{(D_i)}$, where $D_i \geq \aleph$. Let $P(M) \cong \bigoplus_{i=1}^{k}(e_i R)^{(F_i)}$.

Since there exists an epimorphism $\Psi : R^{(|I|)} \to I$, $P(I)$ is isomorphic to a summand of $R^{(|I|)} \cong \bigoplus_{i=1}^{k}(e_i R)^{(n_i |I|)}$. Therefore, $\alpha_i \leq n_i \mid I \mid \leq n_i \mid R^n \mid = n_i n \mid R \mid < \aleph$ for each i. The decompositions $P(N) = P(M) \oplus P(K)$ and $P(N) = P \oplus Ker\ \hat{\varphi}$ imply that $\bigoplus_{i=1}^{k}(e_i R)^{(\alpha_i \cup \beta_i)} \cong \bigoplus_{i=1}^{k}(e_i R)^{(D_i \cup F_i)}$. Since each $\alpha_i < \aleph$, while $\mid D_i \cup F_i \mid \geq \aleph$, we must conclude that $\mid D_i \cup F_i \mid = \beta_i$. So, $Ker\ \hat{\varphi} \cong P(N)$ and one can think of $\hat{\varphi}$ as the projection $p : P(N) \times P(I) \to P(I)$. It then follows that $Ker\ \varphi \cong P(N) \times Ker\ \pi$. Now, N is a homomorphic image of $P(N)$ and, by definition of K, there exists a submodule $N' \subset N$ such that $I \oplus N' = N$. So, there exists a submodule $K' \subset P(N)$ such that $P(N)/K' \cong N'$.

Let $X = K' \times Ker\ \pi \subset Ker\ \varphi$. Then $P(N)/X = [P(N) \times P(I)]/[K' \times Ker\ \pi] \cong N' \times I \cong N$, as desired.

□

ACKNOWLEDGEMENT

S.R. López-Permouth would like to thank the Ministry of Education and Science, Spain as well as the Centre de Recerca Mathemàtica, Barcelona for the support and hospitality during his visit at the Centre.

The authors would like to express their sincere thanks to the referee for his comments and valuable suggestions–in particular, his comments on weakening the hypothesis in Proposition 2.11.

REFERENCES

1. A.H. Al-Huzali, S.K. Jain and S.R. López-Permouth, *Rings whose cyclics have finite Goldie dimension*, to appear in J. Algebra.
2. F.W. Anderson and K.R. Fuller, *Rings and Categories of Modules*, Springer-Verlag, New York, Berlin, Heidelberg, 1974.
3. C. Faith, *Algebra II, Ring Theory*, Springer-Verlag, New York, Berlin, Heidelberg, 1976.
4. J.S. Golan and S.R. López-Permouth, *QI-filters and tight modules*, Comm. Alg. **19**(8) (1991), 2217-2229.
5. S.K. Jain and S.R. López-Permouth, *Rings whose cyclics are essentially embeddable in projectives*, J. Algebra **129** (1990) 257-269.
6. S.K. Jain, S.R. López-Permouth and S. Singh, *On a class of QI-rings*, Glasgow Math. J. **34** (1992), 75-81.
7. S.K. Jain, S.R. López-Permouth, K. Oshiro and M.A. Saleh, *Weakly projective and weakly injective modules*, preprint.
8. Kasch, *Modules and Rings*, Academic Press, London, New York, 1982.
9. S.R. López-Permouth, *Rings characterized by their weakly injective modules*, Glasgow Math. J. **34** (1992), 349-353.
10. L.E.T. Wu and J.P. Jans, *On quasi projectives*, Illinois J. Math. **11** (1967), 439-448.

MODULES SUITABLE FOR THE EXCHANGE OF PROPERTIES WITH THEIR ENDOMORPHISM RINGS

SOUMAYA MAKDISSI KHURI
Mathematics Department, East Carolina University
Greenville, NC 27858

ABSTRACT

The characterization of properties of a module, ${}_RM$, in terms of properties of its endomorphism ring, E, is often obtained by establishing bijections between certain submodules of M and certain left or right ideals of E. In this paper, some important bijections which occur frequently in the literature are classified in terms of conditions on M–such as nondegeneracy, nonsingularity, self-generation and retractability–which are necessary or sufficient for their establishment.

1. Introduction and Preliminaries

When characterizing properties of the endomorphism ring, $E = End_R M$, of a module, ${}_RM$, in terms of properties of M, one frequently used technique is that of establishing certain bijections between those ideals of E which characterize the property under investigation and certain submodules of M. This technique is useful in the characterization of many different types of properties of E and for many different types of modules, and it is interesting to try to determine, for each particular property of E being investigated, some minimal basic conditions on M which ensure that the appropriate bijections can be established. In doing so, one comes across some very interesting general types of modules which are worth investigating in their own right, and in this–mainly expository–paper, we compare these general modules and classify them in terms of which of the different properties of E are transferable for each module.

1.1 Types of Properties of E

The types of properties of E whose characterization in terms of properties of M we shall be classifying, are properties of E defined in terms of either the classes of annihilator left or right ideals of E (denoted by $\mathcal{A}^l$ or $\mathcal{A}^r$), or the classes of finitely generated left or right ideals of E (denoted by $\mathcal{F}^l$ or $\mathcal{F}^r$) or the class of complement left ideals of E (denoted by $\mathcal{C}^l$). Examples of such properties are: E is a *Baer* ring iff every $I \in \mathcal{A}^l$ (or $\mathcal{A}^r$) is a direct summand (d.s.) in E, and E is left (resp. right) upper *Levitzki* iff E satisfies the a.c.c. on $\mathcal{A}^l$ (resp. $\mathcal{A}^r$); E is a (Von Neumann) *regular* ring iff every $I \in \mathcal{F}^l$ (or $\mathcal{F}^r$) is a d.s. in E, and E is left (resp. right) *noetherian* iff E satisfies the a.c.c. on $\mathcal{F}^l$ (resp. $\mathcal{F}^r$); E is a left *CS* ring iff every $I \in \mathcal{C}^l$ is a d.s. in E, and E is a left *Goldie* ring iff E satisfies the a.c.c. on $\mathcal{C}^l$ and $\mathcal{A}^l$.

These properties of E are easily characterized in terms of properties of M as soon as there are bijections between the annihilator or complement or finitely generated ideals which define the property and certain classes of submodules of M. Among such classes, we have the class of *annihilator-closed submodule* of M, defined as $\mathcal{M}^a = \{U \subseteq M : U = \bigcap_{f_\alpha \in H} \ker f_\alpha$, for $H \subseteq E\}$, and its subclass $\mathcal{M}_1^a = \{U \subseteq M : U = \ker f,\ f \in E\}$ of *kernel submodules* of M; or the class of *self-generating submodules* of M, defined as $\mathcal{M}^* = \{U \subseteq M : U = \sum_{f_\alpha \in H} Mf_\alpha$, for $H \subseteq E)$, and its subclass $\mathcal{M}_1^* = \{U \subseteq M : U = Mf,\ f \in E\}$ of *image submodules* of M; and the class $\mathcal{M}^c$ of *complement submodules* of M (U is a complement submodule, or a *complement* in M, if U has no proper essential extension in M).

Notation: $U <'_R M$ denotes that U is an *essential* R-submodule of M, i.e., that U has nonzero intersection with each nonzero submodule of M.

1.2 The Main Bijections

The specific bijections which we shall be discussing are obtained either by mapping a right ideal, I, in E to the submodule $\bigcap_{f \in I} \ker f$ in $\mathcal{M}^a$ (call this mapping G1), or by mapping a left ideal, I, in E to the submodule $\sum_{f \in I} Mf$ in $\mathcal{M}^a$ (call this G2).

Whenever we are able to use G1 to establish bijections between $\mathcal{A}^r$ and $\mathcal{M}^a$, we shall say that we have a_r; if we are able to use G2 to establish bijections between $\mathcal{A}^\ell$ and $\mathcal{M}^*$, we shall say that we have a_ℓ; and if we can use G2 to establish bijections between $\mathcal{C}^\ell$ and $\mathcal{M}^c$, we shall say that we have c_ℓ.

Some properties of E, such as E's being a right or left *Rickart* ring, are defined in terms of the classes of right or left annihilators of elements of E, namely the subclasses of $\mathcal{A}^r$ or $\mathcal{A}^l$:

$$\mathcal{A}_1^r = \{I \subseteq E : I = r(f),\ f \in E\} \subseteq \mathcal{A}^r;$$

$$\mathcal{A}_1^l = \{I \subseteq E : I = l(f),\ f \in E\} \subseteq \mathcal{A}^l;$$

("r" and "l" denote right and left annihilators in E.) Other properties of E, such as E's being left or right *perfect*, or even regularity of E, are defined in terms of the classes, $\mathcal{P}^r$ or $\mathcal{P}^l$, of principal right or left ideals of E. As a consequence, some other very useful and frequently established bijections are:

a_r^1 between $\mathcal{A}_1^r$ and $\mathcal{M}_1^*$, via G1, a_l^1 between $\mathcal{A}_1^l$ and $\mathcal{M}_1^a$, via G2

p_r between $\mathcal{P}^r$ and $\mathcal{M}_1^a$, via G1, and p_l between $\mathcal{P}^l$ and $\mathcal{M}_1^*$, via G2.

To illustrate the usefulness and desirability of establishing these bijections, we list some known results, along with their left-right symmetric analogues:

[1] If a_r exists, then E is Baer iff every $U \in \mathcal{M}^a$ is a d.s. in M, and E is right upper (resp. lower) Levitzki iff M satisfies the d.c.c. (resp. a.c.c.) on $\mathcal{M}^a$[(4) Corollaries 2.6 and 2.7].

[2] If a_l exists, then E is Baer iff every $U \in \mathcal{M}^s$ is a d.s. in M, and E is left upper (resp. lower) Levitzki iff M satisfies the a.c.c. (resp. d.c.c.) on $\mathcal{M}^s$ (left analogue of [1]).

[3] If a_r^1 exists, then E is right Rickart iff every $U \in \mathcal{M}_1^s$ is a d.s. in M (right analogue of [4]).

[4] If a_l^1 exists, then E is left Rickart iff every $U \in \mathcal{M}_1^a$ is a d.s. in M[(4) Proposition 2.9].

[5] If p_r exists, then E is regular iff every $U \in \mathcal{M}_1^a$ is a d.s. in M, and E is a left perfect ring iff M satisfies the a.c.c. on $\mathcal{M}_1^a$[(5) Theorem 3.1(i) and Proposition 4.1(i)].

[6] If p_l exists, then E is regular iff every $U \in \mathcal{M}_1^s$ is a d.s. in M, and E is a right perfect ring iff M satisfies the d.c.c. on $\mathcal{M}_1^s$[(5) Theorem 3.1(ii) and Proposition 4.1(ii)].

[7] If c_l exists, then E is left CS iff every $U \in \mathcal{M}^c$ is a d.s. in M, and E has finite left Goldie dimension iff M satisfies the a.c.c. on $\mathcal{M}^c$[(4) Corollaries 3.8 and 3.11].

1.3 Some Conditions on M Necessary and/or Sufficient for the Establishment of the Above Bijections

The conditions on M most often encountered in trying to establish the main bijections are conditions such as nondegeneracy, nonsingularity, self-generation and retractability; other general types of modules that are suitable for the establishment of the bijections are continuous, trace-accessible and CS modules. All of these conditions will be defined in Section 2, along with some of their generalizations such as e-retractable and CSG modules and a-self-generators.

After giving the definitions, we will compare these conditions on M to each other. We have, for example, that if M is nondegenerate or a self-generator then M is both retractable and an a-self-generator; on the other hand, we give counterexamples to show that: "nondegenerate" and "self-generator" do not imply one another and neither do "retractable" and "a-self-generator". Other examples of these inter-relationships between the conditions are: a trace-accessible M is nondegenerate iff it is retractable; and a nonsingular e-retractable M is an a-self-generator.

1.4 Classification of Conditions on M in Terms of the Seven Bijections

When we come to the classification of conditions on M in terms of the seven bijections listed above, we make use of the relationships in the previous section and develop others by combining conditions on M. For example: (i) either nondegeneracy alone or a combination of nonsingularity with e-retractability will allow us to establish the bijection a_r, (ii) a combination of nondegeneracy with nonsingularity or with c-self-generation will allow us to establish c_l, (iii) either self-generation alone or a combination of non-singularity with c-self-generation will allow us to establish a_l^1. Consequently, the conclusions of [1] hold whenever M is either nondegenerate, a self-generator, or nonsingular and e-retractable; the conclusion of [4] holds whenever M is either a self-generator or a nonsingular c-self-generator; and the conclusions of [7] hold whenever M is either nondegenerate and nonsingular or nondegenerate and c-self-generating.

The preceding examples are of modules for which two of the seven bijections hold. Going one step higher, we have, for example: If M is nonsingular and CS, then we can establish a_r, a_l^1 and p_l, so that the conclusions of [1], [4] and [6] hold; and if M is a trace-accessible self-generator, then a_r, a_l^1 and c_l are establishable, so that the conclusions of [1], [4] and [7] hold. Finally, if M is nonsingular and continuous, then we have a_r, a_l^1, p_r and p_l, so that the conclusions of [1], [4], [5] and [6] hold; if M is free, we have a_r, a_l^1, p_l and c_l, so we have the conclusions of [1], [4], [6] and [7]; and if M is semisimple, all seven bijections and results hold.

The diagram at the end of this paper represents this classification. The seven bijections are placed at the base of the diagram. Then at the first level above the base, one finds those types of modules for which at least one of the bijections is establishable. At the second level, one has those modules for which we can establish at least two bijections, and so on. Consequently, semisimple modules are at the top of the diagram. We can use this diagram in conjuction with [1]-[7] to infer some well-known results:

- If M is semisimple, then E is Baer, regular and left CS.
- If M is free, then E is Baer iff every $U \in \mathcal{M}^a$ is a *d.s.* in M, E is left Rickart iff every kernel submodule is a *d.s.* in M, E is regular iff every image submodule is a *d.s.* in M and E is left CS iff M is CS.
- If M is nonsingular and continuous, then E is Baer and regular.

2. The Main Types of Modules

We will now give the definitions of ten general conditions on M which, alone or combined, are necessary and/or sufficient for the establishment of one or more of the seven main bijections from Section 1.2. The capital letters in front of each definition will make it easier to indicate the relationships between these conditions in the diagram.

1. ND: ${}_RM$ is *nondegenerate* if $T\,m \neq 0$ for each nonzero $m \in M$, where $T = \sum_{f \in Hom_R(M,R)} Mf$ is the *trace* of M in R.
2. SG: ${}_RM$ is a *self-generator* if every submodule, U, of M satisfies $U = M\ Hom_R(M,U)$.

Notation: For $U \subseteq M$, let $a_E(U) = \{f \in E : Uf = 0\}$; and for $H \subseteq E$, let $k_M(H) = \{m \in M : mH = 0\}$. Note that: $U \in \mathcal{M}^a$ iff $U = k_M(H)$, for $H \subseteq E$, iff $U = k_M\, a_E(U)$; when $U \in \mathcal{M}^a$ we will call U an annihilator-closed or *a-closed submodule* of M, and for any $V \subseteq M$, we call $k_M\, a_E(V) = V^a$ the *a-closure* of V.

3. a-SG: ${}_RM$ is an *a-self-generator* if every a-closed submodule, $U, \in \mathcal{M}^a$ satisfies $U = [M\ Hom_R(M,U)]^a$.
4. RE: ${}_RM$ is *retractable* if, for every nonzero submodule, U, of M we have $Hom_R(M,U) \neq 0$.
5. e-RE: ${}_RM$ is *e-retractable* if, for every nonzero complement, C, of M we have $Hom_R(M,C) \neq 0$.
6. NS: ${}_RM$ is *nonsingular* if $Im \neq 0$ for each nonzero $m \in M$, whenever I is an essential left ideal of R.
7. CS: ${}_RM$ is a *CS module* if every complement is a direct summand in M.
8. CSG: ${}_RM$ is a *CSG module* if every complement is self-generating, i.e., if, for every complement, C, in M, we have $C = M\ Hom_R(M,C)$.
9. CT: ${}_RM$ is a *continuous* module if: (a) M is CS and (b) any isomorphic image in M of a direct summand of M is a direct summand in M.
10. TA: ${}_RM$ is *trace-accessible* if $TM = M$, where T is the trace of M in R, or, equivalently, if $TN = M\ Hom_R(M,N)$ for any left R-module ${}_RN$.

When comparing these conditions to each other, it is convenient to have the following:

Notation: For $U \subseteq M$, let $i_E(U) = \{f \in E : Mf \subseteq U\}$; and for $H \subseteq E$, let $s_M(H) = \left\{\sum_{j=1}^n m_j h_j : m_j \in M\, ,\, h_j \in H\, \right\} = MH$.

The notations $s_M(H)$ and MH will be used interchangeably, and we shall identify $i_E(U)$ and $Hom_R(M,U)$. Recall also the notation for the *standard Morita context*, $(R, {}_RM_E\, ,\, {}_EM^*_R, E)$ for M, where $M^* = Hom_R(M,R)$ and we have the $R-R$-bimodule homomorphism $(,) : M \otimes_E M^* \to R$ given by $(m,f) = mf$, for $m \in M$ and $f \in M^*$, and the $E-E$-bimodule homomorphism $[,] : M^* \otimes_R M \to E$ given by $n[f,m] = (n,f)m$, for $n, m \in M$ and $f \in M^*$.

The following equivalent characterizations of nondegeneracy will be useful to us in the sequel:

2.1 Proposition: For any ${}_RM$, the following are equivalent:

a) M is nondegenerate.
b) For each $m \in M$, $[M^*, m] = 0$ implies $m = 0$.
c) For each submodule U of M, $TU <'_R U$, where T is the trace of M in R.

Proof:
$(a) \Leftrightarrow (b) : (M, M^*)m = 0 \Leftrightarrow M[M^*, m] = 0 \Leftrightarrow [M^*, m] = 0$.
$(a) \Rightarrow (c)$: If $0 \neq u \in U$, then $0 \neq Tu \subseteq TU \cap Ru$.
$(c) \Rightarrow (a)$: If $0 \neq m \in M$, then $Tm <' Rm \Rightarrow Tm \neq 0$.

□

It is easy to see that any free module, in fact any generator, is both nondegenerate and a self-generator. Also, it follows directly from the definitions that a self-generator is an a-self-generator; and, if M is a self-generator, then any submodule, U, of M satisfies $U = M\ Hom_R(M,U)$, so that if $0 \neq U$, we have $Hom_R(M,U) \neq 0$, so a self-generator is retractable. We now show that nondegeneracy, like self-generation, implies both a-self-generation and retractability.

2.2 Proposition: If ${}_RM$ is nondegenerate, then M is retractable and an a-self-generator.

Proof: Let M be nondegenerate. If U is a nonzero submodule of M and $0 \neq u \in U$, then by Proposition 2.1(*b*), $[M^*, u] \neq 0$, From $M[M^*, u] = (M, M^*)u \subseteq Ru$; we see that $[M^*, u] \subseteq Hom_R(M,U)$, so $Hom_R(M,U) \neq 0$ and M is retractable.

Let $U = k_M\, a_E(U)$ be an a-closed submodule of M. In order to show that M is an a-self-generator, it will suffice to show that, for such a $U \in \mathcal{M}^a$, we have $a_E(U) = a_E\, s_M\, i_E(U)$, for then it will follow that $U = k_M\, a_E(U) = k_M\, a_E\, s_M\, i_E(U) = [s_M\, i_E(U)]^a$. Clearly, $s_M\, i_E(U) = M i_E(U) \subseteq U$; also, it is easy to see that a_E is order-reversing, so that it follows that $a_E(U) \subseteq a_E\, s_M\, i_E(U)$. For the reverse inclusion, let $f \in a_E\, s_M\, i_E(U)$, so that $0 = [s_M\, i_E(U)]f = [M i_E(U)]f$; then we have $[i_E(U)]f = 0$. If u is any nonzero element of U, then, by nondegeneracy, we have $[M^*, u] \neq 0$. But, as we saw in the first paragraph of this proof, $[M^*, u] \subseteq Hom_R(M,U) = i_E(U)$. Hence, we have $[M^*, uf] = [M^*, u]f \subseteq [i_E(U)]f = 0$; and this implies by nondegeneracy again, that $uf = 0$. Hence $Uf = 0$ and $f \in a_E(U)$, completing the proof.

□

So far we have shown that the relationships between nondegenerate, retractable, self-generating and a-self-generating modules may be summarized diagrammatically as follows:

We will now give some examples to show that no other implications hold in the preceding diagram. We use the symbol " $\not\Rightarrow$ " to indicate the cases where a counter-example is given to show that an implication does not hold.

2.3 Examples:

a) Let R be the ring of all Cauchy sequences in Q with component-wise multiplication and let M be the ideal of zero sequences; then M is nondegenerate but not a self-generator. Thus
$$\text{ND} \not\Rightarrow \text{SG}\,.$$

b) On the other hand, the $\mathbf{Z}$-module $\mathbf{Z}/p^n\mathbf{Z}$ is a self-generator which is not nondegenerate. Thus
$$\text{SG} \not\Rightarrow \text{ND}\,.$$

Since a self-generator is an a-self-generator, b) also gives

$$\text{a-SG} \not\Rightarrow \text{ND}.$$

And since a self-generator is retractable, b) implies

$$\text{RE} \not\Rightarrow \text{ND}.$$

Similarly, since a nondegenerate module is both retractable and an a-self-generator, a) also gives

$$\text{RE} \not\Rightarrow \text{SG}$$

and

$$\text{a-SG} \not\Rightarrow \text{SG}.$$

c) Let R be the ring $\mathbf{Z}$ of integers and let M be the R-module $\left(\bigoplus_{p\in P} \mathbf{Z}_p\right) \oplus \left(\bigoplus_{p\in P} \mathbf{Z}_{p^\infty}\right)$ where P is the set of primes.

Then the socle of M, $Soc(M) = \left(\bigoplus_p \mathbf{Z}_p\right) \oplus \left(\bigoplus_p \mathbf{Z}_p\right)$ is an essential submodule of M and $Hom_R(M, X) \neq 0$ for every nonzero submodule X of $Soc(M)$, whence M is retractable. On the other hand, M is a cogenerator so every submodule of M is a-closed. Therefore, M is an a-self-generator if and only if M is a self-generator. But M is not a self-generator since for any prime p, all the nonzero endomorphisms of $\mathbf{Z}_{p^\infty}$ are surjective. Hence

$$\text{RE} \not\Rightarrow \text{a-SG}.$$

d) Let K be any subfield of the complex numbers, C, and let R be the ring of 2×2 lower triangular matrices with (1,1)-entry in K and with other entries in C :

$$R = \left\{ \begin{pmatrix} k & 0 \\ a & b \end{pmatrix} : k \in K,\ a, b \in \mathbf{C} \right\}.$$

Take $M = Re$, where $e = \begin{pmatrix} 1 & 0 \\ 0 & 0 \end{pmatrix}$. R is a left nonsingular, left CS ring and M is a direct summand of R, hence also CS and so e-retractable. We have $E = End_R M = eRe = K$. Let $\tilde{R}$ be the injective hull of R; then $\tilde{R}$ is the ring of 2×2 matrices over C; also $\tilde{M} = \tilde{R}e$. Because a left R-endomorphism of $\tilde{R}$ is also a left $\tilde{R}$-endomorphism, we get $End_R(\tilde{M}) = End_R(\tilde{R}e) = e\tilde{R}e = \mathbf{C}$. Thus K has nonzero intersection with every nonzero left ideal of C, but K is essential as a K-submodule of C only if $K = \mathbf{C}$. It is known (cf. (4)) that when M is a nonsingular, *retractable* module, then its endomorphism ring E is essential as an E-submodule of the endomorphism ring $\tilde{E}$ of the injective hull $\tilde{M}$ of M. Hence, in this example, M is not retractable unless $K = \mathbf{C}$. On the other hand, M is nonsingular and e-retractable, hence an a-self-generator (as we will show shortly, Proposition 2.4). Thus,

$$\text{a-SG} \not\Rightarrow \text{RE}.$$

In view of the examples in 2.3, we may complete our diagram as follows:

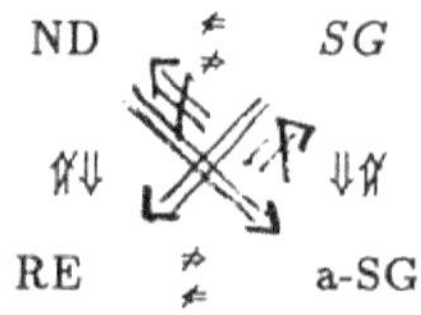

Example d) is due to A.W. Chatters, and Example c) is due to J.L. Garcia Hernandez.

In Example 2.3.d), we used the following result which we now prove for completeness.

2.4 Proposition: Let ${}_RM$ be a nonsingular e-retractable module. Then M is an a-self-generator.

Proof: Let $U = k_M\, a_E(U)$ be an a-closed submodule of M. We will show first that, since M is nonsingular, U is a complement. In order to do this, we will make use of the following well-known property of essential extensions: If $N <'_R K$ and $0 \neq k \in K$, then the left ideal ${}_RI$, or R defined by ${}_RI = [N : k] = \{r \in R : rk \in N\}$ is essential in ${}_RR$. Also recall that, when M is nonsingular, then, to any submodule, Y, of M, there corresponds a unique complement, Y^c, in M such that $Y <' Y^c$. Now let $V = U^c$, so that, in particular, $U <' V$; then, clearly $a_E(V) \subseteq a_E(U)$. Conversely, let $f \in a_E(U)$; then we will show that $f \in a_E(V)$. Let $v \in V$; then the left ideal ${}_RI = [U : v] = \{r \in R : rv \in U\} <'_R R$, and we have $Ivf = 0$. Since ${}_RM$ is nonsingular, this implies $vf = 0$, i.e., $f \in a_E(V)$, and $a_E(U) = a_E(V)$. We have $V \subseteq k_M\, a_E(V) = k_M\, a_E(U) = U$, hence $U = V$, and U is a complement in M.

In the preceding paragraph, we show that, if U and V are any two submodules of M such that $U <'_R V$, then $a_E(U) = a_E(V)$. We will show next that, since M is e-retractable, we have $s_M\, i_E(C) <'_R C$, for any complement submodule, C, of M. Let $0 \neq c \in C$; then we have $0 \neq Y = (Rc)^c \subseteq C$ and $0 \neq i_E(Y) \subseteq i_E(C)$. Let $0 \neq f \in i_E(Y)$; then $0 \neq Mf \subseteq Y$ implies $0 \neq Mf \cap Rc \subseteq s_M\, i_E(C) \cap Rc$, proving that $s_M\, i_E(C) <' C$. Now, if U is any a-closed submodule of M, then U is a complement and we have $s_M\, i_E(U) <'_R U$. Hence $a_E[s_M\, i_E(U)] = a_E(U)$, and $U = k_M\, a_E(U) = k_M\, a_E\, s_M\, i_E(U)$, proving that M is an a-self-generator.

□

Note that in Example 2.3.d), M is e-retractable but not retractable, i.e., we have

$$\text{RE} \;\overset{\not\Leftarrow}{\Rightarrow}\; e\text{-RE}\,.$$

Also in 2.3.d), M is nonsingular but not nondegenerate since a nondegenerate module is retractable, so we have

$$\text{NS} \;\not\Rightarrow\; \text{ND}\,.$$

On the other hand, if R is a ring which is not left nonsingular, then ${}_RR$ is nondegenerate but not nonsingular, as is any free left R-module, so

$$\text{ND} \;\not\Rightarrow\; \text{NS}.$$

In the first part of the proof of Proposition 2.2, we showed that, for any ${}_R M$, if U is a submodule of M, then $[M^*, U] \subseteq Hom_R(M,U)$. Hence, for any M with submodule U, we have $TU = (M, M^*)U = M[M^*, U] \subseteq M\ Hom_R(M,U) = s_M\, i_E(U) \subseteq U$.

2.5 Proposition

a) M is retractable if and only if $s_M\, i_E(U) <'_R U$ for each $0 \neq U < M$.

b) M is nondegenerate if and only if M is retractable and $TU <' \ s_M\, i_E(U)$ for each submodule U of M.

c) A trace-accessible M is nondegenerate if and only if it is retractable.

Proof:

a) Assume that M is retractable and let U be a nonzero submodule of M. If $0 \neq u \in U$, then there is $0 \neq f \in i_E(Ru)$. Hence, since $i_E(Ru) \subseteq i_E(U)$, we have $0 \neq Mf \subseteq Ru \cap s_M\, i_E(U)$, so that $s_M\, i_E(U) <'_R U$. Conversely, if $s_M\, i_E(U) <'_R U$ and $U \neq 0$, then this implies that $i_E(U) \neq 0$.

b) From $TU \subseteq s_M\, i_E(U) \subseteq U$, we have that $TU <' \ U$ if and only if $TU <' \ s_M\, i_E(U)$ and $s_M\, i_E(U) <' \ U$. Since $TU <' \ U$ is equivalent to nondegeneracy by Proposition 2.1, and $TU <' \ s_M\, i_E(U)$ is equivalent to retractability by a), the conclusion follows.

c) If M is trace-accessible then, in particular, $TU = M\ Hom_R(M,U)$ for each submodule U of M, so by *b*), M is nondegenerate if and only if M is retractable.

□

3. Classification of the Principal Bijections in Terms of the Main Modules for Which They Can be Established

We begin by giving necessary and sufficient conditions for the establishment of the bijections: a_r, between $\mathcal{A}^r$ and $\mathcal{M}^a$, and a_l, between $\mathcal{A}^l$ and $\mathcal{M}^*$. For $H \subseteq E$, let $r(H) = \{f \in E : Hf = 0\}$ and $l(H) = \{f \in E : fH = 0\}$. Recall that:

$\mathcal{A}^r = \{I \subseteq E : I = r(H), \text{ for } H \subseteq E\} = \{I \subseteq E : I = rl(I)\}$.

$\mathcal{A}^l = \{I \subseteq E : I = l(H), \text{ for } H \subseteq E\} = \{I \subseteq E : I = lr(I)\}$.

$\mathcal{M}^a = \{U \subseteq M : U = k_M\, a_E(U)\} = \{U \subseteq M : U = k_M(I),\ I \subseteq E\}$.

$\mathcal{M}^* = \{U \subseteq M : U = s_M\, i_E(U)\} = \{U \subseteq M : U = s_M(I),\ I \subseteq E\}$.

3.1 Remark: It follows easily from the definitions that, for $Y \subseteq E$, $a_E\, s_M(Y) = r(Y)$ and $i_E\, k_M(Y) = l(Y)$.

3.2 Theorem:

a) The maps $U \to a_E(U)$ and $I \to k_M(I)$ determine bijections between $\mathcal{M}^a$ and $\mathcal{A}^r$ if and only if $\mathcal{M}^a = k_M(\mathcal{A}^r)$.

b) The maps $U \to i_E(U)$ and $I \to s_M(I)$ determine bijections between $\mathcal{M}^*$ and $\mathcal{A}^l$ if and only if $\mathcal{M}^* = s_M(\mathcal{A}^l)$.

Proof:

a) If a_E and k_M determine bijections between $\mathcal{M}^a$ and $\mathcal{A}^r$ then, for any $U \in \mathcal{M}^a$, we have $a_E(U) = r(Y)$, for some $Y \subseteq E$; hence $U = k_M\, a_E(U) = k_M\, r(Y)$ is in $k_M(\mathcal{A}^r)$. This shows that $\mathcal{M}^a \subseteq k_M(\mathcal{A}^r)$; since, always, $k_M(\mathcal{A}^r) \subseteq \mathcal{M}^a$, we have equality.

Conversely, assume that $\mathcal{M}^a = k_M(\mathcal{A}^r)$. Then, if $U \in \mathcal{M}^a$, we have $U = k_M(r(Y)) = k_M(a_E\, s_M(Y))$ by Remark 3.1, so that $a_E(U) = a_E\, k_M\, a_E\, s_M(Y) = a_E\, s_M(Y) = r(Y) \in \mathcal{A}^r$; that is, a_E maps $\mathcal{M}^a$ to $\mathcal{A}^r$. We have: $U \to a_E(U) \to k_M\, a_E(U) = U$. On the other hand, for $I \in \mathcal{A}^r$, $k_M(I)$ clearly belongs to $\mathcal{M}^a$, and we have: $I = r(Y) \to k_M(I) = k_M(r(Y)) = k_M(a_E\, s_M(Y))$ by Remark 3.1 and hence $a_E\, k_M(I) = a_E\, k_M\, a_E\, s_M(Y) = a_E\, s_M(Y) = r(Y) = I$. So the maps a_E and k_M are inverses of each other and hence determine bijections between $\mathcal{M}^a$ and $\mathcal{A}^r$.

□

The proof of b) is similar.

3.3 Corollary: If M is nondegenerate, a self-generator or a nonsingular e-retractable module, then the bijections a_r can be established between $\mathcal{M}^a$ and $\mathcal{A}^r$.

Proof: We only need to note that the condition $\mathcal{M}^a = k_M(\mathcal{A}^r)$ is equivalent to "M is an a-self-generator", for the latter gives: for any $U \in \mathcal{M}^a$, $U = [s_M\, i_E(U)]^a = k_M\, a_E\, s_M\, i_E(U) = k_M\, r(i_E(U)) \in k_M(\mathcal{A}^r)$; and $\mathcal{M}^a = k_M(\mathcal{A}^r)$ gives: for any $U \in \mathcal{M}^a$, $U = k_M\, r(Y) = k_M\, a_E\, s_M(Y)$, so that $k_M\, a_E\, s_M\, i_E(U) = k_M\, a_E\, s_M\, i_E(k_M\, a_E\, s_M(Y)) = k_M\, rlr(Y) = k_M\, r(Y) = U$.

The Corollary then follows by Propositions 2.2 and 2.4 and the definition of self-generator.

□

Remark: Theorem 3.2.a) is essentially the equivalence between (1) and (2) of Theorem 2.5 of (4), except that "M is an a-self-generator" is used in Theorem 2.5, while $\mathcal{M}^a = k_M(\mathcal{A}^r)$ is used here in Theorem 3.2.

The definition of a self-generator may be re-phrased as: M is a self-generator if every submodule U satisfies $U = s_M\, i_E(U)$, i.e., if every submodule belongs to $\mathcal{M}^s$. Dually, define M to be a *self-cogenerator* if every submodule U satisfies $U = k_M\, a_E(U)$, i.e., if every submodule belongs to $\mathcal{M}^a$. It is easy to show that any cogenerator is a self-cogenerator, and that the bijections a_l can be established for M a self-cogenerator.

We now look at the bijections: a_r^1 between $\mathcal{A}_1^r$ and $\mathcal{M}_1^s$ and a_l^1 between $\mathcal{A}_1^l$ and $\mathcal{M}_1^a$. Recall that:

$\mathcal{A}_1^r = \{I \subseteq E : I = r(f),\ f \in E\} \subseteq \mathcal{A}^r$.

$\mathcal{A}_1^l = \{I \subseteq E : I = l(f),\ f \in E\} \subseteq \mathcal{A}^l$.

$\mathcal{M}_1^a = \{U \subseteq M : U = \ker(f) = k_M(f),\ f \in E\} \subseteq \mathcal{M}^a$.

$\mathcal{M}_1^s = \{U \subseteq M : U = Mf = s_M(f),\ f \in E\} \subseteq \mathcal{M}^s$.

3.4 Theorem:

a) k_M and a_E determine bijections between $\mathcal{A}_1^r$ and $\mathcal{M}_1^s$ if and only if $\mathcal{M}_1^s \subseteq \mathcal{M}^a$.

b) s_M and i_E determine bijections between $\mathcal{A}_1^l$ and $\mathcal{M}_1^a$ if and only if $\mathcal{M}_1^a \subseteq \mathcal{M}^s$.

Proof: a) One direction is clear: if k_M and a_E determine bijections between $\mathcal{A}_1^r$ and $\mathcal{M}_1^s$, then for any $U \in \mathcal{M}_1^s$ we have $U = k_M a_E(U) \in \mathcal{M}^a$. Conversely, assume that $\mathcal{M}_1^s \subseteq \mathcal{M}^a$, so that for each $U \in \mathcal{M}_1^s$, we have $U = k_M a_E(U)$. Then, for any $I \in \mathcal{A}_l^r$, we have $I = r(f)$, for some $f \in E$, hence $I = a_E s_M(f)$, by Remark 3.1, and consequently $k_M(I) = k_M a_E s_M(f) = s_M(f)$ since $s_M(f) \in \mathcal{M}_1^s \subseteq \mathcal{M}^a$; in other words, k_M takes $\mathcal{A}_1^r$ to $\mathcal{M}_1^s$. Also, for any $U \in \mathcal{M}_1^s$, we have $U = s_M(f)$, for some $f \in E$, hence $a_E(U) = a_E s_M(f) = r(f)$, by 3.1, that is, a_E takes $\mathcal{M}_1^s$ to $\mathcal{A}_1^r$. We have, for $I \in \mathcal{A}_1^r$, $I = r(f) = a_E s_M(f) \to k_M(I) = k_M a_E s_M(f) \to a_E k_M(I) = a_E k_M a_E s_M(f) = a_E s_M(f) = I$, and, for $U \in \mathcal{M}_1^s$, $U \to a_E(U) \to k_M a_E(U) = U$. Hence, k_M and a_E are inverses of each other and determine bijections between $\mathcal{A}_1^r$ and $\mathcal{M}_1^s$.

The proof of b) is similar.

□

Remark: Theorem 3.4 is a generalization of Propositions 2.8 and 2.9 of (4).

3.5 Corollary: If M is a self-generator or a nonsingular CSG module, then the bijections a_l^1 can be established between $\mathcal{A}_1^l$ and $\mathcal{M}_1^a$. If M is a self-cogenerator, then the bijections a_r^1 can be established between $\mathcal{A}_1^r$ and $\mathcal{M}_1^s$.

Proof: If M is a self-generator, then every submodule is in $\mathcal{M}^s$, hence by Theorem 3.4.b), we have the bijections a_l^1. Let M be a nonsingular CSG module. In the proof of Proposition 2.4, we showed that, when M is nonsingular, every a-closed submodule is a complement. Since, in a CSG module every complement is in $\mathcal{M}^s$, we have that $\mathcal{M}^a \subseteq \mathcal{M}^s$, so in particular $\mathcal{M}_1^a \subseteq \mathcal{M}^s$, giving the bijections a_l^1.

If M is a self-cogenerator, then every submodule is in $\mathcal{M}^a$, so in particular $\mathcal{M}_1^s \subseteq \mathcal{M}^a$, hence by Theorem 3.4.a) we have the bijections a_r^1.

□

3.6 Proposition: If M is a nonsingular CSG module or a trace-accessible retractable module, then we have the bijections a_r.

Proof: A CSG module is e-retractable since if C is a nonzero complement then $C = s_M i_E(C)$ implies $i_E(C) \neq 0$. Hence, M nonsingular $CSG \Rightarrow M$ nonsingular e-retractable $\Rightarrow$ the bijections a_r are establishable by Corollary 3.3. If M is trace-accessible and retractable, then, by Proposition 2.5, M is nondegenerate, hence by Corollary 3.3, we have a_r.

□

In order to classify the bijections c_ℓ between the complement left ideals, $\mathcal{C}^\ell$, of E and the complement submodules, $\mathcal{M}^c$, of M, we quote, without proof, the following results (cf. (4), Theorems 3.7 and 3.10):

3.7 If M is nondegenerate and CSG or nondegenerate and nonsingular, then the bijections c_ℓ can be established.

Finally, for the bijections: p_r between the principal right ideals, $\mathcal{P}^r$, of E and the kernel submodules, $\mathcal{M}_1^a$, of M; and the bijections p_l between the principal left ideals, $\mathcal{P}^l$, of E and the image submodules, $\mathcal{M}_l^s$, of M, we have

3.8 Proposition [(5), Proposition 2.3]

a) k_M and a_E determine bijections between $\mathcal{P}^r$ and $k_M(\mathcal{P}^r) = \mathcal{M}_1^a$ if and only if every $I \in \mathcal{P}^r$ satisfies $I = a_E\, k_M(I)$.

b) s_M and i_E determine bijections between $\mathcal{P}^l$ and $s_M(\mathcal{P}^l) = \mathcal{M}_1^s$ if and only if every $I \in \mathcal{P}^l$ satisfies $I = i_E\, s_M(I)$.

Proof:

a) If k_M and a_E determine bijections between $\mathcal{P}^r$ and $k_M(\mathcal{P}^r)$ then, for any $I \in \mathcal{P}^r$ we have $I = a_E\, k_M(I)$.
Conversely, assume that $I = a_E\, k_M(I)$ for each $I \in \mathcal{P}^r$. Then for $I = fE \in \mathcal{P}^r$, we have: $I \to k_M(I) = k_M(fE) = k_M(f) \in \mathcal{M}_1^a \to a_E\, k_M(I) = I$; and for $U = k_M(f) = k_M(fE) \in \mathcal{M}_1^a = k_M(\mathcal{P}^r)$, we have: $U \to a_E(U) = a_E\, k_M(fE) = fE \in \mathcal{P}^r \to k_M\, a_E(U) = U$, the last equality since $\mathcal{M}_1^a \subseteq \mathcal{M}^a$. Hence the maps a_E and k_M are inverses of each other and so determine bijections between $\mathcal{P}^r$ and $\mathcal{M}_1^a$.
The proof of b) is similar.

□

In connection with p_r and p_l, we shall need to make use of the following result from (5), which we quote without proof:

3.9 [(5), Theorem 3.3]

a) If every $U \in \mathcal{M}_1^s$ is a direct summand in M, then every $I \in \mathcal{P}^r$ satisfies $I = a_E\, k_M(I)$.

b) If every $U \in \mathcal{M}_1^a$ is a direct summand in M, then every $I \in \mathcal{P}^l$ satisfies $I = i_E\, s_M(I)$.

We can now combine results 3.3-3.9 to obtain examples of modules for which three of the main bijections hold.

3.10 Proposition: If M is a trace-accessible self-generator, then the bijections a_r, a_l^1 and c_l can be established. If M is nonsingular and CS, then a_r, a_l^1 and p_l can be established.

Proof: Let M be a trace-accessible self-generator. Then by self-generation, we have a_r by Corollary 3.3 and a_l^1 by Corollary 3.5. Since a self-generator is retractable, M is trace-accessible and retractable, hence nondegenerate by Proposition 2.5.c); and since a self-generator is, in particular, a CSG module, M is a nondegenerate CSG module, hence we have c_l by 3.7.

Let M be nonsingular and CS. Since a CS module is a CSG module, we have a_r by Proposition 3.6 and we have a_l^1 by Corollary 3.5. Since M is nonsingular, we have $\mathcal{M}^a \subseteq \mathcal{M}^c$, as we showed in the proof of Proposition 2.4, hence, since M is CS, every $U \in \mathcal{M}^a$ is a direct summand in M. Since $\mathcal{M}_1^a \subseteq \mathcal{M}^a$, we have that every $U \in \mathcal{M}_1^a$ is a direct summand in M, hence, by 3.9.a), every $I \in \mathcal{P}^l$ satisfies $I = i_E s_M(I)$, which, by Proposition 3.8.b) implies that we have the bijections p_l.

□

Recall that M is *quasi-injective* (QI) if any homomorphism from a submodule of M into M extends to an endomorphism of M; and that M is *quasi-projective* (QP) if, for each epimorphism p from M onto a module M' and for each $g \in Hom_R(M, M')$, there exists an endomorphism of M such that $g = hp$.

It is shown in (1) that: If M is quasi-injective, then every $I \in \mathcal{P}^r$ (in fact, every $I \in \mathcal{F}^r$) satisfies $I = a_E k_M(I)$; and if M is quasi-projective, then every $I \in \mathcal{P}^l$ (in fact, every $I \in \mathcal{F}^l$) satisfies $I = i_E s_M(I)$. These results help to give the next Corollary.

3.11 Corollary:

a) If M is quasi-injective, then we have the bijections p_r.
b) If M is quasi-projective, then we have the bijections p_l.
c) If M is projective and retractable, then we have p_l and a_r.

Proof: A combination of the remarks preceding the Corollary with Proposition 3.8 gives a) and b). Let M be projective and retractable; then, since a projective module is quasi-projective, we have p_l by b), and since a projective module is trace-accessible, M is, in particular, trace-accessible and retractable, hence we have a_r by Proposition 3.6.

□

We now have some examples of modules with four of the main bijections available.

3.12 Proposition: If M is nonsingular and continuous–in particular, if M is nonsingular and quasi-injective–then we have the bijections $a_r a_l^1$, p_l and p_r. If M is a free module, then we have a_r, a_l^1, c_l and p_l.

Proof: Let M be nonsingular and continuous. Since a continuous module is CS, we have a_r, a_l^1 and p_l by Proposition 3.10. Let $U \in \mathcal{M}_1^*$, so that $U = Mf$, for $f \in E$. Since M is nonsingular, the submodule $k_M(f)$, being a-closed, is also a complement; hence, since M is CS, there is a submodule, V, of M such that $M = k_M(f) \oplus V$. Let f_1 denote the restriction of f to V; then $f_1 : V \to Vf_1$ is an isomorphism and we have $Mf = Vf = Vf_1$. Hence, since V is a direct summand and M is continuous, $Vf_1 = Mf$ is a direct summand in M. Now, by 3.9, since every $U \in \mathcal{M}_1^*$ is a direct summand in M, we have $I = a_E k_M(I)$ for every $I \in \mathcal{P}^r$, which by Proposition 3.8.a), implies that we have the bijections p_r.

Let M be a free module; then M is projective and a generator, hence quasi-projective, trace-accessible and a self-generator. Therefore, we have p_l by Corollary 3.11b), and we have a_r, a_l^1 and c_l by Proposition 3.10.

□

Finally, an example where we have all seven of the main bijections:

3.13 Proposition: If M is a semisimple module, then we have the bijections a_r, a_l, a_r^1, a_l^1, c_l, p_r and p_l.

Proof: Let M be semisimple. Then, since M is a self-generator, we have a_r and a_l^1, and since M is a self-cogenerator, we have a_l and a_r^1. Since M is both quasi-injective and quasi-projective, we have p_r and p_l, and since M is trace-accessible and a self-generator, we have c_l.

□

This classification is summarized in the following diagram.

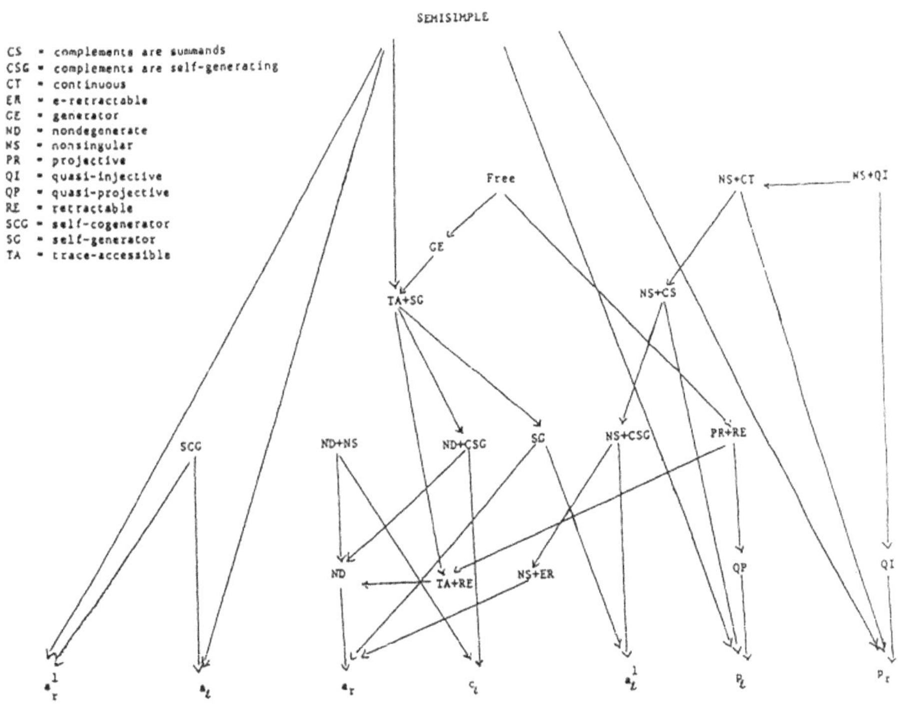

REFERENCES

1. T. Albu and C. Nastasescu, *Relative Finiteness in Module Theory*, Dekker, New York, 1984.
2. A.W. Chatters and S.M. Khuri, *Endomorphism rings of modules over nonsingular CS rings*, J. London Math. Soc. **21**(2) (1980), 434-444.
3. C. Faith, *Lectures on injective modules and quotient rings*, Lecture Notes in Mathematics, Vol. **246**, Spring-Verlag, Heidelberg, 1972.
4. S.M. Khuri, *Correspondence theorems for modules and their endomorphism rings*, J. Algebra **122** (1989), 380-396.
5. S.M. Khuri, *Modules with regular, perfect, noetherian or artinian endomorphism rings*, in Non-Commutative Ring Theory, Lecture Notes in Mathematics, Vol. **1448**, eds. S.K. Jain and S.R. López-Permouth, p. 7-18.
6. B. Zimmerman-Huisgen, *Endomorphism rings of self-generators*, Pacific J. Math. **61** (1975), 587-602.

CLASSICAL KRULL DIMENSION AND THE SECOND LAYER CONDITION

KARL A. KOSLER
Department of Mathematics, University of Wisconsin-Waukesha
Waukesha, Wisconsin 53188

ABSTRACT

This paper examines an exact, symmetric extension of classical Krull dimension to one-sided modules over a Noetherian ring with the second layer condition. Techniques that used regular Krull dimension for examining Krull symmetric Noetherian rings are applied, using classical Krull dimension, to yield some parallel results about ring decompositions and the quotient ring problem.

1. Introduction and Definitions

Aside from investigations concerning localization at collections of prime ideals (e.g., see [12]), interest in Noetherian rings with the second layer condition has focused on generalizing results about fully bounded Noetherian (FBN) rings. Research in this direction originated with Jategaonkar in [11] where certain results from [10] for FBN rings and their modules are shown to carry over to rings with the second layer condition and their tame modules. Some examples are the 'Jordan-Holder' theorem for the basic series of a tame module, Jacobson's conjecture and the invariance of classical Krull dimension across bimodules (see [11; 1.6, 1.8(1), 2.1]). A more recent account of progress in this area can be found in [3].

The methods used to obtain the results in [11] were adapted from techniques developed for FBN rings in [10] which relied heavily on the symmetry of regular (i.e., Rentschler-Gabriel) Krull dimension. This was accomplished by showing that bimodule Krull dimension is an extension of classical Krull dimension to Noetherian bimodules over a Noetherian ring satisfying the left second layer condition (see [11; 1.7]). Subsequently, in [11] classical Krull dimension served as a replacement for regular Krull dimension in applying the methods of [10] to this class of rings.

In this paper, we expand the scope of this tactic by showing that classical Krull dimension has an extension to one-sided modules that admits some of the techniques used to examine Krull symmetric Noetherian rings. For a left Noetherian ring R that satisfies the left second layer condition, the *classical Krull dimension* of a left R-module M, denoted $\mathrm{Cldim}(M)$, is defined to be the deviation of the set $\{K \subseteq M \mid K \text{ is a submodule and } M/K \text{ is tame}\}$ partially ordered by inclusion. In

Section 2, Cldim( ) is shown to be an exact dimension function that takes on the usual value of classical Krull dimension when applied to R and its factor rings. In case R is Noetherian and satisfies the second layer condition on both sides, Cldim( ) is symmetric in the sense that Cldim $({}_R B)$ = Cldim (B_R) for a Noetherian $R - R$ bimodule B.

Section 3 is devoted to examining modules that are critical with respect to classical Krull dimension. Jategaonkar's basic series from [11] is shown to be a straightforward classical Krull dimension analogue of a Krull composition series. This is used to show that a stronger version of Jacobson's conjecture holds for Noetherian rings with the second layer condition on both sides. Viewing the prime ideals associated with a basic series for ${}_R R$ and R_R (called the *basic primes*) as versions of Muller's Krull primes from [15], we characterize the regular elements of R and determine when R has an Artinian classical quotient ring. This is done in a fashion that is virtually identical to the way Krull primes were used to accomplish this in [15] for Krull symmetric Noetherian rings. In fact, if R is Krull symmetric, then the basic prime ideals coincide with the Krull primes.

While the results of Section 3 may be of interest in their own right, their purpose is to provide the necessary machinery for Section 4 which examines the β -homogeneous ideals of a Noetherian ring R that satisfies the second layer condition. A tame module M is called β*- homogeneous* for some ordinal β, provided Cldim(N) = Cldim $(M) = \beta$ for all submodules $0 \neq N \subseteq M$. From [11; 2.8] and 2.1, if R is homogeneous, then R has an Artinian classical quotient ring. One purpose of this section is to generalize this result to the non-homogeneous case. Another purpose is to determine conditions under which R decomposes into a product of homogeneous rings. In both cases, we obtain the classical Krull dimension analogues of the results in [6] and [17] (as proved in [2]) that addressed these problems for Krull symmetric Noetherian rings. The advantage of using classical Krull dimension is illustrated by one of the main results of this section: if R has an Artinian classical quotient ring, then the largest ideal V with Cldim$(R/\ell(V)) = 0$ is an ideal direct summand of R. In our setting this improves upon the well known result of [4] that the largest Artinian ideal is an ideal direct summand of R.

Throughout, a condition that is unmodified with 'left' or 'right' will mean that the condition holds on both sides. Unless specified otherwise, all modules will be left R-modules. If M is a module and $S \subseteq M$ is a non-empty subset, then $\ell(S) = \{r \in R \mid rS = 0\}$. For a right R-module, and non-empty subset $S \subseteq M$, $rt\ (S) = \{r \in R \mid Sr = 0\}$. If $I \subseteq R$ is an ideal and M is a module, then ann ${}_M(I) = \{x \in M \mid Ix = 0\}$ is a submodule. If $N \cap K \neq 0$ for all submodules $0 \neq K \subseteq M$, then N is *essential* in M and we will write $N \leq_e M$. A submodule $N \subseteq M$ is *closed* in M provided N has no proper essential extensions in M.

If U is a uniform module, then P = ass $(U) = \{r \in R \mid rU' = 0$ for some submodule $0 \neq U' \subseteq U\}$ is a prime ideal. A prime ideal is called a *left associated prime* for a module M provided P = ass(U) for some uniform submodule

$0 \neq U \subseteq M$. The collection of all associated primes of a module M is denoted Ass $({}_R M)$. Right associated primes and Ass (M_R) are defined similarly.

A uniform left R-module U over a left Noetherian ring R is called *tame* provided the injective hull $E(U)$ is isomorphic to a direct $E(R/$ ass $(U))$. A module M is called *tame* if M contains an essential direct sum of tame uniform submodules. If every uniform submodule of $E(R/P)/\operatorname{ann}_{E(R/P)}(P)$ is tame for every prime ideal P of R, then R is said to satisfy the *left second layer* condition. The *right second layer condition* is defined similarly. From [11], FBN rings, enveloping algebras of solvable Lie algebras and group rings of polycyclic-by-finite groups over commutative rings satisfy the second layer condition on both sides.

Finally, $| M |$ will denote the *Krull dimension* of a module M as defined in [9]. Critical modules and critical composition series are defined in [9] and [8] and will be referred to here as *Krull critical* modules and *Krull composition series* , respectively.

2. Classical Krull Dimension of Modules

For a module M, let $\mathcal{T}(M) = \{K \subseteq M \mid K$ is a submodule with M/K tame$\}$ partially ordered by inclusion. Define the *classical Krull dimension* of M, denoted Cldim(M), to be the deviation of $\mathcal{T}(M)$ as defined in [14; 6.1.2]. Thus, Cldim$(M) = -1$ if and only if no factor M/K is tame for any proper submodule $K \subseteq M$. It follows that Cldim$(M) = 0$ if Cldim$(M) \neq -1$ and any chain $M = K_0 \supseteq K_1 \supseteq K_2 \supseteq \cdots$ with $K_i \in \mathcal{T}(M)$ for all i is eventually constant. In general, if $\alpha > 0$ is an ordinal and Cldim$(M) \not< \alpha$, then Cldim$(M) = \alpha$ provided for any descending chain $M = K_0 \supseteq K_1 \supseteq \cdots$ where $K_i \in \mathcal{T}(M)$ for all i, Cldim $(K_i/K_{i+1}) < \alpha$ for all but finitely many i. In case M is finitely generated, Cldim(M) always exists and Cldim$(M) \leq | M |$ by [16; 3.1.4].

The first part of this section is devoted to showing that Cldim( ) is an exact dimension function for modules over a left Noetherian ring with the left second layer condition. Unfortunately, $\mathcal{T}(M)$ is not *a-priori* a modular lattice and so exactness does not follow from arguments such as those found in [16].

Throughout R will always denote a left Noetherian ring that satisfies the left second layer condition. A module M will be called *finitely annihilated* provided $\ell(M) = \ell(x_1, \cdots, x_n)$ for some $x_1, \cdots, x_n \in M$. Note that in this case $| R/\ell(M) | = | M |$. Heavy use will be made of the following result from [13] and its corollary.

2.1 Proposition [13; 2.8]. A module M is tame iff all finitely generated submodules of M are finitely annihilated.

2.2 Corollary. Let M be a finitely generated module.
(a) If $N \subseteq M$ is a tame submodule and M/N is tame, then M is tame.
(b) If M is tame, then $M/\operatorname{ann}_M(I)$ is tame for all ideals I of R.

Proof:
(a) Let $0 \neq X \subseteq M$ be a prime critical submodule. If $X \subseteq N$, then X is tame. If $X \not\subseteq N$, then $0 \neq X/X \cap N \cong X + N/N \subseteq M/N$ and hence, by 2.1,

$X/X \cap N$ is finitely annihilated. Clearly, $X \cap N$ is finitely annihilated. Let $A = \ell(X \cap N)$ and $B = \ell(X/X \cap N)$. Then $| R/A | = | X \cap N |$ and $| R/B | = | X/X \cap N |$. Also, $AB \subseteq \ell(X)$ and so $| R/AB | \geq | R/\ell(X) |$. Now, $| R/AB | = \sup\{| R/A |, | R/B |\} = \sup \{| X \cap N |, | X/X \cap N |\} = | X |$. Thus, $| X | \geq | R/\ell(X) |$. It follows that $| X | = | R/\ell(X) |$. Since $\ell(X)$ is a prime ideal, X is tame.

(b) Let $0 \neq \bar{X} = X + \text{ann}_M(I)/\text{ann}_M(I) \subseteq M/\text{ann}_M(I)$ be a submodule. Then $\ell(X) = \ell(x_1, \cdots, x_n)$ for some $x_1, \cdots, x_n \in X$. If $rx_i \in \text{ann}_M(I)$ for all i, then $Irx_i = 0$ for all i and hence $IrX = 0$. Thus, $r \in \ell(\bar{X})$. This establishes the inclusion $\ell(\bar{x}_1, \cdots, \bar{x}_n) \subseteq \ell(\bar{X})$ where $\bar{x}_i = x_i + \text{ann}_M(I)$ for each i. Since the reverse inclusion is trivial, it follows that $\ell(\bar{X}) = \ell(\bar{x}_1, \cdots, \bar{x}_n)$. The conclusion now follows from 2.1.

We will frequently use the following alternate definition of classical Krull dimension. From 2.2(a), if M is a finitely generated module, then $\text{Cldim}(M) = \alpha$ provided $\text{Cldim}(M) \not< \alpha$ and for any descending chain $M = M_0 \supseteq M_1 \supseteq \cdots$ with M_i/M_{i+1} tame for all i, $\text{Cldim}(M_i/M_{i+1}) < \alpha$ for all but finitely many i.

2.3 Proposition. If M is a finitely generated module and $N \subseteq M$ is a submodule, then

(a) $\text{Cldim}(M/N) \leq \text{Cldim}(M)$

(b) $\text{Cldim}(N) \leq \text{Cldim}(M)$

Proof:

(a) Define a map $\mathcal{T}(M/N) \to \mathcal{T}(M)$ by $K/N \to K$. The conclusion follows from [16; 3.1.4].

(b) Proceed by transfinite induction on the ordinal $\beta = \text{Cldim}(M)$. If $\beta = -1$, then no proper factor of M is tame. Suppose that $\text{Cldim}(N) \neq -1$. Then N/H is tame for some proper submodule $H \subseteq N$. If $N/H \leq_e M/H$, then M/H is tame which is a contradiction. It follows that $N/H \oplus X/H \leq_e M/H$ for some proper closed submodule $0 \neq X/H \subseteq M/H$. Thus, $N/H \cong N + X/X \leq_e M/X$ and hence M/X is tame which is impossible. Therefore, $\text{Cldim}(N) = -1$.

Assume that $\beta > 0$ and that the proposition is true for all finitely generated tame modules with classical Krull dimension less than β. Suppose that $\text{Cldim}(N) > \beta$ for some submodule $N \subset M$. Then $N \neq 0$ and according to [14; 6.1.7], there is a chain $N = N_0 \supset N_1 \supset \cdots$ such that $N_i \in \mathcal{T}(N)$ and $\text{Cldim}(N_i/N_{i+1}) = \beta$ for all i.

If M/N_i is tame for all but finitely many i, then for the chain $M \supset N_0 \supset N_1 \supset \cdots$, $\text{Cldim}(N_i/N_{i+1}) < \beta$ for all but finitely many i, which is a contradiction. Suppose that it is not the case the M/N_i is tame for all but finitely many i. If one quotient, say M/N_t, is tame, then $M/N_t \cong (M/N_{t+1})/(N_t/N_{t+1})$ tame together with N_t/N_{t+1} tame implies, by 2.2, that M/N_{t+1} is tame. A simple induction argument yields that all of the modules M/N_k are tame for $k \geq t$. Since this contradicts the supposition, it must be true that none of the quotients M/N_i are tame. In particular, N/N_1 is not an essential submodule of M/N_1. Thus, $N/N_1 \oplus X_1/N_1 \leq_e M/N_1$ for some closed submodule $X_1/N_1 \neq 0$. Note that N/N_1 is isomorphic to an essential submodule of M/X_1 and so M/X_1 is tame.

Since $N/N_1 \oplus X_1/N_1 \leq_e M/N_1$, $N/N_2 + X_1/N_2 \leq_e M/N_2$. Suppose that $N_1/N_2 = N/N_2 \cap X_1/N_2 \leq_e X_1/N_2$. Then X_1/N_2 is tame. Also, $(N/N_2 + X_1/N_2)/(X_1/N_2) \cong N + X_1/X_1 \cong N/N \cap X_1 = N/N_1$ is tame. Therefore, by 2.2, $N/N_2 + X_1/N_2$ is tame. But since $N/N_2 + X_1/N_2 \leq_e M/N_2. M/N_2$ is tame which is a contradiction. It follows that $N_1/N_2 \oplus X_2/N_2 \leq_e X_1/N_2$ for some closed submodule $X_2/N_2 \neq 0$. Also, since $N_1/N_2 \neq 0$, $X_2 \subset X_1$. Furthermore, $N_1/N_2 \cong N_1 + X_2/X_2 \leq_e X_1/X_2$ and so X_1/X_2 is tame. Continuing in this fashion yields a strict descending chain $M \supset X_1 \supset X_2 \supset \cdots$, where for each i, X_i/X_{i+1} is tame and there is a 1-1 map $N_i/N_{i+1} \to X_i/X_{i+1}$.

Now, $\text{Cldim}(X_i/X_{i+1}) < \beta = \text{Cldim}\ (M)$ for all but finitely many i. Thus, by hypothesis, $\text{Cldim}(N_i/N_{i+1}) \leq \text{Cldim}\ (X_i/X_{i+1}) < \beta$. However, this contradicts the choice of the chain $N = N_0 \supset N_1 \supset \cdots$. Therefore, $\text{Cldim}(N) \leq \text{Cldim}\ (M)$.

2.4 Theorem. If M is a finitely generated module and $N \subseteq M$ is a submodule, then $\text{Cldim}(M) = \sup\{\text{Cldim}(N), \text{Cldim}(M/N)\}$.

Proof: Let $\alpha = \sup\{\text{Cldim}(N), \text{Cldim}(M/N)\}$. By 2.3, $\alpha \leq \text{Cldim}\ (M)$. To show that $\text{Cldim}(M) \leq \alpha$, we proceed by transfinite induction on α. If $\alpha = -1$, then no proper factor of N or M/N is tame. It's easy to see that no proper factor of M is tame. Therefore, $\text{Cldim}(M) = -1$. Assume that $\alpha \geq 0$ and that the result holds true for all ordinals $\beta < \alpha$. The remainder of the proof utilizes the following.

Sublemma.

(a) If $K_1, \cdots, K_n$ are finitely generated modules with $\text{Cldim}(K_i) < \alpha$, then $\text{Cldim}(K_1 \oplus \cdots \oplus K_n) = \text{Cldim}\ (K_j)$ for some j.

(b) If H is a finitely generated module, then H contains a unique largest submodule K with $\text{Cldim}(K) < \alpha$.

(c) If K is a tame module with $\text{Cldim}(K) < \alpha$, then $\text{Cldim}(R/\ell(K)) = \text{Cldim}\ (K)$.

(d) If H is a finitely generated tame module and K is the unique largest submodule of H with $\text{Cldim}(K) < \alpha$, then $K = \text{ann}_H\, \ell(K)$.

Proof of Sublemma:

(a) The conclusion of (a) follows from the transfinite induction hypothesis together with an easy induction argument on the index n.

For (*b*), let K be the sum of all submodules of H with classical Krull dimension less than α. Since K is finitely generated, $K = K_1 + \cdots + K_n$ for some submodules $K_1, \cdots, K_n$ of H where $\text{Cldim}(K_i) < \alpha$ for all i. The conclusion now follows from (a) and 2.3(a).

(c) Let $A = \ell(K)$. From 2.1, there is a 1-1 map $R/A \to K^{(m)}$ for some natural number m. By (a) and 2.3(*b*), $\text{Cldim}(R/A) \leq \text{Cldim}\ (K) < \alpha$. Further, since there is an onto map $(R/A)^{(k)} \to K$ for some natural number k, $\text{Cldim}(K) \leq \text{Cldim}\ ((R/A)^{(k)})$ by 2.3(a). Also, $\text{Cldim}(R/A) < \alpha$ implies that

$\text{Cldim}(R/A)^{(k)} = \text{Cldim }(R/A)$ by (a). Thus, $\text{Cldim}(K) \leq \text{Cldim }(R/A)$. Therefore, $\text{Cldim}(R/A) = \text{Cldim }(K)$.

(d) Let $A = \ell(K)$. Clearly, $K \subseteq \text{ann}_H(A)$. Since $\text{ann}_H(A)$ is a finitely generated R/A-module, there is an onto map $(R/A)^{(k)} \to \text{ann}_H(A)$ for some natural number k. By (a), (c) and 2.3(a), $\text{Cldim }(\text{ann}_H(A)) \leq \text{Cldim }(R/A) = \text{Cldim }(K) < \alpha$. Thus, $\text{ann}_H(A) \subseteq K$. Therefore, $K = \text{ann}_H(A)$.

To complete the proof of the theorem, suppose that $\text{Cldim}(M) > \alpha$. Choose $K \subseteq M$ a submodule maximal with respect to $\text{Cldim}(M/K) > \alpha$. Then there exists a chain $M/K = X_0/K \supset X_1/K \supset \cdots$, such that for all i, $X_i \in \mathcal{T}(M)$ and $\alpha = \text{Cldim }(X_i/X_{i+1}) < \text{Cldim }(M/K) \leq \text{Cldim }(M)$. By the choice of K, $\text{Cldim}(M/X_i) \leq \alpha$ for all i. Now, by 2.3(b), $\text{Cldim}(X_i/X_{i+1}) \leq \text{Cldim}(M/X_{i+1})$. Thus, $\text{Cldim}(X_i/X_{i+1}) = \text{Cldim}(M/X_{i+1}) = \alpha$ for all i.

Let $Z_0 = M$. Choose $Z_1/X_1 \in \mathcal{T}(Z_0/X_1)$ maximal with respect to $\text{Cldim}(Z_0/Z_1) = \alpha$. Then $X_1/X_2 \subseteq Z_1/X_2 \subset Z_0/X_2$ and hence $\text{Cldim}(Z_1/X_2) = \alpha$. Choose $Z_2/X_2 \in \mathcal{T}(Z_1/X_2)$ maximal with respect to $\text{Cldim}(Z_1/Z_2) = \alpha$. Continue in this fashion to obtain a strict descending chain $M = Z_0 \supset Z_1 \supset \cdots$ with each factor Z_i/Z_{i+1} tame, $\text{Cldim}(Z_i/Z_{i+1}) = \alpha$ and with the property that for all $Y/Z_{i+1} \in \mathcal{T}(Z_i/Z_{i+1})$, $\text{Cldim}(Z_i/Y) < \alpha$.

Consider the chain $N \supseteq N \cap Z_1 \supseteq N \cap Z_2 \supseteq \cdots$. Since $\text{Cldim}(N) \leq \alpha$, there exists a natural number m such that for all $j \geq m$, $\text{Cldim}(N \cap Z_j/N \cap Z_{j+1}) < \alpha$. Suppose that $N \cap Z_j/N \cap Z_{j+1} \neq 0$ for some $j \geq m$. Since there is a 1-1 map $N \cap Z_j/N \cap Z_{j+1} \to Z_j/Z_{j+1}$, Z_j/Z_{j+1} contains a non-zero submodule with classical Krull dimension less than α. By (b) of the sublemma, $Z_j/Z_{j+1} \supset V/Z_{j+1}$ the unique largest submodule of Z_j/Z_{j+1} with $\text{Cldim}(V/Z_{j+1}) < \alpha$. Let $U = Z_j/Z_{j+1}$, $W = V/Z_{j+1}$ and let $B = \ell(W)$. By (d) of the sublemma, $W = \text{ann}_U(B)$. Since $Z_j/V \cong U/W$, Z_j/V is tame by 2.2(b) whence $\text{Cldim}(Z_j/V) < \alpha$. Invoking the inductive hypothesis, we have $\text{Cldim}(Z_j/Z_{j+1}) \leq \sup\{\text{Cldim}(V/Z_{j+1}), \text{Cldim}(Z_j/V)\} < \alpha$ which is a contradiction. Thus, $N \cap Z_j = N \cap Z_{j+1}$ for all j. Consequently, the canonical map $Z_j/Z_{j+1} \to Z_j + N/Z_{j+1} + N$ is an isomorphism and each factor $Z_j + N/Z_{j+1} + N$ is tame. Consider the chain of submodules in $\mathcal{T}(Z_m + N/N) : Z_m + N/N \supset Z_{m+1} + N/N \supset \cdots$. For all but finitely many indices $k \geq m$, $\text{Cldim}(Z_k + N/Z_{k+1} + N) < \text{Cldim }(Z_m + N/N) \leq \text{Cldim }(M/N) \leq \alpha$. Again, this yields $\text{Cldim}(Z_k/Z_{k+1}) < \alpha$ which is a contradiction.

2.5 Corollary. If M is a finitely generated tame module, then $\text{Cldim}(R/\ell(M)) = \text{Cldim }(M)$. In particular, this holds true for a Noetherian $R-S$ bimodule M where S is any ring.

Proof. The proof of the first statement is identical to the proof of part (c) of the sublemma in 2.4. For the second statement, let $0 \neq F \subseteq M$ be a R-submodule. Then FS is finitely generated as a right S-module and hence $FS = x_1S + \cdots + x_nS$ for some $x_1, \cdots, x_n \in F$. It follows that $\ell(F) = \ell(x_1, \cdots, x_n)$. Thus, by 2.1, M is a tame R-module. The conclusion now follows from the first statement.

The next set of results address the problem of computing Cldim(R) by utilizing the ordinals Cldim(R/P) where P is a prime ideal. The reader should be aware that despite the terminology, it has not yet been determined that Cldim( ) coincides with the usual notion of classical Krull dimension when applied to rings.

2.6 Proposition. Let N denote the prime radical of R. Then Cldim(R) = Cldim (R/N) = sup Cldim (R/P) | P is a minimal prime ideal of R}.

Proof. Same as [8, 13.5] with Cldim( ) in place of Krull dimension.

2.7 Proposition. If R is a prime ring, then Cldim(R/P) < Cldim(R) for all non-zero prime ideals P of R.

Proof. Suppose that Cldim(R/P) = Cldim (R) for some non-zero prime ideal P. Let $X_0 = R$ and let $X_1 = P \in \mathcal{T}(R)$. Since R is a prime ring, X_1 contains a regular element c. Then $X_0c/X_1c \subseteq X_1/X_1c$ and Cldim(X_0/X_1) = Cldim (X_0c/X_1c) = Cldim (R). Choose $X_2/X_1c \subseteq X_1/X_1c$ a submodule maximal with respect to $X_0c/X_1c \oplus X_2/X_1c \leq_e X_1/X_1c$. Then $X_0/X_1 \cong X_0c/X_1c \cong X_0c + X_2/X_2 \leq_e X_1/X_2$. Thus, X_1/X_2 is tame and $X_1c \subseteq X_2$. Also, Cldim(R) = Cldim(X_0c/X_1c) $\leq$ Cldim(X_1/X_2). Therefore, Cldim(X_1/X_2) = Cldim(R). Continuing in this fashion yields a strict descending chain $R = X_0 \supset X_1 \supset \cdots$ with Cldim(X_i/X_{i+1}) = Cldim(R) and X_i/X_{i+1} tame for all i. However, no such chain is supposed to exist. Therefore, Cldim(R/P) <Cldim (R).

2.8 Corollary. If $P \subset Q$ are prime ideals of R, then Cldim(R/P) > Cldim (R/Q).

We are now in a position to show that Cldim(R) coincides with the usual value of the classical Krull dimension of R. In the following, we use dim(R) to denote this ordinal. Recall that dim(R) is the least ordinal γ such that $\mathcal{X}_\gamma$ =Spec(R) where the sets $\mathcal{X}_\alpha$ are defined inductively as follows. Let $\mathcal{X}_{-1}$ be the empty set. If $\alpha \geq 0$ and $\mathcal{X}_\beta$ has been defined for all $\beta < \alpha$, then $\mathcal{X}_\alpha$ is the set of prime ideals P of R such that all prime ideals properly containing P belong *to* $\cup \{\mathcal{X}_\beta \mid \beta < \alpha\}$.

2.9 Theorem. Cldim(R) = dim(R).

Proof. To show that dim(R) $\leq$ Cldim(R), we use transfinite induction on α = Cldim (R). Suppose that $\alpha = 0$. Let P be a prime ideal of R. If $Q \supset P$ is a prime ideal, then by 2.8, Cldim(R/Q) < Cldim(R/P) $\leq$ Cldim(R) = 0 which is impossible. Therefore, P is a maximal ideal. It follows that dim(R) = 0.

Assume that $\alpha > 0$ and that the result holds true for all ordinals less than α. Let $P \subset Q$ be prime ideals. Then, by 2.8, β = Cldim (R/Q) < Cldim(R/P) $\leq \alpha$. Also, R/Q is a left Noetherian ring that satisfies the left second layer condition. Thus, by the inductive hypothesis, dim(R/Q) $\leq \beta$ and hence $Q \in \mathcal{X}_\beta$. This demonstrates that for all prime ideals P, if $Q \supset P$ for some prime ideal Q, then $Q \in \mathcal{X}_\beta$ for some ordinal $\beta < \alpha$. It follows that $P \in \mathcal{X}_\alpha$. Therefore, dim(R) $\leq \alpha$.

For the reverse inequality, proceed by transfinite induction on α. From above, the result is true for $\alpha = 0$. Assume that $\alpha > 0$ and that the result holds true for all ordinals less than α. By 2.6, $\alpha =$ Cldim (R/P) for some prime ideal P and so it suffices to show that $\dim(R/P) \geq \alpha$. Thus, without loss of generality, we may assume that R is a prime ring.

Let $R = I_0 \supset I_1 \supset \cdots$ be a descending chain where $I_i \in \mathcal{T}(R)$ for all i. For each i, let $A_i = \ell(I_i/I_{i+1})$. Then for all but finitely many i, $\text{Cldim}(I_i/I_{i+1}) =$ Cldim $(R/A_i) < \alpha$ by 2.5. In this case, $A_i \neq 0$ and by the induction hypothesis, $\text{Cldim}(R/A_i) \leq \dim(R/A_i)$. Now, since R is a prime ring, $\dim(R/A_i) < \dim(R)$ whenever $A_i \neq 0$ by [8; 12A(c)]. Thus, for all but finitely many i, $\text{Cldim}(I_i/I_{i+1}) < \dim(R)$. Therefore, $\text{Cldim}(R) \leq \dim(R)$.

The last result of this section demonstrates that Cldim( ) is a symmetric dimension function if R is Noetherian and satisfies the second layer condition on both sides.

2.10 Proposition. Let R be a Noetherian ring that satisfies the second layer condition. If B is a Noetherian $R-R$ bimodule, then $\text{Cldim}({}_R B) =$ Cldim (B_R).

Proof. Let $S = R/\ell(B)$ and let $T = R/rt(B)$. Then, B, S and T satisfy the hypothesis of [8; 12.5] and hence $\text{Cldim}(S) =$ Cldim (T). The result follows from 2.5.

3. Critical Modules and Prime Ideals

Let R be a Noetherian ring that satisfies the left second layer condition. A non-zero finitely generated module M is called *classically critical* , or *Cl-critical* for short, provided M is tame and $\text{Cldim}(M/N) <$ Cldim (M) for all $0 \neq N \in \mathcal{T}(M)$. If the ordinal $\alpha =$ Cldim (M) needs to be specified, then M will be called α - *Cl-critical*.

The purpose of this section is to use classically critical modules together with Jategaonkar's basic series from [11] and the corresponding collection of assassinator primes, to characterize the regular elements of R and to determine when R has an Artinian classical quotient ring. From [11], a uniform submodule U of a finitely generated module M is called a *basic submodule of M* provided U is maximal among those uniform submodules $V \subseteq M$ that satisfy $\ell(V) =$ ass (V) and $\text{Cldim}(R/\ell(V)) = \min\{\text{Cldim}(R/P) \mid P \in \text{Ass } (M)\}$ (see [11; 1.8(2) and pg. 36]). A *basic series* for M is a series $0 = B_0 \subset B_1 \subset \cdots \subset B_n = B$ such that B_i/B_{i-1} is a basic submodule of M/B_{i-1}.

In the following, it is shown that basic submodules are precisely the maximal classically critical submodules of minimal dimension. Throughout, R *will always denote a Noetherian ring with the left second layer condition.*

3.1 Lemma. Let X be a Cl-critical module. Then

(a) X is uniform.

(b) $\text{Cldim}(X) =$ Cldim (Y) for all non-zero submodules $Y \subseteq X$.

(c) $\ell(X) =$ass (X).

Proof.

(a) If X is not uniform, then $X_1 \oplus X_2 \leq_e X$ where $X_i \subseteq X$ is a closed submodule for $i = 1, 2$. Thus, since X/X_i is tame, $\text{Cldim}(X/X_i) < \text{Cldim}(X)$ for each i. However, the canonical 1-1 map $X \to X/X_1 \oplus X/X_2$ yields that $\text{Cldim}(X) \leq \sup\{\text{Cldim}(X/X_i) \mid i = 1, 2\}$ which is impossible. Therefore, X is uniform.

(b) Let $W = \text{ann}_X \ell(Y)$. Then $W \in \mathcal{T}(X)$ by 2.2 and hence $\text{Cldim}(X/W) < \text{Cldim}(X)$. It follows from 2.4 that $\text{Cldim}(X) = \text{Cldim}(W)$. Now, $\ell(W) = \ell(Y)$. Thus, from 2.5, $\text{Cldim}(W) = \text{Cldim}(R/\ell(W)) = \text{Cldim}(R/\ell(Y)) = \text{Cldim}(Y)$. Therefore, $\text{Cldim}(X) = \text{Cldim}(Y)$.

(c) Let X be a Cl-critical module such that $X/L \neq 0$ where $L = \text{ann}_X(P)$ and $P = \text{ass}(X)$. By 2.2, $L \in \mathcal{T}(X)$ and so $\text{Cldim}(X/L) < \text{Cldim}(X)$. Choose $Q \in \text{Ass}(X/L)$. Then $Q = \ell(Y/L)$ for some submodule $0 \neq Y/L \subseteq X/L$. From 2.2(b), [7; 1.3] and [12; 8.2.7], $\text{Cldim}(R/P) \leq \text{Cldim}(R/Q)$. Now, 2.5 and (b) above yield that $\text{Cldim}(R/Q) = \text{Cldim}(Y/L) < \text{Cldim}(X) = \text{Cldim}(Y)$. Thus, $\text{Cldim}(Y) = \sup \text{Cldim}(L), \text{Cldim}(Y/L)\} = \text{Cldim}(L) = \text{Cldim}(R/P) > \text{Cldim}(R/Q)$ which is a contradiction. Therefore, $X = L$.

3.2 Proposition. Let M be a finitely generated tame module and let $0 \neq B \subseteq M$ be a uniform submodule with $\text{Cldim}(M) = \beta$. Then B is a basic submodule of M iff B is a maximal β-Cl-critical submodule of M and $\beta = \min\{\text{Cldim}(N) \mid 0 \neq N \subseteq M$ is a submodule$\}$. In this case $B = \text{ann}_A(P)$ where A is any maximal essential extension of B in M and $P = \ell(B) = \text{ass}(B)$.

Proof. Let $\varphi = \min\{\text{Cldim}(N) \mid 0 \neq N \subseteq M$ is a submodule of $M\}$ and let $\varsigma = \min\{\text{Cldim}(R/P) \mid P \in \text{Ass}(M)\}$. Choose $0 \neq N \subseteq M$ a submodule with $\varphi = \text{Cldim}(N)$. Let $P \in \text{Ass}(N)$. Then $P = \ell(Y)$ for some submodule $0 \neq Y \subseteq N$ and by 2.5, $\text{Cldim}(R/P) = \text{Cldim}(Y) = \varphi$. Thus, $\varsigma \leq \varphi$. Since any $P \in \text{Ass}(M)$ satisfies $\text{Cldim}(R/P) = \text{Cldim}(K)$ for some submodule with $K \subseteq M$, $\varsigma \geq \varphi$. Therefore, $\varsigma = \varphi$.

Let $V \subseteq M$ be a uniform submodule with $\ell(V) = \text{ass}(V) = Q$. Let $W \in \mathcal{T}(V)$. Since V is finitely annihilated, V is a torsion-free R/Q-module. Thus, since $W \leq_e V$, and V/W is finitely annihilated, $Q \subset \ell(V/W) \subseteq R$. By 2.5 and [8; 12A(c)], $\text{Cldim}(R/\ell(V/W)) = \text{Cldim}(V/W) < \text{Cldim}(R/Q) = \text{Cldim}(V)$ and hence V is Cl-critical. Conversely, if V is Cl-critical, then by 3.1, V is uniform and $\ell(V) = \text{ass}(V)$.

The first statement now follows from the definition of a basic submodule of M. The last statement follows from the maximality of B and the equivalence established in the last paragraph.

For the next result, the *classical dimension sequence* of a finitely generated tame module M, $\text{Cldim}(M) = \alpha_1 > \cdots > \alpha_n \geq 0$ is the (necessarily) finite set of ordinals that occur as the classical Krull dimension of a non-zero submodule of M. From the exactness of Cldim( ) established in 2.4, the submodule

$M_i = \sum\{K \subseteq M \mid K$ is a submodule with $\text{Cldim}(K) \leq \alpha_i\}$ is the unique largest submodule with $\text{Cldim}(M_i) = \alpha_i$.

3.3 Theorem. Let M be a finitely generated tame module with classical dimension sequence $\text{Cldim}(M) = \alpha_1 > \cdots > \alpha_n \geq 0$. Then M has at least one basic series

$$M = B_m \supset \cdots \supset B_1 \supset B_0 = 0$$

such that

(a) $\beta_m \geq \cdots \geq \beta_1$ where $\beta_j = \text{Cldim}\ (B_j/B_{j-1})$ for $j = 1, \cdots, m$.
(b) $\{\beta_1, \cdots, \beta_m\} = \{\alpha_1, \cdots, \alpha_n\}$.
(c) B_j/B_{j-1} is β_j -Cl-critical for $j = 1, \cdots, m$.
(d) for each $i = 1, \cdots, n$, $M_i = B_j$ for some j.

If $M = C_k \supset \cdots \supset C_1 \supset C_0 = 0$ is any other basic series for M, then $k = m$ and there exists a permutation π of $\{1, \cdots, m\}$ such that for all j, B_j/B_{j-1} and $C_{\pi(j)}/C_{\pi(j)-1}$ contain isomorphic copies of each other.

Furthermore, if R satisfies the right second layer condition and if M is uniform, then $\text{Cldim}(M) = \text{Cldim}\ (N)$ for all submodules $0 \neq N \subseteq M$.

Proof: The existence and uniqueness statements come from [11; 2.1]. Parts (a)-(c) follow from 3.2 and the definition of a basic series. For (d), an easy induction argument using (a) establishes that $\text{Cldim}(B_j) = \beta_j$ if B_j/B_{j-1} is β_j-Cl-critical. By (b), $B_j \subseteq M_i$ for some i. Since M_i is Noetherian, that part of the series contained in M_i must terminate with M_i.

For M uniform, let $P = \text{ass}\ (M)$ and for $j = 1, \cdots, m$, let $P_j = \ell(B_j/B_{j-1})$. By the second statement of 3.2, [7; 1.3] and [11; 1.6, 1.7], $\text{Cldim}(R/P_j) =$ $\text{Cldim}\ (R/P)$ for all j. It follows from (b) that the only ordinal in the classical dimension sequence of M is $\text{Cldim}(R/P)$.

If R satisfies the second layer condition on both sides, then according to [8; 12.8], $\bigcap_{n=1}^{\infty} I^n = 0$ where I is the intersection of all the maximal ideals of R. Using essentially the same proof and 3.3, more is true.

3.4 Corollary. Let R be a Noetherian ring that satisfies the second layer condition and let $\text{Cldim}(R) = \alpha_1 > \cdots \geq \alpha_n \geq 0$ be the classical dimension sequence of R. For any ordinal $\beta \leq \alpha_n$, let $J = \cap\{P \in \text{Spec}(R) \mid \text{Cldim}(R/P) = \beta\}$. Then $\bigcap_{n=1}^{\infty} J^n = 0$.

Proof. Let $0 \neq I \subseteq R$ be a left ideal and let $A = \ell(I)$. Then $\text{Cldim}(R/\ell(I)) =$ $\text{Cldim}(I) = \alpha_i$ for some i. By [8; 12.2], there exists a prime ideal $P \supseteq \ell(I)$ such that $\text{Cldim}(R/P) = \beta$. There is a 1-1 map $R/\ell(I) \to I^{(k)}$ for some k. The canonical map $R/\ell(I) \to R/P$ extends to a map $I^{(k)} \to E(R/P)$. Thus, $\text{Hom}_R\ (\text{I}, E(R/P)) \neq 0$.

Let $E = \oplus\{E(R/P) \mid P \in \text{Spec}(R)$ and $\text{Cldim}(R/P) = \beta\}$ and let $K =$ $\bigcap\{\ker(f) \mid f \in \text{Hom}_R(R, E)\}$. From above, it must be the case that $K = 0$ and hence, there is a 1-1 map $R \to \Pi_S E$ for some set S. Thus, it suffices to show that for every prime P with $\mathrm{Cldim}(R/P) = \beta$, every element of $E(R/P)$ is annihilated by some power of J.

Choose $0 \neq x \in E(R/P)$ and let $Rx = B_m \supset \cdots \supset B_0 = 0$ be a basic series for Rx with $P_j = \ell(B_j/B_{j-1})$ for $j = 1, \cdots, m$. Then $P_1 P_2 \cdots P_m x = 0$. By 3.3, each P_j satisfies $\mathrm{Cldim}(R/P_j) = \beta$. Therefore, $J \subseteq P_j$ for all j and hence $J^n x = 0$.

For the remainder of this paper, $\mathrm{Cldim}(R) = \alpha_1 > \cdots \geq \alpha_n \geq 0$ will denote the classical dimension sequence of ${}_R R$, and for each $i = 1, \cdots, n$, T_i will denote the unique largest left ideal of R with $\mathrm{Cldim}(T_i) = \alpha_i$.

Since R is a Noetherian ring, then from 2.5, R is a tame left R-module. Thus, by 3.3, ${}_R R$ has a basic series $R = B_m \supset \cdots \supset B_0 = 0$. By 3.1, each $P_j = \ell(B_j/B_{j-1})$ is a prime ideal and by 3.3(c), (b) and 2.5, $\mathrm{Cldim}(R/P_j) = \alpha_i$ for some i. Such a prime ideal will be called a *left* α_i*- basic prime* of R. If R also satisfies the right second layer condition, then by 2.10, the classical dimension sequences of ${}_R R$ and R_R coincide and in this case, the *right* α_i*- basic primes* are defined similarly. The collection of all prime ideals that are left or right α_i-basic primes for some i will be denoted by Bprime(R). Propositions 3.6, 3.9 and 3.10 below makes it clear that the basic primes are analogues of Muller's Krull primes introduced in [15]. Whether or not the two sets coincide in our setting is an open question (see the comment at the end of this section).

3.5 Lemma. For each i, T_i is a two sided ideal of R and the classical dimension sequence of R/T_i is $\mathrm{Cldim}(R/T_i) = \alpha_1 > \cdots > \alpha_{i-1}$.

Proof. If $0 \neq r \in R$, then $T_i r$ is a homomorphic image of T_i and so $\mathrm{Cldim}(T_i r) \leq \mathrm{Cldim}(T_i) = \alpha_i$. Thus, $T_i r \subseteq T_i$.

If $0 \neq X/T_i \subseteq R/T_i$, then $\mathrm{Cldim}(X) = \alpha_j > \alpha_i$ for some j. Also, $\mathrm{Cldim}(X) = \sup\{\mathrm{Cldim}(X/T_i), \mathrm{Cldim}(T_i)\}$. Therefore, $\mathrm{Cldim}(X/T_i) = \alpha_j$.

If $1 \leq j \leq i-1$, then there exists a left ideal Y of R with $\mathrm{Cldim}(Y) = \alpha_j > \alpha_i$. Repeating the argument above, $\mathrm{Cldim}(Y + T_i/T_i) = \mathrm{Cldim}\ (Y) = \alpha_j$.

3.6 Proposition. Let R be a Noetherian ring that satisfies the second layer condition.

(a) If P is a minimal prime ideal of R, then P is a left and right α_i-basic prime for some α_i.

(b) A prime P is a left α_i-basic prime iff P is minimal over $\ell(T_i/T_{i+1})$.

(c) Ass $({}_R R) \cup$ Ass$(R_R) \subseteq$ Bprime (R).

Proof.

(a) Let $R = X_m \supset \cdots \supset X_0 = 0$ be a basic series for ${}_R R$ with left basic primes $P_i = \ell(X_i/X_{i-1})$ for $i = 1, \cdots, m$. Then $0 = P_1 \cdots P_m R = P_1 \cdots P_m \subseteq P$ and hence, $P_j \subseteq P$ for some j. Since P is minimal, $P = P_j$. Similarly, P is a right α_k-basic prime for some k. By 2.10, $\alpha_i = \alpha_k$.

(b) Let $A = \ell(T_i/T_{i+1})$. If P is a left α_i-basic prime, then $\mathrm{Cldim}(R/P) = \alpha_i$ and $P = \ell(X_j/X_{j-1})$ for some j where $R = X_m \supset \cdots \supset X_0 = 0$ is a basic series for ${}_R R$. By 3.3, $T_i \supseteq X_j \supset X_{j-1} \supseteq T_{i+1}$ and it follows that $A \subseteq P$. Also, Cldim(R/A) =Cldim$(T_i/T_{i+1}) = \alpha_i$. By 2.8, P/A is a minimal prime ideal of R/A.

Let $P/A \subseteq R/A$ be a minimal prime ideal. By 3.3, $T_i = X_k \supset \cdots \supset X_t = T_{i+1}$ for some k and t. Thus, in R/A, the product of the left α_i-basic primes is zero. Repeating the argument used to establish (a) yields the desired conclusion.

(c) If $X \subseteq R$ is a α_i -Cl-critical left idea, then $X \cap T_{i+1} = 0$ and so there is a 1-1 map $X \to T_i/T_{i+1}$. Thus, in the notation of (b), $A \subseteq \ell(X) = P$. Note that by 3.1(c), P =ass (X). Since Cldim(R/P) =Cldim $(R/A) = \alpha_i$, P/A is a minimal prime ideal of R/A by 2.8. The result follows from (b).

3.7 Lemma. Let R be a Noetherian ring that satisfies the second layer condition. Suppose that Ass$({}_R R) \cup$ Ass(R_R) consists of minimal primes. Then for each $i = 1, \cdots, n$, R contains a α_i -Cl-critical left ideal.

Proof. Let β be an ordinal and let $\mathcal{A} = \{D \subseteq R \mid D$ is a left ideal and Cldim$(R/D) < \beta\}$. Using 2.4, it is easy to verify that $\mathcal{A}$ is an idempotent filter in the sense of [5; pg. 25]. Thus, by [5, 3.3], $\mathcal{A}$ defines a torsion theory τ on left R-modules given by $\tau(M) = \sum\{Rx \subseteq M \mid$ Cldim$(Rx) < \beta\}$. It follows from 2.10 and [1; 1.4, 1.1] that $\tau(R)$ is a closed left ideal of R. Now, if $\beta = \alpha_i$, then $\tau(R) = T_{i+1}$. Thus, $T_{i+1} \oplus K \leq_e R$ for some left ideal K. Since T_{i+1} is closed, $K \cong T_{i+1} \oplus K/T_{i+1} \leq_e R/T_{i+1}$. By 3.5, α_i is the smallest ordinal in the classical dimension sequence of R/T_{i+1} and so by the first statement of 3.3 and by 3.2, R/T_{i+1} contains a α_i -Cl-critical submodule. It follows that K must contain a α_i -Cl-critical submodule.

3.8 Corollary. Let R be a Noetherian ring that satisfies the second layer condition. Suppose that Ass$({}_R R) \cup$ Ass(R_R) consists of minimal primes. Then for each i, $T_i \cap \ell(T_i) = T_i \cap rt(T_i) = 0$. In particular, $\ell(T_i) = rt(T_i)$ for each i.

Proof. If $T_i \cap \ell(T_i) \neq 0$, then by 3.7, $T_i \cap \ell(T_i) \supseteq X$ a α_i -Cl-critical right ideal. Since $XT_i = 0$, $T_i \subseteq rt(X) = P$ which is a minimal prime ideal by 3.1(c) and the hypothesis. By 2.5, Cldim$(R/P) = \alpha_i$. Since P/T_i is also a minimal prime ideal of R/T_i, Cldim$(R/P) = \alpha_i$ is an ordinal in the classical dimension sequence of R/T_i by 3.6(a). However, this contradicts 3.5. Therefore, $T_i \cap \ell(T_i) = 0$. Similarly, $T_i \cap rt(T_i) = 0$. The last statement easily follows from these two results.

For an ideal I, let $\mathcal{C}(I) = \{c \in R \mid rc \in I$ or $cr \in I$ implies that $r \in I\}$. For a collection S of ideals, we use $\mathcal{C}(S)$ to denote the *set* $\cap\{\mathcal{C}(I) \mid I \in S\}$.

It is an easy consequence of the last statement of 3.8 that $\mathcal{C}(0) \subseteq \mathcal{C}(\ell(T_i)) = \mathcal{C}(rt(T_i))$ for each $i = 1, \cdots, n$. Using this fact, the proofs of the following results are verbatim copies of Muller's proofs of [15; 9, 21] with classical Krull dimension in place of regular Krull dimension.

3.9 Corollary. Let R be a Noetherian ring that satisfies the second layer condition. Suppose that Ass$({}_R R) \cup$ Ass(R_R) consists of minimal primes. Then

$\mathcal{C}(0) = \mathcal{C}$ (Bprime(R)).

3.10 Corollary. Let R be a Noetherian ring that satisfies the second layer condition. Then R has an Artinian classical quotient ring iff Bprime(R) consists of minimal primes.

By 3.1 and [8; 6.19], any Cl-critical module X imbeds into $R/\ell(X) =$ $R/$ass (X). Thus, Cl-critical modules are critical with respect to Krull dimension. However, basic series and Krull composition series are not the same. For example, let $R = S \oplus T$ where S and T are simple Noetherian domains with $| S |= 2$ and $| T |= 1$, and the operations are performed coordinate-wise. The series $R \supset S \times 0 \supset 0$ is a basic series but is not a Krull composition series since the Krull dimension of the factors appear in the wrong order. Note though that $R \supset 0 \times T \supset 0$ is a Krull composition series and the annihilators of its factors are the same as the basic primes obtained from the first series.

The prime ideals ass(U) where U is a factor from a Krull composition series for ${}_R R$ or R_R are called it Krull primes and the collection of all Krull primes is denoted Kprime(R). Using 3.3 and the Krull dimension version of 3.3 (e.g., [8; 13.9]), its not hard to see that Bprime(R) = Kprime (R) if one assumes that factors of a Krull composition series for both ${}_R R$ and R_R have prime annihilators. For the converse, choose a maximal Krull critical submodule $X \subseteq {}_R R$ of minimal Krull dimension, and choose a basic series for X, $0 = B_0 \subset \cdots \subset B_m = X$. If $B_1 \neq X$, then the annihilator Q of the top factor satisfies both $| R/Q |<| X |$ and $Q \in$ Bprime(R) =Kprime (R) which is impossible (since the latter implies that $| R/Q |$ must now be an ordinal in the Krull dimension sequence of ${}_R R$). An easy induction using [8; 13.9] completes the proof.

As noted in [15], ass $(U) = \ell(U)$ for any factor U from a Krull composition series of R if R is Krull symmetric. Thus, whether or not Bprime(R) = Kprime(R) is an open question.

4. The Homogeneous Components

A tame module M will be called *β-homogeneous* , for an ordinal β, provided Cldim(N) = β for all non-zero finitely generated submodules $N \subseteq M$. Jategaonkar in [11] showed that a Noetherian ring R satisfying the second layer condition has an Artinian classical quotient ring if R possesses a finitely generated faithful homogeneous module. From 2.1, this requirement is equivalent to R being homogeneous.

One purpose of this section is to generalize the notion of homogeneity to a set of conditions which are equivalent to the existence of an Artinian classical quotient ring for any Noetherian ring R that satisfies the second layer condition. Another purpose is determine conditions under which R decomposes into a product of α_i-homogeneous ideals where Cldim(R) $= \alpha_1 > \cdots > \alpha_n \geq 0$ is the classical dimension sequence of R. Our goal here is to obtain the classical Krull dimension analogues of the results in [17] (as proved in [2]) that addressed the Krull dimension version of this problem for Krull symmetric Noetherian rings.

We begin by showing that when associated primes are minimal, R contains a unique maximal non-zero α_i-homogeneous ideal for each i. *Throughout this section, it will be assumed that R is a Noetherian ring that satisfies the second layer condition.*

4.1 Proposition. Suppose that Ass $({}_R R)$ consists of minimal primes. If $S, T \subseteq E(R)$ are β-homogeneous submodules, then $S+T$ is β-homogeneous.

Proof. First, note that by [2; 2.1], every submodule of $E(R)$ is finitely annihilated. Thus, without loss of generality, we may assume that both S and T are finitely generated.

Let $0 \neq X \subseteq S+T$ be a Cl-critical submodule. By 3.1(c), $P = \ell(X) \in \mathrm{Ass}({}_R R)$ and so P is a minimal prime. Now, $\ell(S) \cdot \ell(T) \subseteq \ell(S) \cap \ell(T) \subseteq \ell(X) = P$. Thus, $\ell(S) \subseteq P$ or $\ell(T) \subseteq P$. If $\ell(S) \subseteq P$, then $P/\ell(S) \subseteq R/\ell(S)$ is a minimal prime. Since S is β-homogeneous, it is an easy consequence of 2.5 that $R/\ell(S)$ is also β-homogeneous. By 3.6, $\mathrm{Cldim}(R/P) = \beta$. Also, $\mathrm{Cldim}(R/P) = \mathrm{Cldim}(X)$. Therefore, X is β-Cl-critical. It follows that $S+T$ is β-homogeneous.

4.2 Proposition. Suppose that Ass $({}_R R) \cup \mathrm{Ass}(R_R)$ consists of minimal primes. Let H_i be the sum of the α_i-homogeneous left ideals of R for $i = 1, \cdots, n$. Then for all i,

(a) H_i is the unique maximal non-zero α_i-homogeneous left ideal of R as well as the unique maximal α_i-homogeneous right ideal of R.
(b) $E(H_i)$ is α_i-homogeneous.
(c) H_i is closed in R both as a left ideal and a right ideal.
(d) The classical dimension sequence of R/H_i is $\mathrm{Cldim}(R/H_i) = \alpha_i > \cdots > \alpha_{i-1} > \alpha_{i+1} > \cdots > \alpha_n \geq 0$.
(e) $\ell(T_i/T_{i+1}) = \ell(H_i)$.

Proof. Since $H_i \neq 0$ by 3.7, (a) Follows from 4.1 and 2.10.

(b) For each i, $E(H_i) = \bigoplus_{k=1}^{t} E(U_k)$ where $U_k \subseteq H_i$ is a uniform submodule for $k = 1, \cdots, t$. By 3.3, every finitely generated submodule of $E(U_k)$ is α_i-homogeneous.
(c) Statement (c) is an immediate consequence of (b).
(d) Let $H = \bigoplus_1^n H_i$. Then $H <_e R$ and by (c), $\bigoplus_{j \neq i} H_j \cong H/H_i \leq_e R/H_i$. Therefore, $E(R/H_i) \cong \bigoplus_{j \neq i} E(H_j)$. The result is now an easy consequence of (b).
(e) The proof of (e) is just the proof of [2; 3.1(b)] with classical Krull dimension in place of Krull dimension.

4.3 Corollary. Suppose that $\mathrm{Ass}({}_R R) \cup \mathrm{Ass}(R_R)$ consists of minimal primes. Then for all i,

(a) $\bigoplus_{j \neq i} H_j \subseteq \ell(H_i)$.
(b) $H_i \cap \ell(H_i) = 0$.

(c) $\bigoplus_1^n H_i$ is faithful as a left and right ideal.
(d) $\ell(H_i) = rt(H_i)$.

Proof.
(a) If $j \neq i$, then $H_i \cdot H_j \subseteq H_i \cap H_j = 0$.
(b) Let $K = \bigoplus_{j \neq i} H_j$. By (a), $K \subseteq \ell(H_i)$. Let $0 \neq X \subseteq \ell(H_i)$ be a Cl-critical it right ideal. Then $H_i \subseteq rt(X) = P \in Ass(R_R)$ by 3.1(c). Thus, P is a minimal prime whence P/H_i is a minimal prime ideal of R/H_i. By 3.6 and 4.2(d), $\text{Cldim}(R/P) = \alpha_j$ for some $j \neq i$. Therefore, $\text{Cldim}(X) = \text{Cldim}(R/P) = \alpha_j$. Consequently, $X \subseteq H_j$. It follows that $K \leq_e \ell(H_i)$ as right ideals. Therefore, since $K \cap H_i = 0$, $\ell(H_i) \cap H_i = 0$.
(c) If $A = \ell(\bigoplus_1^n H_i) = \bigcap_1^n \ell(H_i) \neq 0$, then A contains a α_j -Cl-critical submodule for some j. Then $H_j \cap \ell(H_j) \supseteq H_j \cap A \neq 0$ which contradicts (b).
(d) Statement (d) is an easy consequence of (b) and its right-hand counterpart.

According to 3.6, every minimal prime ideal P is a α_i-basic prime for some α_i. In the following, such a prime will be called a it minimal α_i- it prime. As with Krull primes in [15], the relationship between the ideals H_i and the α_i-basic primes determines when R has an Artinian classical quotient ring.

4.4 Theorem. Suppose that Ass $({}_R R) \cup \text{Ass}(R_R)$ consists of minimal primes. If P is a minimal α_i-prime, then $H_i \not\subseteq P$. Furthermore, R has an Artinian classical quotient ring iff for all i, $H_i \not\subseteq P$ for all α_i-basic prime ideals P.

Proof. If P is a minimal α_i-prime and $H_i \subseteq P$, then P/H_i is a minimal prime ideal of R/H_i. By 3.6(a), $\text{Cldim}(R/P) = \alpha_i$ is an ordinal in the classical dimension sequence of R/H_i. However, this contradicts 4.2(d).

The proof of the second statement is the same as the proof of [2; 3.7] with 3.9 in place of [15; 21].

4.5 Corollary. Suppose that Ass $({}_R R) \cup \text{Ass}(R_R)$ consists of minimal primes. Then R has an Artinian classical quotient ring iff $\bigoplus_1^n H_i$ contains a regular element.

Proof. Let $H = \bigoplus_1^n H_i$ and let N be the prime radical of R. If R has an Artinian classical quotient ring, then $\mathcal{C}(0) = \mathcal{C}(N)$. For any minimal prime P, $H_i \not\subseteq P$ for some i by 4.4. Thus $\phi \neq H_i \cap \mathcal{C}(P) \subseteq H \cap \mathcal{C}(P)$. By [15; 1], $H \cap \mathcal{C}(N) = H \cap \mathcal{C}(0) \neq \phi$.

Conversely, suppose that $H \cap \mathcal{C}(0) \neq \phi$. Let P be an α_i-classical prime. By 3.6(b), 4.2(e) and 4.3(a), $\bigoplus_{j \neq i} H_j \subseteq \ell(H_i) \subseteq P$. Thus, if $H_i \subseteq P$, then $H \subseteq P$ and P contains a regular element, say c. From 4.3(d), it's easy to see that $c \in \mathcal{C}(\ell(H_i)) = \mathcal{C}(rt(H_i))$. However, since $P/\ell(H_i)$ is a minimal prime ideal of $R/\ell(H_i)$, then $c \in \mathcal{C}(\ell(H_i)) \subseteq \mathcal{C}(P)$ which is impossible. Therefore, $H_i \not\subseteq P$. By 4.4, R has an Artinian classical quotient ring.

In [6], it is shown that a Krull symmetric Noetherian ring R with minimal associated primes has an Artinian classical quotient ring iff for certain ideals K_1, $\cdots$, K_m, $| R/K_i | < \beta_i$ where $0 \leq \beta_m < \cdots < \beta_1 = | R |$ is the Krull dimension sequence of R. A similar result involving classical Krull dimension is obtained for Noetherian rings that satisfy the second layer condition.

4.6 Corollary. Suppose that Ass $({}_R R) \cup \text{Ass}(R_R)$ consists of minimal primes. Then R has an Artinian classical quotient ring iff $\text{Cldim}(R/H_i \oplus \ell(H_i)) < \alpha_i$ for all i.

Proof. Note that from 2.5, $\text{Cldim}(R/H_i \oplus \ell(H_i)) \leq \alpha_i$ for each i. It is a consequence of 2.6, 3.6(b) and 4.2(e) that equality holds if and only if there exists $P \in \alpha_i\text{-Bprime}(R)$ such that $H_i \oplus \ell(H_i) \subseteq P$. The result now follows from 4.4

We are now in position to determine when R decomposes into a product of homogeneous ideals. As with Warfield's result [16; 10], this depends on R/N possessing such a decomposition.

4.7 Lemma. Suppose that R has an Artinian classical quotient ring. Then for all $i, H_i + N/N$ is α_i-homogeneous.

Proof. Let $Q_1, \cdots, Q_m$ denote the minimal α_i-primes for some fixed i. Let P be a minimal α_j-prime for some $j \neq i$. By 4.4, $H_j \not\subseteq P$. Since $H_j \subseteq \ell(H_i)$, then $\ell(H_i) \not\subseteq P$ and hence $H_i \subseteq P$. Consequently, $H_i \cap N = H_i \cap Q_1 \cap \cdots \cap Q_m$. Thus, $H_i + N/N \cong H_i/H_i \cap Q_1 \cap \cdots \cap Q_m \cong H_i + (Q_1 \cap \cdots \cap Q_m)/(Q_1 \cap \cdots \cap Q_m) \subseteq R/(Q_1 \cap \cdots \cap Q_m)$ which is α_i-homogeneous.

4.8 Theorem. Suppose that R has an Artinian classical quotient ring. Then H_i is an ideal direct summand of R iff the α_i-homogeneous component of R/N is an ideal direct summand of R/N. In this case, $H_i + N/N$ is the α_i-homogeneous component of R/N.

Proof. Let $K = \bigoplus_{j \neq i} H_j$ and let $H = \bigoplus_1^n H_i$. By 4.5, H contains a regular element c.

If $R = H_i \oplus Y$ for some ideal Y, then $K \subseteq Y$ and $R/N = H_i + N/N \oplus Y + N/N$. By 4.7, $H_i + N/N$ is α_i-homogeneous. Now, if $0 \neq X/N \subseteq Y + N/N$ is a α_i-homogeneous submodule, then from 4.7, $KX \subseteq N$. Also, $H_i X \subseteq H_i Y = 0$. Thus, $HX \subseteq N$ and hence $cX \subseteq N$. Since $c \in \mathcal{C}(N)$, $X \subseteq N$ which is a contradiction. It follows that $H_i + N/N$ is the α_i-homogeneous component of R/N.

Conversely, suppose that $R/N = T/N \oplus J/N$ where T/N is the α_i-homogeneous component of R/N and J/N is an ideal. Then there exist orthogonal idempotents $e_1, e_2 \in R$ such that $e_1 + e_2 = 1$, $e_1 + N$ and $e_2 + N$ are central in R/N and $T/N = Re_1 + N/N$ and $J/N = Re_2 + N/N$.

It follows from 4.7 that $H_i + N/N \subseteq T/N$ and $K + N/N \subseteq J/N$. Now, $c = c_1 + c_2$ where $c_1 \in H_i$ and $c_2 \in K$. Then $c + N = (c_1 + N) + (c_2 + N) = (c_1 e_1 + N) + (c_2 e_2 + N)$. Since $(c_2 e_2)e_1 = 0$, $c_2 e_2 \in H_i e_1^{-1} = \{r \in R \mid re_1 \in H_i\}$. Also, $c_1 e_1 \in H_i \subseteq H_i e_1^{-1}$. Thus, $c + N \in H_i e_1^{-1} + N/N$. Similarly, $c_1 e_1 \in K e_2^{-1}$ and $c_2 e_2 \in K e_2^{-1}$ and hence $c + N \in K e_2^{-1} + N/N$.

Since $c + N \in H_i e_1^{-1} + N/N$, $c + N = d + N$ where $de_1 \in H_i$. Since

$c \in \mathcal{C}(0) = \mathcal{C}(N)$, $d \in \mathcal{C}(N) = \mathcal{C}(0)$. Thus, $H_i \supseteq de_1R \cong e_1R$ and hence $e_1R \subseteq H_i$. In particular, $e_1 \in H_i$ and this establishes the inclusion $Re_1 + N/N \subseteq H_i + N/N$. Therefore, $H_i + N/N = Re_1 + N/N$.

It follows from $c + N \in Ke_2^{-1} + N/N$ that $c + N = d' + N$ where $d'e_2 \in K$ and $d' \in \mathcal{C}(0)$. Now, $d'e_2 \in K$ yields that $d'e_2H_i = 0$. Thus, $e_2 \in \ell(H_i)$. From 4.3(b), $H_i \cap e_2R = 0$. Thus, by modularity, $H_i = e_1R \oplus (H_i \cap e_2R) = e_1R$. Using 4.3(b) and modularity again, $\ell(H_i) = \ell(H_i) \cap (H_i \oplus e_2R) = e_2R \oplus (\ell(H_i) \cap H_i) = e_2R$. Therefore, $R = H_i \oplus \ell(H_i)$.

4.9 Corollary. Suppose that R has an Artinian classical quotient ring. Let V be the largest ideal with $\text{Cldim}(V) = 0$. Then V is an ideal direct summand of R.

Proof. According to 4.8, it suffices to show that the largest ideal of R/N with classical Krull dimension 0 is an ideal direct summand of R/N. Thus, without loss of generality, we may assume that R is a semiprime ring.

Suppose that $V \neq 0$. Let M be a maximal ideal of R. Since M is a prime ideal, $V \subseteq M$ or $\ell(V) \subseteq M$. If $\ell(V) \subseteq M$, then $\text{Cldim}(R/M) = \text{Cldim}(R/\ell(V)) = \text{Cldim}(V) = 0$. From 2.8, $M/\ell(V)$ is a minimal prime of $R/\ell(V)$. Thus, by 3.6(b), M is a 0-classical prime ideal, and by 3.10, M is a minimal prime ideal of R.

Let $M_1, \cdots, M_k$ be the distinct maximal ideals of R that contain $\ell(V)$. Let S be the intersection of the remaining minimal prime ideals. Then $M_1 \cap \cdots \cap M_k \cap S = 0$. Since M_i is a maximal ideal, $M_i \oplus \left[\left(\bigcap_{j \neq i} M_j\right) \cap S\right] = R$ for all i. Thus, M_1 is an ideal direct summand of R. Suppose that $M_1 \cap \cdots \cap M_t \oplus J = R$ for some ideal J where $t < k$. Since $(M_1 \cap \cdots \cap M_t)J = 0$ and $M_1 \cap \cdots \cap M_t \not\subseteq M_{t+1}$, $J \subseteq M_{t+1}$. Thus, by modularity, $M_{t+1} = M_{t+1} \cap (J \oplus M_1 \cap \cdots \cap M_t) = J \oplus M_1 \cap \cdots \cap M_{t+1}$. Also, $R = M_{t+1} \oplus \left[\left(\bigcap_{j \neq t+1} M_j\right) \cap S\right]$. Thus, $M_1 \cap \cdots \cap M_{t+1}$ is an ideal direct summand of R. Eventually, $R = M_1 \cap \cdots \cap M_k \oplus T$ for some ideal T.

Now, $T \cong R/(M_1 \cap \cdots \cap M_k)$ implies that $\text{Cldim}(T) = 0$. Thus, $T \subseteq V$. Let P be a minimal prime ideal of R that is not a maximal ideal. Then $\text{Cldim}(R/P) > 0$ and hence, by 3.6(a), (b), $\ell(V)$ is not contained in P. It follows that $V \subseteq P$. Therefore, $V \subseteq S$ and hence $V \cap M_1 \cap \cdots \cap M_k = 0$. Using modularity again, $V = V \cap (T \oplus M_1 \cap \cdots \cap M_k) = T \oplus (V \cap M_1 \cap \cdots \cap M_k) = T$. Consequently, $R = M_1 \cap \cdots \cap M_k \oplus V$.

ACKNOWLEDGEMENT

The author gratefully acknowledges the support provided by a UW-Centers Summer 1991 Research Grant under which this work was completed.

REFERENCES

1. J.A. Beachy, *Stable torsion radicals over FBN-rings*, J. Pure Appl. Algebra **24** (1982), 235-244.

2. A.K. Boyle and K.A. Kosler, *Krull symmetric Noetherian rings and their homogeneous components*, J. Pure Appl. Algebra **41** (1986), 139-148.
3. K.A. Brown and R.B. Warfield, *The influence of ideal structure on representation theory*. J. Algebra **116** (1988), 294-315.
4. S.M. Ginn and P.B. Moss, *A decomposition theorem for Noetherian orders in Artinian rings*, Bull. London Math Soc. **9** (1977), 177-181.
5. J.S. Golan, *Localization of Noncommutative Rings*, Marcel Dekker, New York, 1975.
6. A. Goldie and G. Krause, *Artinian quotient rings of ideal invariant Noetherian rings*, J. Algebra **63** (1980), 374-388.
7. K.R. Goodearl, *Linked injectives and Ore localizations*, J. London Math. Soc. **37** (1988), 404-420.
8. K.R. Goodearl and R.B. Warfield, *An Introduction to Noncommutative Rings*, London Math. Soc. Student Texts **16**. Cambridge University Press, Cambridge, 1989.
9. R. Gordon and J.C. Robson, *Krull dimension*, AMS Memoirs **133** (1973).
10. A.V. Jategaonkar, *Jacobson's conjecture and modules over fully bounded Noetherian rings*, J. Algebra **30** (1974), 103-121.
11. ———, *Solvable Lie Algebras, polycyclic by finite groups and bimodule Krull dimension*, Comm. in Algebra **10** (1982), 361-366.
12. ———, *Localization in Noetherian Rings*, London Math Soc. Lecture Note Series, **98**, Cambridge University Press, London/New York, 1985.
13. K.A. Kosler, *Module extensions and the second layer condition*, Comm. in Algebra **20** (1992), 69-91.
14. J.C. McConnell and J.C. Robson, *Noncommutative Noetherian Rings*, John Wiley and Sons, Chichester, 1987.
15. B.J. Mueller, *The quotient problem for Noetherian rings*, Canad. J. Math. **33** (1981), 734-748.
16. C. Nastasescu and F. van Oystaeyen, *Dimensions of Ring Theory*, D. Reidel Publishing, Dordrecht, 1987.
17. R.B. Warfield, *Quotient rings and localization for Noetherian rings*, J. Algebra **72** (1981), 166-182.

THE EXCHANGE PROPERTY FOR QUASI-CONTINUOUS MODULES

SAAD H. MOHAMED
Department of Mathematics, Kuwait University, Kuwait 13060
and
BRUNO J. MÜLLER
Department of Mathematics, McMaster University
Hamilton, Ontario, Canada L8S 4K1

ABSTRACT

The exchange property has been demonstrated, among others, for injective [10], quasi-injective [3] and continuous [4] modules. In contrast, quasi-continuous modules do not always enjoy even the finite exchange property. The questions, already posed in the fundamental paper [2], whether the finite exchange property always implies the full exchange property, is still open. Here we establish this implication for nonsingular quasi-continuous modules. We also provide a new proof of the exchange property for continuous modules, using a different approach. Many of our lemmas apply to arbitrary quasi-continuous modules, and might prove useful in extending the implication.

1. Let M be a right module over an arbitrary ring R. We denote the endomorphism ring of M by S, the Jacobson radical of S by J, and the ideal of S consisting of all endomorphisms with essential kernel by Δ. A family of endomorphisms $\phi_i \in S(i \in I)$ is called summable if for each $m \in M$, $\phi_i(m) = 0$ holds for almost all $i \in I$. This means that $\sum_{i\in I} \phi_i$ is convergent in the finite topology. The ideal Δ is always closed in this topology ([5], 3.23(3)). [A.L.S. Corner has drawn to our attention the easily verified fact that, for a module with the finite exchange property, the Jacobson radical J of S is closed as well.]

Proposition and Definition 1 ([5], [12]). *M is said to have the (finite) exchange property if it satisfies the following equivalent conditions, for an (finite) index set I:*

(1) Whenever $M \oplus N = \bigoplus_{i\in I} A_i$, then there exist submodules $B_i \leq A_i$ such that $M \oplus N = M \oplus \left(\bigoplus_{i\in I} B_i\right)$;
(2) The special instance of (1) where all A_i are isomorphic to M;
(3) For every summable family $\phi_i (i \in I)$ with $\sum \phi_i = 1$ there exist orthogonal idempotents $e_i \in S\phi_i$ such that $\sum e_i = 1$.

It is well known that M has the finite exchange property if and only if S has the (finite) exchange property, as left or as right module over itself. Such S is called an exchange ring [11], or suitable [6].

For background information on continuous and quasi-continuous modules, and the conditions (C_1), (C_2) and (C_3), see ([5], Chapter 2). In particular, any quasi-continuous module is the direct sum of a quasi-injective and a squarefree quasi-continuous module ([5], 2.37; a module is squarefree if it does not contain a direct sum of two nonzero isomorphic submodules). As the exchange property is preserved under finite direct sums, we may assume that our quasi-continuous module M is squarefree. Then all idempotents of S are central modulo Δ([5], 3.4).

Not every quasi-continuous module has the finite exchange property: for instance, every uniform module is quasi-continuous but has the (finite) exchange property if and only if the endomorphism ring is local. The following characterization extends this observation:

Proposition 2. A squarefree quasi-continuous module M has the finite exchange property if and only if $\Delta \leq J$, and S/Δ has local Pierce stalks (i.e., $\bar{S}/m\bar{S}$ is a local ring, for every maximal ideal m of the Boolean algebra of idempotents of $\bar{S} = S/\Delta$).

Proof. As observed above, all idempotents of $\bar{S}$ are central. By ([6], 1.5, 1.9 and 2.1) and ([5], 3.7), S is suitable if and only if S/Δ is suitable and $\Delta \leq J$. By ([1], 3.4), S/Δ is an exchange ring if and only if all its stalks are local.

□

2. The aim of this section is to establish the exchange property for nonsingular quasi-continuous modules with the finite exchange property, using the characterization of Proposition 1(3). The subsequent lemmas are deliberately stated with weaker assumptions than needed, in the expectation that they might be utilized for a proof of a more general result.

We fix a summable family $\sum_{i \in I} \phi_i = 1$ in S. We define $\phi_F = \sum_{i \in F} \phi_i$ and $K_F = \bigcap_{i \notin F} ker\ \phi_i$, for all finite subsets F of I. Note that $\phi_F(x) = x$ holds for all $x \in K_F$. For any submodule N of M, N^* denotes a closure of N in M.

Lemma 3. Let M have (C_1). If the idempotents of S commute, then the closure N^* of any submodule N of M is unique. In particular, $K_G^* \leq K_F^*$ holds whenever $G \subseteq F$.

Proof. By (C_1), every closure N^* of a submodule N is a direct summand, i.e., $N^* = eM$ for some $e = e^2 \in S$. If fM is another closure of N, then $N \leq eM \cap fM = efM$. Since ef is again idempotent, we have $eM = efM = fM$. In particular, as $G \subseteq F$ implies $K_G \leq K_F$, we have $K_G^* \leq K_F^*$ under the present assumptions.

□

Lemma 4. Let M be quasi-continuous and squarefree. If there are closures K_F^*, for all finite F, such that $K_G^* \leq K_F^*$ whenever $G \subseteq F$, then there exist submodules M_F such that $M = \bigoplus_F M_F$, and $\phi_F(x) = x$ for all x in some essential submodule of M_F.

Proof. By way of induction over n, we assume that submodules M_G of M have been defined for all $|G| < n$, such that $\sum M_G$ is direct and $K_G^\bullet = \sum_{H\subsetneq G} \oplus M_H$.

For any F with $|F| = n$, and any $G \subset F$, we have $M_G \subseteq K_G^\bullet \subseteq K_F^\bullet$, and, therefore, $\sum_{G\subset F} \oplus M_G \subseteq K_F^\bullet$. Since this finite direct sum is a direct summand of M by (C_3), we obtain $\sum_{G\subset F} \oplus M_G \oplus X = K_F^\bullet$. We define $M_F = X$.

We have to show that $\sum_{|F|\leq n} M_F$ is direct. Suppose not, and select a non-direct subsum $\sum_{a=1}^m M_{F_a}$ with a minimal number m of terms.

If $\cup_{a=2}^m F_a \subseteq F_1$, then $F_a \subset F_1$ for $a = 2,\cdots,m$; hence $\sum_{a=1}^m M_{F_a}$ is direct as part of $\sum_{G\subset F_1} \oplus M_G \oplus M_{F_1} = K_{F_1}^\bullet$. Thus the supposition implies $\cup_{a=2}^m F_a \not\subseteq F_1$, and we have an index $k \notin F_1$ with $k \in F_b$ for some $2 \leq b \leq m$.

We consider the endomorphism ϕ_k. We select arbirtary projections $\pi_a : M \to M_{F_a}$. Since M is squarefree, idempotents of endo M are central modulo Δ ([5], 3.4), which means that $\delta_a = \phi_k\pi_a - \pi_a\phi_k$ has essential kernel.

This yields essential submodules $X_a = M_{F_a} \cap \ker\delta_a \cap K_{F_a} \subseteq' M_{F_a}$. In consequence, the sum $\sum_{a=1}^M X_a$ is also non-direct, and we can find $x_a \in X_a$, not all zero, such that $\sum_{a=1}^m x_a = 0$. The minimality of m implies $x_a \neq 0$ for *all* a.

Now $x_1 \in K_{F_1}$ and $k \notin F_1$ yield $\phi_k(x_1) = 0$ and, therefore, $0 = \sum_{a=2}^m \phi_k(x_a)$. Moreover, $x_a \in \ker\delta_a$ implies $0 = \delta_a(x_a) = \phi_k\pi_a(x_a) - \pi_a\phi_k(x_a)$, hence $\phi_k(x_a) = \phi_k\pi_a(x_a) = \pi_a\phi_k(x_a) \in M_{F_a} (a = 2,\cdots,m)$. Consequently, the minimality of m forces $\phi_k(x_a) = 0$ for all a.

In particular, we have $\phi_k(x_b) = 0$. We recall $k \in F_b$ and $0 \neq x_b \in K_{F_b} = \cap_{i\notin F_b} \ker\phi_i$. Thus $x_b \in \cap_{i\notin F_b - k} \ker\phi_i = K_{F_b-\{k\}} \subseteq K_{F_b-\{k\}}^\bullet = \sum_{G\subsetneq F_b-\{k\}} \oplus M_G$, hence $x_b \in \sum_{G\subset F_b} \oplus M_G \cap M_{G_b} = 0$. This contradiction establishes the directness of $\sum_{|F|\leq n} M_G$, and completes the inductive construction.

Since $\phi(x) = x$ holds for all $x \in K_F$, and since K_F is essential in $K_F^\bullet$, $\phi(x) = x$ holds on the essential submodule $K_F \cap M_F$ of M_F. By summability, we have $M = \sum_F K_F$. Therefore, $M = \sum_F K_F^\bullet = \sum_F \bigoplus_{G\subseteq F} M_G = \bigoplus_G M_G$.

□

Lemma 5. Let M have the finite exchange property, and let idempotents of S be central modulo Δ. Assume the conclusions of Lemma 4. Then there exist orthogonal idempotents $e_i \in S\phi_i$ such that $\sum_{i\in I} e_i = 1$.

Proof. Let $\pi_F : M = \bigoplus_F M_F \to M_F$ be the projections; they form a summable family with $\sum_F \pi_F = 1$.

$\phi_F\pi_F - \pi_F$ vanishes on an essential submodule of M_F, as well as on $\bigoplus_{G\neq F} M_G$. Thus it has essential kernel, and we have $\phi_F\pi_F - \pi_F = \delta_F \in \Delta$. We deduce $\delta_F = \delta_F\pi_F$; hence the family δ_F is summable. We define $\delta = \sum_F \delta_F$ and obtain $\delta\,\pi_F = \delta_F$. Since Δ is closed in the finite topology, we have $\delta \in \Delta \leq J$. Now $\phi_F\pi_F = \pi_F + \delta\,\pi_F = (1+\delta)\pi_F$ and, therefore, $\pi_F = \pi_F(1+\delta)^{-1}\phi_F\pi_F$.

The finite exchange property is inherited by $\pi_F S\,\pi_F$ ([6], 1.10). We obtain orthogonal idempotents $e_{Fi} (i \in F)$ such that $\sum_i e_{Fi} = \pi_F$ and $e_{Fi} \in (\pi_F S\,\pi_F)$ $(\pi_F(1+\delta)^{-1}\phi_i\pi_F)$. Then the family e_{Fi} is summable (over all i and F), and

$\sum_F \sum_i e_{Fi} = \sum_F \pi_F = 1$. The e_{Fi} are orthogonal (since $e_{Fi}e_{Fj} = 0$) by construction if $i \neq j$, and $e_{Fi}e_{Gj} = e_{Fi}\pi_F \cdot \pi_G e_{Gj} = 0$ if $F \neq G$). Write $e_{Fi} = s_{Fi}\phi_i\pi_F$ where $s_{Fi} \in \pi_F S\, \pi_F (q+\delta)^{-1}$.

As the π_F are central module Δ, we have $\phi_i\pi_F = \pi_F\phi_i + \delta_{Fi}$ with $\delta_{Fi} \in \Delta$. The family δ_{Fi} is clearly summable; we define $\tau = \sum_F \sum_i e_{Fi}s_{Fi}\delta_{Fi}$. Then $e_{Fi}\tau = e_{Fi}s_{Fi}\delta_{Fi}$. Thus $e_{Fi} = e_{Fi}s_{Fi}\phi_i\pi_F = e_{Fi}s_{Fi}\pi_F\phi_i + e_{Fi}s_{Fi}\delta_{Fi} = e_{Fi}s_{Fi}\pi_F\phi_i + e_{Fi}\tau$ hence $e_{Fi}(1-\tau) = e_{Fi}s_{Fi}\pi_{Fi}\phi_i$. We construct orthogonal idempotents $e_i = (1-\tau)^{-1}\sum_F e_{Fi}(1-\tau) = (1-\tau)^{-1}(\sum_F e_{Fi}s_{Fi}\pi_F)\phi_i \in S\phi_i$. Then $\sum_i e_i = (1-\tau)^{-1}\sum_i\sum_F e_{Fi}(1-\tau) = 1$.

□

Theorem 6. Any nonsingular quasi-continuous module, with the finite exchange property, has the exchange property.

Proof. As noted before, we may assume that M is squarefree. Nonsingularity implies $\Delta = 0$ and, therefore, idempotents of S are central. Thus the Lemmas 3, 4 and 5 apply, in sequence.

□

Remarks.

(1) A different situation, where the finite exchange property implies the exchange property, arises if the quasi-continuous module is the direct sum of indecomposables (as is always the case if the ring is right-noetherian; cf. ([5], 2.19)): the finite exchange property implies that all the indecomposables have local endomorphism rings ([9],
Proposition 1); the condition lsTn holds automatically ([5], 2.24); thus the full exchange property follows ([12]; cf. [5], 2.25).

(2) Our proof works verbatim if every submodule of N has a unique closure. This property was recently characterized for quasi-continuous modules in ([7], Section 2), and in general in [8]: it is slightly more general than nonsingularity. The nonsingular case enjoys the additional simplification that the K_F are closed and hence coincide with the $K_F^\bullet$.

(3) The property which makes our proof work is the possibility to select closures $K_F^\bullet$ in such a way that $K_G^\bullet \subseteq K_F^\bullet$ holds whenever $G \subseteq F$ (cf. Lemma 4). The following example shows that there are obstructions to an inductive selection of such closure, due to their non-uniqueness in the general case

Example 7. Consider a module B with a proper essential submodule A. Let $\bar{B} = B/A$, with natural map $\gamma(x) = \bar{x}$. Define $M = B \oplus \bar{B} = (1+\gamma)B \oplus \bar{B}$. Pick the endomorphisms $\phi_1(x,y) = (0,\bar{x})$, $\phi_2(x,y) = (x,0)$, $\phi_3(x,y) = (0, y-\bar{x})$ of M. Then $\phi_1 + \phi_2 + \phi_3 = 1$, $ker\ \phi_1 = A \oplus \bar{B}$, $ker\ \phi_2 = \bar{B}$, $ker\ \phi_3 = (1+\gamma)B$. Therefore, $K_1 = ker\ \phi_2 \cap ker\ \phi_3 = 0$, $K_2 = ker\ \phi_1 \cap ker\ \phi_3 = A$ and $K_{12} = ker\ \phi_3 = (1+\gamma)B$. Thus we have $K_1^\bullet = 0$, and we may choose $K_2^\bullet = B$ or $= (1+\gamma)B$. If the first choice is made (inductively, over $|\ F\ |$), then $K_2^\bullet \not\subseteq K_{12} = K_{12}^\bullet$.

/

To obtain a module $M = B \oplus \bar{B}$ which is quasi-continuous with the exchange property, we take $B = R$ to be a local V-domain, with maximal ideal A.

3. In this section we approach the problem using the characterization (2) of Proposition 1. We obtain a new proof for the exchange property of continuous modules. All the preliminary lemmas hold for squarefree modules with (C_1).

Lemma 8. Let K be an arbitrary module. If a submodule L of K is squarefree and essential in K, then K is squarefree.

Proof. Suppose K is not squarefree. Then there exist nonzero submodules X and Y of K such that $X \cong Y$ and $X \cap Y = 0$. Let $f : X \to Y$ be an isomorphism; let $X_1 = X \cap L$ and $Y_1 = fX_1$. Then $0 \neq X_1 \cong Y_1$. Let $Y_2 = Y_1 \cap L$ and $X_2 = f^{-1}Y_2$. Then $0 \neq Y_2 \cong X_2$. Now $X_2 \leq L$, $Y_2 \leq L$ and $X_2 \cap Y_2 \leq X \cap Y = 0$. This is a contradiction, as L is squarefree.

□

Lemma 9. If K is a squarefree module, then every monomorphism $g : K \to K$ has essential image.

Proof. Assume $gK \cap N = 0$ for some submodule N of K. Then $N \cong gN \leq gK$. Since K is squarefree, $N = 0$.

□

Modules A and B are called orthogonal if they have no nonzero isomorphic submodules (cf. [5], p. 13).

Lemma 10. Let $X = \bigoplus_{i \in I} X_i$, where X_i and X_j are orthogonal for $i \neq j$. If a module Y contains an isomorphic copy Y_i of each X_i, then the sum $\sum_{i \in I} Y_i$ is direct, and hence Y contains a copy of X.

Proof. It is enough to consider a finite subset F of I. Let $H \subseteq F$ be maximal such that $\sum_{i \in H} Y_i$ is direct. Assume $H \neq F$, and consider $j \in F - H$. Then $(\bigoplus_{i \in H} Y_i) \cap Y_j \neq 0$. Then by ([5], 1.31), Y_j and Y_k are not orthogonal for some $k \in H$. This implies that X_j and X_k are not orthogonal, in contradiction to the assumption.

□

Let M be a squarefree module with (C_1). Let $A = M \oplus N = \bigoplus_{i \in I} A_i$ with $A_i \cong M$. By Zorn's Lemma, there exists a submodule $B \leq A$ maximal with respect to the properties $B = \bigoplus_{i \in I} B_i$ with $B_i \leq A_i$, and $M \cap B = 0$. The exchange property for M follows if we prove $A = M \oplus B$.

Clearly each B_i is a closed submodule of A_i, and hence a direct summand of A_i. We write $A_i = B_i \oplus C_i$ and $C = \bigoplus_{i \in I} C_i$; then $A = B \oplus C$.

Let $f : A = B \oplus C \to C$ be the natural projection. As $M \cap B = 0$, the restriction $f \mid M$ is a monomorphism. Using the maximality of B, one can check that fM is essential in C. Since M is squarefree, C is squarefree by Lemma 8. Hence C_i is orthogonal to C_j for all $i \neq j$. Let $g_i : A_i \to M$ be isomorphisms.

Since $C_i \cong g_i C_i \leq M$, C embeds into M by Lemma 10. If $h : C \to M$ is such an embedding, then $M \xrightarrow{f|M} C \xrightarrow{h} M$ is a monomorphism and, therefore, hfM is essential in M by Lemma 9.

An easy argument yields now a new proof of ([4], Theorem 5):

Theorem 11. Every continuous module has the exchange property.

Proof. We may again assume that the continuous module M is squarefree. By (C_2), hfM is a direct summand of M, and hence equal to M. Consequently $C = fM \oplus ker\ h = fM$. Therefore, $A = M + ker\ f = M \oplus B$.

□

REFERENCES

[1] W.D. Burgess and W. Stephenson, *Rings all whose Pierce stalks are local*, Canad. Math. Bull **22** (1979), 150-164.

[2] P. Crawley and B. Jonsson, *Refinements for infinite direct decompositions of algebraic systems*, Pacific J. Math. **14** (1964), 797-855.

[3] L. Fuchs, *On quasi-injective modules*, Annali Scoula Normal Sup. Pisa **23** (1969), 541-546.

[4] S.H. Mohamed and B.J. Müller, *Continuous modules have the exchange property*, Abelian Group Theory (Perth, 1987), Contemp. Math. **87** (1989), 285-289.

[5] S.H. Mohamed and B.J. Müller, *Continuous and discrete modules*, London Math. Soc. Lecture Notes Series **147**, Cambridge Univ. Press (1990).

[6] W.K. Nicholson, *Lifting idempotents and exchange rings*, Trans. Amer. Math. Soc. **229** (1977), 269-278.

[7] B.L. Osofsky, *Non-quasi-continuous quotients of finitely generated quasi-continuous modules* (preprint).

[8] P.F. Smith, *Modules for which every submodule has a unique closure* (preprint).

[9] R.B. Warfield, *A Krull-Schmidt Theorem for infinite sums of modules*, Proc. Amer. Math. Soc. **22** (1969), 460-465.

[10] R.B. Warfield, *Decomposition of injective modules*, Pacific J. Math. **31** (1969), 263-276.

[11] R.B. Warfield, *Exchange rings and decompositions of modules*, Math. Ann. **199** (1972), 31-36.

[12] B. Zimmermann-Huisgen and W. Zimmermann, *Classes of modules with the exchange property*, J. Algebra **88** (1984), 416-434.

RING THEORETIC PROPERTIES OF THE COORDINATE RINGS OF QUANTUM SYMPLECTIC AND EUCLIDEAN SPACE *

IAN M. MUSSON
Mathematical Sciences Department, University of Wisconsin-Milwaukee
Milwaukee, Wisconsin 53201

ABSTRACT

We study some ring theoretic properties of the coordinate rings of quantum symplectic and euclidean space. If A is one of these coordinate rings, we show that A is an iterated Ore extension and all prime ideals of A are completely prime. We show that A has a finite dimensional filtration such that the associated graded ring is the coordinate ring of a quantum m-space. We also identify the division ring of fractions of A.

Let A be the coordinate ring of quantum symplectic space or of quantum euclidean space. We obtain a simpler set of defining relations for A than those given for example in [2] or [9]. This result is stated without proof in [10, Prop. 5] and [11, Prop. 5.1] for the euclidean case. We work throughout over a field K containing an element q that is not a root of unity.

We show that A is an iterated Ore extension and that prime ideals of A are completely prime. This last result is also obtained for the symplectic case in [4] by a different method.

An algebra with generators $X_1, \cdots, X_m$ and relations $X_i X_j = q_{ij} X_j X_i$ for some $q_{ij} \in K$ whenever $i < j$ will be called the coordinate ring of a quantum m-space. We show that the division ring of fractions of A is isomorphic to the division ring of fractions of some quantum m-space.

Also there is a finite dimensional filtration of A for which the associated graded ring is the coordinate ring of a quantum m-space.

1. Quantum Symplectic Space

1.1 The coordinate ring of quantum symplectic 2n dimensional space is the algebra $A = \mathcal{O}_q(spK^{2n})$ generated over K by $X_1, \cdots, X_{2n}$ with relations

$$X_i X_j = q X_j X_i \qquad 1 \leq i < j \leq 2n\ , \quad j \neq i' \tag{0}$$

* This research was partially supported by National Security Agency grant number 91H0017.

and

$$X_{i'} X_i = X_i X_{i'} + (q^2 - 1) \sum_{j=1}^{i'-1} q^{\rho(i') - \rho(j)} \epsilon_{i'} \epsilon_j X_j X_{j'} \tag{1.i}$$

for $1 \leq i \leq n$, [2, definition 14], [9, p. 158].

Here

$$i' = (2n+1) - i \ , \ \epsilon_i = 1 \quad \text{if} \quad 1 \leq i \leq n,$$

$$\epsilon_i = -1 \quad \text{if} \quad n+1 \leq i \leq 2n$$

and

$$(\rho(1), \cdots, \rho(2n)) = (n, n-1, \cdots, 1, \text{-}1, \cdots, -n) \ .$$

We can rewrite the relation (1.i) in the form

$$X_{i'} X_i = X_i X_{i'} - (q^2 - 1) \sum_{j=1}^{n} q^{i+j-2n-2} X_j X_{j'} \tag{2.i}$$

$$+(q^2 - 1) \sum_{j=n+1}^{i'-1} q^{i+j-2n-1} X_j X_{j'} \ .$$

Now set $a_1 = 0$ and for $2 \leq i \leq n+1$,

$$a_i = \sum_{k=1}^{i-1} q^{k-i} X_k X_{k'} \ .$$

We show that the relations $(2.j)$ are equivalent to the relations

$$X_i X_{i'} - q^2 X_{i'} X_i = (q^2 - 1) a_i \ . \tag{3.i}$$

Note that (3.i) involves only $i+1$ terms $X_k X_{k'}$ whereas (2.i) involves $i'+1$ such terms. We derive (3.i) from the relations $(2.j)$ by reverse induction on i. Set

$$b_{i,k} = \sum_{j=2n-k+2}^{2n-i} q^{i+j-2n-1} X_j X_{j'} \quad \text{for} \quad i+2 \leq k \leq n+1,$$

and $b_{i,i+1} = 0$. Then (2.i) can be written

$$X_{i'} X_i = X_i X_{i'} - (q^2 - 1)(q^{i-n-1} a_{n+1} - b_{i,n+1}) \ .$$

Note that

$$a_{\ell+1} = q^{-1}a_\ell + q^{-1}X_\ell X_{\ell'}$$

and

$$b_{i,\ell+1} = b_{i,\ell} + q^{i-\ell}X_{\ell'}X_\ell \ .$$

If $i = n$ this gives

$$X_{n'}X_n = X_nX_{n'} - (q^2-1)q^{-1}a_{n+1} = X_nX_{n'} - (q^2-1)q^{-2}(a_n + X_nX_{n'})$$

and (3.n) follows. If $i < n$, then using (3.n), $\cdots$, (3.i+1)

$$q^{i-n-1}a_{n+1} - b_{i,n+1} = q^{i-n-2}a_n - b_{i,n} + q^{i-n-2}(X_nX_{n'} - q^2X_{n'}X_n)$$

$$= q^{i-n}a_n - b_{i,n} = \cdots = q^{-1}a_{i+1} - b_{i,i+1} = q^{-2}(a_i + X_iX_{i'}) \ .$$

Hence

$$X_{i'}X_i = X_iX_{i'} - (1-q^{-2})(a_i + X_iX_{i'})$$

$$= q^{-2}X_iX_{i'} - (1-q^{-2})a_i$$

and (3.i) follows.

1.2 We draw some consequences from the relations (0) and (3.i). First observe that X_1 and $X_{1'}$ are normal elements of A such that $A/X_1A + X_{1'}A \cong \mathcal{O}_q(spK^{2(n-1)})$. Also the subalgebra B of A generated by X_1, $X_{1'}$, $\cdots$, X_{n-1}, $X_{(n-1)'}$ is isomorphic to $\mathcal{O}_q(spK^{2(n-1)})$.

Let R be a K-algebra, σ an automorphism of R and δ a σ-derivation of R. The algebra $S = R[X;\sigma,\delta]$ generated by R and the indeterminate X subject to the relations $rX = X\sigma(r) + \delta(r)$ for $r \in R$, is called a skew polynomial algebra over R; see [7, 1.2.4] for more details. We show, by induction on n that $A = \mathcal{O}_q(spK^{2n})$ is an iterated skew polynomial algebra over K.

To build A we order the generators X_i by setting $X_1 < X_{1'} < X_2 < X_{2'} < \cdots < X_n < X_{n'}$. Our assertion amounts to showing that the monomials $X_1^{a_1}X_{1'}^{b_1} \cdots X_n^{a_n}X_{n'}^{b_n}$ are linearly independent. We use Bergman's diamond lemma [1].

The relations (0) and (3.i) give a set of reductions for any product uv with $u, v \in \{X_1, X_{1'}, \cdots, X_n, X_{n'}\}$ and $u > v$. Hence the monomials above span A. To show the result it is enough to show that the overlap $(uv)w = u(vw)$ can be reduced to a common expression using these reductions.

By induction $C = K[X_1, X_{1'}, \cdots . X_{n-1}, X_{(n-1)'}]$ is an iterated skew polynomial algebra. Thus it suffices to check only overlaps of the form $(X_n X_\ell)X_i = X_n(X_\ell X_i)$, $(X'_n X_\ell)X_i = X_{n'}(X_\ell X_i)$ and $(X_{n'} X_n)X_i = X_{n'}(X_n X_i)$. This is routine and is left to the reader.

For completeness we give the automorphisms and skew derivations involved in obtaining A from C.

Theorem . Set $C = K[X_1, X_{1'}, \cdots, X_{n-1}, X_{(n-1)'}]$, $B = C[X_n]$ and $A = B[X_{n'}] = \mathcal{O}_q(spK^{2n})$.

1) $B = C[X_n;\ \sigma, \delta]$ is a skew polynomial algebra over C with $\sigma(X_i) = qX_i$, $\sigma(X_{i'}) = q^{-1}X_{i'}$ for $1 \leq i \leq n-1$ and $\delta = 0$.
2) $A = B[X_{n'};\ \sigma', \delta']$ is a skew polynomial algebra over B with $\sigma'(X_i) = qX_i$, $\sigma(X_{i'}) = q^{-1}X'_i$ for $1 \leq i \leq n-1$, $\sigma'(X_n) = q^2 X_n$, and $\delta'(X_n) = (q^2-1)a_n$.

Proof. This follows easily from the relations obtained in 1.1.

Corollary. If q is not a root of unity, then prime ideals of $\mathcal{O}_q(spK^{2n})$ are completely prime.

Proof. This follows immediately from [3, Theorem 2.3] once we observe that in 2) $\delta'\sigma' = q^2\sigma'\delta'$. Note that a_n is a linear combination of terms $X_\ell X_{\ell'}$ with $1 \leq \ell \leq n-1$, so that $\sigma'(a_n) = a_n$, hence $\delta'\sigma'(X_n) = q^2(q^2-1)a_n = q^2\sigma'\delta'(X_n)$.

We remark that a complete classification of primitive ideals of A is given in [4], and from this it is deduced that all prime ideals are completely prime.

1.3 Next we set for $1 \leq i \leq n$,

$$v_i = a_i + X_i X_{i'} = \sum_{k=1}^{i} q^{k-i} X_k X_{k'} \ .$$

By (0) we have

$$X_j X_k X_{k'} = X_k X_{k'} X_j \qquad \text{if} \quad 1 \leq k < j \leq n \tag{4}$$

$$= q^2 X_k X_{k'} X_j \qquad \text{if } 1 \leq j < k \leq n. \tag{5}$$

On the other hand, by (3.i)

$$v_i = q^2(X_{i'} X_i + a_i) \ .$$

Now (4) gives $X_i a_i = a_i X_i$, hence

$$X_i v_i = q^2(X_i X_{i'} X_i + a_i X_i) = q^2 v_i X_i \ . \tag{6}$$

Also by (4),

$$X_j v_i = v_i X_j \qquad \text{if} \quad j > i. \tag{7}$$

Furthermore for $j < i$ we have

$$v_i = q^{j-i} v_j + \sum_{k=j+1}^{i} q^{k-i} X_k X_{k'}$$

whence by (5) and (6)

$$X_j v_i = q^2 v_i X_j \ . \tag{8}$$

Finally if $j \neq i$ we have $X_j X_{j'} X_i X_i = X_i X_{i'} X_j X_{j'}$ and hence

$$v_i v_j = v_j v_i \ . \tag{9}$$

We now identify the division ring of fractions Fract(A) for $A = \mathcal{O}_q(spK^{2n})$.

Let $\epsilon = (\epsilon_1, \epsilon_2, \cdots, \epsilon_n) \in \mathbf{Z}_2^n$. We define 2^n subalgebras A_ϵ of A as follows. For $1 \leq i \leq n$, set $Y_i = X_i$ if $\epsilon_i = 1$ and $Y_i = X_{i'}$ if $\epsilon_1 = -1$. Then A_ϵ is the subalgebra of A generated by $Y_1, \cdots, Y_n, v_1, \cdots, v_n$.

Theorem. For any $\epsilon \in \mathbf{Z}_2^n$, $\text{Fract}(A) = \text{Fract}(A_\epsilon)$ and A_ϵ is the coordinate ring of a quantum 2n-space.

Proof. For ease of notation we assume $\epsilon = (1, 1, \cdots, 1)$. Let $A_0 = A$ and for $1 \leq i \leq n$, let A_i be the subalgebra of A generated by $X_1, \cdots, X_n$ and $v_1, \cdots, v_i, X_{(i+1)'}, \cdots, X_{n'}$. Then $A = A_0 \supset A_1 \supset A_2 \supset \cdots \supset A_n$. The element X_i is normal in A_i, so $\{X_i^n\}$ is Ore in A_i. The localization $(A_i)_{X_i}$ contains $X_{i'} = X_i^{-1}(v_i - q^{-1} v_{i-1})$ and hence contains A_{i-1}. Any element of $(A_i)_{X_i}$ can be written in the form $X_i^{-r} a$ with $r \geq 0$ and $a \in A_i \subseteq A_{i-1}$. It follows that $\{X_i^r\}$ is Ore in A_{i-1} and $(A_i)_{X_i} = (A_{i-1})_{X_i}$. Set $X = X_n X_{n-1} \cdots X_1$. Then $(A_n)_X = (A_0)_X$ and hence $\text{Fract}(A) = \text{Fract}(A_n)$. From relations (0) and (6)-(9) $A_n = A_\epsilon$ satisfies the defining relations of the coordinate ring of a quantum $2n$-space. To show that A_ϵ is actually isomorphic to this coordinate ring, we must show that the ordered monomials in $X_1, v_1, \cdots, X_{n-1}, v_{n-1}, X_n, v_n$ are linearly independent. We use induction on n. Let B and C be the subalgebras of A as in Theorem 1.3. Note that $X_1, v_1, \cdots, X_{n-1}, v_{n-1}, X_n$ belong to B. If m is any monomial in these elements, then the degree of mv_n^d considered as a polynomial in $X_{n'}$ with coefficients in B is equal to d. Thus it is enough to show linear independence of the ordered monomials in $X_1, v_1, \cdots, X_{n-1}, v_{n-1}, X_n$. Similarly using the fact that $B = C[X_n]$ is a skew polynomial algebra over C, we reduce to showing linear independence of the ordered monomials in $X_1, v_1, \cdots, X_{n-1}, v_{n-1}$, but this is our induction hypothesis.

1.4 We show that $A = \mathcal{O}_q(spK^{2n})$ has a finite dimension filtration such that the associated graded ring Gr(A) is the coordinate ring of a quantum 2n-space.

The filtration of A is defined by giving X_i and $X_{i'}$ degree i for $1 \leq i \leq n$. More precisely, since A is an iterated Ore extension, it has a basis over K consisting of monomials

$$X = X_1^{a_1} X_{1'}^{b_1} X_2^{a_2} X_{2'}^{b_2} \quad \cdots \quad X_n^{a_n} X_{n'}^{b_n} .$$

We let F_r be the span of such monomials X with $\sum_{i=1}^n i(a_i + b_i) \leq r$. From the relations it follows easily that $F_r X_i$, $F_r X_{i'}$ are contained in F_{r+i} for each i and hence that $F_r F_s \subseteq F_{r+s}$. Furthermore the factor F_r / F_{r-1} is spanned by the images $\bar{X}$ of monomials X with deg $X = r$.

Theorem. The associated graded ring $GrA = \bigoplus_{r \geq 0} F_r / F_{r-1}$ is generated by the elements $\bar{X}_1, \cdots, \bar{X}_{n'}$ subject to relations $\bar{X}_i \bar{X}_j = \lambda_{ij} \bar{X}_j \bar{X}_i$ where

$$\lambda_{ij} = q \qquad \text{if} \quad i < j \neq i$$

$$\lambda_{ii'} = q^2 \qquad \text{if} \quad i < i'$$

$$\lambda_{ii} = 1 \qquad \text{and} \qquad \lambda_{ij} \lambda_{ji} = 1 \qquad \text{if} \quad i \neq j.$$

Proof. This follows easily from the fact that the monomials X as above with deg $X = r$ are linearly independent modulo F_{r-1}.

Corollary.

1) A satisfies the Nullstellensatz over K, that is the endomorphism ring of any simple module is equal to K, and any prime ideal is the intersection of primitive ideals.
2) Gel'fand-Kirillov dimension is exact for A.

Proof. This follows immediately from [6, Theorem 3.8].

2. Quantum Euclidean Space

2.1 To define the coordinate ring of quantum Euclidean space $O_q(soK^n)$ we need some preliminary notation. For n even set $n = 2k$ and for n odd $n = 2k + 1$. For $1 \leq i \leq n$, let $i' = n + 1 - i$, $\rho(i) = n/2 - i$ and $\rho(i') = i - n/2$. If n is odd set $\rho(k + 1) = 0$. In the relations which follow, the "numerator" N and "denominator" D are defined by $N = q^2 - 1$, $D = 1 + q^{n-2}$.

Then as in [2, definition 12] or [9, p. 159], $O_q(soK^n)$ may be defined as the algebra over K generated by $X_1, \cdots, X_n$ subject to the relations

$$X_i X_j = q X_j X_i \qquad 1 \leq i < j \leq n \quad , \quad j \neq i' \tag{0}$$

and

$$X_{i'}X_i = X_iX_{i'} + N/D\left(q^{n-2}\sum_{\ell=1}^{i'-1} q^{\rho(i')-\rho(\ell)}X_\ell X_{\ell'} - \sum_{\ell=i'}^{n} q^{\rho(i')-\rho(\ell)}X_\ell X_{\ell'}\right) . \quad (1.i)$$

To simplify the relations (1.i) we introduce

$$Z_i = X_{i'}X_i - X_iX_{i'} \ , \ u_i = \sum_{\ell=i+1}^{i'-1} q^{i+n/2-2-\rho(\ell)}X_\ell X_{\ell'}$$

and

$$D_i = D + N(1+q^2+\cdots+q^{2(i-1)}) = q^{2i}(1+q^{n-2i-2}) .$$

We claim that the relations (1.i) are equivalent to the relations

$$D_iZ_i = Nu_i \qquad 1\le i\le k . \quad (2.i)$$

Assume relations (1.i) for $1 \le i \le k$. To derive (2.i) we assume inductively that (2.j) holds for $j < i$. Then we use a second induction on j to show that

$$D_{j-1}Z_i = N\{-q^{i+j-2}Z_j + u_i + \sum_{\ell=j+1}^{i}(q^{\ell+i-2}X_\ell X_{\ell'} - q^{2j-2+i-\ell}X_{\ell'}X_\ell)\} \quad (3.i.j)$$

for $1 \le j \le i$.

Note that (3.i.i) will then yield

$$D_{i-1}Z_i = N(-q^{2i-2}Z_i + u_i)$$

and hence (2.i).

Note that (1.i) can be rewritten

$$DZ_i = N\left\{\sum_{\ell=1}^{i'-1} q^{n/2+i-2-\rho(\ell)}X_\ell X_{\ell'} - \sum_{\ell=i'}^{n} q^{i-n/2-\rho(\ell)}X_\ell X_{\ell'}\right\}$$

$$= N\left\{q^{i-1}Z_1 + u_i + \sum_{\ell=2}^{i}(q^{\ell+i-2}X_\ell X_{\ell'} - q^{i-\ell}X_{\ell'}X_\ell)\right\}$$

and this is (3.i.1).

Now we show that for $j < i$, (2.j) and (3.i.j) imply (3.i.j+1). By (2.j) and (3.i.j)

$D_{j-1} Z_i$

$$= N \left[-q^{i+j-2}(N/D_j)u_j + u_i + \sum_{\ell=j+1}^{i} (q^{\ell+i-2} X_\ell X_{\ell'} - q^{2j-2+i-\ell} X_{\ell'} X_\ell) \right] \quad (4)$$

Next by examining the coefficients of $X_\ell X_{\ell'}$ occurring in (4) we obtain

$$D_{j-1} Z_i = N(D_{j-1}/D_j) \left[-q^{i+j-1} Z_{j+1} + u_i \right.$$

$$\left. + \sum_{\ell=j+2}^{i} (q^{\ell+i-2} X_\ell X_{\ell'} - q^{2j-\ell+i} X_{\ell'} X_\ell) \right] . \quad (5)$$

For example, for $j+1 \leq \ell \leq i$, the coefficient of $X_\ell X_{\ell'}$ in (4) is

$$N\{q^{\ell+i-2} - q^{i+j-2}(N/D_j) \text{ (coefficient of } X_\ell X_{\ell'} \text{ in } u_j)\}$$

$$= Nq^{\ell+i-2}(1 - q^{2j-2}(N/D_j)) = q^{\ell+i-2} N D_{j-1}/D_j$$

and the coefficient of $X_{\ell'} X_\ell$ is

$$-N(q^{2j-2+i-\ell} + q^{i+j-2} q^{n+j-\ell-2} N/D_j)$$

$$= -Nq^{2j-\ell+i-2}(1 + q^{n-2}(N/D_j)) = -Nq^{2j-\ell+i} D_{j-1}/D_j \; .$$

Now (3.i.j+1) follows directly from (5).

2.2 We can now make a further simplification to the defining relations for $A = \mathcal{O}_q(soK^n)$. By reverse induction on i, it can be shown that relations (2.i) $1 \leq i \leq k$ are equivalent to the relations

$$Z_i = N \sum_{\ell=i+1}^{k} q^{\ell-i-2} X_\ell X_{\ell'} \qquad \text{if } n = 2k \text{ is even} \quad (6)$$

$$Z_i = (q-1) q^{(2k-2i-1)/2} X_{k+1}^2 + N \sum_{\ell=i+1}^{k} q^{\ell-i-2} X_\ell X_{\ell'} \qquad \text{if } n = 2k+1 \text{ is odd} . \quad (7)$$

We mention some inductive techniques that can be used in studying A. Firstly the subalgebra of A generated by the elements X_i, for $2 \leq i \leq 2'$ is isomorphic of $\mathcal{O}_q(soK^{n-2})$. Also if $n = 2k+1$ is odd, the element X_{k+1} is normal in A and

$A/X_{k+1}A \cong \mathcal{O}_q(soK^{n-1})$. If $n = 2k$ is even then for any a, $b \in K$ not both zero, the element $X = aX_k + bX_{k'}$ is normal in A and $A/XA \cong \mathcal{O}_q(soK^{n-1})$.

The proofs of the above statements are all straightforward. For example, if $n = 2k$ and we set $X_{k'} = \lambda X_k$ in (6) we obtain

$$Z_i = (q-1)q^{(2k-2i-3)/2}Y^2 + N\sum_{\ell=i+1}^{k-1} q^{\ell-i-2}X_\ell X_{2k+1-\ell}$$

where $Y = (\lambda(q^{1/2} + q^{-1/2}))^{1/2}X_k$.

Now if we replace Y by X_k and $X_{2k+1-\ell}$ by $X_{2k-\ell}$ for $i+1 \le \ell \le k-1$, we obtain (7) with $k-1$ in place of k. This shows that $A/(\lambda X_k - X_{k'})A \cong \mathcal{O}_q(soK^{n-1})$.

2.3 Theorem. Set $C = K[X_2, \cdots, X_{2'}]$, $B = C[X_1]$ and $A = B[X_{1'}] = K[X_1, X_2, \cdots, X_{2'}, X_{1'}] = \mathcal{O}_q(soK^n)$. Then

a) A is an iterated skew polynomial algebra over K. Moreover $B = C[X_1, \sigma, \delta]$ where $\delta = 0$ and $\sigma(X_j) = q^{-1}X_j$ for $2 \le j \le 2'$, and $A = B[X_{1'}, \sigma', \delta']$ where $\sigma'(X_j) = qX_j$, $\delta'(X_j) = 0$ for $2 \le j \le 2'$, $\sigma'(X_1) = X_1$ and

$$\delta'(X_1) = -N\sum_{\ell=2}^{k} q^{\ell-3}X_\ell X_{\ell'} \quad \text{if} \quad n = 2k \quad \text{is even}$$

$$\delta'(X_1) = -(q-1)q^{(2k-3)/2}X_{k+1}^2 - N\sum_{\ell=2}^{k} q^{\ell-3}X_\ell X_{\ell'} \quad \text{if} \quad n = 2k+1 \quad \text{is odd}.$$

b) Prime ideals of A are completely prime.

Proof. This follows as in the proof of the theorem and corollary in 1.2.

2.4 We identify the division ring of fractions of $A = \mathcal{O}_q(soK^n)$.

For $1 \le i \le k$, $Z_i = X_{i'}X_i - X_iX_{i'}$ can be written as a linear combination of $X_\ell X_{\ell'}$ with $\ell > i$, by (6) and (7). Hence by (0); $X_iZ_i = q^2Z_iX_i$ and $X_{i'}Z_i = q^{-2}Z_iX_{i'}$. Therefore, if we set $\Omega_i = X_iX_{i'} - q^2X_{i'}X_i$ we obtain

$$X_i\Omega_i = \Omega_iX_i, \; X_{i'}\Omega_i = \Omega_iX_{i'}. \tag{8}$$

On the other hand, (0) gives

$$\begin{aligned} X_j\Omega_i &= q^2\Omega_iX_j \qquad \text{if } j < i \\ X_j\Omega_i &= q^{-2}\Omega_iX_j \qquad j > i' \\ X_j\Omega_i &= \Omega_iX_j \qquad i < j < i'. \end{aligned} \tag{9}$$

Hence

$$\Omega_i \Omega_j = \Omega_j \Omega_i \ . \tag{10}$$

Now for $\epsilon = (\epsilon_1, \cdots, \epsilon_k) \in \mathbf{Z}_2^k$ we set $Y_i = X_i$ if $\epsilon_i = 1$ and $Y_i = X_{i'}$ if $\epsilon_i = -1$. Let A_ϵ be the subalgebra of A generated by $Y_1, \cdots, Y_k, \Omega_1, \cdots, \Omega_k$ if $n = 2k$ is even and by $Y_1, \cdots, Y_k, X_{k+1}, \Omega_1, \cdots, \Omega_k$ if $n = 2k+1$ is odd.

Theorem. For any $\epsilon \in \mathbf{Z}_2^k$, Fract$(A)$ =Fract(A_ϵ) and A_ϵ is the isomorphic to the coordinate ring of a quantum n-space.

Proof. The element $Y = Y_1 \cdots Y_k$ is normal in A_ϵ and we can see as in the proof of Theorem 1.3 that $(A_\epsilon)_Y = A_Y$, and that A_ϵ is isomorphic to the coordinate ring of quantum n-space.

2.5 We define a finite dimension filtration on $A = \mathcal{O}_q(soK^n)$ such that Gr(A) is the coordinate ring of a quantum n-space. If $n = 2k$ the filtration is obtained by setting $deg\ X_i = deg\ X_{i'} = k + 1 - i$. If $n = 2k+1$ we set $deg\ X_{k+1} = 1$ and $deg\ X_i = deg\ X_{i'} = k + 2 - i$ for $1 \le i \le k$.

Theorem. The associated graded ring Gr(A) of the above filtration is generated by $\bar{X}_1, \cdots, \bar{X}_n$ subject to the relations $\bar{X}_i \bar{X}_j = \lambda_{ij} \bar{X}_j \bar{X}_i$ where

$$\lambda_{ij} = q \qquad \text{if } i < j \neq i'$$

$$\lambda_{ii} = \lambda_{ii'} = 1 \quad \text{and} \quad \lambda_{ij}\lambda_{ji} = 1 \qquad \text{if } i \neq j \ .$$

2.6 We denote the center of the algebra R by $Z(R)$. It is shown in [4] that $Z(\mathcal{O}_q(spK^{2n})) = K$. On the other hand, when $A = \mathcal{O}_q(soK^n)$, Ω_1 is central in A by (8) and (9). In fact, we have

Theorem.

a) $Z(A) = K[\Omega_1]$.

b) If $n = 2k+1$ is odd, then $Z(\text{Fract}(A)) = K(\Omega_1)$.

If $n = 2k$ is even, then $Z(\text{Fract }(A)) = K(\Omega_1, X_k^{-1} X_{k'})$.

Proof. We assume n is even. The odd case is slightly easier. Let $F = K(\Omega_1)$, $L = F(X_k^{-1} X_{k'})$. By Theorem 2.4, A has the same division ring of fractions as the algebras $F[X_1^{\pm 1}, \cdots, X_k^{\pm 1}, \Omega_2^{\pm 1}, \cdots, \Omega_k^{\pm 1}] = R$ and $L[X_1^{\pm 1}, \cdots, X_k^{\pm 1}, \Omega_2^{\pm 1}, \cdots, \Omega_{k-1}^{\pm 1}] = S$.

Now set $T_i = X_i$ for $1 \le i \le k$ and $T_i = \Omega_{2k-1}$ for $k+1 \le i \le 2k-2$. The Ω_{2k-1} generators T_i for the L-algebra S satisfy relations $T_i T_j = q^{m(i,j)} T_j T_i$ for suitable integers $m(i,j)$. From Eqs. (8)-(10) the matrix $\log_q m(i,j)$ has zeros below the antidiagonal, (that is $m(i,j) = 1$ if $i + j \ge n$) and the product of the antidiagonal entries is nonzero. Thus the matrix is nonsingular and so by [8, Proposition 1.3] $Z(S) = L$. By embedding Fract(A) into a suitable Malcev-Neumann group algebra over L, as in [8, §2] it is now easy to show that $Z(\text{Fract}(A)) = L$.

It follows that $Z(R) = R \cap Z(S) = F[X_k^{-1} X_{k'}]$ and hence that $Z(A) = A \cap Z(R) = K[\Omega_1]$.

Remark. In $A = \mathcal{O}_q(soK^4)$, set $a = X_4$, $b = -X_2$, $c = X_3$ and $d = X_1$. Then $A = K[a, b, c, d]$ is the coordinate ring of quantum 2×2 matrices with relations as in [5, p. 6] and $\Omega_1 = (1 - q^2)\det_q$, with $\det_q = da - qcb$ the quantum determinant.

REFERENCES

[1] G. Bergman, *The diamond lemma for ring theory*, Adv. Math. **29** (1978), 178-218.

[2] L.D. Faddeev, N. Yu. Reshetikhin and L.A. Takhtadzhyan, *Quantization of Lie groups and Lie algebras*, Leningrad Math. J. (1990), 193-225.

[3] K.R. Goodearl and E.S. Letzter, *Prime factor algebras of the coordinate ring of quantum matrices*, preprint.

[4] T.J. Hodges and S.-Q. Oh, *Primitive ideals of the coordinate ring of quantum symplectic space*, preprint.

[5] Yu. I. Manin, *Quantum groups and non-commutative geometry*, Centre de Recherches Math. Université de Montréal, 1988.

[6] J.C. McConnell, Quantum groups, filtered rings and Gel'fand-Kirillov dimension, Lecture Notes in Mathematics, **1448**, Springer-Verlag.

[7 J.C. McConnell and J.C. Robson, *Noncommutative noetherian rings*, Wiley-Interscience, New York, 1987.

[8] J.C. McConnell and J.J. Pettit, *Crossed products and multiplicative analogues of Weyl algebras* J. London Math. Soc. **38** (1988), 47-55.

[9] S.P. Smith, *Quantum groups: An introduction and survey for ring theorists*, in Noncommutative Rings M.S.R.I. Publ. **24**, Springer-Verlag, New York, 1992, pp. 131-178.

[10] M. Takeuchi, *Quantum orthogonal and symplectic groups and their embedding into quantum* GL_n, Proc. Japan Acad **65** (1989), 55-58.

[11] M. Takeuchi, *Matric bialgebras and quantum groups*, Israel J. Math. **72** (1990), 232-251.

NON-QUASI-CONTINUOUS QUOTIENTS OF FINITELY GENERATED QUASI-CONTINUOUS MODULES

BARBARA L. OSOFSKY *

Department of Mathematics, Rutgers—The State University, New Brunswick, NJ 08904

ABSTRACT

Let M be a finitely generated quasi-continuous module with endomorphism ring Λ, $\{e_i | \; i \in \omega\}$ an infinite set of orthogonal idempotents $\subseteq \Lambda$, and $K = \sum_{i \in \omega} e_i M$. Then M/K is not quasi-continuous if for every direct sum decomposition $M/K = U \oplus V$, the projection π of M/K to U lifts to a map (not necessarily idempotent) from M to M. Some conditions and an example are given which imply that the quotient M/K cannot even contain a quasi-continuous direct summand. There is a deliberate attempt to make the hypotheses on M as far from implying that Λ is von Neumann regular as we possibly can and still get the results.

In [7] and [8], the author showed that the quotient module of a self injective von Neumann regular ring modulo an ideal generated by an infinite set of orthogonal idempotents is not injective. The purpose of showing this proposition was to prove the theorem that if every cyclic R-module is injective then the ring R is semisimple artinian.

The argument used to prove the proposition is intuitively compelling. Start with an injective cyclic module M and an infinite set of orthogonal idempotents $\{e_i \,|\, i \in \omega\}$ in $\Lambda = End(M)$. Pick injective hulls $E_A M$ of infinite subsums $\sum_{\alpha \in A} e_\alpha M$ where A runs over some maximal set of almost disjoint subsets of ω, and show there is no projection of $M / \sum_{i=0}^{\infty} e_i M$ onto a summand containing a countably infinite subset of the $E_A M$ with kernel containing the others. If the module M is a von Neumann regular ring itself, there are lots of things going on that can assist in the proof, such as projectivity and unique maximal essential extensions of submodules. How much can be proved without such extra properties? In this paper we weaken both of the mentioned properties significantly in a way that still allows the result to be proved.

*The author wishes to thank the University of Glasgow for its hospitality while this paper was being written. She also wishes to thank Professor Patrick F. Smith for his significant assistance in this research.

The proposition or theorem of [7] has been used by many authors, almost always by quoting rather than improving the argument. The argument itself has been modified to apply to different cases only twice to the best of my knowledge. The first time was by Damiano [2], who looked at a finitely generated cyclic module rather than the ring itself. The second time was by Osofsky and Smith [10] where a new approach enabled the argument to be generalized. Their approach did not indicate which of three possible candidates for non-*CS* module was the actual culprit preventing a quotient modulo an infinite direct sum from being non-*CS*. Then Dung and Smith [3] used the original result rather than the argument for it to show that a quasi-projective module M has the property that every non-zero subquotient contains a non-zero injective submodule if and only if M is a semi-artinian V-module. This paper was inspired by the Dung-Smith result.

In Section 1 of this paper we find a small set of hypotheses for which the idea of the original proof in [7] works. We next include a Section 2 discussing when submodules of a quasi-continuous module M have unique maximum essential extensions in M. This discussion, interesting in its own right, will be necessary in the proof of one of our main theorems in the third section. In the fourth section, we obtain a strengthening of the Dung–Smith result, and get rid of the categorical equivalence used in [3] to transfer the original theorem from the case of the quasi-projective module to its endomorphism ring. The final section explores an example where the theorems of Sections 3 and 4 do not apply, but we can still show that an appropriate quotient of the finitely generated injective module in question does not have a quasi-continuous direct summand.

1 The major hypotheses for our theorems

In this section we discuss the hypotheses we need and indicate intuitively why we need each of them to prove our theorems. Let us first look at the main result in [7].

Proposition 1 (Osofsky, 1964) *Let R be a self-injective von Neumann regular ring, $\{e_i \mid i \in \omega\}$ a set of non-zero orthogonal idempotents of R. Then $R/(\oplus_{i=0}^{\infty} e_i R)$ is not injective.*

The purpose of this Proposition is to study cyclic modules, and we keep that idea. However, the fact that M is an R-module is irrelevant. We might as well be working in an arbitrary $\mathcal{A}B5$ category. But there cyclic makes no sense. So instead we assume

(A1) M is finitely generated. The relevant categorical property is that any map $\varphi : M \longrightarrow \oplus_{i=0}^{\infty} e_i M$ factors through a finite subsum, that is, $\varphi = (\sum_{i=0}^{n} e_i)\varphi$ for some natural number n.

This assumption also insures that, for any infinite set $\{e_i \mid i \in \omega\}$ of orthogonal idempotents in Λ, $\oplus_{i=0}^{\infty} e_i M$ is not a direct summand of M. In the case of modules over a ring, for $n < \infty$, a Morita equivalence takes an n-generated R-module to a cyclic R_n-module where R_n is $n \times n$ matrices over R. So unless you are studying a particular ring R as in [7], there is effectively no difference between cyclic and finitely generated.

For arbitrary modules M, there is no reason to expect $M/(\oplus_{i=0}^{\infty} e_i M)$ to have any particular properties. If you start with an infinite rng (ring without 1) direct sum of copies of a field F and adjoin $1 \times F$, then this ring with identity has an injective quotient modulo the direct sum. So we want to be able to say more about direct summands of M so that a large number of them exist. Injectivity is too strong. But we at least want submodules to be essential in direct summands. This is the *CS* property of [1] and [10]. The problem with this hypothesis is that a submodule K maximal with respect to $K \cap N = 0$ for some direct summand N is a direct summand of M but $N \oplus K$ need not equal M. Patrick Smith has pointed out a classical example of this. Let $M = \mathbb{Z} \oplus \mathbb{Z}$ and $N = \langle 1, 2 \rangle \mathbb{Z}$, $K = \langle 1, 4 \rangle \mathbb{Z}$. This example has the additional property that an intersection of direct summands is a direct summand. As noted above, the paper [10] had to use a considerably modified argument and did not show exactly which of three modules was not *CS*. So we will strengthen *CS* to

(A2) M is quasi-continuous. That is, every submodule of M is essential in a direct summand of M and if U and V are direct summands of M with $U \cap V = 0$, then $U \oplus V$ is a direct summand of M.

This is a step closer than *CS* to quasi-injective and injective, which is where the main applications lie. Even closer is continuity. A module M is continuous provided it is quasi-continuous and every submodule isomorphic to a direct summand is a direct summand. Reference [6] is an excellent reference for these properties, and contains proofs of facts we state and/or use here without proof.

If we are to connect injectivity or non-injectivity of $M/(\oplus_{i=0}^{\infty} e_i M)$ with properties of M, we need some connection between direct summands of $M/(\oplus_{i=0}^{\infty} e_i M)$ and direct summands of M. In the cases of [7], [8], [2], and [3], the modules in question had von Neumann regular endomorphism rings. That implies two particularly useful properties. One is that any submodule λM for $\lambda \in \Lambda$ is a direct summand. The second is that any submodule of M is essential in precisely one direct summand. Rather surprisingly, both of these properties which play a prominent role in the proof of Proposition 1 in [7] can be considerably weakened without destroying the argument.

The properties (A1) and (A2) are part of the essence of our Theorems, and not just a part of the proof. For the proof of our generalization we need another property which has three different forms. Unlike the other properties, this may hold for some submodules $K \subseteq M$ but not others. This is a restricted form of quasi-projectivity.

(A3) Assume $M/K = U \oplus V$ with π the projection of M/K to U with kernel V followed by inclusion of U into M/K, and ν the natural map from M to M/K. Then there is a $\lambda \in \Lambda$ with $\nu\lambda = \pi\nu$.

Pictorially we have

$$\begin{array}{ccccc} & M & \xrightarrow{\nu} & M/K & \\ & | & & \downarrow & | \\ \exists\lambda & | & & U & \pi \\ & \downarrow & & \downarrow & \downarrow \\ & M & \xrightarrow[\nu]{} & M/K & \end{array}$$

(A3)′ This is (A3) with the additional assumption that the image of λ is actually a direct summand of M.

There is a third variant on this theme which is on the surface even stronger than (A3)′. We list it here because it seems more intuitively satisfying than (A3)′.

(A3)″ Let $M/K = U \oplus V$. Then there exist submodules A and B of M such that $M = A \oplus B$ and $\nu A = U$, $\nu B = V$. That is, direct sum decompositions of M/K lift to direct sum decompositions of M.

Note that (A3)″ $\Longrightarrow$(A3)′ $\Longrightarrow$(A3) since assuming (A3)″ we can decompose any $x \in M$ into a sum $\pi_A(x) + (1 - \pi_A)(x)$ and $\pi\nu(x) = \nu\pi_A(x)$ where π_A is the projection of M onto A with kernel B. It is easy to see that in general (A3) does not imply (A3)′ by looking at the ring $\mathbb{Z}$.

We make some observations about these variants of property (A3).

1. If M is quasi-projective then (A3) holds for any $K \subseteq M$.

2. If M is finitely generated and (A3) holds for $K = \oplus_{i=0}^{\infty} e_i M$, since $\nu\lambda^2 = \pi\nu\lambda = \pi^2\nu = \pi\nu = \nu\lambda$ there is an n such that for $e = (1 - \sum_{i=1}^{n} e_i) \in \Lambda$, $e\lambda^2 = e\lambda$. But that is not the same thing as saying $e\lambda e\lambda = e\lambda$. Thus (A3)′ may possibly be stronger than (A3) even for this special K. If you know something additional, such as $e\lambda(1-e) = 0$, then $e\lambda$ is an idempotent in Λ, and it also satisfies $\nu e\lambda = \nu\lambda = \pi\nu$.

3. If Λ is von Neumann regular then for every $\lambda \in \Lambda$, $\lambda\Lambda = e\Lambda$ so $\lambda M = eM$ for some $e = e^2 \in \Lambda$. Thus λM is a direct summand of M. This has nothing to do with continuity or number of generators of M.

4. If Λ is von Neumann regular then (A3)$\Longrightarrow$(A3)$''$. Lift both U and V via λM and μM respectively. The lifted modules are direct summands of M and so is their intersection. Replace one by a complement of the intersection in it.

5. If M is a finitely presented R-module, then (A3)$''$ holds for $K = \sum_{i=0}^{\infty} e_i M$. One way to see this follows. By means of the Morita equivalence between modules over R and modules over $n \times n$ matrices over R induced by $\otimes_R \left(\bigoplus_{j=1}^{n} R_j\right)$, we reduce to the case that $M = mR$ is cyclic, $U = msR + K$, $V = mtR + K$. If the relations on M are generated by $\{mr_i = 0 \mid 1 \leq i \leq n\}$, then some finite sum $\sum_{j=0}^{k} e_j M$ contains $\{msr_i\} \cup \{mtr_i\} \cup \{m(s^2 - s), m(t^2 - t), mst, mts\}$ and then $M = \left(1 - \sum_{j=0}^{k} e_j\right) msR \oplus \left(1 - \sum_{j=0}^{k} e_j\right) mtR \oplus \left(\sum_{j=0}^{k} e_j\right) mR$ gives the required liftings. Alternatively, one could use the fact that the sequence

$$0 \longrightarrow K \longrightarrow M \longrightarrow M/K \longrightarrow 0$$

is pure exact, so for M finitely presented, $\mathrm{Hom}(M, \cdot)$ preserves the exactness.

6. It is easy to find examples where M is a finitely generated quasi-continuous module and (A3) holds for all $K = \sum_{j=0}^{\infty} e_j M$ but M is not quasi-projective. Let R be your favorite non-artinian self-injective commutative von Neumann regular ring, and $M = R \oplus S$, where S is a simple R-module not isomorphic to a direct summand. Any $K = \sum_{j=0}^{\infty} e_j M$ is of the form $K' \oplus \bar{S}$ where $K' \subseteq K$ and $\bar{S} \subseteq S$ and direct summands of the quotient M/K look like direct summands of R/K' with perhaps a summand isomorphic to S. Yet M does not even have unique injective hulls, as $\langle 1, 0\rangle R$ and $\langle 1, \bar{1}\rangle R$ are both injective hulls of the annihilator of S. As remarked in [9], the injective submodules of this module are precisely the finitely generated submodules.

7. We will modify the above example in our final section.

8. I believe it to be very difficult to construct and verify an example of a finitely generated quasi-continuous module M where (A3)$'$ fails for $K = \sum_{j=0}^{\infty} e_j M$.

The properties (A1), (A2), and (A3) are all that are necessary to prove a generalization of Proposition 1. To generalize the Dung–Smith result, we will need additional properties which are related to but may be weaker than unique maximum essential extensions.

2 Unique essential extensions

We wish to look at a situation when $M \Big/ \left(\sum_{j=0}^{\infty} e_j M\right)$ cannot even contain a quasi-continuous direct summand. A hypothesis that works is unique maximum essential extensions, but we can weaken that. To appreciate the problems that arise, we first discuss the property that every submodule of M has a maximum essential extension containing all essential extensions in a quasi-continuous module M. This is essentially a specialization of the kind of things done in [4] and [5].

One standard way to ensure that M contains unique maximum essential extensions of submodules of M is to put on conditions that imply $\Lambda = End(M_R)$ is von Neumann regular. That the regularity of Λ implies unique essential extensions has been known for 30 years. Yet it is not necessary for unique maximum essential extensions as the $\mathbb{Z}$-module $\mathbb{Q}/\mathbb{Z}$ shows.

We will let the notation $N \subseteq' M$ mean N is an essential submodule of the module M.

Proposition 2 *Let M be a quasi-continuous module with endomorphism ring Λ. Then the following are equivalent:*

(i) For all submodules $K \subseteq M$, there exists a direct summand $E(K)$ of M which is an essential extension of K and contains every essential extension of K in M.

(ii) For all $\lambda \in \Lambda$ with $\ker(\lambda)$ essential in M, if $\epsilon = \epsilon^2 \in \Lambda$ has $Image(\lambda)$ essential in ϵM, then $\lambda \circ (1 - \epsilon) = 0$.

(iii) If $M = A \oplus B$ and $N \subseteq M$, then $N \cap A$ has no proper essential extension in N.

(iv) If $M = A \oplus B = C \oplus D$, then $A \cap C$ is a direct summand of M.

Proof: $(i) \Longrightarrow (ii)$. (This works for any CS module M.) Let $\lambda \in \Lambda$ have $Image(\lambda) \subseteq' \epsilon M$ and $\ker(\lambda) \subseteq' M$. Set $K = \ker(\lambda) \cap (1-\epsilon)M$. Then $K \subseteq' (1-\epsilon)M$ and $K \subseteq' \{x + \lambda(x) \mid x \in (1-\epsilon)M\}$. By (i), $(1-\epsilon)M \supseteq \{x + \lambda(x) \mid x \in (1-\epsilon)M\}$, so $Image(\lambda) = 0$.

$(ii) \Longrightarrow (iii)$. Let $M = A \oplus B$ and $N \subseteq M$. Let C be any maximal essential extension of $N \cap A$ in A, and $A = C \oplus D$. Let ϵ project M to C with kernel $B \oplus D$. Let F be any closure of $N \cap A$ in M. Define $\lambda \in \Lambda$ by $\lambda \restriction_{(1-\epsilon)M} = 0$ and $\lambda \restriction_F = (1-\epsilon) \restriction_F$. (This uses the full strength of quasi-continuity as opposed to CS to get λ defined on all of M and not just an essential submodule.) Then $\ker(\lambda)$ is essential in M and $Image(\lambda) \subseteq (1-\epsilon)M$ and so is essential in a direct summand of $(1-\epsilon)M$. By (ii), $\lambda \circ \epsilon = 0$, so $\lambda = 0$ and $F \subseteq \epsilon M \subseteq A$. Thus $F \cap N \subseteq A$ so $F \cap N = N \cap A$ and $N \cap A$ is closed in N.

$(iii) \Longrightarrow (iv)$. Any closed submodule of a direct summand of M is a direct summand of M.

$(iv) \Longrightarrow (i)$. Any two maximal essential extensions of a submodule $N \subseteq M$ are direct summands of M and so by (iv) their intersection is a direct summand of M. Hence they must coincide. □

For the quasi-continuous module M, set

$$\Delta(\Lambda) = \{\lambda \in \Lambda \mid \ker(\lambda) \textit{ is essential in } M\}.$$

This is the Jacobson radical of Λ if M is continuous. Proposition 2 contains the following:

Corollary 3 *Let M be a quasi-continuous module. Then every submodule of M has a unique maximum essential extension in M if and only if for all $\epsilon = \epsilon^2 \in \Lambda$, $\epsilon\,\Delta(\Lambda)\,(1-\epsilon) = 0$.*

We digress for a short comment on quasi-injective modules.

Corollary 4 *Let M be a quasi-injective module. Then $M \oplus M$ has unique maximum essential extensions of submodules if and only if $End(M)$ is von Neumann regular.*

Proof: We note that $M \oplus M$ is quasi-injective if and only if M is quasi-injective. If $End(M)$ is von Neumann regular, then so is the Morita equivalent ring $M_2\,(End(M)) = End(M \oplus M)$, and this implies unique maximum essential extensions. For the only if direction, condition (ii) of Proposition 2 shows that there are no homomorphisms from $(M, 0)$ to $(0, M)$ with kernel essential in $(M, 0)$, so the endomorphism ring of the continuous module M must be von Neumann regular.

We are now ready to introduce our fourth property, that will be used to show that $M \big/ \sum_{j=0}^{\infty} e_j M$ cannot even contain a quasi-continuous direct summand in certain cases.

(A4) Let $M = A \oplus B = C \oplus D$. Then $A + C$ is a direct summand of M.

The definition of quasi-continuity is CS plus the special case of (A4) when A and C have $A \cap C = 0$.

If every submodule of a quasi-continuous module M has a unique maximum essential extension in M, then for A and C direct summands of M we have $A \cap C$ is a direct summand of M and hence of C. Then $A + C = A \oplus \iota(C/A \cap C)$ is a direct summand of M where ι is a direct sum inclusion. However, (A4) does not imply unique maximum essential extensions as the example in Observation 6 shows. That M is an injective, finitely generated R-module for which a sum of two injective submodules is injective, but the intersection $\langle R, 0\rangle \cap \langle 1, \bar{1}\rangle\, R$ is not a direct summand of M. References [4], [5], and [9] are sources for information on an intersection of summands being a summand, a sum of summands being a summand, and the relationships between these properties. For quasi-continuous modules about all we can say is the intersection property implies the sum property but not conversely.

3 The theorems

With these preliminaries out of the way, we state and prove our general theorems. The seemingly elaborate statement of the first theorem will prove very useful in the last two sections of the paper.

Theorem A *Let M be a quasi-continuous, finitely generated module (or object in an $\mathcal{A}B5$ category) with endomorphism ring Λ. Let $\{e_i \mid i \in \omega\}$ be a set of non-zero orthogonal idempotents of Λ, and $\{f_i \mid i \in \omega\}$ idempotents of Λ such that $e_i f_i = f_i e_i = f_i$ for all $i \in \omega$. Then for any submodule K of M such that M satisfies (A3) for K and*

$$\oplus_{i=0}^{\infty} f_i M \subseteq K \subseteq \oplus_{i=0}^{\infty} e_i M,$$

we have M/K is not quasi-continuous.

Proof: We do essentially what was done in [7].

For any subset $A \subseteq \omega$, let $E_A \in \Lambda$ be a projection onto a maximal essential extension of $\oplus_{i\in A} f_i M$ in M with kernel containing $(\oplus_{i\in A}(e_i - f_i)M) \oplus (\oplus_{j\in\omega\setminus A} e_j M)$. Note that $E_A e_i = f_i \chi_A(i)$ where χ_A is the characteristic function of A. Note also that for any finite subset $\mathcal{F} \subseteq \omega$, $E_A M \cap \sum_{i\in\mathcal{F}} e_i M = \sum_{i\in\mathcal{F}\cap A} f_i M$.

Now let $\mathcal{A} = \{A_i \mid i \in \omega\}$ be a family of pairwise disjoint infinite subsets of ω and $\mathcal{B}$ a family $\{B_\alpha \mid \alpha \in \mathcal{I}\}$ of subsets of ω such that $S = \mathcal{A} \cup \mathcal{B}$ is a maximal family of almost disjoint subsets of ω. That is, any pair of sets in $\mathcal{A} \cup \mathcal{B}$ has finite intersection, and any infinite subset of ω has infinite intersection with some set in S.

We show that the family $\{E_s \mid s \in S\}$ is independent modulo K, that is, the sum $\sum_{s\in S} E_s M + K$ is direct in M/K. Assume some $\sum_{i=1}^{n} E_{S_i} m_i \in K$. Let $X = \{\alpha \mid \alpha \in S_j \cap S_k \text{ } for \text{ } some \text{ } j \neq k\}$. Then X is finite, and since $f_\alpha M \subseteq K$, we have $\sum_{i=1}^{n}(1 - \sum_{\alpha\in X} f_\alpha) E_{S_i} m_i \in K$. In this new sum, the sets $\{S_i \mid 1 \le i \le n\}$ are effectively disjoint. But then every module $(1 - \sum_{\alpha\in X} f_\alpha) E_{S_i} m_i R$ is essential over its intersection with $\sum_{j\in S_i\setminus X} e_j M$. Since $K \subseteq \sum_{j=0}^{\infty} e_j M$, for some n,

$$\left(1 - \sum_{\beta=1}^{n} e_\beta\right) \sum_{i=1}^{n} \left(1 - \sum_{\alpha\in X} f_\alpha\right) E_{S_i} m_i = 0.$$

If $0 \neq \left(1 - \sum_{\beta=1}^{n} e_\beta\right)(1 - \sum_{\alpha\in X} f_\alpha) E_{S_i} m_i R$, that module is essential over its non-zero intersection with $\sum_{j\in S_i\setminus X} e_j M$, a module having zero intersection with

$$\sum_{j\neq i} \left(1 - \sum_{\beta=1}^{n} e_\beta\right) \left(1 - \sum_{\alpha\in X} f_\alpha\right) E_{S_j} m_j R.$$

We conclude that $0 = \left(1 - \sum_{\beta=1}^{n} e_\beta\right)(1 - \sum_{\alpha\in X} f_\alpha) E_{S_i} m_i R$, so each summand is in $\sum_{\beta=1}^{n} e_\beta M$. But that forces each summand to be in $\sum_{\beta=1}^{n} f_\beta M$ and so in K.

Assume M/K is quasi-continuous. Then there is a direct sum decomposition $M/K = U \oplus V$ where $U + K$ is an essential extension of $\sum_{A\in\mathcal{A}} E_A M + K$ and $V \supseteq \sum_{B\in\mathcal{B}} E_B M + K$. Let ν, π, and λ be as in property (A3). Then for all $A \in \mathcal{A}$ and $m \in M$, $\pi\nu E_A m = \nu E_A m$ so $\lambda E_A m - E_A m \in K$. By property (A1), there exists an $n \in \omega$ with $(1 - \sum_{i=0}^{n} e_i)(\lambda E_A - E_A)M = 0$. Hence for almost all $i \in A$, $e_i \lambda E_A = e_i E_A$ and then $e_i \lambda e_i = e_i \lambda E_A e_i = e_i E_A e_i = e_i$. Similarly, for all $B \in \mathcal{B}$, $\pi\nu E_B m = 0$ so $\lambda E_A m \in K$ and for almost all $j \in B$, $e_j \lambda e_j = 0$.

Let C be the image of a choice function on

$$\{\{i \in A \mid e_i \lambda e_i = e_i\} \mid A \in \mathcal{A}\}.$$

By maximality of $\mathcal{S}$, C must have infinite intersection with some $s \in \mathcal{S}$, and s cannot be in $\mathcal{A}$, so it must be in $\mathcal{B}$, contradicting the fact that for almost all $j \in s \in \mathcal{B}$, $e_j \lambda e_j = 0$. □

Theorem A and our above observations is sufficient to prove the Dung–Smith result [3, Theorem 18] which inspired this paper. However, we are also in a position to prove a stronger result. The proof is essentially an adaptation of the ideas in [3, Section 1] using Theorem A instead of Proposition 1.

Theorem B *Let M be a quasi-continuous, finitely generated module such that either*

(i) (A3)′ holds for submodules of M of the form $\sum_{j=0}^{\infty} e_j M$ and a sum of two direct summands of M is a direct summand, or

(ii) (A3) holds for submodules of M of the form $\sum_{j=0}^{\infty} e_j M$ and no nonzero element of Λ has essential kernel.

Then for any set of orthogonal idempotents in Λ with $\sum_{j=0}^{\infty} e_j M$ essential in M, $M \big/ \left(\sum_{j=0}^{\infty} e_j M\right)$ cannot contain a non-zero quasi-continuous direct summand.

Proof: Let us assume that (ii) holds. By Proposition 2 an intersection of direct summands is a direct summand. Hence we have a sum of direct summands is a direct summand, i.e. condition (A4), in both case (i) and case (ii).

Assume $M \big/ \left(\sum_{j=0}^{\infty} e_j M\right)$ has a quasi-continuous direct summand U with

$$M \Big/ \left(\sum_{j=0}^{\infty} e_j M\right) = U \oplus V.$$

Lift the projection π onto V to a map λ from M to M. Then there is an idempotent $\epsilon \in \Lambda$ with λM essential in ϵM. Let $K = \lambda M + \sum_{j=0}^{\infty} e_j M$. Then $M/K \simeq U$ is quasi-continuous.

By condition (A4), the chain

$$\epsilon M \subseteq \epsilon M + e_0 M \subseteq \epsilon M + e_0 M + e_1 M \subseteq \cdots \subseteq \epsilon M + \sum_{i=0}^{n} e_i M \subseteq \cdots$$

is a chain of direct summands of M. Thus $\epsilon M + \sum_{i=0}^{n+1} e_i M = \epsilon M + \sum_{i=0}^{n} e_i M \oplus f_{n+1} M$ where $f_{n+1}^2 = f_{n+1}$. Let $\epsilon' M$ be an essential extension of $\sum_{j=0}^{\infty} f_j M$. Then $\epsilon M \oplus \epsilon' M = M$ since it is a direct summand containing the essential submodule $\sum_{j=0}^{\infty} e_j M$. Moreover

$$U \simeq M/K \simeq (\epsilon M/\lambda M) \oplus \left(\epsilon' M \Big/ \sum_{j=0}^{\infty} f_j M\right)$$

so $\epsilon' M \big/ \sum_{j=0}^{\infty} f_j M$ is quasi-continuous. If there are an infinite number of non-zero f_i, this contradicts Theorem A. Hence we need only rule out the possibility that $M = \epsilon M + \sum_{i=0}^{n} e_i M$ for some n. In this case, quasi-continuity implies that there is an $N \subseteq \sum_{i=0}^{n} e_i M$ such that $M = \epsilon M \oplus N$ since a complement of $\epsilon M \cap \sum_{j=0}^{n} e_j M$ in $\sum_{j=0}^{n} e_j M$ will serve as such an N.

If (i) holds, then $\epsilon M = \lambda M$, and

$$\begin{aligned} U &= (1-\pi)\,\nu\,(\epsilon M \oplus N) \simeq (1-\pi)\,\nu\,(\lambda M \oplus N) \\ &= (1-\pi)\,\nu\lambda M + (1-\pi)\,\nu N = (1-\pi)\,\pi\nu M + 0 = 0. \end{aligned}$$

If (ii) holds, by finite generation (A1) and Observation 2, there is an integer n such that, for $e = 1 - \sum_{j=0}^{n} e_j$, $e\,(\lambda^2 - \lambda) = 0$. Since no element of Λ has essential kernel in this case, $e\,(1-\lambda)\,\epsilon = 0$. Then

$$\begin{aligned} U &= (1-\pi)\,\nu\,(\epsilon M \oplus N) \simeq \nu\,(1-\lambda)\,\epsilon M + (1-\pi)\,\nu N \\ &= \nu e\,(1-\lambda)\,\epsilon M + \nu\,(1-e)\,(1-\lambda)\,\epsilon M + 0 = 0. \end{aligned}$$

In either case we get a contradiction. □

4 An application

In this section, we state and improve the Dung-Smith result.

Proposition 5 (Dung-Smith 1991) *Let M be a quasi-projective module. Then the following are equivalent:*

(i) M is a semi-artinian module with every simple quotient of M M-injective.

(ii) Every non-zero cyclic subquotient of M contains a non-zero M-injective submodule.

Sketch of Proof: One direction is clear. Now assume every non-zero cyclic subquotient of M contains a non-zero M-injective submodule. Let K be the union of the ascending socle series for M. Then if M/K is not zero it contains a non-zero cyclic quasi-injective direct summand N which is socle-free. Since K is invariant in M and M is quasi-projective, M/K and hence N is quasi-projective. This implies that (A3) holds for all submodules of N. Since the image of any nonzero map in End(N) contains a direct summand which must be N-projective, no element of this endomorphism ring has essential kernel. The injectivity of N implies that the endomorphism ring of N is von Neumann regular and so (A3)′ holds. By Theorem B a cyclic submodule of N cannot contain an infinite sum of N-injective submodules so it must be of finite Goldie dimension and hence semisimple. But the only semisimple submodule of N is zero, so N and hence M/K are both 0. □

Theorem C *Let M be a finitely generated module such that every quotient $M' = M/K$ satisfies (A3) for submodules of the form $M' / \sum_{j=0}^{\infty} e_j M'$ for $\{e_i \mid i \in \omega\}$ orthogonal idempotents in End(M'). Assume also either submodules of M' have unique maximum essential extensions, or M' satisfies (A3)′ for submodules of the form $\sum_{j=0}^{\infty} e_j M'$ and also (A4). Then the following are equivalent:*

(i) Every quotient of M has a quasi-continuous direct summand.

(ii) Every quotient of M contains a uniform direct summand.

Proof: Since a uniform module is quasi-continuous, (ii) implies (i) is clear.

For (i) implies (ii), assume this is false. Then some quotient of M contains a quasi-continuous direct summand which does not have finite Goldie dimension. Hence it contains an infinite direct sum of direct summands. Since the hypotheses of Theorem B hold, the quotient cannot contain a quasi-continuous direct summand. □

Since quasi-projective implies (A3) for all submodules, the Dung-Smith proposition that the endomorphism ring of a quasi-projective module such that any submodule contains a direct summand cannot contain elements with essential kernel is all that is needed to derive their result from our Theorems. What this paper shows is that quasi-projective is too strong a property. The result is much more general. Exactly what the best 'minimal' set of hypotheses is is still not clear.

5 An example

In the previous section, we looked at the case where a quotient of a finitely generated quasi-continuous module M modulo a sum $\sum_{j=0}^{\infty} e_j M$ could not even contain a quasi-continuous direct summand. We needed some special hypotheses closely related to unique maximum essential extensions. In this section, we construct an example where these hypotheses do not hold. We then show that the conclusion of Theorem B is still true in this case.

Let p be a prime natural number, and n and integer > 1. Let

$$R = \prod_{i=0}^{\infty} \left(\mathbb{Z}/p^2\mathbb{Z}\right)_i .$$

Each factor $\mathbb{Z}/p^2\mathbb{Z}$ is a self-injective ring, and also injective as an R-module. Thus R_R is a product of R-injective modules and so R is a self-injective ring.

To simplify the description of our next step, we will use the concept of an ultrafilter. For the purposes of this paper, one may think of an ultrafilter on a set $\mathcal{A}$ as follows. The set of all subsets of $\mathcal{A}$ forms a boolean algebra under the operations

$$A + B = (A \cup B) \smallsetminus (A \cap B)$$
$$A * B = A \cap B.$$

Identify this power set with $2^{\mathcal{A}}$ which is the set of all functions from $\mathcal{A}$ to $2 = \{0, 1\}$. These operations then become the usual operations in $\prod_{\alpha \in \mathcal{A}} \mathbb{Z}/2\mathbb{Z}_\alpha$. Take a maximal ideal $\mathfrak{M}$ in this boolean algebra. Set

$$\mathcal{U}_{\mathfrak{M}} = \{A \subseteq \mathcal{A} \,|\, \textit{there is a } B \in \mathfrak{M} \textit{ with } A = \textit{ the set of zeros of } B\}.$$

Such a $\mathcal{U}_{\mathfrak{M}}$ has the properties

(*i*) $A \in \mathcal{U}_{\mathfrak{M}}$ and $B \in \mathcal{U}_{\mathfrak{M}} \Longrightarrow A \cap B \in \mathcal{U}_{\mathfrak{M}}$.

(*ii*) $A \in \mathcal{U}_{\mathfrak{M}}$ and $B \supseteq A \Longrightarrow B \in \mathcal{U}_{\mathfrak{M}}$.

(*iii*) The empty set $\emptyset \notin \mathcal{U}_{\mathfrak{M}}$.

(*iv*) For all $A \subseteq \mathcal{A}$, either $A \in \mathcal{U}_{\mathfrak{M}}$ or $\mathcal{A} \smallsetminus A \in \mathcal{U}_{\mathfrak{M}}$.

The first three of these properties define a filter, the fourth makes it an ultrafilter. A principal ultrafilter is $\{A \subseteq \mathcal{A} \,|\, n \in A\}$ for some fixed $n \in \omega$. Take $\mathcal{U} = \mathcal{U}_{\mathfrak{M}}$ for some non-principal maximal ideal $\mathfrak{M}$. Given an ultrafilter $\mathcal{U}$, you can take a product $\prod_{i \in \omega} \mathfrak{A}_i$ of algebraic objects with the same operations and identify any two infinite sequences which agree on a set in the ultrafilter. The corresponding equivalence classes form a similar algebraic object $\prod_{i \in \omega} \mathfrak{A}_i/\mathcal{U}$ called an ultraproduct. Any property expressible in terms of first order logic holds in the ultraproduct if and only if it holds in a set of factors indexed by a member of the ultrafilter.

Let $\mathcal{A}$ be an infinite subset of ω and let $\mathcal{U}$ be a non-principal ultrafilter on $\mathcal{A}$. The ring

$$\hat{R} = \prod_{\alpha\in\mathcal{A}} \left(\mathbb{Z}/p^2\mathbb{Z}\right)_\alpha \Big/ \mathcal{U}$$

has all the first order properties of $\mathbb{Z}/p^2\mathbb{Z}$ which is a finite ring, and so $\hat{R}$ must be isomorphic to $\mathbb{Z}/p^2\mathbb{Z}$ as a ring. Dividing a product of rings by an ultrafilter is the same as dividing the product ring by the ideal generated by the characteristic functions of the complements of sets in the ultrafilter. Since the ideal you are dividing out by is generated by idempotents in the commutative ring R, any map from an ideal of R to $\hat{R}$ is zero on that ideal, so $\hat{R}$ is actually an injective R-module.

Now partition ω into a countable family $\{\mathcal{A}_j|\ j\in\omega\}$ of disjoint infinite sets. For each j form a ring $\hat{R}_j$ by dividing $\prod_{\alpha\in\mathcal{A}_j}\mathbb{Z}/p^2\mathbb{Z}_\alpha$ by a non-principal ultrafilter $\mathcal{U}_j$ on $\mathcal{A}_j$. The quotient of R obtained by dividing out by the ideal generated by characteristic functions of the $\{\{A_j\setminus U|\ U\in\mathcal{U}_j\}|\ j\in\omega\}$ contains a copy of $\prod_{j\in\omega}\hat{R}_j$ which is an injective R-module. Since it is a direct summand of a cyclic R-module, it is a cyclic R-module.

This process can be iterated an infinite number of times. Set $R^{(0)} = R$ and $R^{(1)} = \prod_{j\in\omega}\hat{R}_j$. Divide ω into a countable disjoint family of infinite sets and perform the same construction on $\prod_{j\in\omega}\hat{R}_j$ as was done on R. The result is another cyclic injective R-module, call it $R^{(2)}$, that looks like a product of copies of $\mathbb{Z}/p^2\mathbb{Z}$. Repeat to get $R^{(3)}$, $R^{(4)}$, $\cdots$. Now let

$$M_\ell = \textstyle\sum_{i=0}^{\ell} R^{(i)} \qquad\qquad 1\le \ell<\infty\ .$$

Fix the integer $\ell\ge 1$ and let $M = M_\ell$. Then M is a finitely generated injective R-module.

We remark that we could replace $\mathbb{Z}/p^2\mathbb{Z}$ by $\mathbb{Z}/p^n\mathbb{Z}$ or by any uniserial ring with a finite composition series without changing the relevant properties of M.

We now look at some of these relevant properties of the module M.

(a) A sum of direct summands of M need not be injective. For example, the submodule $\left\langle 1,0,\vec{0}\right\rangle R + \left\langle 1,p,\vec{0}\right\rangle R$ is not injective, but each summand is.

(b) Let K be any finitely generated submodule of $R^{(i)}$. We show that K is a direct sum $K = \eta_1 R^{(i)}\oplus\eta_2 pR^{(i)}$, where the η_j are orthogonal idempotents. Note that the R and $R^{(i)}$ submodules of $R^{(i)}$ coincide, so there is no ambiguity in this statement. To see this, first let $S^1 = R^{(i)}/pR^{(i)}$. Then S^1 is a von Neumann regular ring, so K/pK is of the form $\tilde{\eta}S^1$ for some idempotent $\tilde{\eta}\in K/pK\subseteq S^1$. Since $\tilde{\eta}pR^{(i)}$ is nilpotent, $\tilde{\eta}$ lifts to an idempotent $\eta_1\in K$. Then $K=\eta_1R^{(i)}\oplus(1-\eta_1)K$, where $(1-\eta_1)K\subseteq pR^{(i)}$. $(1-\eta_1)K$ is finitely generated, so if η_2 is an idempotent generating its injective hull, $K=\eta_1K\oplus\eta_2pK$.

(c) Let K be any countably generated submodule of M and let $\varphi \in End(M/K)$. Then $\varphi\left(R^{(j)}+K\right) \subseteq \sum_{i=j}^{\ell} R^{(i)}+K$. To see this, let $\varphi\left(R^{(j)}+K\right) = \left\langle\vec{\alpha}, \vec{\beta}\right\rangle + K$ where $\vec{\alpha} \in \sum_{i=0}^{j-1} R^{(i)}$ and $\vec{\beta} \in \sum_{i=j}^{\ell} R^{(i)}$. Then $\vec{\alpha}\cdot\left(0:1^{(i)}\right) \subseteq K$. But in constructing $R^{(i)}$ from $R^{(i-1)}$ we have to throw into the kernel any characteristic function of a set which has finite intersection with each of the A_j. Hence no countably generated submodule of $\sum_{i=0}^{j-1} R^{(i)}$ can contain $\vec{\alpha}\cdot\left(0:1^{(i)}\right)$ without also containing $\vec{\alpha}$ so the projection of K to $\sum_{i=0}^{j-1} R^{(i)}$ must contain $\vec{\alpha}$ as asserted.

(d) M satisfies (A3) for countably generated submodules K. Indeed, let $\varphi \in End(M/K)$. By property (c), $\varphi\left(1^{(i)}\right) = \left(0:\vec{\beta}\right)$ where the map $1^{(i)} \longmapsto \vec{\beta}$ is a well defined map from $R^{(i)}$ to M. The direct sum of these maps lifts φ.

(e) Let $0 \neq x \in M$. Let ϵ_i denote the projection of M to $R^{(i)}$ with kernel the product of the rest of the $R^{(j)}$s. There is a smallest integer i such that $\epsilon_i(x) \neq 0$. Call this i the **order** of x. By the way $R^{(i)}$ was constructed, the injective hulls of simple submodules of $R^{(i)}$ are quotients of idempotent projections onto disjoint countable subsets of the original product $R = \prod_{j\in\omega} \mathbb{Z}/p^2\mathbb{Z}$. We call the set of all elements of ω which are in one of the disjoint sets corresponding to a simple submodule of $\epsilon_i(x)R$ the **support** of x.

(f) M satisfies (A3)″ for countably generated submodules K. To see this, let $\varphi \in End(M/K)$ be idempotent. By property (c), $\varphi\left(1^{(i)}\right) = \beta_i + K = \epsilon_i\beta_i + \gamma_i + K$, where γ_i has order $> i$. By property (b), $\epsilon_i\beta_i R^{(i)} = \eta_1^{(i)} R^{(i)} + \eta_2^{(i)} pR^{(i)}$, where ϵ_i is as in property (e). Now $\epsilon_i\beta_i$ determines an element in $R^{(i)}$ and we will identify it with that element. Then $\varphi^2\left(1^{(i)}\right) = \varphi\left(\epsilon_i\beta_i + \gamma_i + K\right) = \left(\epsilon_i\beta_i\right)^2 +$ terms of higher order $+ K = \eta_1^{(i)}u_i +$ terms of higher order $+ K$. Also, since

$$\left[\varphi^2\left(1^{(i)}\right)u_i^{\,2} - \varphi\left(1^{(i)}\right)u^{i}\right]\cdot u_i^{\,-1} \in K,$$

without loss of generality we may take $\epsilon_i\beta_i = \eta_1^{(i)}$. Now the matrix μ of the lifting λ of φ is

$$\mu = \begin{bmatrix} \eta_1^{(0)} & 0 & \cdots & 0 \\ \mu_{1,0} & \eta_1^{(1)} & 0 & \vdots \\ \vdots & \vdots & \ddots & 0 \\ \mu_{\ell,0} & \mu_{\ell,1} & \cdots & \eta_1^{(\ell)} \end{bmatrix}.$$

Notice that $\left(\mu^2-\mu\right)^{\ell+1} = 0$. Hence there is an idempotent $\mu' \in \mathbb{Z}/p^2\mathbb{Z}\,[\mu]$ with $\mu' \equiv \mu$ modulo $\left(\mu^2-\mu\right)\cdot\mathbb{Z}/p^2\mathbb{Z}\,[\mu]$. For example, if $\ell = 1$, the idempotent may be taken as $3\mu^2 - 2\mu^3$. But $(\mu^2 - \mu)(M) \subseteq K$, so the map corresponding to μ' also lifts φ.

(g) Let e and f be orthogonal idempotents in Λ. Assume $eM + fM$ contains a simple submodule $S \subseteq R^{(i)}$. Then $S \subseteq eM$ or $S \subseteq fM$ since the only isomorphic copy of S in M is S itself. If $S \subseteq eM$, then eM contains the unique injective hull of S in M.

(h) Let $\{e_i \mid i \in \omega\}$ be an infinite set of orthogonal idempotents of Λ, and ϵ any non-zero idempotent. Assume $\sum_{j=0}^{\infty} e_j M$ is essential in M and $M \big/ \left(\sum_{j=0}^{\infty} e_j M + \epsilon M\right)$ is quasi-continuous. We observe that there is an $n \in \omega$ with $\sum_{j=0}^{n} e_j M + \epsilon M$ essential in M. Assume not. Then we have a sequence $\left\{\mathcal{E}\left(\sum_{j=0}^{n} e_j M + \epsilon M\right) \middle| \, n \in \omega\right\}$ with an infinite number of terms, where $\mathcal{E}(N)$ denotes an injective hull of N. Then

$$\bigcup_{n\in\omega} \mathcal{E}\left(\sum_{j=0}^{n} e_j M + \epsilon M\right) = \epsilon M \oplus \sum_{j=0}^{\infty} \tilde{e}_j M$$

for some infinite set of orthogonal idempotents $\{\epsilon, \tilde{e}_i\}$ and each of the $\tilde{e}_i M$ which is non-zero contains some injective hull of a simple which also lies in some $e_j M$. Thus we have $K = \sum_{j=0}^{\infty} e_j M + \epsilon M$ lying between $\epsilon M \oplus \sum_{j=0}^{\infty} \tilde{e}_j M$ and a sum of injective hulls of simples. Since (A3) holds for K, Theorem A shows this cannot happen.

(i) Property (h) shows exactly why Theorem A was stated the way it was. Given a finitely generated quasi-continuous module $\overline{M}$ which has (A3)$'$ for submodules of the form $\sum_{j=0}^{\infty} e_j \overline{M}$ and (A3) for countably generated submodules, and such that every nonzero intersection of direct summands contains a direct summand of $\overline{M}$, we can conclude that any quasi-continuous quotient of the form $\overline{M} \big/ \left(\epsilon\overline{M} + \sum_{j=0}^{\infty} e_j \overline{M}\right)$ with $\sum_{j=0}^{\infty} e_j \overline{M}$ essential has $\epsilon\overline{M} + \sum_{j=0}^{n} e_j \overline{M}$ essential for some n, which, by adding orthogonal idempotents, might as well be taken as 0.

(j) We finally get to our last result. Even though we cannot apply Theorem B directly to our module M of this section, property (h) plus a chain condition on kernels of products of idempotents in Λ is enough to show that no quotient of M modulo an essential infinite direct sum $\sum_{j=0}^{\infty} e_j M$ can contain a quasi-continuous direct summand. For assume some such quotient $M \big/ \left(\sum_{j=0}^{\infty} e_j M\right)$ does contain such a summand U. By property (f), there exists an idempotent $\epsilon \in \Lambda$ lifting a projection onto U. Among all such ϵ, pick one with the property that $\epsilon M \subseteq \sum_{j=i}^{\ell} R^{(j)}$ for i as large as possible. By property (h), $(1-\epsilon) M + \sum_{i=1}^{n} e_i M$ is essential in M. Then $\sum_{i=1}^{n} e_i M$ must contain the socle of ϵM by property (g), and by taking an injective hull of that socle in $e_0 M$, we may assume that $e_0 M$

and ϵM both have the same socle. Then $(1-e_0)\epsilon M$ is a direct summand of M also lifting U, and it must be contained in $\sum_{j=i+1}^{\ell} R^{(j)}$ because $(1-e_0)$ kills the socle of $\epsilon M \cap R^{(i)}$, does not multiply any element by p, and so cannot have any projection to $R^{(i)}$, and property (c) shows it moves every $R^{(j)}$ to something with order $\geq j$ since 0 is countably generated. This contradicts the maximality of i.

(k) It would be very nice to be able to take $\ell = \infty$ and somehow keep the property that M is finitely generated by looking at it as a module over some subring of Λ rather than R. Attempts to do this seem to destroy the delicate structure of M and lead to unique injective hulls of submodules of M over the new ring.

(l) This example shows that the hypotheses of Theorem B are still too strong. It can generate other hypotheses that will work to get the conclusions of Theorem B, but they are highly specialized if the module does not have even a sum of direct summands a direct summand. If we are going to use our Theorem A, we will need some hypothesis which forces enough direct summands to exist inside appropriate kernels. A not unreasonable such hypothesis weakening the Dung-Smith one on cyclics containing injectives is insisting that at least a non-zero intersection of direct summands contains some non-zero direct summand. Perhaps further examples might shed some light on whether or not the following interesting statement is true or false:

> Let M be finitely generated quasi-continuous with (A3)′ (or (A3)″) for infinite sums $\sum_{j=0}^{\infty} e_j M$, (A3) for countably generated submodules, and a non-zero intersection of direct summands of M contains a non-zero direct summand. Then $M \big/ \sum_{j=0}^{\infty} e_j M$ cannot contain a quasi-continuous direct summand if $\sum_{j=0}^{\infty} e_j M$ is essential.

References

[1] Chatters, A. W. and Harjavanis, C.R., *Rings in which every complement right ideal is a direct summand*, Quart. J. Math. Oxford (2) **28**(1977), 61–80.

[2] Damiano, R. F., *A right pci ring is right noetherian*, Proc. Amer. Math. Soc. **77**(1979), 11–14.

[3] Dung, N. V. and Smith, P. F., *On semi-artinian V-modules*, J. Pure and Applied Algebra **82**(1992), 27–37.

[4] Garcia, J. L., *Properties of direct summands of modules*, Comm in Alg **17**(1989), 73–92.

[5] Hausen, J., *Modules with the summand intersection property*, Comm in Alg **17**(1989), 135–144.

[6] Mohamed, S. H. and Müller, B. J., *Continuous and discrete modules*, London Math. Soc. Lecture Note Series nr. 147, Cambridge Univ. Press, Cambridge, 1990.

[7] Osofsky, B. L., *Rings all of whose finitely generated modules are injective*, Pacific J. Math **14**(1964), 645–650.

[8] Osofsky, B. L., *Noninjective cyclic modules*, Proc. Amer. Math. Soc. **19**(1968), 1383–1384.

[9] Osofsky, B. L., *Rings whose idempotent generated ideals form a lattice with ACC*, Rings, Modules, and Radicals (Hobart 1987), Pitman Res. Notes Math 204, Longman Sci. Tech, Harlow (1989), 118-124.

[10] Osofsky, B. L. and Smith, P. F., *Cyclic modules whose quotients have all complement submodules direct summands*, J. Algebra **139**(1991), 342–354.

WHEN DIRECT SUMS OF SINGULAR INJECTIVES ARE INJECTIVE

STANLEY S. PAGE and YIQIANG ZHOU
Department of Mathematics, The University of British Columbia
Vancouver, Canada V6T 1Z2

Abstract . Let R be an associative ring with identity. We show that every direct sum of singular injective right R-modules is injective iff any one of the following equivalent conditions holds. i) Every singular injective right R-module is a direct sum of uniform singular injective right R-modules; ii) Every singular injective right R-module has a decomposition that complements direct summands; iii) There exists a cardinal number c such that every singular injective right R-module is a direct sum of c-generated modules; iv) Every ascending chain of right ideals of R: $I_1 \subseteq I_2 \subseteq \cdots \subseteq I_n \subseteq \cdots$ such that the injective hulls of all I_{i+1}/I_i are singular, terminates; v) For any countable set of right ideals of R: $\{I_i : E(R/I_i)$ is singular, $i \in \mathbf{N}\}$, the chain $\cap_{i\geq 1} I_i \subseteq \cap_{i\geq 2} I_i \subseteq \cdots$ terminates; vi) Every ascending chain of right ideals of R: $I_1 \subseteq I_2 \subseteq \cdots \subseteq I_n \subseteq \cdots$ such that all $E(R/I_i)$ are singular, terminates. A classification of uniform singular injective modules is given for any commutative ring which satisfies any one of the above equivalent conditions.

Introduction. It is well known that a ring R is right Noetherian if and only if all direct sums of injective right R-modules are injective. Teply characterized rings for which every direct sum of non-singular injective right R-modules is injective [5]. These rings turn out to be rings for which the Goldie torsion theory is a Noetherian torsion theory. Another characterization of these rings is that they have finite torsionfree width, i.e. there is a bound on the number of summands in any right ideal which is the direct sum of torsionfree modules. But there is no characterization for those rings for which direct sums of singular injective right modules are injective. A result of Page and Yousif states that if R is a right nonsingular ring, then R is a right almost Noetherian ring if and only if all direct sums of singular injective right R-modules are injective [3, Coro.2.9]. One can provide (1) an example of almost Noetherian ring which is not nonsigular, but for which all direct sums of singular injective right modules are injective and (2) an example of ring which is not right almost Noetherian, but for which direct sums of singular injective right modules are injective. So it is natural for one to consider the general question: When is every direct sum of singular injective right modules injective? This is the subject of this paper. To begin with, we will observe that the class of singular modules with singular injective hulls does not, in general, form a torsion class. We will show that rings for which direct sums of singular injective right modules are injective are characterized by any one of the following equivalent conditions. i) Every singular injective right R-module is a direct sum of uniform singular injective right R-modules; ii) Every singular injective right R-module has a decomposition which complements direct summands; iii) There exists a cardinal number c such that every singular injective right R-module is a direct sum of c-generated modules; iv) Every ascending chain of right ideals of R: $I_1 \subseteq I_2 \subseteq \cdots \subseteq I_n \subseteq \cdots$ such that the injective hulls of all I_{i+1}/I_i are singular, terminates; v) For any countable set of right ideals of R: $\{I_i : E(R/I_i)$ is singular, $i \in \mathbf{N}\}$, the chain $\bigcap_{i\geq 1} I_i \subseteq \bigcap_{i\geq 2} I_i \subseteq \cdots$ terminates; vi) Every ascending chain of right ideals of R: $I_1 \subseteq I_2 \subseteq \cdots \subseteq I_n \subseteq \cdots$ such that all $E(R/I_i)$ are singular, terminates. As an application, we give a classification of uniform singular injective modules for commutative ring for which direct sums of singular injective modules are injective.

Throughout all rings are associative with identity and all modules are unital. For a right R-module M, $E(M_R)$, $Soc(M_R)$ and $Z(M_R)$ will denote the injective hull, the socle and the singular submodule respectively. For a submodule N of M, the notation $N \subseteq^{\oplus} M$ denotes that N is a summand of M, and $N \subset M$ means that N is a proper submodule of M. For a ring R, we let $H(R) = \{I \subseteq R_R : E(R/I)$ is singular $\}$ and $D_0 = \bigoplus\{E(R/I) : I \in H(R)\}$. We denote the set of positive integers and the set of integers by $\mathbf{N}$ and $\mathbf{Z}$ respectively.

The main results. Proposition 1. Let M_R be a module. Then $E(M)$ is singular if and only if no non-zero submodules of M can be embedded in R.

Proof. Suppose $E(M)$ is singular. If $0 \neq N \subseteq M$, and $f : N \to R$ is a R-module monomorphism, then $E(N) \cong E(f(N))$ is singular and $E(R) = E(f(N)) \oplus X$ for some X. Write $1 = q+w$, $q \in E(f(N))$ and $w \in X$. Then $q^{\perp}$ is essential in R, implying that $q^{\perp} \cap f(N) \neq 0$. Hence $0 \neq q^{\perp} \cap f(N) = (q+w)(q^{\perp} \cap f(N)) = w(q^{\perp} \cap f(N)) \subseteq f(N) \cap X$. A contradiction.

Conversely, if $E(M)$ is not singular, then $x^{\perp}$ is not essential in R for some $0 \neq x \in E(M)$. There exists nonzero right ideal I of R such that $x^{\perp} \cap I = 0$. Thus $I \cong xI$, and so xI can be embedded in R, implying that $0 \neq xI \cap M$ can be embedded in R. □ **Corollary 2.** (*a*) $Soc(R_R)$ is nonsingular if and only if every singular simple right R-module has a singular injective hull;

(*b*) $Z(R_R) = 0$ if and only if every cyclic singular right R module has a singular injective hull if and only if every singular right R-module has a singular injective hull. □

It is easy to note that $H(R)$ contains all dense right ideals of R. It is of some interest to look for the largest idempotent filter in $H(R)$ which contains all dense right ideals of R. It turns out that such an idempotent filter must be the set of all dense right ideals of R as the following proposition shows:

Proposition 3. Let R be a ring, and $H(R)$ as above. Then the following hold:

(a) If $I \in H(R)$, then $(I : a) \in H(R)$ for all $a \in R$;

(b) Any intersection of finite elements in $H(R)$ is in $H(R)$;

(c) Let H_1 be a nonempty subset of $H(R)$ which contains all dense right ideals of R. Then H_1 forms an idempotent filter iff H_1 is the set of all dense right ideals of R.

Proof. (a)&(b). These are easy.

(c). Let H_1 be as above and $I \in H_1$. If $0 \neq f \in Hom(J/I, R)$, where $I \subseteq J \subseteq R$, then $ker f = K/I$ for some $K \subseteq J$, and J/K is embeddable in R. Then $E(R/K)$ is not singular by Proposition 1. Hence $K \notin H_1$, contradicting the fact that H_1 is an idempotent filter. So $Hom(J/I, R) = 0$ for all $I \subseteq J \subseteq R$. Thus I is a dense right ideal of R. □

The following proposition gives a necessary and sufficient condition for the class of right modules with singular injective hulls to form a torsion class or a torsionfree class. **Proposition 4.** Let $Z_0 = \{r \in R : rI = 0 \text{ for some } I \in H(R)\}$, and $S_0 = \bigcap\{I : I \in H(R)\}$. Then

(a) Z_0 and S_0 both are ideals of R;

(b) The class of right R-modules with singular injective hulls forms a torsion class iff $Z_0 = (0)$;

(c) The ideal S_0 is an essential right ideal of R iff the class of right R-modules with singular injective hulls forms a torsionfree class iff $H(R)$ is closed under arbitrary intersections.

Proof. (a). Suppose $r_1, r_2 \in Z_0$, $a \in R$. Then there exist $I_1, I_2 \in H(R)$ such that $r_1 I_1 = 0$, $r_2 I_2 = 0$. Hence $(ar_1)I_1 = 0$, $(r_1 - r_2)(I_1 \cap I_2) = 0$ and $(r_1 a)(I_1 : a) = 0$. Since $I_1 \cap I_2$ and $(I_1 : a)$ both are in $H(R)$, by Proposition 3, we have $ar_1, r_1 - r_2$ and $r_1 a$ all are in Z_0. Hence Z_0 is an ideal.

The set S_0 is the intersection of the annihilators of all singular injective right R-modules, hence is an ideal.

(b). First we note that the class of right R-modules with singular injective hulls forms a torsion class iff $H(R)$ is an idempotent filter iff, by Proposition 3, $H(R)$ is the set of all dense right ideals of R. Suppose that $0 \neq x \in Z_0$. Then $I \subseteq x^{\perp} \subset R$ for some $I \in H(R)$. Since $R/x^{\perp} \cong xR$, we have $\mathrm{Hom}(R/x^{\perp}, R) \neq 0$. Hence $x^{\perp}$ is not a dense right ideal of R. It follows that I is not a dense right ideal. For the converse, suppose $Z_0 = (0)$. If $I \in H(R)$, then $(I : a) \in H(R)$ for any $a \in R$ by Proposition 3. So, ${}^{\perp}(I : a) = (0)$, since $Z_0 = (0)$. Hence I is a dense right ideal of R by [4, Prop.6.4,

P149]. (c). Suppose S_0 is an essential right ideal of R. Let $M = \prod_t M_t$ with every $E(M_t)$ singular. Then $E(M) = \prod_t E(M_t)$. For any $x \in E(M)$, write $x = (x_t)$ with $x_t \in E(M_t)$ for all t. We have $x_t^{\perp} \in H(R)$ since $E(R/x_t^{\perp}) \cong E(x_tR) \subseteq E(M_t)$ and $E(M_t)$ is singular. Then $S_0 \subseteq x_t^{\perp}$ for all t, implying $xS_0 = (0)$. It follows that $E(M)$ is singular. Therefore we have shown that the class of right R-modules with singular hulls is closed under products, and hence the class is a torsionfree class.

Next suppose that the class of right R-modules with singular injective hulls is a torsionfree class. Then it is closed under products. Let $\{I_t : t \in A\} \subseteq H(R)$. Since $E(R/\bigcap_t I_t) \hookrightarrow \prod_{t\in A} E(R/I_t)$ and $\prod_{t\in A} E(R/I_t)$ is singular, then $\bigcap_{t\in A} I_t \in H(R)$. Hence, $H(R)$ is closed under intersections.

Finally, if $H(R)$ is closed under arbitrary intersections, then $S_0 \in H(R)$, showing S_0 is an essential right ideal of R. □

Remarks : i) For any ring R, $Z_0 \subseteq Z(R_R)$. If $R = \mathbf{Z}/4\mathbf{Z}$, then $Z(R_R) = 2R$, but $Z_0 = (0)$. These examples show that if R is a right non-singular ring, then the class of right R-modules with singular injective hulls forms a torsion class, but the converse is not true in general. ii) Example 13 (1) of this article provides a ring R for which $Z_0 = Z(R_R) = Soc(R_R) \neq 0$. Hence the class of right modules with singular injective hulls does not form a torsion class in general.

iii) If $R = \mathbf{Z}$, then $H(R)$ is the set of all nonzero ideals of R. And we have $S_0 = (0)$ is not essential in R. Hence the class of right modules with singular injective hulls does not form a torsionfree class in general.

iv) From the proof of (c) of Proposition 3, we note that the class of right R-modules with singular injective hulls forms a torsion class iff $H(R)$ is closed under super sets, i.e. if $I \in H(R)$ and $I \subseteq J$, then $J \in H(R)$. **Lemma 5**. Let $\{M_\alpha : \alpha \in A\}$ be a collection of right R-modules and B a set. Then $(\bigoplus_A M_\alpha)^{(B)} \cong \bigoplus_A ({M_\alpha}^{(B)})$.

Proof. This is well-known. □

Proposition 6. The following are equivalent for a ring R:

(a) Every direct sum of singular injective right R-modules is injective;

(b) $M^{(\mathbf{N})}$ is injective for every singular injective right R-module M;

(c) ${D_0}^{(\mathbf{N})}$ is injective.

Proof. (a) $\Rightarrow$ (b). Obvious. (b) $\Rightarrow$ (c). Let $E = E(D_0)$. E is singular, for otherwise, there exists a nonzero cyclic submodule xR of D_0 such that xR can be embedded in R by Proposition 1. But $xR \subseteq \bigoplus_{I\in F} E(R/I)$ for a finite subset F of $H(R)$. It follows that $\bigoplus_{I\in F} E(R/I)$ is not singular, a contradiction. So E is singular.

and then $E^{(\mathbf{N})}$ is injective by (b). Therefore $E^{(H(R))}$ is injective by [1, 25.1, P289]. But for each $I \in H(R)$, $E(R/I)$ is a summand of E. Thus $D_0 = \bigoplus_{I\in H(R)} E(R/I)$ is a summand of $E^{(H(R))}$. Therefore we have D_0 is injective. Then D_0 is a singular injective module. Hence ${D_0}^{(\mathbf{N})}$ is injective by (b). $(c) \Rightarrow (a)$. First we claim that if $M = \oplus_A E_\alpha$ such that each $E_\alpha \cong E(R/I)$ for some $I \in H(R)$, then M is injective. To show this, we may write $M = \oplus\{E_\alpha : \alpha \in A_t, t \in T\}$ such that if $\alpha_1 \in A_{t_1}$, $\alpha_2 \in A_{t_2}$, then $E_{\alpha_1} \cong E_{\alpha_2}$ iff $t_1 = t_2$. For each $t \in T$, choose one $\alpha_t \in A_t$, and one $I_t \in H(R)$ such that $E(R/I_t) \cong E_{\alpha_t}$. Then we have $M = \oplus\{E_\alpha : \alpha \in A_t, t \in T\} \cong \bigoplus_{t\in T} E_{\alpha_t}^{(A_t)} \cong \bigoplus_{t\in T} E(R/I_t)^{(A_t)} \subseteq^{\oplus} \bigoplus_{t\in T} E(R/I_t)^{(A)} \subseteq^{\oplus} \bigoplus_{I\in H(R)} E(R/I)^{(A)} \cong [\bigoplus_{I\in H(R)} E(R/I)]^{(A)}$ [by Lemma 5]$\cong {D_0}^{(A)}$. Since ${D_0}^{(\mathbf{N})}$ is injective by (c), ${D_0}^{(A)}$ is injective by [1, 25.1, P289]. It follows that M is injective, establishing our claim. Now to show (a), it suffices to show that every singular injective module is a direct sum of modules of form $E(R/I)$ with $I \in H(R)$ by the above. So we suppose M is a singular injective right R-module. Since $xR \cong R/x^{\perp}$ for any $x \in M$, then $E(xR)$ is isomorphic to $E(R/I)$ for some $I \in H(R)$. Set $\mathbf{F} = \{X : X$ is an independent set of nonzero cyclic submodules of $M\}$. We define a partial order $\leq$ on $\mathbf{F}$ by $X_1 \leq X_2$ iff $X_1 \subseteq X_2$. Then $(\mathbf{F}, \leq)$ is an inductive set. By Zorn's Lemma, there exists a maximal element, X say, in $\mathbf{F}$. Then $\sum_{N\in X} E(N) = \bigoplus_{N\in X} E(N)$. By the above, $\bigoplus_{N\in X} E(N)$ is injective. Thus $M = (\bigoplus_{N\in X} E(N)) \oplus Y$ for some Y. But the maximality of X implies that $Y = 0$. Therefore $M = \bigoplus_{N\in X} E(N)$, with each $E(N)$ isomorphic to $E(R/I)$ for some $I \in H(R)$. □ **Definition**. Let H be any family of right ideals of R. For any countable subset $\{I_i : i \in \mathbf{N}\}$ of H, let $H_j = \bigcap_{i\geq j} I_i$. Then $H_1, H_2, \cdots, H_n, \cdots$ forms an ascending sequence. We call any sequence constructed in this manner a type $A(H)$ sequence.

Theorem 7. The following are equivalent for a ring R:

(a) Every direct sum of singular injective right R-modules is injective; (b) For any chain of right ideals of R: $I_1 \subseteq I_2 \subseteq \cdots$ such that $E(I_{i+1}/I_i)$ is singular for each i, there exists an m such that $I_i = I_m$ for each $i \geq m$;

(c) Every sequence of type $A(H(R))$ becomes stable.

Proof. $(a) \Rightarrow (b)$. Suppose there exists a strictly ascending chain of right ideals of R: $I_1 \subset I_2 \subset \cdots \subset I_n \subset \cdots$ such that $E(I_{i+1}/I_i)$ is singular for each i. Let $I = \cup I_i$, and $E = \bigoplus_{i=1}^{\infty} E(I_{i+1}/I_i)$. Then E is injective by (a). Since $E(I_{i+1}/I_i)$ is injective, the inclusion map from I_{i+1}/I_i to $E(I_{i+1}/I_i)$ can be extended to a map $\eta_i : R/I_i \rightarrow E(I_{i+1}/I_i)$. We define a map $f : I \longrightarrow \bigoplus_{i=1}^{\infty} E(I_{i+1}/I_i)$ via $\pi_i(f(a)) = \eta_i(a + I_i)$, where π_i is the projection of E on $E(I_{i+1}/I_i)$. It is easy to see that f is well-defined. Hence there is $x \in E$ such that $f(a) = xa$ for all $a \in I$. Choose k such that $x \in \bigoplus_{i=1}^{k} E(I_{i+1}/I_i)$. Then, if $i > k$ and $a \in I_{i+1}$, we have $a + I_i =$

$\eta_i(a + I_i) = \pi_i(f(a)) = \pi_i(xa) = \pi_i(x)a = 0 \cdot a = 0$, impling $a \in I_i$. Thus $I_{i+1} = I_i$, a contradiction.

$(b) \Rightarrow (c)$. If $H_1, H_2, \cdots, H_n, \cdots$ is a sequence of type $A(H(R))$, then there exist $I_i \in H(R), i = 1, 2, \cdots$, such that $H_j = \bigcap_{i \geq j} I_i$ for all j. For each j, $H_{j+1}/H_j = H_{j+1}/(I_j \cap H_{j+1}) \cong (H_{j+1} + I_j)/I_j \subseteq R/I_j$. Since $I_j \in H(R)$, then $E(R/I_j)$ is singular, and hence $E(H_{j+1}/H_j)$ is singular. This shows that the sequence $\{H_i\}$ satisfies the condition (b) and hence that $H_j = H_m$ for all $j \geq m$, for some fixed m. $(c) \Rightarrow (a)$. By Proposition 6, it suffices to show that $M^{(\mathbb{N})}$ is injective whenever M is a singular injective module. So we suppose M is a singular injective right R-module. Write $M^{(\mathbb{N})} = \bigoplus_{i=1}^{\infty} M_i$ with $M_i \cong M$. Let $f : I \longrightarrow \bigoplus_{i=1}^{\infty} M_i$, where I is a right ideal of R. For each i, let π_i be the projection of $\bigoplus_{i=1}^{\infty} M_i$ onto M_i. Since M_i is injective, there exists a homomorphism $f_i : R \rightarrow M_i$ that extends $\pi_i \circ f$. Now let $q_i = f_i(1) \in M_i$. Then $q_i^{\perp} \in H(R)$ for all i. Let $H_j = \bigcap_{i \geq j} q_i^{\perp}$. Then $H_1, H_2, \cdots, H_n, \cdots$ is a sequence of type $A(H(R))$. So by (c), there exists some m such that $H_j = H_m$ for all $j \geq m$. If $r \in I$, then $\pi_i \circ f(r) = 0$ for all but finite i, i.e. r is contained in some H_k. It follows that $r \in H_m$. Hence $I \subseteq H_m$. Therefore we have $f(I) \subseteq \bigoplus_{i=1}^{m-1} M_i$. Since $\bigoplus_{i=1}^{m-1} M_i$ is injective, we can apply the Baer Injective Lemma to obtain the result. □

We next show that every direct sum of singular injective right R-modules is injective iff R has a.c.c. on $H(R)$. We establish two lemmas first.

Lemma 8. Suppose that $I_1 \subseteq I_2 \subseteq \cdots$ is a chain of right ideals of R such that all $E(I_{i+1}/I_i)$ are singular. Then $E(I/I_i)$ is singular for every i, where $I = \bigcup_i I_i$.

Proof. If $E(I/I_i)$ is not singular for some i, then by Proposition 1, there exists a nonzero submodule X/I_i of I/I_i ($X \subseteq I$) such that X/I_i is embeddable in R. Noting that $I_i \subset X = X \cap I = I_i \cup [\bigcup_{j>i}(X \cap I_j)]$, we have $I_i \subset X \cap I_j$ for some $j > i$. Since $X \cap I_i = I_i$, we may assume that $X \cap I_{j-1} = I_i$. Then $((X \cap I_j) + I_{j-1})/I_{j-1} \cong (X \cap I_j)/(X \cap I_{j-1}) = (X \cap I_j)/I_i \subseteq X/I_i$. This shows that I_j/I_{j-1} has a nonzero submodule which is embeddable in R. Hence $E(I_j/I_{j-1})$ is not singular by Proposition 1, a contradiction. □

Lemma 9. Let $I_1 \subseteq I_2 \subseteq \cdots$ be a chain as in Lemma 8 and $I = \bigcup I_i$. Then there exists $K \in H(R)$ such that $I_1 \subseteq K$ and $(I_{i+1} + K)/(I_i + K) \cong I_{i+1}/I_i$ for all $i > 1$.

Proof. If $I_1 \in H(R)$, then let $K = I_1$ and we are done. Assume $I_1 \notin H(R)$. There exists a submodule X/I_1 of R/I_1 $(X \subseteq R)$ which is maximal with respect to $(I/I_1) \cap (X/I_1) = 0$. Let $K = X$. Then clearly $K \cap I = I_1$. And it follows that $K \cap I_i = I_1$ for all i. Hence $(I_{i+1} + K)/(I_i + K) \cong I_{i+1}/(I_{i+1} \cap (I_i + K)) = I_{i+1}/(I_i + (I_{i+1} \cap K)) = I_{i+1}/(I_i + I_1) = I_{i+1}/I_i$. Note that K/I_1 is a closed submodule of R/I_1. By using [2, Prop.1.4, P18], we have that I/I_1 can embed in R/K as an essential submodule. Hence $E(I/I_1) \cong E(R/K)$. It follows that $K \in H(R)$ by Lemma 8. □ **Theorem 10**. The following are equivalent for a ring R:

(a) Every direct sum of singular injective right R-modules is injective;

(b) R has a.c.c. on $H(R)$.

Proof. $(a) \Rightarrow (b)$. By Theorem 7.

$(b) \Rightarrow (a)$. It suffices to show (b) of Theorem 7. Suppose there exists a strictly ascending chain of right ideals of R: $I_1 \subset I_2 \subset \cdots \subset I_n \subset \cdots$ such that $E(I_{i+1}/I_i)$ is singular for every i. We obtain a contradiction by constructing a strictly ascending chain $K_1 \subset K_2 \subset \cdots$ with every $K_i \in H(R)$. By Lemma 9, there exists $K_1 \in H(R)$ such that $K_1 \cap I_i = I_1$ and $(I_{i+1} + K_1)/I_i \cong I_{i+1}/I_i$ for all i. Then $K_1 \subset I_2 + K_1$ and $I_2 + K_1 \subset I_3 + K_1 \subset \cdots$ is a strictly ascending chain with all $E((I_{i+1} + K_1)/(I_i + K_1))$ singular for $i \geq 2$. Suppose we have constructed $K_1, K_2, \cdots, K_n$ such that all K_i are in $H(R)$, $K_1 \subset K_2 \subset \cdots \subset K_n$, $K_n \subset I_{n+1} + K_n$ and $I_{n+1} + K_n \subset I_{n+2} + K_n \subset \cdots$ is a strictly ascending chain with $E((I_{i+1} + K_n)/(I_i + K_n))$ singular for all $i \geq n+1$. Applying Lemma 9 to the chain $I_{n+1} + K_n \subset I_{n+2} + K_n \subset \cdots$, we have a $K_{n+1} \in H(R)$ such that $I_{n+1} + K_n \subseteq K_{n+1} \subset I_{n+2} + K_{n+1}$ and $K_{n+1} + I_{n+2} \subset K_{n+1} + I_{n+3} \subset \cdots$ is a strictly ascending chain with $E((I_{i+1} + K_{n+1})/(I_i + K_{n+1}))$ singular for all $i \geq n+2$. The induction principle implies that there exists a sequence $\{K_i : i \in \mathbf{N}\}$ with all K_i in $H(R)$ and $K_1 \subset K_2 \subset \cdots K_n \subset \cdots$ is a strictly ascending chain. The theorem is proved. □ **Corollary 11**. If R is a right almost Noetherian ring, then every direct sum of singular injective right R-modules is injective. The converse is true if $Soc(R_R)$ is non-singular.

Proof. Suppose R is a right almost Noetherian ring. Then R satisfies a.c.c. on essential right ideals of R by [3, Coro.2.9], and hence R has a.c.c. on $H(R)$.

For the converse, suppose direct sums of singular injective right R-modules are injective. If $I_1 \subset I_2 \subset \cdots$ is a strictly ascending chain of essential right ideals of R, then for each i, there exist K_i and J_i such that $I_i \subseteq K_i \subset J_i \subseteq I_{i+1}$ and K_i is a maximal submodule of J_i. Since each R/K_i is singular, we have J_i/K_i is a singular simple module. If $Soc(R_R)$ is non-singular, then all $E(J_i/K_i)$ are singular by Corollary 2. Hence $E = \bigoplus_i E(J_i/K_i)$ is injective. Let $f_i : R/K_i \longrightarrow E(J_i/K_i)$ be a homomorphism that extends the inclusion from J_i/K_i to $E(J_i/K_i)$ and let $I = \cup I_i$. We define a map $f : I \longrightarrow E$ by $\pi_i \circ f(a) = f_i(a + K_i)$, where π_i is the projection of E on $E(J_i/K_i)$. Since E is injective, there is a $x \in E$ such that $f(a) = xa$ for all $a \in I$. Then there exists m such that $\pi_i \circ f = 0$ if $i > m$. If $i > m$ and $a \in J_i$, then $0 = \pi_i \circ f(a) = f_i(a + K_i) = a + K_i$, implying $J_i = K_i$. This is a contradiction. □

A module M is said to be c-generated if there is an indexed set $(x_t)_{t \in C}$ that spans

M with $c = card\ C$.

Theorem 12. The following are equivalent for a ring R:

(a) Every direct sum of singular injective right R-modules is injective;

(b) Every singular injective right R-module is a direct sum of uniform singular injective right R-modules;

(c) Every singular injective right R-module has a decomposition that complements direct summands;

(d) There exsits a cardinal number c such that every singular injective right R-module is a direct sum of c-generated right R-modules.

Proof. $(a) \Rightarrow (b)$. We claim that if R satisfies (a), then every singular injective right R-module has a uniform submodule. To see this, let M be such a module. Then if $0 \neq z \in M$, $E(R/z^{\perp}) \cong E(zR)$ is a summand of M, implying $E(R/z^{\perp})$ is singular and so $z^{\perp} \in H(R)$. We note that if R satisfies (a), then R has acc on $H(R)$ by Theorem 10. Therefore there exists a maximal element, $z^{\perp}$ say, in $\{y^{\perp} : 0 \neq y \in M\}$. We show next that zR is a uniform module. Suppose R/H, where $H = z^{\perp}$, is not uniform. Then $E(R/H) = X \oplus Y$, for some nonzero modules X and Y . Write $e = 1 + H = x + y$, $x \in X$ and $y \in Y$. Since R/H is essential in $E(R/H)$, then $x \neq 0$ and $y \neq 0$. Let p_1, p_2 be the projections of $E(R/H)$ onto X and Y, respectively. Then $p_1(R/H) = xR$, $p_2(R/H) = yR$. Since $p_1((R/H) \cap Y) = p_2((R/H) \cap X) = 0$, $ker(p_1|_{R/H}) \neq 0$, and $ker(p_2|_{R/H}) \neq 0$. Then $x^{\perp}$ contains H properly and $y^{\perp}$ contains H properly too. This contradicts the maximality of $z^{\perp} = H$, and our claim holds. Now let M be an arbitrary singular injective right R-module. By the above claim, there exists a uniform submodule of M. By using Zorn's Lemma, there exists a maximal independent set W of uniform submodules of M. Then $\sum_{N \in W} E(N) = \bigoplus_{N \in W} E(N)$ and each $E(N)$ is uniform. By (a), $\bigoplus_{N \in W} E(N)$ is injective. Hence $M = (\bigoplus_{N \in W} E(N)) \oplus P$. If $P \neq 0$, then P has a uniform submodule P_1 by the above claim, and then $W \cup \{P_1\}$ would be an independent set of uniform submodules, contradicting the maximality of W. Hence $P = 0$ and $M = \bigoplus_{N \in W} E(N)$.

$(b) \Rightarrow (c)$. By [1, 25.5, P291].

$(c) \Rightarrow (a)$. Let D be a direct sum of singular injective right R- modules. Since every singular injective module has a decomposition that complements direct summands, then by [1,Ex.12.2, P148] every singular injective module is a direct sum of uniform singular injective modules. Therefore we can write $D = \bigoplus\{E'_{\alpha} : \alpha \in B_t, t \in T\}$ such that all E'_{α} are uniform singular injective, and if $\alpha_1 \in B_{t_1}$, $\alpha_2 \in B_{t_2}$, then $E_{\alpha_1} \cong E'_{\alpha_2}$ if and only if $t_1 = t_2$. For each t, choose one $\alpha_t \in B_t$ and let $F = \{\alpha_t : t \in T\}$

and $D_1 = \bigoplus_{\alpha \in F} E'_\alpha$. Let $E = E(D_1{}^{(\mathbf{N})})$. E is singular, for otherwise, there exists a nonzero cyclic submodule xR of $D_1{}^{(\mathbf{N})}$ which can be embedded in R by Proposition 1. But xR can be embedded in $(\bigoplus_{\alpha \in F_1} E'_\alpha)^{(n)}$ for some finite subset F_1 of F and some n. By Proposition 1, we have $(\bigoplus_{\alpha \in F_1} E'_\alpha)^{(n)}$ is not singular, a contradiction. So E is singular. Write $E = \oplus_A E_\alpha$, a decomposition that complements direct summands. For each $\alpha \in F$, let $A(\alpha) = \{\beta \in A : E_\beta \cong E'_\alpha\}$. Now for each $n > 0$, the injective module $E'^{(n)}_\alpha$ is isomorphic to a direct summand of E. So by [1, 12.2, P142], $Card(A(\alpha)) \geq n$. And hence $A(\alpha)$ is infinite. Let $B = \bigcup_{\alpha \in F} A(\alpha)$. Then $D_1^{(\mathbf{N})}$ is isomorphic to a summand of the direct summand $\oplus_B E_\alpha$ of E. Therefore $D_1^{(\mathbf{N})}$ is injective. Let $\bar{B}$ be the disjoint union of $\{B_t : t \in T\}$. Then by [1, 25.1, P289], $D_1^{(\bar{B})}$ is injective. By using Lemma 5, we have $D \cong \bigoplus\{E'_\alpha : \alpha \in B_t,\ t \in T\} \cong \bigoplus_{t \in T} E'^{(B_t)}_{\alpha_t} \subseteq^{\oplus} \bigoplus_{t \in T} E'^{(\bar{B})}_{\alpha_t} \cong (\bigoplus_{t\in T} E'_{\alpha_t})^{(\bar{B})} \cong D_1^{(\bar{B})}$. Then D is injective. $(a) \Rightarrow (d)$. Since the set $\{E(R/I) : I \in H(R)\}$ contains an isomorphic copy of each uniform singular injective right R-module, then there exists a cardinal number c such that every uniform singular injective right R-module is c-generated. Now the equivalence of (a) and (b) implies that every singular injective module is a direct sum of c-generated modules. $(d) \Rightarrow (a)$. Assume that R satisfies (d). To show (a), it suffices to show that $M^{(\mathbf{N})}$ is injective whenever M is a singular injective module by Proposition 6. But the necessary argument has been given by Faith and Walker (see the proof of [1, 25.8, P293]). □ Next we turn to the case where R is a commutative ring with a.c.c. on $H(R)$. We can present a classification of uniform singular injective modules over such a ring.

Theorem 13. Let R be a commutative ring with a.c.c. on $H(R)$. Then

(a) M is a uniform singular injective R-module iff $M \cong E(R/I)$ for some prime ideal I of R with $I \in H(R)$;

(b) M is a singular injective R-module iff $M = \oplus_I E(R/I)$, where each I is a prime ideal of R with $I \in H(R)$;

(c) $I \longmapsto E(R/I)$ provides a bijection between the set of prime ideals in $H(R)$ of R and the set of isomorphism classes of uniform singular injective R-modules. The converse map is given by $E \longmapsto x^\perp$, where $x^\perp$ is a (or the) maximal element in $\{x^\perp : 0 \neq x \in E\}$.

Proof. (a). If $I \in H(R)$ and I is a prime ideal of R, then $E(R/I)$ is singular and R/I is a uniform R-module. Thus $E(R/I)$ is a uniform singular injective R-module. Conversely, if M is a uniform singular injective module, by the assumption on R, there is a maximal element, $x^\perp$ say, in $\{y^\perp : 0 \neq y \in M\}$. Let $ab \in x^\perp$, $a \notin x^\perp$. Since $x^\perp \subseteq (xa)^\perp$, we have $(xa)^\perp = x^\perp$ by the choice of $x^\perp$. Then $b \in (xa)^\perp = x^\perp$. So $x^\perp$

is a prime ideal of R. And it is easy to see that $x^{\perp} \in H(R)$. Now the fact that M is uniform implies that $M = E(xR) \cong E(R/x^{\perp})$.

(*b*). By (*a*) and Theorem 12.

(*c*). By (*a*), the map $f : I \longmapsto E(R/I)$ is a surjection between the set of prime ideals in $H(R)$ of R and the set of uniform singular injective R-modules. To show that f is an injection, we let $E(R/I_1)elh\cong E(R/I_2)$, where $I_1, I_2 \in H(R)$ and both are prime ideals of R . We take $0 \neq \bar{r} = r + I_2 \in h(R/I_1) \cap (R/I_2)$. Then $I_1 \subseteq \bar{r}^{\perp}$. If $x \in \bar{r}^{\perp}$, then $rx \in I_2$, implying $x \in I_2$ since I_2 is a prime ideal. Hence $\bar{r}^{\perp} \subseteq I_2$. It follows that $I_1 \subseteq I_2$. By symmetry, $I_2 \subseteq I_1$. □

Finally we present examples as promised in the introduction.

Example 14. Let A be a ring, $\{M_\alpha : \alpha \in I\}$ a set of $A - A$ bimodules. Let $R = A \oplus (\oplus_{\alpha \in I} M_\alpha)$. R will become a ring under the following ' + ', and ' $\circ$ ':

$$(a; \cdots, x_\alpha, \cdots) + (b; \cdots, y_\alpha, \cdots) = (a + b; \cdots, x_{h\alpha} + y_\alpha, \cdots)$$
$$(a; \cdots, x_\alpha, \cdots) \circ (b; \cdots, y_\alpha, \cdots) = (ab; \cdots, ay_\alpha + x_\alpha b, \cdots).$$

We take $A = \mathbf{Z}$, and $M_i = \mathbf{Z}/(p_i)^{n_i}$, where p_i is the ith prime number, n_i is a positive number, and $i = 1, 2, \cdots$. It is easy to check that $Soc(R_R) = \oplus_{i=1}^{\infty} p_i^{n_i - 1} M_i$ is an essential right ideal of R, and $Z(R_R) = \oplus_{i=1}^{\infty} M_i$. (1) We choose $n_i = 1$ for every i. Then $Z(R_R) = Soc(R_R) \neq (0)$, and $H(R) = \{Soc(R_R), R\}$. R is clearly a right almost Noetherian ring. (2) We choose $n_i = 2$ for all i. Then $Soc(R_R) = \oplus_{i=1}^{\infty} p_i M_i$. Since $(\oplus M_i)/(\oplus p_i M_i) \cong \oplus(M_i/p_i M_i)$ and each $M_i/p_i M_i \neq 0$, R is not an almost right Noetherian ring. If $Soc(R_R) \subseteq I$, and $M_i \not\subseteq I$, then $(M_i + I)/I \cong$

$p_i M_i$ as right R-modules. By Proposition 1, $I \notin H(R)$. On the other hand, if $I = (n) \oplus (\oplus_{i=1}^{\infty} M_i)$ with $n \neq 0$, then $R/I \cong \mathbf{Z}/(n)$ as right R-modules. But it is easy to see the latter has a nonzero submodule embeddable in $Soc(R_R)$, and hence I is not in $H(R)$ by Proposition 1. Therefore we have $H(R) = \{\oplus M_i, R\}$. By Theorem 10, every direct sum of singular injective right R-modules is injective.

REFERENCES

[1] F.W.Anderson and K.R.Fuller, Rings and Categories of Modules, Springer-Verlag, 1974.

[2] K.R.Goodearl, Ring Theory: Nonsingular Rings and Modules, Marcel Dekker, Inc., 1976.

[3] S.S.Page and M.F.Yousif, Relative injectivity and chain conditions, *Comm. in Algebra* , **17**(4), 899-924, 1989.

[4] B.Stenström, Rings of Quotients, Springer-Verlag, 1975.

[5] M.Teply, Some aspects of Goldie's torsion theory, *Pacific J. Math.* **29**(1969), 447-459.

POLYNOMIALS OVER DIVISION RINGS, AND THEIR APPLICATIONS

LOUIS H. ROWEN
Bar-Ilan University
Ramat-Gan 52900, Israel

ABSTRACT

A survey of factorization of polynomials over division rings, focusing on Wedderburn's method, as extended by Jacobson and Lam-Leroy. Applications are given to the structure theory of finite dimensional division algebras, and brief consideration is given at the end to polynomials in several indeterminates.

Introduction.

As in the familiar commutative situation, polynomials are a key tool in the study of algebraicity in noncommutative rings. The serious study of polynomials over division rings started with Wedderburn's innovative paper [**27**] which proved that every division algebra D of degree 3 (i.e. dimension 9 over its center F) has an element $a \notin F$ such that $a^3 \in F$. In the course of the proof he showed more generally that any polynomial f irreducible over F with a root in D, actually splits into linear factors over D. In case $\deg(f) = 3 = \deg(D)$ he showed actually that

$$f = (\lambda - a^2da^{-2})(\lambda - ada^{-1})(\lambda - d), \tag{1}$$

from whence one concludes $a^3 \in F$.

After Wedderburn's method had been applied successfully also to division algebras of degree 4, cf. [**1**], Jacobson [**9**] succeeded in applying the theory of modules to refine and improve Wedderburn's method. Jacobson also showed how this theory yields straightforward proofs of basic facts such as the Skolem-Noether Theorem and Hilbert's Theorem 90. Despite various flurries of activity, the theory lay largely dormant until the papers of Treuer [**24**],[**25**],[**26**] and Lam-Leroy [**11**],[**12**],[**13**], the latter taking up the gauntlet and carrying the Wedderburn-Jacobson theory to its full generalization for Ore extensions. In the interim there have been a myriad of applications to algebraic elements of division algebras, as we shall describe. Much of the theory can be extended to polynomials over simple Artinian rings, cf. [**20**], but we shall not discuss that here.

So far, very little has been ascertained about polynomials in several indeterminates over a division ring D; the basic tool that $F[\lambda_1, \lambda_2]$ is a unique factorization domain is not available in the noncommutative situation. Some scant information which is known will be discussed in §5.

This short survey describes some of the research concerning polynomials over division rings, and its application to basic questions concerning algebraicity. I shall try to show how this subject includes a variety of applications, by covering the following topics:

1. Wedderburns's method (elements) and its applications (The Skolem-Noether Theorem and Hilbert's Theorem 90)
2. Jacobson's method (modules) and its applications
3. The Lam-Leroy theory
4. The structure of finite dimensional division algebras of low degree
5. The structure of the ring $D[\lambda_1, \ldots, \lambda_t]$

Certain of these topics have already been the subjects of expositions, and in fact one of the goals here is to show how these treatments fit together. The other goal is to penetrate quickly to some striking results. Special attention is paid to Jacobson's method, since it can be developed quite rapidly but I do not know of any short exposition in print.

Throughout, D will be a division ring with center F, and λ is a commuting indeterminate. Polynomials are written as left polynomials, i.e. as $\sum a_i\lambda^i$ $(a_i \in D)$.

§1. Wedderburn's Method.

We say g divides f in $D[\lambda]$ (written $g|f$) if $f = hg$ for suitable h in $D[\lambda]$. Note for any $d \neq 0$ in D that $g|df$ iff $g|f$. If $f = \sum a_i\lambda^i$ write $f(d)$ for $\sum a_i d^i$. By the Euclidean algorithm, for any d in D,

$$f(\lambda) = g(\lambda)(\lambda - d) + f(d) \quad \text{for suitable} \quad g(\lambda) \text{ in } D[\lambda]. \tag{2}$$

In particular, $f(d) = 0$ iff $(\lambda - d)|f$. A conjugate of d is an element of the form ada^{-1} for suitable $a \neq 0$ in D. Wedderburn [**27**] proved the following remarkable theorem.

THEOREM A. *If $f \in F[\lambda]$ is monic irreducible and has a root d_1 in D then*

$$f = (\lambda - d_n)\ldots(\lambda - d_1)$$

where each d_i is conjugate to d_1. (In other words, if f has a root in D then f splits over D.)

Wedderburn's proof can also be found in [**10**, chapter 2], [**19**, pp.178ff], [**20**], and actually contains a computation of the d_i: if $f_i(\lambda) = (\lambda - d_i)\ldots(\lambda - d_1)$ and d is a conjugate of d_1 which is not a root of f_i then d_{i+1} may be taken to be $f_i(d)df_i(d)^{-1}$. (Furthermore, any factorization $f = (\lambda - d_n)\ldots(\lambda - d_1)$ can be obtained in this way, cf. [**8**, §1].) This follows at once from the basic observation, which is a consequence of the definitions:

REMARK 1.1: If $f = gh$ and $h(d) \neq 0$ then $f(d) = g(h(d)dh(d)^{-1})h(d)$. In particular, if d is a root of f but not of h then $h(d)dh(d)^{-1}$ is a root of g.

NOTE: In the proof of theorem A one shows that $f \in D[\lambda] \setminus \{0\}$ has minimal degree with respect to the property that every conjugate of d is a root of f. This turns out to be the key property in generalizations of theorem A.

Furthermore, theorem A has a partial converse, also a consequence of Remark 1.1:

THEOREM B. *If $f(\lambda) = (\lambda - d_n) \dots (\lambda - d_1) \in D[\lambda]$ and $f(d) = 0$ then d is a conjugate of some d_i.*

These results already give us a basic structure theorem for division algebras.

APPLICATION 1.2. (Special case of the Skolem-Noether theorem) *If $K \subset D$ is isomorphic (as F-algebra) to a finite dimensional subfield $F(a)$ of D then K is conjugate to $F(a)$.*

PROOF: Let $\varphi : F(a) \to K$ be the isomorphism and $b = \varphi(a)$. Then b satisfies the minimal polynomial of a, so is conjugate to a, by theorem B. But clearly $K = F(b)$. □

Lam-Leroy [**11**] extend these results to Ore extensions $D[\lambda; \sigma, \delta]$ where σ is an endomorphism of D and δ is a σ-derivation (i.e. $\delta(ab) = \sigma(a)\delta(b) + \delta(a)b$); multiplication is given by $\lambda d = \sigma(d)\lambda + \delta(d)$. The trick is to define $f(d)$ such that (2) still holds; the key to computing $f(d)$ is the following observation:

REMARK 1.3: Suppose $f = \sum a_i\lambda^i$. Then $f(d) = g(d)$ where $g = a_0 + \sum_{i \geq 1} a_i\lambda^{i-1}d$ is rewritten as a left polynomial.

For example if $\delta = 0$ then $\lambda a = \sigma(a)\lambda$, so $g = a_0 + \sum_{i \geq 1} a_i\sigma^{i-1}(d)\lambda^{i-1}$, and by induction $f(d) = a_0 + \sum_{i \geq 1} a_i\sigma^{i-1}(d) \dots \sigma(d)d$. In other words, writing $N_0(d) = 1$ and $N_i(d) = \sigma^{i-1}(d) \dots \sigma(d)d$, we conclude $f(d) = \sum_{i \geq 0} a_i N_i(d)$. Furthermore, $\sigma(a)(\lambda - d)a^{-1} = \lambda - \sigma(a)da^{-1}$, motivating us to consider the notion of σ-conjugate of d, which is $\sigma(a)da^{-1}$ for suitable $a \neq 0$ in D.

In general (for $a \neq 0$) one can define the (σ, δ)-conjugate d^a of d to be $(\sigma(a)d + \delta(a))a^{-1} = g(d)a^{-1}$ where $g = \lambda a$. Then $(d^a)^b = d^{ab}$ for all $a, b \neq 0$ in D; a direct but more elaborate compuation yields

GENERAL REMARK 1.4: [**11**, "Product Theorem 2.7"] If $f = gh$ and $h(d) \neq 0$ then $f(d) = g(d^{h(d)})h(d)$.

Using this in place of remark 1.1, we can copy over the statements and proofs of Theorems A,B to a more general situation, replacing "conjugate" by "(σ, δ)-conjugate". To wit, given d in D, write Δ_d for $\{d^a : 0 \neq a \in D\}$.

THEOREM A'. *Suppose $0 \neq f \in D[\lambda; \sigma, \delta]$ has minimal degree with respect to $f(\Delta_d) = 0$. Then $f = (\lambda - d_n) \dots (\lambda - d_1)$, where each d_i is (σ, δ)-conjugate to d_1.*

THEOREM B'. *If $f(\lambda) = (\lambda - d_n)\dots(\lambda - d_1) \in D[\lambda;\sigma,\delta]$ and $f(d) = 0$ then d is a (σ,δ)-conjugate of some d_i.*

REMARK 1.5: For any d in D, $\{f \in T : f(\Delta_d) = 0\} \triangleleft T$, by remark 1.4.

Application: $\delta = 0$. Let us consider the case where $\delta = 0$; i.e., we work in $T = D[\lambda;\sigma]$. Given d in D, note that $d^d = \sigma(d) \in \Delta_d$, and $\sigma^{-1}(d) = d^{\sigma^{-1}(d^{-1})} \in \Delta_d$. Let $I_d = \{f \in T : f(\Delta_d) = 0\}$. Take $f \in I_d$ of minimal degree; the principal task here is to determine what form such f can take. Write $f = \sum_{i=0}^{t} d_i\lambda^i$.

First we claim that $d_0 \neq 0$. Otherwise $f = h\lambda$ for some g in $D[\lambda]$; remark 1.4 implies

$$h(d^a) = f(\sigma^{-1}(d^a))(d^a)^{-1} = 0$$

for all $a \neq 0$ in D, so $h \in I_d$, contrary to minimality of $\deg f$.

Now dividing through by d_0, we may assume $d_0 = 1$. For any $a \in D$, the polynomial $af - fa \in I_d$ has degree $\leq \deg f$ but has constant term 0. Hence, by the previous paragraph, $af = fa = \sum_{i=0}^{t} d_i\sigma^i(a)\lambda^i$, so $a_i d_i = d_i\sigma^i(a)$ for each $i \leq t-1$; implying conjugation by d_i^{-1} produces the automorphism σ^i, for each i such that $d_i \neq 0$. In particular σ has finite period n, for some $n \geq 1$, and taking b in D such that $\sigma^n(a) = b^{-1}ab$, we can write $f = g(b\lambda^n)$ for some $g \in F[\lambda]$. Furthermore, noting that $\lambda f - f\lambda = \lambda^2 h$ for suitable h, we argue as before to show $h \in I_d$ and thus $h = 0$, from which we conclude that each coefficient of g is fixed by σ, i.e., $g \in F^\sigma[\lambda]$.

The minimality of f in I_d is equivalent to g being irreducible. Thus we have

THEOREM A''. *Suppose $n \geq 1$ is the smallest positive number such that σ^n is inner, and σ^n is given by conjugation by b. Suppose $f = g(b\lambda^n)$ where $g \in F^\sigma[\lambda]$ is irreducible, such that $f(d) = 0$. Then $f = (\lambda - d_n)\dots(\lambda - d_1)$, where each d_i is σ-conjugate to d_1.*

THEOREM 1.6. (Hilbert's Theorem 90, cf. [9, theorem 3.27]) *Assume σ is an automorphism of D of period n, where no smaller power of σ is inner, and write $N(d)$ for $\sigma^{n-1}(d)\dots\sigma(d)d$. Suppose $\gamma = N(d) \in F$. Then $N(d') = \gamma$ iff $d' = \sigma(a)da^{-1}$ for suitable $a \neq 0$ in D.*

PROOF: ($\Leftarrow$) Take $g = \lambda - \gamma$ and $b = 1$ in Theorem A'', i.e. $f = \lambda^n - \gamma$. Then $f(d') = 0$ iff $N(d') = \gamma$. Hence by theorem A'' and theorem B', $f(d') = 0$ iff d' is σ-conjugate to d, which is the assertion here. □

Taking $\gamma = d = 1$ we have

COROLLARY 1.7: $N(d) = 1$ iff $d = \sigma(a)a^{-1}$ for suitable $a \neq 0$.

This application will be generalized in §§3; the discussion before theorem A'' will be handled more consisely and more generally in remark 3.3. Note that the assumption in theorem 1.6 was a trivial instance of the hypothesis in theorem A''; one can get other versions by using different examples of g.

§2. Jacobson's method.

In his pioneering book [**9**], Jacobson reworked Wedderburn's theory in terms of modules, using the structure of $D[\lambda]$ and $D[\lambda,\sigma]$; recently he has written a lucid presentation in [**10**], to which we shall refer repeatedly. Lam-Leroy have pushed this much farther in a recent series of papers [**11**],[**12**],[**13**],[**14**], (the latter with Leung and Matzuk), and we shall also point to their results when they do not involve extra complications. (Their more technical results are discussed in §3.)

Write T for $D[\lambda;\sigma,\delta]$. First recall that T satisfies the Euclidean algorithm; thus T is a principal left ideal domain. We shall assume that σ is an automorphism; then by symmetry T is also a principal right ideal domain. Furthermore any two polynomials $0 \neq p,q \in T$ satisfy $Tp \cap Tq \neq 0$ since T is an Ore domain, and thus p,q have a least common (left) multiple.

Suppose we have arbitrary $f \neq 0$ in T. Then T/Tf is a cyclic T-module which is finite dimensional over D, and thus has a composition series

$$T/Tf \supset T_1/Tf \supset \ldots \supset T_n/Tf = 0$$

i.e. $Tf = T_n \subset T_{n-1} \subset \ldots \subset T_1 \subset T$. Write $T_1 = Tf_1$ for suitable f_1 dividing f. Then $T_2 = Tf_2f_1, \ldots, T_n = Tf_n \ldots f_1$, for suitable f_i such that $f_n \ldots f_1 = f$.

Note $T/Tf_i \approx T_{i-1}/T_i$ under the map $g + Tf_i \mapsto gf_{i-1}\ldots f_1 + Tf_if_{i-1}\ldots f_1$ (which is an isomorphism since T is a domain). Thus each f_i is irreducible. Furthermore, since $T/Tf_1, \ldots, T/Tf_n$ are the composition factors of T/Tf, the Jordan-Holder Theorem instantly gives us the following sort of uniqueness:

We say polynomials g and h are (left) <u>similar</u> if $T/Tg \approx T/Th$ as T-modules.

THEOREM 2.1. ([**9**]) *Any factorization of $f \in D[\lambda]$ into irreducibles is "unique", up to similarity and permutation of the factors.*

This theorem whets our appetite as to when two polynomials are similar, and the following straightforward criterion is useful.

PROPOSITION 2.2. ([**9**, p.33]) *Polynomials g,h are similar iff there are u,v such that v,g are relatively prime and $ug = hv$. Note that one can take $\deg u < \deg h$.*

COROLLARY 2.3. (i) *g,h are (left) similar iff they are right similar (i.e. $T/gT \approx T/hT$ as right T-modules).*

(ii) *If g,h are similar then $\deg g = \deg h$ (and thus $\deg u = \deg v$).*

(iii) *If h is monic and similar to $\lambda - a$ then $h = \lambda - b$ where a,b are conjugate.*

Suppose $f = f_n \ldots f_1$ is factored into irreducibles, and g is similar to f_1. Then we write $ug = f_1v$ for v relatively prime to g. Although g need not necessarily divide f in general, it does in one important case.

Definition 2.4: $f \in T$ is 2-sided if $fT = Tf$.

Remark 2.5: (i) If f is 2-sided then $Tf \triangleleft T$.

(ii) Conversely if $I \triangleleft T$ then I has a unique monic element f of minimal degree, and $I = Tf = fT$ by the Euclidean algorithm. The map $Tf \mapsto f$ defines a multiplicative monoid isomorphism from {ideals of T} to {monic 2-sided elements of T}.

(iii) If $f = hg$ and two of these elements are 2-sided then so is the third. (For example, suppose f, g are 2-sided. Then $Thg = Tf = fT = hgT = hTg$, so $Th = hT$.)

(iv) If a 2-sided element $f = gh$ then g divides f. (Indeed $fh \in Tf = Tgh$ so $f \in Tg$.)

2-sided elements satisfy the following property akin to primeness.

Proposition 2.6. *Suppose $hg \in Tf$ where f is 2-sided, but $h \notin Tf$. Then one can factor $g = g_1 \dots g_k$ into irreducibles, such that some g_i divides f. (In particular, if g is irreducible then g divides f.)*

Proof: (motivated by [**3**, lemma A]) By induction on $\deg g$. Write $gT + fT = pT$. Then p is not a unit, since otherwise $h = hpp^{-1} \in hgT + hfT \subseteq Tf + hTf = Tf$. On the other hand $gT \subset pT$ since otherwise $g|f$ by remark 2.5(iv). Hence $g = pq$ where $\deg p, \deg q < \deg g$; then $(hp)q \in Tf$, so by induction we are done unless $hp \in Tf$, and we conclude by applying induction once more. □

We are ready to describe factorizations of 2-sided elements, to generalize theorems A,B. Let us say f is 2-sided irreducible if one cannot find a factorization of f into 2-sided noninvertible elements. Clearly any 2-sided element has a factorization into 2-sided irreducibles. By remark 2.5(iii), f is 2-sided irreducible iff Tf is a maximal ideal. Thus the factorization of a 2-sided element into 2-sided irreducibles corresponds to the factorization of an ideal as a product of maximal ideals of T. To this end Jacobson proved

Theorem 2.7. ([**9**, theorem 3.9 and its preceding lemmas]) *If $I_1, I_2 \triangleleft T$ then $I_1 I_2 = I_2 I_1$. Equivalently, all 2-sided elements of T commute. Consequently, the factorization of 2-sided elements into irreducible 2-sided elements is unique.*

(The proof is based on the reduction to maximal ideals.) Thus in studying the factorization of a 2-sided element f, we may assume f is 2-sided irreducible.

Theorem 2.8. (Jacobson) *Suppose f is 2-sided irreducible. Then f can be factored into irreducibles $f = f_n \dots f_1$, and in any such factorization $f_1, \dots, f_n$ are similar (and any irreducible dividing f is similar to f_1.) Conversely any irreducible polynomial similar to f_1 divides f.*

Proof: (sketch) The ring $\overline{T} = T/Tf$ is simple and finite dimensional as a D-module, so is simple Artinian. Hence $\overline{T}$ is a finite direct sum of irreducible $\overline{T}$-modules which are unique up to isomorphism, so by the considerations preceding theorem 2.1, f is a product of irreducibles, and they are all similar. The last assertion follows from proposition 2.6. □

NOTE: (cf. [**10**, theorem 1.5]) $T/Tf = M_n(\Delta)$ where n is the number of irreducibles in the factorization $f_n \dots f_1$. Also $\Delta = \mathrm{End}_T(T/Tf_1)^{\mathrm{op}}$, which can be identified with S/Tf where $S = \{s \in T : f_1 s \in Tf_1\}$, the idealizer of Tf_1.

Clearly theorem 2.8 contains theorems A and B and more.

COROLLARY 2.9. *If f is 2-sided irreducible and of prime degree p, then either f is irreducible (in T) or f is a product of linear factors .*

PROOF: If $f = f_n \dots f_1$ then $p = \deg f = n \deg f_1$, so either $n = 1$ or $n = p$. □

Let us examine the ideals of T, which already are seen to be the key to our discussion.

LEMMA 2.10. *Suppose $I = Tf \triangleleft T$ with f monic. Then*
(i) *$fd = \sigma^n(d)f$ for all d in D.*
(ii) *$\lambda f = f \cdot (\lambda + d)$ for suitable d in D.*

PROOF: (i) $fd - \sigma^n(d)f = (\lambda^n + \dots)d - \sigma^n(d)(\lambda^n + \dots) \in I$ of lower degree, so is 0.
(ii) $\lambda f \in fT$ so $\lambda f = f\lambda'$ for some monic polynomial λ' of degree 1 (seen by comparing leading terms). Thus $\lambda' = \lambda + d$ for some d. □

Let C denote the center of T. Of course any element of C is 2-sided. Although the converse is not true in general one does have

PROPOSITION 2.11. (cf. [**10**, theorems 1.2 and 1.3]) *Suppose f is monic 2-sided. If σ is the identity then $f \in C$.*

PROOF: Let $I = Tf \triangleleft T$. $\lambda f - f\lambda \in I$ has lower degree than $\deg f$ so must be 0; likewise $df - fd = 0$. □

Although proposition 2.11 is rather special, one often can reduce to it.

PROPOSITION 2.12. (cf. [**10**, proposition 1.3]) *If σ is inner, i.e. given by $d \mapsto udu^{-1}$ for $0 \neq u \in D$ then $D[\lambda; \sigma, \delta] \approx D[\lambda'; 1, \delta']$ where $\lambda' = u^{-1}\lambda$ and $\delta' = u^{-1}\delta$.*

PROOF: $\lambda' d = u^{-1}\lambda d = u^{-1}(\sigma(d)\lambda + \delta(d)) = d\lambda' + \delta'(d)$. □

COROLLARY 2.13. *If $[D : F] < \infty$ and σ fixes F then any ideal of $D[\lambda; \sigma, \delta]$ is generated by an element of C.*

PROOF: σ is inner by the Skolem-Noether theorem, so we can write $D[\lambda; \sigma, \delta]$ as $D[\lambda'; 1, \delta']$ and apply proposition 2.11. □

DIGRESSION: One gets analogous results when instead δ is σ-inner, i.e. of the form $d \mapsto ad - \sigma(d)a$ for suitable a in D. Then $D[\lambda; \sigma, \delta] = D[\lambda'; \sigma]$ where $\lambda' = \lambda - a$ and $\sigma(\lambda') = \lambda - \sigma(a) = \lambda' - (\sigma(a) - a)$, thereby reducing to the case $\delta = 0$.

If $\delta = 0$ then any 2-sided element $f = d\lambda^k g$ for suitable k, suitable d in D, and suitable g in C. (Indeed one takes d to be the lowest-order coefficient of f; then g is 2-sided and $g(0) = 1$, implying $g \in C$. Compare with the discussion in §1)

Actually there is a general result, due to Cauchon: Any 2-sided element f must have the form $dh^k g$ where $g \in C$, and $h \in T$ is a nonconstant monic polynomial of least degree such that $hT \subseteq Th$. The finer points of this fact and its applications to the subject at hand are discussed in [**12**].

This trick of exploiting a "nicer" form of the Ore extension is studied in depth by Lam-Leroy [**12**],[**13**], under the name "cv-polynomials." ("cv" is short for "change of variable").

Let F^σ denote $\{\alpha \in F : \sigma(\alpha) = \alpha\}$. It follows easily by matching coefficients in lemma 2.10(i) that if C has some non-constant element then taking monic $f \in C$ of minimal degree $n \geq 1$, one has $\sigma^n = 1$. Conversely, for $\delta = 0$, if σ^n is inner, say $\sigma^n : d \mapsto udu^{-1}$, then clearly C contains $F^\sigma[\lambda']$, where F^σ denotes $\{\alpha \in F : \sigma(\alpha) = a\}$ and $\lambda' = u^{-1}\lambda^n$, and in this case $[T : C] \leq n^2$ (since $[F : F^\sigma] \leq n$ by Galois theory). When $\delta \neq 0$ one has to contend with the possibility that $C \subset D$, but then of course T is simple (by proposition 2.11(i)). For more detailed results see [**10**, theorem 1.2] and [**13**], in which this turns out to be one of the key observations.

Applications to finite dimensional division algebras.

We are ready to start the machine rolling. Assume $\delta = 0$. Suppose D is a finite dimensional division algebra over F, and let $T = D[\lambda; \sigma]$ and $C = \text{Cent}(T)$. Suppose $f \in T$ is irreducible. Then $L = Tf$ is a maximal left ideal of T.

Now let us also assume σ^n is inner on D, for suitable n. As just noted, $[T : C] \leq n^2$ so T is a ring with polynomial identity, implying by Kaplansky's theorem that L contains a maximal 2-sided ideal of T, which in turn contains a monic polynomial $\tilde{f}$ of minimal degree. By corollary 2.13, we see $\tilde{f} \in C$. As a special case, if $f = \lambda - a$ (which obviously is irreducible) then $\tilde{f}$ is the minimal polynomial of a over C. In particular, we have proved such a polynomial exists, but this is just the beginning. An easy application of theorem 2.8 and the subsequent note yields

THEOREM 2.14. ([**10**, corollary 1.1]) *$D[\lambda]/D[\lambda]f \approx M_n(\Delta)$, where $n = \frac{\deg \tilde{f}}{\deg f}$. In case $f = \lambda - d$ then $n = \deg \tilde{f}$, and Δ is the centralizer of d.*

Here is a nice application.

THEOREM 2.15. [**10**, theorem 1.7] (Wedderburn's criterion) *Suppose the smallest number p such that σ^p is an inner automorphism of D is prime, and $\sigma^p : x \to uxu^{-1}$. Then $T/T(\lambda^p - u)$ is a division ring iff u is* <u>not</u> *a norm, i.e. $u \neq N_p(d)$ for all d in D.*

PROOF: $f = \lambda^p - u$ is 2-sided irreducible, by direct calculation, so T/Tf is simple, and is a division ring iff f is irreducible in T. But this is false iff f has a linear factor $\lambda - d$; i.e., iff $0 = f(d) = N_p(d) - u$. $\square$

Appendix to §2: The roots of a polynomial.

Much of our attention has been focused on using the minimal polynomial of an algebraic element d to study D in terms of d. Conversely, one might ask how one can study the roots of an arbitrary polynomial f in $T = D[\lambda; \sigma, \delta]$ of degree n. Clearly theorems A and B imply that all the roots of f lie in at most n (σ, δ)-conjugacy classes; indeed, if d is a root then writing $f = g(\lambda)(\lambda - d)$ one knows any root of f not (σ, δ)-conjugate to d is a root of g, so one concludes by induction. Even when the roots lie in a single conjugacy class, one might wonder how $d_1, \dots, d_n$ of theorem A compare, i.e. if they are distinct, and if perhaps f is the least common multiple of $\lambda - d_1, \dots, \lambda - d_n$. Such a factorization exists when f is separable and $\deg D$ is a prime power; for arbitrary f of degree > 2 one can at least find a nonrepeating factorization. Haile-Rowen [8, proposition 3.2 and theorem 3.8] obtains these facts and explores related questions.

On the other hand, the natural question arises as to how one can adjoin roots of an irreducible polynomial f. If f is 2-sided then as in the commutative situation one could take $D_1 = T/Tf$, in which $\lambda + Tf$ is a root of f in D_1; however, if Tf is not an ideal the situation becomes much more complex. A good treatment is found in Treuer [**26**], who uses coproducts.

§3. The Lam-Leroy Theory.

Many of the results described in §2 rely on fortuitous circumstances such as the existence of nonconstant polynomials in $C = \mathrm{Cent}(D[\lambda; \sigma, \delta])$, since otherwise one cannot even speak of algebraic elements in §2. In particular one often needs to assume $\delta = 0$. One of the achievements of Lam-Leroy was to free themselves of this restriction in [**11**], [**12**], in the context of an arbitrary automorphism σ and a σ-derivation δ (on D). They thereby succeded also in unifying this theory with another theory of Jacobson and Amitsur which dealt only with derivations.

DEFINITION 3.1: A set $\Delta \subseteq D$ is algebraic if there exists $0 \neq f \in T = D[\lambda; \sigma, \delta]$ such that $f(\Delta) = 0$; the degree of Δ is the minimal possible degree of such f.

By this definition any singleton $\{d\}$ is an algebraic set, satisfying the linear polynomial $\lambda - d$. Lam-Leroy appeal to a basic idea of algebraic geometry.

DEFINITION 3.2: An algebraic set Δ is full if the following property holds:

(P1) If $f(d) = 0$ for all $f \in T$ such that $f(\Delta) = 0$ then $d \in \Delta$.

Note by theorems A and B that property P1 implies a is (σ, δ)-conjugate to some element of Δ. (Recall $d^a = (\sigma(a)d + \delta(a))a^{-1}$.)

Conversely we have

REMARK 3.3: [**12**, lemma 5.2] Suppose Δ is closed under (σ, δ)-conjugation. Then $I = \{g \in T : g(\Delta) = 0\}$ is an ideal of T, as in remark 1.5, and thus contains a monic 2-sided element, by remark 2.5.

In this way we see that the results of §2 are applicable immediately. We can push §2 even further by reexamining the T-module $\overline{T}_d = T/T(\lambda - d)$, given d in D. Given h in T write $\overline{h}$ for $h + T(\lambda - d)$; then $\overline{h} = \overline{h(d)}$ by (2), and clearly $h(d)$ is the unique representative of $\overline{h}$ in D. Identifying $\overline{T}_d$ with D via $\overline{a} \leftrightarrow a$ for a in D, we can describe the T-module action as

$$g\overline{h} = g(d^{h(d)})h(d)$$

(which is taken to be 0 if $h(d) = 0$). Note by remark 1.4 that $g\overline{h} = f(d)$, where $f = gh$. For example, letting $a = h(d)$,

$$(\lambda - b)a = \lambda a - ba = \sigma(a)\lambda + \delta(a) - ba = \sigma(a)d + \delta(a) - ba = (d^a - b)a.$$

Then $\operatorname{Ann} \overline{T}_d = \{g : g(d^a) = 0\}$ for all $a \neq 0$ in D; we conclude $\operatorname{Ann} \overline{T}_d \neq 0$ iff Δ_d is algebraic, where Δ_d denotes the set of (σ, δ)-conjugates of d. This yields

PROPOSITION 3.4. *If T is not primitive then every (σ, δ)-conjugacy class in D is algebraic.*

These considerations lead us to define the centralizer $C_{\sigma,\delta}(d) = \{0\} \cup \{b \in D : d^b = d\}$, which can be checked to be a division subring of D; then one has the following generalization of theorem 2.8:

THEOREM 3.5. *If f is the minimal polynomial of Δ_d then $T/Tf \approx \operatorname{End}_{C_{\sigma,\delta}(d)} \overline{T}_d \approx M_n(C_{\sigma,\delta}(d))$, where $n = \deg f$.*

$\overline{T}_d$ is also naturally a *right* $C_{\sigma,\delta}(d)$-module, since

$$(\lambda - d)a = (\lambda - d^a)a = \lambda a - \delta(a) - (d^a a - \delta(a)) = \lambda a - \delta(a) - \sigma(a)d = \sigma(a)(\lambda - d)$$

for $a \neq 0$ in $C_{\sigma,\delta}(d)$. Another description of the action is $\overline{(\sum d_i \lambda^i)}a = \sum d_i \sigma^i(a) N_i(d)$ for a in $C_{\sigma,\delta}(d)$.

Now for arbitrary g in T we define the transformation L_g on $\overline{T}_d$ to be left multiplication by g. Clearly L_g is a right $C_{\sigma,\delta}(d)$-module map; the dimenson of its kernel is at most $\deg g$, as seen via the Jacobson-Chevalley Density Theorem. Taking this all together yields

THEOREM 3.6. *(cf. [**12**, theorem 5.10]) Δ_d is algebraic iff D is finite dimensional as a right $C_{\sigma,\delta}(d)$-module, in which case this dimension equals the degree of Δ_d.*

This topic is developed in considerable detail in [**12**],[**13**] with somewhat different notation, under the name of "λ-transform," cf. [**12**, corollary 5.23ff]. We turn now to a sublime generalization of Hilbert's theorem 90. Rather than describe the proofs of these results here, let me refer the reader to the very well-written paper [**13**], and merely quote its main results, from [**13**, §5].

Suppose $\Delta = \Delta_d$ is algebraic, satisfying the 2-sided polynomial $p_0(\lambda)$ in $T = D[\lambda; \sigma, \delta]$. Let $p \in T$ be a monic nonconstant polynomial of minimal degree such that $pT = Tp$. Such a polynomial exists, because $p_0T = Tp_0$; furthermore σ has finite period. Let $n = \deg p$, and take k minimal such that $(\sigma^n)^k$ is inner, say $(\sigma^n)^k : d \mapsto udu^{-1}$. Write $N(a)$ for $N_k(a)$, taken with respect to the automorphism σ^n and derivation 0.

THEOREM 3.7. ([**13**, theorem 5.4]) *Notation as above, the following are equivalent for any d' in D :*

(1) $d' \in \Delta$;
(2) $p(d')$ *is* σ^n*-conjugate to* $p(d)$;
(3) $u^{-1}N(p(d'))$ *is conjugate (in the usual meaning) to* $u^{-1}N(p(d))$;
(4) $u^{-1}N(p(d'))$ *is algebraic over* F, *with the same minimal polynomial as* $u^{-1}N(p(d))$.

For example, if $\delta = 0$ then $p(\lambda) = \lambda$ and $n = 1$, so $N(d) = \sigma^{k-1}(d)\dots\sigma(d)d$; we see from the theorem that d' is σ-conjugate to d iff $u^{-1}N(d') = v^{-1}u^{-1}N(d)v$ for suitable v. Thus we have theorem 1.6 as a special case (when $u = 1$). Consequently, for $d' = 1$, one has the result that d has the form $\sigma(a)a^{-1}$ iff $N(d)$ is a multiplicative commutator (i.e., $uvu^{-1}v^{-1}$).

[**13**, example 6.4] describes the situation when $\sigma = 1$, rederiving a theorem of Jacobson concerning algebraic derivations. In this way, one sees that [**13**, theorem 5.4] unifies a variety of versions of Hilbert's theorem 90. Lam-Leroy [**11**] also use this theory to interpret the Dieudonne determinant of Vandermonde and Wronskian matrices.

§4. The structure of finite dimensional division algebras of low degree.

In this short section we shall describe briefly some of the applications of factorizations of polynomials, spread over 70 years. Since many of the proofs depend on computing the correct element, we shall often provide a short sketch of the proof. Throughout, D is a finite dimensional division algebra over its center F. Recall the *degree* of D is defined as $\sqrt{[D:F]}$, and is the integer n such that $D \otimes \overline{F} \approx M_n(\overline{F})$, where $\overline{F}$ is the algebraic closure of F. Furthermore viewing $D \subseteq M_n(\overline{F})$ in this manner enables us to define the reduced norm $N_{red}(d)$ (resp. reduced trace $T_{red}(d)$) of an element d to be its determinant (resp. trace), viewing d as a matrix.

Degree 3. When $d \in D \setminus F$ satisfies a polynomial $f(\lambda)$ of degree 3 over F, one can write f as in (1), cf. [**27**]. Thus $a^3d = da^3$, yielding the following consequences:

Fact 4.1 (Wedderburn): D is cyclic. (Although there are other proofs of this fact, this approach gives a generalization to arbitrary division algebras containing an algebraic element of degree 3, cf. [**20**, proposition 5.2].)

Fact 4.2: If $N_{red}(d) = 1$ then d is a multiplicative commutator. To see this, note that (1) implies $1 = a^2da^{-1}da^{-1}d$ so $a^{-3} = (a^{-1}d)^3$; hence by the Skolem-Noether theorem $a^{-1} = b^{-1}(a^{-1}d)b$ for suitable b in D, so $d = aba^{-1}b^{-1}$.

Fact 4.3 [**2**]: If $T_{red}(d) = 0$ then, by direct computation, $3d$ is the additive commutator $ab - ba$ where $b = da^{-1} - a^{-1}d$. (One can continue this argument to show that d is itself an additive commutator.)

Degree 4. Whenever $d \in D$ satisfies an irreducible monic polynomial $f(\lambda) = \lambda^4 + \sum_{i=0}^{3} \alpha_i\lambda^i$ of degree 4 over F then f splits into linear factors $f = (\lambda - d_4)(\lambda - d_3)(\lambda - d_2)(\lambda - d_1)$ over D; pairing these off into quadratic factors permits us to write

$$f = (\lambda^2 + a\lambda + b)(\lambda^2 + a'\lambda + b'). \tag{3}$$

Suppose $T_{red}(d) = 0$, i.e. $\alpha_3 = 0$. Then matching coefficients of λ^3, λ^0, and λ^2 respectively yields the following:

$$a' = -a, \quad b' \in Fb^{-1}, \text{ and } a^2 \text{ commutes with } b.$$

Thus (3) is a factorization of f over $C_D(a^2)$, and one can conclude a^2 has degree 1 or 2. This easily yields

FACT 4.4: ([**1**], also cf. [**19**, theorem 3.28]) If $\deg D = 4$ then D has a maximal subfield Galois over F with Galois group $\mathbf{Z}_2 \times \mathbf{Z}_2$.

Suppose moreover that $\alpha_1 = \alpha_3 = 0$. Then f can be written in the form $g(\lambda^2)$, which enables us to assume $f = (\lambda + d')(\lambda - d')(\lambda + d)(\lambda - d)$ for suitable d' conjugate to d; since f is central one also has $f = (\lambda - d)(\lambda + d')(\lambda - d')(\lambda + d)$, leading to the following result.

FACT 4.5: (Rowen-Saltman [**21**]) For $\alpha_1 = \alpha_3 = 0$, the following are equivalent: (i) α_0 (the reduced norm of d) is a square in F; (ii) $F(d)/F$ is Galois with group $\mathbf{Z}_2 \times \mathbf{Z}_2$. (In fact one can choose d' in $F(d)$ such that $F(d) \approx F(dd') \otimes F(d + d')$.)

An immediate consequence is the theorem of Knus-Parimala-Sridharan, that if D (of degree 4) has a transpose-type involution (*) of first kind and if the reduced norm of an antisymmetric element is a square, then D is isomorphic to the tensor product of two (*)-invariant quaternion subalgebras.

When D has an involution and has degree 8, then one can show using the general structure theory that D contains an element of degree 4, and using (3) one can then show that D has a maximal subfield Galois over the center with Galois group $\mathbb{Z}_2 \times \mathbb{Z}_2 \times \mathbb{Z}_2$; Tignol then used this fact to prove $M_2(D)$ is a tensor product of quaternion algebras. Details are given in [**10**] or [**19**].

Another application is an unpublished result of Amitsur about splitting fields. Suppose $\deg D = n$. Since $D[\lambda]$ is a PI-domain, we could view $D[\lambda]$ as an order in $D(\lambda) \approx D \otimes_F F(\lambda)$, also a division algebra of degree n over $F(\lambda)$. If $f \in D[\lambda]$ is irreducible over D, then $\mathrm{N}_{\mathrm{red}} f$ is a polynomial in $F[\lambda]$ which is a multiple of f and thus of $\tilde{f}$, notation as in §2. On the other hand, $\tilde{f}$ is a product of polynomials similar to f, so we see $\tilde{f} = \mathrm{N}_{\mathrm{red}} f$ iff $\deg \tilde{f} = n \deg f$, iff $D[\lambda]/\langle \tilde{f} \rangle$ is a direct sum of n irreducible modules. But $D[\lambda]/\langle \tilde{f} \rangle \approx D \otimes_F F[\lambda]/F[\lambda]\tilde{f}$, so the latter condition holds iff $K = F[\lambda]/F[\lambda]\tilde{f}$ is a splitting field for D; reversing this argument yields

THEOREM 4.6. (Amitsur, unpublished) *Suppose h is an irreducible polynomial over F. Then $F[\lambda]/F[\lambda]h$ is a splitting field for D iff h is the reduced norm of an irreducible polynomial over D.*

Finally, let us recall that Albert found another neat application of Wedderburn's method, to prove that every ordered finite dimensional division algebra is commutative. (Indeed, otherwise take a positive element d of reduced trace 0. By theorem A, we would have 0 a sum of conjugates of d, which all must be positive - contradiction!)

§5. The structure of the ring $D[\lambda_1, \ldots, \lambda_t]$.

Inspired by the module-theoretic applications in §2, several authors have delved into the algebraic structure of $D[\lambda_1, \ldots, \lambda_t]$, where $\lambda_1, \ldots, \lambda_t$ are commuting indeterminates over D.

Unfortunately, many features of $D[\lambda]$ do not carry over to $D[\lambda_1, \ldots \lambda_t]$, and in fact $D[\lambda_1, \ldots, \lambda_t]$ fails to have many of the nice properties of $F[\lambda_1, \ldots, \lambda_t]$ when $t > 1$. (For example, the polynomial ring in 2 indeterminates over the algebra of quaternions has nonfree projective modules.) Although some deep positive results have been obtained (cf. [**22**],[**23**] for example), our concern here is the ring-theoretic structure of $D[\lambda, \ldots, \lambda_t]$, in particular its left ideals L (so that we could study the cyclic modules $D[\lambda_1, \ldots, \lambda_t]/L$ in analogy with §2.) Unfortunately, not much is known yet, but there are certain facts which should be mentioned. The starting point is the result of Amitsur-Small:

[3, LEMMA C]. *$L \cap D[\lambda_i] \neq 0$ for each maximal left ideal L of $D[\lambda_1, \ldots, \lambda_t]$ and each i.*

It follows readily that maximal left ideals of $D[\lambda_1, \ldots, \lambda_t]$ have finite codimension over D; in fact this was Amitsur-Small's goal. Here are some other consequences, the first of which generalizes some of the considerations in §2:

[3, THEOREM 2]. *$D[\lambda_1, \ldots \lambda_t]$ is primitive iff for some n, $M_n(D)$ contains a subfield of* *transcendence degree t over F. Consequently, $D_n[\lambda_1, \ldots, \lambda_t]$ is primitive iff $n \geq t$, where D_n denotes the n-th Weyl division algebra.*

[18, PROPOSITION 1.8]. *Any maximal left ideal L of $D[\lambda_1, \ldots \lambda_t]$ is generated by $\leq t$ elements; there are examples when fewer than t elements suffice.*

In case $[D : F] < \infty$, as noted earlier, any maximal left ideal L of $R = D[\lambda_1, \ldots, \lambda_t]$ intersects $F[\lambda_1, \ldots, \lambda_t]$ at a maximal ideal I, from which it follows (by passing to R/RI) that $L \cap D[\lambda_1]$ is the intersection of at most n maximal left ideals of $D[\lambda_1]$. However, $L \cap D[\lambda_1]$ need not be maximal, and in fact Amitsur-Small [4] have shown there is always an example of L such that $L \cap D[\lambda_1]$ is not maximal, for each finite dimensional noncommutative D except Hamilton's algebra of quaternions over the reals. Nonetheless, $L \cap D[\lambda_1]$ is generated by an element which is the lcm of similar irreducibles, cf. §2.

REFERENCES

1. Albert, A.A., *A determination of all normal division algebras in 16 units*, Trans. Amer. Math. Soc. **32** (1929), 253-260.
2. Amitsur S.A. and Rowen, L., *Elements of reduced trace 0*, Israel J. Math. (submitted).
3. Amitsur S.A. and Small, L., *Polynomials over division rings*, Israel J. Math. **31** (1978), 353-358.
4. ——————, *in preparation.*
5. Cauchon, G., *Idéaux bilatères et centre des anneaux de polynômes de Ore sur les anneaux quasi-simples*, in "Springer Lecture Notes," **740**, 1979, pp. 397-407.
6. Cohn, P.M., "Skew Field Constructions," LMS Lecture Notes, **27**, 1977.
7. Gordon, B. and Motzkin, T., *On the zeros of polymomials over division rings*, Trans. Amer. Math. Soc. **116** (1965), 218-226; correction **122** (1966), p. 547.
8. Haile D. and Rowen L.H., *Factorizations of polynomials*, preprint.
9. Jacobson, N., "The Theory of Rings," Amer. Math. Soc. Surveys II, 1943.
10. Jacobson, N. and Saltman, D., forthcoming book on finite dimension division algebras.
11. Lam, T.Y. and Leroy, A., *Vandermonde and Wronskian matrices over division rings*, J. Algebra **118** (1988), 308-336.
12. ——————, *Algebraic conjugacy classes and skew polynomial rings*, in "Perspectives in Ring Theory," F. van Oystaeyen and L. De Bruyn eds., Kluwer, 1988, pp. 153-203.
13. ——————, *Hilbert 90 theorems over division rings*, Trans. Amer. Math. Soc (to appear).
14. Lam, T.Y., Leung, K.H., Leroy, A. and Matzuk, J., *Invariant and semi-invariant polynomials in skew polynomial rings*, in "Ring Theory 1989 in honor of S.A. Amitsur," L.H. Rowen ed., Israel Mathematical Conference Proceedings **1**, 1989, pp. 247-261.
15. Lawrence, J. and Simmons, G., *Equations in division rings: A survey*, Amer. Math. Monthly **96** (1989), 220-237.
16. Leung, K.G., *On a generalization of Albert's Theorem*, Israel J. Math..
17. Ore, Oyestein, *Theory of noncommutative polynomials*, Ann. of Math. **34** (1933), 480-508.
18. Resco,R., *A reduction theorem for the primitivity of the tensor product*, Math. Z. **170** (1980), 65-76.
19. Rowen, L.H., "Polynomial Identities in Ring Theory," Pure and Applied Math. **83**, Academic Press, Boston, 1980.
20. ——————, *Wedderburn's method and algebraic elements of simple Artinian rings*, in "Azumaya Algebras, Actions and Modules in honor of G. Azumaya," ed. D. Haile, Contemporary Math. **124**, 1992, pp. 179-202.
21. Rowen, L.H. and Saltman, D., *The discriminant of an involution*, preprint.

22. Stafford, T., *Projective modules over polynomial extensions of division rings*, Invent. Math. **59** (1980), 105-117.
23. Suslin, A.A., *Structure of projective modules over rings of polynomials in the case of a noncommutative ring of coefficients*, Trudy Mat. Inst. Steklov **148** (1978), 233-252 (Russian).
24. Treuer, J., *On duality for field extensions*, J. Algebra **119** (1988), 1-22.
25. ———, *Separate zeros and Galois extensions of skew fields*, J. Algebra **120** (1989), 392-405.
26. ———, *Noncommutative splitting fields*, J. Algebra **129** (1990), 367-379.
27. Wedderburn, J.H.M., *On division algebras*, Trans. Amer. Math. Soc. **22** (1921), 129-135.

MODULES FOR WHICH EVERY SUBMODULE HAS A UNIQUE CLOSURE

PATRICK F. SMITH
Department of Mathematics, University of Glasgow
Glasgow G12 8QW, Scotland, UK

ABSTRACT

A module M is called UC-module if every submodule has a unique closure (i.e., maximal essential extension) in M. The paper establishes a large number of conditions equivalent to this property.

1. Introduction

Throughout, R will denote a general associative ring with identity and M a unital right R-module. A submodule K of M is closed (in M) provided K has no proper essential extension in M. It is well known that, for any submodule N of M, there exists a closed submodule K of M such that N is essential in K, and we call K a closure of N (in M). In this note we survey and investigate modules with the property that every submodule has a unique closure (in the module). Such modules we shall call UC-modules ("UC" for "unique closures").

A submodule K of a module M will be called R-closed provided $K \cap mR$ is not essential in mR for all $m \in M \backslash K$. (In [10], Renault calls R-closed submodules simply "closed" and, like Goldie [4], calls our closed submodules "complements", but Goodearl [5] uses the terminology adopted here.) Following [4] and [10], we define the following equivalence relation ρ on the lattice of submodules of M; given submodules K, L of M we define $K \rho L$ whenever the submodule $K \cap L$ is essential in both K and L. It is easy to see that $K \rho L$ if and only if the set of submodule H of M such that $K \cap H = 0$ is the same as the set of submodules H of M such that $L \cap H = 0$. For any element m in M we denote the annihilator in R of m by $r(m)$, i.e.,

$$r(m) = \{r \in R : mr = 0\} .$$

We are now ready to give necessary and sufficient conditions, some of which are new and some well known, for a module to be a UC-module.

Theorem. Let R be any ring. Then the following statements are equivalent for a right R-module M.

(1) M is a UC-module.

(2) Given submodules $K \subseteq K'$, $L \subseteq L'$ of M such that K is essential in K' and L is essential in L', then $K + L$ is essential in $K' + L'$.

(3) Given submodules K, K', L, L' of M such that $K\rho K'$ and $L\rho L'$, then $(K + L)\rho(K' + L')$.

(4) Given submodules K, L of M with $K \cap L$ essential in K, then L is essential in $K + L$.

(5) Every essential submodule N of M has the property that $(N \cap K) + (N \cap L)$ is essential in $K + L$, for all submodules K, L of M.

(6) There exists an essential submodule N of M such that N is a UC-module and $(N \cap K) + (N \cap L)$ is essential in $K + L$ for all submodules K, L of M.

(7) If K is a closed submodule of M, then $K \cap N$ is a closed submodule of N, for any submodule N of M.

(8) The intersection of any two closed submodules of M is also closed.

(9) The intersection of any collection of closed submodules of M is closed.

(10) If $K_\lambda \subseteq L_\lambda (\lambda \in \Lambda)$ are submodules of M such that K_λ is closed in L_λ for all λ in Λ, then $\cap_\Lambda K_\lambda$ is closed in $\cap_\Lambda L_\lambda$.

(11) Every (2-generator) submodule of M is a UC-module.

(12) M/K is a UC-module for every closed submodule K of M.

(13) Given any submodule N of M and homomorphism $\varphi : N \to M$ such that $N \cap \varphi(N) = 0$, then $ker\ \varphi$ is a closed submodule of N.

(14) Given any submodule N of M and homomorphism $\varphi : N \to M$ such that $N \cap \varphi(N) = 0$, then $\varphi^{-1}(K)$ is a closed submodule of N for every closed submodule K of $\varphi(N)$.

(15) Given elements m, m' in M with $mR \cap m'R = 0$, $r(m) \subseteq r(m')$ and $r(m')/r(m)$ essential in $R/r(m)$, then $m' = 0$.

(16) There does not exist a right R-module X and a proper essential submodule Y of X such that the module $X \oplus (X/Y)$ embeds in M.

(17) Given submodules $K \subseteq K'$, $L \subseteq L'$ of M such that $K' \cap L' = 0$, $K'/K \cong L'/L$ and K is essential in K', then L is essential in L'.

(18) Given submodules $K \subseteq K'$, $L \subseteq L'$ of M such that $K' \cap L' = 0$, $K'/K \cong L'/L$ and K is closed in K', then L is closed in L'.

(19) Every closed submodule of M is R-closed.

(20) For every submodule N of M, $N^* = \{m \in M : N \cap mR \text{ is essential in } mR\}$ is a submodule of M.

In this case, for each submodule N of M, $N^* = \{m \in M : N \text{ is essential in } N + mR\}$ and N^* is the unique closure of N.

Among examples of UC-modules, we would mention uniform modules and semisimple modules. In addition, Johnson [8, Theorem 6.4] proved that nonsingular modules are UC-modules (see also [10, p. 45, Corollaire 2] and [4, Theorem 2.2]). Renault [10, Chapitre 3]) discusses UC-modules at some length and proves the equivalence of (1), (3), (8), (9), (15) and (19) of the theorem. (For example, see

[10, Theoremes 3.3, 3.4 and 3.5, Proposition 3.4]). In addition, his paper gives a lot of information about (what we call) closed submodules. Camillo and Zelmanowitz [2, Proposition 1, Corollary 2] prove the equivalence of (13) and (16) in the theorem. (Note that in [2], UC-modules are called *dimension modules,* for a reason that will become apparent–see Proposition 14 below; but see also Example 15.) At this point we could also mention the work of Garcia [3], Hausen [7] and Wilson [12] on modules for which the intersection of any two direct summands is a direct summand, and of Osofsky [9] because, as she points out, UC-modules are related to modules satisfying her condition (A4).

Before proceeding to the proof of the theorem, we wish to make some remarks concerning it. First consider statement (2). It is clear that a module M satisfies (2) if and only if the following statement holds:

(2') Given any index set Λ and submodules $L_\lambda \subseteq N_\lambda$ of M such that L_λ is essential in N_λ for all λ in Λ, then $\sum_\Lambda L_\lambda$ is essential in $\sum_\Lambda N_\lambda$.

Moreover, it is always the case that if K and L are essential submodules of K' and L', respectively, then $K \cap L$ is an essential submodule of $K' \cap L'$. In addition, the corresponding fact for (3) is also true, namely given submodules K, K', L, L' of M, then $K\rho K'$ and $L\rho L'$ together imply $(K \cap L)\rho(K' \cap L')$.

A module M satisfies (3) if and only if the following statement holds:

(3') Given submodules K_λ, $L_\lambda\,(\lambda \in \Lambda)$ of M such that $K_\lambda \rho\, L_\lambda\,(\lambda \in \Lambda)$, then $(\sum_\Lambda K_\lambda)\rho(\sum_\Lambda L_\lambda)$.

For, if (3) holds then so too does (2) (an easy check), and hence so also does (2'). Now this gives that $\sum_\Lambda (K_\lambda \cap L_\lambda)$ is essential in both $\sum_\Lambda K_\lambda$ and $\sum_\Lambda L_\lambda$.

Note further that the converse of (4) is always true, i.e., if K, L are submodules of M such that L is essential in $K + L$, then $K \cap L$ is essential in K. It is clear that a module M satisfies (5) if and only if the following statement holds:

(5') Every essential submodule N of M has the property that $\sum_\Lambda (N \cap L_\lambda)$ is essential in $\sum_\Lambda L_\lambda$, for every non-empty collection of submodules $L_\lambda\,(\lambda \in \Lambda)$ of M.

Recall that a module M is called *distributive* provided $N \cap (K + L) = (N \cap K) + (N \cap L)$, for all submodules N, K, L of M. It is clear from (5) that distributive modules are UC-modules (see [2, Proposition 6]). Even for a UC-module M, (5) need not be true for arbitrary submodules N. For example, consider the $\mathbf{Z}$-module $M = \mathbf{Z} \oplus \mathbf{Z}$. Let $N = \mathbf{Z}(1,0)$, $K = \mathbf{Z}(1,1)$, $L = \mathbf{Z}(0,1)$. Then $N \cap K = N \cap L = 0$, but $K + L = M$.

Now consider (6). It is not the case in general that every essential extension of a UC-module is a UC-module. For example, let p be any prime and let M denote the $\mathbf{Z}$-module $(\mathbf{Z}/\mathbf{Z}p) \oplus (\mathbf{Z}/\mathbf{Z}p)$. Then M is semisimple, and hence M is a UC-module. By the theorem, $M' = (\mathbf{Z}/\mathbf{Z}p^2) \oplus (\mathbf{Z}/\mathbf{Z}p)$ is not a UC-module, nor is any module containing M' as a submodule. In particular, the injective hull of a UC-module is not a UC-module, in general.

It is clear that (8) can be restated thus: the intersection of any finite collection of closed submodules of M is closed. Goldie [4, p. 204] asked if this is the case for modules of finite uniform (Goldie) dimension and he remarks that if it is, then it would have certain nice consequences. Our theorem can be taken as a vindication of Goldie's comment!

Now consider (11). We have just remarked that the $\mathbf{Z}$-module $M' = (\mathbf{Z}/\mathbf{Z}p^2) \oplus (\mathbf{Z}/\mathbf{Z}p)$ is not a UC-module, for any prime p, but every cyclic submodule of M' is uniform and hence UC. It should be clear that not every homomorphic image of a UC-module is a UC-module (see (12)). For example, the $\mathbf{Z}$-module M' above is a homomorphic image of the UC-module $\mathbf{Z} \oplus \mathbf{Z}$.

2. Proof of the Theorem

For convenience, we shall give a complete proof in a series of lemmas. Throughout, R is a ring with identity and M a unital right R-module. The following notation will be used:

$N \leq M$: N is an R-submodule of M

$N \leq_e M$: N is an essential submodule of M (i.e., $N \cap L \neq 0$ for all $0 \neq L \leq M$)

$K \leq_c M$: K is a closed submodule of M

$K \leq_r M$: K is an R-closed submodule of M.

Lemma 1. The following statements are equivalent for a submodule K of a module M.

(i) $K \leq_c M$.

(ii) K is maximal in $\{L \leq M : L \cap N = 0\}$, for some $N \leq M$.

(iii) $K \leq L \leq_e M \Rightarrow L/K \leq_e M/K$.

In this case $K \oplus N \leq_e M$.

Proof. See, for example, [5, Propositions 1.3 and 1.4].

Corollary 2.

(i) $K \leq_c H \leq_c M \Rightarrow K \leq_c M$.

(ii) $K \leq L \leq_c M \Rightarrow L/K \leq_c M/K$.

(iii) $K \leq L$, $K \leq_c M$, $L/K \leq_c M/K \Rightarrow L \leq_c M$.

Proof. (i) See, for example, [5, Proposition 1.5].

(ii), (iii) Easy checks.

Lemma 3. The following statements are equivalent for a module M.

(1) M is a UC-module.

(2) $K \leq_e K' \leq M$, $L \leq_e L' \leq M \Rightarrow K + L \leq_e K' + L'$.

(3) K, L, K', $L' \leq M$, $K\rho K'$ and $L\rho L' \Rightarrow (K + L)\rho(K' + L')$.

(4) K, $L \leq M$, $K \cap L \leq_e K \Rightarrow L \leq_e K + L$.

Proof. (1) $\Rightarrow$ (2). Suppose that $K \leq_e K'$, $L \leq_e L'$. There exists a unique $H \leq_c M$ such that $K + L \leq_e H$. There exists $J \leq H$ such that $K \leq_e J \leq_c H$. By Corollary 2(i), $J \leq_c M$. Because $K \leq_e K'$ it follows that $K' \leq J \leq H$. Similarly $L' \leq H$. Thus $K + L \leq_e K' + L'$.

(2) $\Rightarrow$ (3). Suppose that $K\rho K'$, $L\rho L'$. Then $K \cap K' \leq_e K$ and $L \cap L' \leq_e L$. Thus $N = (K \cap K') + (L \cap L') \leq_e K + L$. Similarly, $N \leq_e K' + L'$. Thus $(K + L)\rho(K' + L')$.

(3) $\Rightarrow$ (4). Suppose that $K \cap L \leq_e K$. Then $(K \cap L)\rho\ K$. Thus $[(K \cap L) + L]\rho(K + L)$. But this means that $L\ \rho(K + L)$ and hence $L \leq_e K + L$.

(4) $\Rightarrow$ (1). Let $N \leq M$. Suppose that $N \leq_e K \leq_c M$ and $N \leq_e K' \leq_c M$. Then $K \cap K' \leq_e K$. By (4), $K' \leq_e K + K'$, and hence $K' = K + K'$. Similarly $K = K + K'$. Thus $K = K'$.

Corollary 4. A module M is a UC-module if and only if every (2-generator) submodule of M is a UC-module.

Proof. It is clear that every submodule of a UC-module is a UC-module, by Lemma 3 (1) $\Longleftrightarrow$ (2). Conversely, suppose that every 2-generator submodule of M is a UC-module. Let $K \leq_e K' \leq M$, $L \leq_e L' \leq M$. Let $0 \neq x \in K' + L'$. Then $x = y + z$ for some $y \in K'$, $z \in L'$. Let $N = yR + zR$. Note that $K \cap N \leq_e K' \cap N$ and $L \cap N \leq_e L' \cap N$. By hypothesis and Lemma 3,

$$(K \cap N) + (L \cap N) \leq_e (K' \cap N) + (L' \cap N) .$$

In particular, because $x = y + z \in (K' \cap N) + (L' \cap N)$ there exists $r \in R$ such that $0 \neq xr \in (K \cap N) + (L \cap N) \subseteq K + L$. It follows that $K + L \leq_e K' + L'$. By Lemma 3, M is a UC-module.

Lemma 5. The following statements are equivalent for a module M.

(1) M is a UC-module.

(5) $N \leq_e M \Rightarrow (N \cap K) + (N \cap L) \leq_e K + L$, for all $K, L \leq M$.

(6) There exists $N \leq_e M$ such that N is a UC-module and $(N \cap K) + (N \cap L) \leq_e K + L$, for all $K, L \leq M$.

Proof. (1) $\Rightarrow$ (6). Clear

(6) $\Rightarrow$ (5). Let $N \leq_e M$ and let $K, L \leq M$. By hypothesis, there exists $H \leq_e M$ such that H is a UC-module and

$$(H \cap K) + (H \cap L) \leq_e K + L .$$

Now $N \cap H \leq_e H$ gives

$$[(N \cap H) \cap (K \cap H)] + [(N \cap H) \cap (L \cap H)] \leq_e (K \cap H) + (L \cap H) ,$$

by Lemma 3, (1) $\Rightarrow$ (2). Thus $(N \cap K) + (N \cap L) \leq_e K + L$.

(5) $\Rightarrow$ (1). Suppose that $K, L \leq M$ and $K \cap L \leq_e K$. By Zorn's Lemma, there exists $P \leq M$ such that P is maximal with respect to the property $P \cap (K \cap L) = 0$, and in this case $Q = P \oplus (K \cap L) \leq_e M$ (Lemma 1). By hypothesis,

$(Q \cap K) + (Q \cap L) \leq_e K + L$. But $P \cap K = 0$ gives $Q \cap K = K \cap L \subseteq L$, and hence $Lleq_e K + L$. By Lemma 3, (4) $\Rightarrow$ (1), M is a UC-module.

Lemma 6. The following statements are equivalent for a module M.

(1) M is a UC-module.
(7) $K \leq_c M$, $N \leq M \Rightarrow K \cap N \leq_c N$.
(8) $K \leq_c M$, $L \leq_c M \Rightarrow K \cap L \leq_c M$.
(12) M/K is a UC-module for any $K \leq_c M$.

Proof. (1) $\Rightarrow$ (7). Let $K \leq_c M$, $N \leq M$. There exists a unique $L \leq_c M$ such that $K \cap N \leq_e L$. There exists $H \leq K$ such that $K \cap N \leq_e H \leq_c K$. By Corollary 2(i), $H \leq_c M$ and hence $H = L$. Now suppose that $K \cap N \leq_e P \leq N$. Then $P \leq L = H \leq K$. Thus $P = K \cap N$. It follows that $K \cap N \leq_c N$.

(7) $\Rightarrow$ (8). Immediate by Corollary 2(i).

(8) $\Rightarrow$ (12). Let $K \leq L \leq M$, $K \leq N \leq M$ such that $L/K \leq_c M/K$ and $N/K \leq_c M/K$. By Corollary 2(iii), $L \leq_c M$ and $N \leq_c M$. By (8), $L \cap N \leq_c M$ and hence, by Corollary 2(ii), $(L/K) \cap (N/K) = (L \cap N)/K \leq_c M/K$. It follows that M/K is a UC-module.

(12) $\Rightarrow$ (1). Clear.

Recall that we call a submodule K of a module M R-closed if $K \cap mR$ is not essential in mR (written $K \cap mR \not\leq_e mR$) for all m in $M \backslash K$, and in this case we write $K \leq_r M$. The next lemma, which should be compared with Corollary 2, lists some properties of R-closed modules. Parts (i) and (v) are taken essentially from [10, Theorem 3.3].

Lemma 7.

(i) $K \leq_r M \Rightarrow K \leq_c M$.
(ii) $K \leq_r H \leq_r M \Rightarrow K \leq_r M$.
(iii) $K \leq L \leq_r M \Rightarrow L/K \leq_r M/K$.
(iv) $K \leq L$, $K \leq_r M$, $L/K \leq_r M/K \Rightarrow L \leq_r M$.
(v) $K_\lambda \leq_r M (\lambda \in \Lambda) \Rightarrow \cap_\Lambda K_\lambda \leq_r M$.
(vi) $N \leq M$, $K \leq_r M \Rightarrow K \cap N \leq_r M$.

Proof.

(i) Suppose $K <_r M$. Suppose further that $K \leq_e L \leq M$. Let $x \in L$. Then $K \cap xR \leq_e xR$, so that $x \in K$. Thus $L = K$. It follows that $K \leq_c M$.
(ii) Suppose that $K \leq_r H \leq_r M$. Let $m \in M \backslash K$. If $m \notin H$ then $H \cap mR \not\leq_e mR$ and hence $K \cap mR \not\leq_e mR$. If $m \in H$ then $K \cap mR \not\leq_e mR$. It follows that $K \leq_r M$.
(iii) Suppose that $L/K \not\leq_r M/K$. Then there exists $m \in M$ such that $m + K \notin L/K$ and $(L/K) \cap [(m + K)R] \leq_e (m + K)R$. Thus $m \notin L$ and $L \cap mR \leq_e mR$. It follows that $L \not\leq_r M$.
(iv) Let $m \in M \backslash L$. Then $m + K \notin L/K$, and hence

$$(L/K) \cap (m+K)R \not\leq_e (m+K)R \ .$$

Thus there exists $r \in R$ such that $mr \notin K$ but $L \cap mrR \subseteq K$. Now $mr \notin K$ implies that $K \cap mrR \not\leq_e mrR$. But $L \cap mrR = K \cap mrR$. Thus $L \cap mrR \not\leq_e mrR$, and hence $L \cap mR \not\leq_e mR$.

(v) Let $K = \cap_\Lambda K_\lambda$. Let $m \in M \backslash K$. Then $m \notin K_\lambda$ for some $\lambda \in \Lambda$. By assumption, $K_\lambda \cap mR \not\leq_e mR$ and hence $K \cap mR \not\leq_e mR$. Thus $K \leq_r M$.

(vi) Let $x \in N \backslash (N \cap K)$. Then $x \notin K$, and thus

$$(K \cap N) \cap xR = K \cap xR \not\leq_e xR.$$

It follows that $K \cap N \leq_r N$.

The converse of Lemma 7(i) is false in general, as the next result shows.

Lemma 8. The following statements are equivalent for a module M.

(1) M is a UC-module.

(19) $K \leq_c M \Rightarrow K \leq_r M$.

(9) $K_\lambda \leq_c M (\lambda \in \Lambda) \Rightarrow \cap_\Lambda K_\lambda \leq_c M$.

Proof. (1) $\Rightarrow$ (19). Suppose that M is a UC-module. Suppose that $K \leq_c M$. Let $m \in M \backslash K$. Then $K \not\leq_e (K + mR)$. By Lemma 3 (1) $\Rightarrow$ (4), $K \cap mR \not\leq_e mR$. Hence $K \leq_r M$.

(19) $\Rightarrow$ (9). By Lemma 7(i), (v).

(9) $\Rightarrow$ (1). By (8) $\Rightarrow$ (1) of Lemma 6.

Corollary 9. A module M is a UC-module if and only if the following condition holds:

(10) $K_\lambda \leq_c L_\lambda \leq M (\lambda \in \Lambda) \Rightarrow \cap_\Lambda K_\lambda \leq_c \cap_\Lambda L_\lambda$.

Proof. Suppose that M is a UC-module. Suppose further that $K_\lambda \leq_c L_\lambda \leq M$ $(\lambda \in \Lambda)$. Let $L = \cap_\Lambda L_\lambda$. Note that L and L_λ $(\lambda \in \Lambda)$ are all UC-modules, by Corollary 4. By (1) $\Rightarrow$ (7) of Lemma 6, $K_\lambda \cap L \leq_c L$ $(\lambda \in \Lambda)$. Hence

$$\cap_\Lambda K_\lambda = \cap_\Lambda (K_\lambda \cap L) \leq_c L \ ,$$

by Lemma 8. The converse is clear by Lemma 6 (or Lemma 8).

Lemma 10. The following statements are equivalent for a module M.

(1) M is a UC-module.

(13) For each $N \leq M$, $ker\ \varphi \leq_c N$, for each $\varphi \in Hom_R(N, M)$ with $N \cap \varphi(N) = 0$.

(15) For all m, $m' \in M$ such that $mR \cap m'R = 0$, $r(m) \subseteq r(m')$ and $r(m')/r(m) \leq_e R/r(m)$, then $m' = 0$.

(16) There does not exist an R-module X and a proper essential submodule Y of X such that the module $X \oplus (X/Y)$ embeds in M.

Proof. (1) $\Rightarrow$ (13). Let $N \leq M$ and $\varphi \in Hom_R(N, M)$ such that $N \cap \varphi(N) = 0$. There exists $L \leq M$ such that $ker\ \varphi \leq_e L \leq_c N$. Using Corollary 2(i), we see that $L \leq_c P = N \oplus \varphi(N)$. Note that P is a UC-module by Corollary 4. Define $\theta : N \to P$ by

$$\theta(x) = x + \varphi(x)(x \in N) \ .$$

Clearly θ is a monomorphism. Thus

$$ker\ \varphi = \theta(ker\ \varphi) \leq_e L \oplus \varphi(L) \leq P \ .$$

Now P being a UC-module gives $L \oplus \varphi(L) \subseteq L$, and hence $\varphi(L) = 0$. Thus $ker\ \varphi = L \leq_c N$.

(13) $\Rightarrow$ (15). Suppose that $m, m' \in M$ satisfy $mR \cap m'R = 0$, $r(m) \subseteq r(m')$ and $r(m')/r(m) \leq_e R/r(m)$. Define $\varphi : mR \to m'R$ by $\varphi(mr) = m'r (r \in R)$. Then φ is a homomorphism and ker $\varphi = mr(m')$. Let $0 \neq x \in mR$. Then $x = mr$ for some $r \notin r(m)$. There exists $s \in R$ such that $rs \notin r(m)$ but $rs \in r(m')$. Thus $0 = m'rs = \varphi(mrs)$, and $0 \neq mrs \in xR \cap ker\ \varphi$. Thus ker $\varphi \leq_e mR$. By (13), it follows that $ker\ \varphi = mR$ and hence $m' = \varphi(m) = 0$.

(15) $\Rightarrow$ (16). Let X be an R-module and $Y \leq_e X$ such that there exists a monomorphism $\alpha : X \oplus (X/Y) \to M$. Suppose that $Y \neq X$. Let $x \in X/Y$. Let $m = \alpha(x)$, $m' = \alpha(x + Y)$. Then it can easily be checked that $m, m' \in M$, $mR \cap m'R = 0$, $r(m) \subseteq r(m')$ and $r(m')/r(m) \leq_e R/r(m)$. By (15), $\alpha(x + Y) = m' = 0$, which is a contradiction. Thus $Y = X$.

(16) $\Rightarrow$ (1). Suppose that M is not a UC-module. There exist $K \leq_e K_1 \leq_c M$, $K \leq_e K_2 \leq_c M$ and $K_1 \neq K_2$. Thus $K_1 \cap K_2 \leq_e K_1$, $K_1 \cap K_2 \leq_e K_2$, but $K_1 \cap K_2 \not\leq_e (K_1 + K_2)$. Let $0 \neq L \leq K_1 + K_2$ such that $K_1 \cap K_2 \cap L = 0$. Clearly $K_1 \cap L = 0$ and $K_2 \cap L = 0$. Now

$$L \cong L/(K_1 \cap L) \cong (L + K_1)/K_1 \leq (K_2 + K_1)/K_1 \cong K_2(K_1 \cap K_2) \ ,$$

and hence $L \cong H/(K_1 \cap K_2)$, for some $K_1 \cap K_2 \leq_e H \leq K_2$. Now $H \cap L = 0$ and $H \oplus L \leq M$. Thus (16) implies (1).

Lemma 11. The following statements are equivalent for a module M.

(1) M is a UC-module.

(17) Given $K \leq_e K' \leq M$, $L \leq L' \leq M$ such that $K' \cap L' = 0$ and $K'/K \cong L'/L$, then $L \leq_e L'$.

(18) Given $K \leq_c K' \leq M$, $L \leq L' \leq M$ such that $K' \cap L' = 0$ and $K'/K \cong L'/L$, then $L \leq_c L'$.

(14) For each $N \leq M$, $\varphi^{-1}(L) \leq_c N$, for each $\varphi \in Hom_R(N, M)$ with $N \cap \varphi(N) = 0$ and each $L \leq_c \varphi(N)$.

Proof. (1) $\Rightarrow$ (17). Suppose that $\alpha : K'/K \to L'/L$ is an isomorphism. Let $y \in L' \backslash L$. Then $y + L = \alpha(x + K)$ for some $x \in K'$. Let $E = \{r \in R : xr \in K\}$. Then E is a right ideal of R. Suppose that $yE = 0$. Then we can define a homomorphism $\theta : (x + K)R \to yR$ by $\theta(xr + K) = yr (r \in R)$. Thus $yR \cong xR/K''$, for some $K \cap xR \leq K'' \leq xR$. Thus $K'' \leq_e xR$ and $xR \oplus yR \leq M$, contradicting Lemma 10 (1) $\Rightarrow$ (16). Thus $yE \neq 0$. Suppose that $e \in E$ satisfies $ye \neq 0$. Then $ye + L = \alpha(xe + K) = 0$, so that $0 \neq ye \in L$. It follows that $L \leq_e L'$.

(17) $\Rightarrow$ (18). Let $\beta : L'/L \to K'/K$ be an isomorphism. Let $L \leq_e N \leq L'$. Then $\beta(N/L) = H/K$, for some $K \leq H \leq K'$. Thus $N/L \cong H/K$ and $N \cap H = 0$. By (17), $K \leq_e H$. But $K \leq_c K'$. Thus $K = H$, and hence $L = N$. It follows that $L \leq_c L'$.

(18) $\Rightarrow$ (14). Given $\varphi : N \to M$ with $N \cap \varphi(N) = 0$, let $L \leq_c \varphi(N)$. Note that φ induces an isomorphism $N/\varphi^{-1}(L) \cong \varphi(N)/L$. By (18), $\varphi^{-1}(L) \leq_c N$.

(14) $\Rightarrow$ (1). By (13) $\Rightarrow$ (1) of Lemma 10.

Lemma 12. A module M is a UC-module if and only if for each submodule N of M, $N^* = \{m \in M : N \cap mR \leq_e mR\}$ is a submodule of M. In this case, $N^* = \{m \in M : N \leq_e (N + mR)\}$ and N^* is the unique closure of N.

Proof. Suppose first that M is a UC-module. Let $N \leq M$. By (1) $\Rightarrow$ (4) of Lemma 3, $m \in N^*$ if and only if $N \leq_e N + mR$. Let $m_1, m_2 \in N^*$. Then $N \leq_e N + m_1 R$ and $N \leq_e N + m_2 R$. By Lemma 3 (1) $\Rightarrow$ (2), $N \leq_e N + m_1 R + m_2 R$. It follows that $N \leq_e N + (m_1 + m_2)R$. Hence $m_1 + m_2 \in N^*$. It now easily follows that $N^* \leq M$.

Conversely, suppose that $N^* \leq M$, for any $N \leq M$. Clearly $N \leq_e N^*$. Suppose that $N \leq_e K \leq_c M$. For any k in K, $N \cap kR \leq_e kR$, and hence $k \in N^*$. Thus $K \leq_e N^*$ and hence $K = N^*$. It follows that N^* is the unique closure of N.

The proof of the theorem is now complete.

3. A Special Case

For certain classes of modules, other equivalent conditions can be given for a module to be a UC-module. We shall mention only one of these. The uniform dimension of a module M will be denoted by $u(M)$. Recall that $u(M) = \infty$ or n, for some non-negative integer n, where n is the maximum number of non-zero summands in a direct sum of submodules of M. We shall say that a module M has *finite uniform dimension locally* if every finitely generated submodule of M has finite uniform dimension. For example, if R is a right Noetherian ring, or, more generally, if R is a ring with right Krull dimension, then every right R-module has finite uniform dimension locally (see [6, Lemma 1.1 and Proposition 1.4]).

Lemma 13. Let $K \leq M$.

(i) $K \leq_c M \Rightarrow u(M) = u(K) + u(M/K)$.

(ii) $u(M) < \infty$, $u(M) = u(K) + u(M/K) \Rightarrow K \leq_c M$.

Proof.

(i) By Lemma 1 there exists $N \leq M$ such that $N \cap K = 0$, $N \oplus K \leq_e M$ and $(N \oplus K)/K \leq_e M/K$. Note that $N \cong (N \oplus K)/K$. Thus

$$u(M) = u(N \oplus K) = u(N) + u(K) = u(M/K) + u(K) .$$

(ii) There exists $K' \leq M$ such that $K \leq_e K' \leq_c M$. Then $K'/K \leq_e M/K$ (Corollary 2(ii)). Now (i) gives

$$u(K) + u(M/K') = u(K') + u(M/K') = u(M) = u(K) + u(M/K)$$

$$= u(K) + u(M/K') + u(K'/K) .$$

Note that $u(K)$ and $u(M/K')$ are both finite. Thus $u(K'/K) = 0$, and hence $K = K' \leq_c M$.

Proposition 14. Let M be a module which has finite uniform dimension locally. Then the following statements are equivalent.

(i) M is a UC-module.

(ii) $u(K + L) + u(K \cap L) = u(K) + u(L)$, for all submodules K, L of M.

(iii) $u(K + L) + u(K \cap L) = u(K) + u(L)$, for all finite dimensional submodules K, L of M.

Proof. (*i*) $\Rightarrow$ (*ii*). Let K, $L \leq M$. If $u(K) = \infty$ or $u(L) = \infty$, then both sides of the equation in (ii) are infinite, and so the equation holds. Suppose now that $u(K) < \infty$ and $u(L) < \infty$. We know that $K \leq_e K' \leq_c M$ and $L \leq_e L' \leq_c M$. By Lemma 6, $K' \cap L' \leq_c M$, and hence $K' \cap L' \leq_c K'$. Now $L' \leq_c K' + L'$ gives, by Lemma 13(i),

$$u(K' + L') - u(L') = u((K' + L')/L') = u(K'/(K' \cap L'))$$

$$= u(K') - u(K' \cap L') ,$$

and, in particular, $u(K' + L') < \infty$. But $u(K) = u(K')$, $u(L) = u(L')$, $u(K \cap L) = u(K' \cap L')$, and, by Lemma 3, $u(K + L) = u(K' + L')$. Thus $u(K + L) + u(K \cap L) = u(K) + u(L)$.

(*ii*) $\Rightarrow$ (*iii*). Clear.

(*iii*) $\Rightarrow$ (*i*). Suppose that M satisfies (iii). Let M' be any 2-generator submodule of M. By assumption, M' has finite uniform dimension. Let $A \leq_c M'$, $B \leq_c M'$. Then, because A and B are finite dimensional, $A \leq_c A + B$ gives

$$u(B/(A\cap B)) = u((A+B)/A) = u(A+B) - u(A) = u(B) - u(A\cap B)\ .$$

By Lemma 13(ii), $A\cap B \leq_c B$, and hence, by Corollary 2(i), $A\cap B \leq_c M'$. By Lemma 6, M' is a UC-module. Finally, by Corollary 4, M is a UC-module.

Proposition 14 is based on [2, Proposition 1 and Corollary 2] (see also [1] and [11]). The condition that M have finite uniform dimension locally in Proposition 14 is necessary because of the following example.

Example 15. Let R be a domain which is not right Ore and let E be a maximal right ideal of R. Let $M = R\oplus(R/E)$. Then the R-module M satisfies:

$$u(K+L) + u(K\cap L) = u(K) + u(L)$$

for all submodules K, L of M, but M is not a UC-module.

Proof. An example of such a domain is the free algebra on two variables over a field F (see, for example, [5, p. 101, ex. 5]). For any non-zero right ideal A of R, R embeds in A, and hence $u(A) = \infty$ (see [5, Theorem 3.30]). It is clear that E is an essential right ideal of R and that if K is a non-zero submodule of M then $K = 0\oplus(R/E)$ or $K\cap(R\oplus 0)\neq 0$. Thus $K = 0\oplus(R/E)$ or $u(K) = \infty$. Now it is immediate that for any submodules K and L of M, $u(K+L)+u(K\cap L) = u(K)+u(L)$. By Lemma 10 (1) $\Rightarrow$ (16), M is not a UC-module.

REFERENCES

1. V.P. Camillo and J.M. Zelmanowitz, *On the dimension of a sum of modules*, Comm. in Algebra **6** (1978), 345-352.
2. V.P. Camillo and J.M. Zelmanowitz, *Dimension modules*, Pacific J. Math. **91** (1980), 249-261.
3. J.L. Garcia, *Properties of direct summands of modules*, Comm. in Algebra **17** (1989), 73-92.
4. A.W. Goldie, *Semi-prime rings with maximum condition*, Proc. London Math. Soc. **(3)10** (1960), 201-220.
5. K.R. Goodearl, *Ring theory: Nonsingular rings and modules* (Dekker, New York, 1976).
6. R. Gordon and J.C. Robson, *Krull dimension*, Mem. Amer. Math. Soc. **133** (1978).
7. J. Hausen, *Modules with the summand intersection property*, Comm. in Algebra **17** (1989), 135-148.
8. R.E. Johnson, *Structure theory of faithful rings II*, Trans. Amer. Math. Soc. **84** (1957), 523-544.

9. B.L. Osofsky, *Non-quasi-continuous quotients of finitely generated quasi-continuous modules*, preprint.
10. G. Renault, *Etude des sous-modules complements dans un module*, Bull. Soc. Math. France Mem. **9** (1967).
11. A del Valle, *Goldie dimension of a sum of modules*, University of Glasgow preprint 92/11.
12. G.V. Wilson, *Modules with the summand intersection property*, Comm. in Algebra **14** (1986), 21-38.

TORSIONFREE B^*-MODULES

MARK L. TEPLY and B. DAVID REDMAN, JR.
Mathematical Sciences Department
University of Wisconsin-Milwaukee
Milwaukee, WI 53201

ABSTRACT

Let R be a ring and let τ be an hereditary torsion theory of left R-modules. As in previous papers, a module M is called a B^*-module if $Ext_R(M, X) = 0$ for every τ-divisible X and every X with τ-bounded order. When the filter for τ has a cofinal subset of finitely generated left ideals, we characterize when every τ-torsionfree module is a B^*-module. This characterization shows that R must be obtained via an essential product of explicit matrix rings and that τ is determined from the Goldie torsion theory for these matrix rings.

1. Preliminaries

In this paper, all rings have an identity element and all modules are unitary. Additionally, τ will always denote a nontrivial (hereditary) torsion theory of left modules with associated filter $\mathcal{L}_\tau$ of left ideals. For a left module M, we let $\tau(M)$ denote the τ-torsion submodule of M. A module M is τ-injective if $Ext_R(T, M) = 0$ for all τ-torsion T. We let $E(M)$ denote the injective hull of a module M; then $E_\tau(M) = \{e \in E(M) \mid Ie \subseteq M \text{ for some } I \in \mathcal{L}_\tau\}$ is τ-injective. A torsion theory τ is called stable if $E(M)$ is τ-torsion whenever M is τ-torsion. Naturally associated with τ is a localization functor $Q_\tau(-)$ given by $Q_\tau(M) = E_\tau(M/\tau(M))$. We note that $Q_\tau = Q_\tau(R)$ is a ring and a right R-module in a natural way. As usual, τ is called perfect if $Q_\tau(M) = Q_\tau \bigotimes_R M$ for each module M. One of the most studied torsion theories is the Goldie torsion theory τ_g, whose torsionfree class is precisely the nonsingular left modules. If R is a left nonsingular ring, then the τ_g-torsion modules are the singular left R-modules. If $R \to R'$ is an onto ring homomorphism, then every R'-module is naturally an R-module; so we say that a torsion theory τ of R-modules induces τ_g on R' if the following condition holds: any R'-module M is τ-torsion as an R-module if and only if M is τ_g-torsion as an R'-module. A submodule N of a module M is called τ-closed (sometimes called τ-pure elsewhere) if M/N is τ-torsionfree; in case $\tau = \tau_g$ we simply say that N is closed in M. For these definitions and more information on torsion theory, see [9].

We will also use the concept of an essential product of rings, which is fully explained in [11] and [12]. We use $pd_R M$ to denote the projective dimension of M and $lgldimR$ for the left global dimension of a ring R; see [2] for more information on homological algebra concepts.

A τ-torsion module T is said to have τ-bounded order if T is contained in a module possessing a set of generators annihilated by some $I \in \mathcal{L}_\tau$. Modules with τ-bounded order appear many places in the literature; for example, see [1], [6], [7], [10], [14], [15] and their references. A torsion theory τ has the Splitting Property, denoted by SP (respectively, the Bounded Splitting Property, denoted by BSP), if any short exact sequence

$$0 \to T \to X \to F \to 0$$

splits whenever F is τ-torsionfree and T is τ-torsion (with τ-bounded order). As in [15], we define a module D to be τ-divisible if D is a homomorphic image of a direct sum of τ-injective modules. A module B is a B^*-module if $Ext_R(B, X) = 0$ for each τ-divisible X and each X with τ-bounded order. Motivated by the study of Baer modules (e.g., see [3]) over a commutative integral domain, Fuchs and Viljoen characterized the B^*-modules over a valuation ring in [4]. The study of B^*-modules for torsion theories over general rings was then begun in [15].

In this paper, we consider the problem of characterizing the torsion theories for which every τ-torsionfree module is a B^*-module. For a torsion theory with a filter $\mathcal{L}_\tau$ that has a cofinal subset of finitely generated left ideals, we solve the problem of determining when every τ-torsionfree module is a B^*-module. Our solution shows that τ must be a specific type of torsion theory over a specialized type of ring. In particular, $R = \tau(R) \bigoplus_R R'$, where τ induces τ_g on $R/\tau(R) \cong R'$ (Theorem 8). R' is a specific type of essential product of finitely many rings R_i with a simple artinian left maximal quotient ring (Theorem 12), and each of these rings R_i has an explicit matrix ring structure (Theorems 13 and 18).

2. The Results

Before we can prove our first main result (Theorem 8), we need several lemmas.

Lemma 1. If every τ-torsionfree module is a B^*-module, then τ is stable.

Proof. Let E be injective. Then $\tau(E)$ is τ-injective by [9, Prop. 8.5]. So by hypothesis, $Ext(E/\tau(E), \tau(E)) = 0$, and hence $\tau(E)$ is a direct summand of E. Thus τ is stable by [9, Prop. 8.6].

Corollary 2. If every τ-torsionfree module is a B^*-module, then $R = \tau(R) \oplus R'$ for some ${}_R R'$.

Proof. This follows from Lemma 1, the hypothesis, and [7, Lemma 3.2].

Lemma 3. If every τ-torsionfree module is a B^*-module, then the localization functor for τ is exact.

Proof. Let $0 \to K \to E \to F \to 0$ be exact with E, F τ-torsionfree and E τ-injective. Let T be τ-torsion. By [9, Prop. 44.1], we need to show that $Ext(T, F) = 0$.

By Corollary 2, we have $R = \tau(R) \bigoplus R'$ for some R'. Since $\tau(R)T$ is an epimorphic image of a direct sum of copies of $\tau(R)$, there exists an exact sequence

$$0 \to G \to \bigoplus \tau(R) \to \tau(R)T \to 0 .$$

This induces the exact sequence

$$0 = Hom(G, F) \to Ext(\tau(R)T, F) \to Ext(\bigoplus \tau(R), F) = 0 .$$

Thus $Ext(\tau(R)T, F) = 0$, and it is sufficient to show that $Ext(T/\tau(R)T, F) = 0$.

There is an exact sequence

$$0 \to L \to \bigoplus R/\tau(R) \to T/\tau(R)T \to 0 .$$

Since K is a τ-closed submodule of E, then K is τ-injective. Since L is τ-torsionfree, then by our hypothesis, $Ext(L, K) = 0$. Since $R/\tau(R) \cong R'$ is projective, we now have an exact sequence

$$0 = Ext(L, K) \to Ext^2(T/\tau(R)T, K) \to Ext^2(\bigoplus R/\tau(R), K) = 0.$$

Hence $Ext^2(T/\tau(R)T, K) = 0$. Using the τ-injectivity of E, we now have the exact sequence

$$0 = Ext(T/\tau(R)T, E) \to Ext(T/\tau(R)T, F) \to Ext^2(T/\tau(R)T, K) = 0 ,$$

which gives $Ext(T/\tau(R)T, F) = 0$ as desired.

Lemma 4. Let $\mathcal{L}_\tau$ have a cofinal subset of finitely generated left ideals. If every τ-torsionfree module is a B^*-module, then Q_τ is a semisimple artinian ring, and hence τ induces the Goldie torsion theory τ_g on $R/\tau(R)$.

Proof. Let M and N be Q_τ-modules. By Lemma 3 and [9, Prop. 45.1], τ is perfect, and hence M and N are τ-torsionfree. Since every Q_τ-module is τ-divisible, then $Ext_R(M, N) = 0$ by hypothesis. Since τ is perfect, $Ext_{Q_\tau}(M, N) \cong Ext_R(M, N) = 0$. Thus Q_τ must be semisimple artinian.

Since the only faithful perfect torsion theory that can have a semisimple localization for the ring is the Goldie torsion theory, then τ induces τ_g on $R/\tau(R)$.

Lemma 5. If $\tau(R) = \tau(R)e$ for some $e = e^2$, then $Tor_n^R(R/\tau(R), M) = 0$ for every $n \geq 1$ and every $R/\tau(R)$-module M.

Proof. This is an easy induction that uses the basic properties *of* $\bigotimes$ and Tor.

Lemma 6. If S has τ-bounded order, then the $R/\tau(R)$-module $Hom_R(R/\tau(R), S)$ has bounded over for the torsion theory induced by τ on $R/\tau(R)$.

Proof. This is a straightforward computation.

Lemma 7. Suppose that $R = \tau(R) \bigoplus {}_R R'$ and that τ has BSP. Then the following statements hold.

(1) If a $R/\tau(R)$-module M has τ-bounded order, then inj $\dim_{R/\tau(R)} M \leq 1$.
(2) If M is an $R/\tau(R)$-module and E is τ-injective, then $Ext^2_R(M, R'\tau(R)E) = 0$.
(3) If E is τ-injective, then $E/R'\tau(R)E$ is also τ-injective.
(4) If $pd_R N \leq 1$, Q_τ is a flat right R-module, $Ext(N, \bigoplus_X \tau(R)) = 0$ for any set X, and $Q_\tau \otimes N$ is a projective Q_τ-module, then $Ext_R(N, D) = 0$ for each τ-divisible module D.

Proof.

(1) Let $N \in R/\tau(R) - mod$, and let $0 \to K \to \bigoplus R/\tau(R) \to N \to 0$ be exact. Then we have the exact sequence

$$Ext_{R/\tau(R)}(K, M) \to Ext^2_{R/\tau(R)}(N, M) \to Ext^2_{R/\tau(R)}(\bigoplus R/\tau(R), M) = 0$$

for a module M with τ-bounded order. By Lemma 5 and [2, VI.4.1.3], we obtain

$$Ext_{R/\tau(R)}(K, M) \cong Ext_{R/\tau(R)}(R/\tau(R) \bigotimes K, M) \cong Ext_R(K, M) = 0 .$$

Hence $Ext^2_{R/\tau(R)}(N, M) = 0$, as desired.

(2) Let $0 \to K \to \bigoplus R/\tau(R) \to M \to 0$ be an exact sequence. Then K is τ-torsionfree. Since $R'\tau(R)E$ has τ-bounded order and $R/\tau(R)$ is projective, then

$$0 = Ext^1_R(K, R'\tau(R)E) \to Ext^2_R(M, R'\tau(R)E) \to$$

$$Ext^2_R(\bigoplus R/\tau(R), R'\tau(R)E) = 0$$

is exact. Hence $Ext^2_R(M, R'\tau(R)E) = 0$.

(3) Let T be τ-torsion, and let E be τ-injective. We have the exact sequence

$$0 = Ext_R(T, E) \to Ext_R(T, E/R'\tau(R)E) \to Ext^2_R(T, R'\tau(R)E) .$$

It is sufficient to show that $Ext^2_R(T, R'\tau(R)E) = 0$. We use (2) to obtain the exact sequence

$$0 = Ext^2_R(T/\tau(R)T, R'\tau(R)E) \to Ext^2_R(T, R'\tau(R)E) \to$$

$$Ext^2_R(\tau(R)T, R'\tau(R)E) \ .$$

Thus it suffices to show that $Ext^2_R(\tau(R)T, R'\tau(R)E) = 0$. By (2) we have the exact sequence

$$Ext^2_R(\tau(R)T/R'\tau(R)T, R'\tau(R)E) \to Ext^2_R(\tau(R)T, R'\tau(R)E) \to$$

$$Ext^2_R(R'\tau(R)T, R'\tau(R)E) = 0 \ .$$

So we need to show that $Ext^2_R(\tau(R)T/R'\tau(R)T, R'\tau(R)E) = 0$. Since $R/\tau(R)$ is projective, as a right $R/R'\tau(R)$ -module, then

$$Tor^R_p(R/\tau(R), \tau(R)T/R'\tau(R)T) \cong Tor^{R/R'\tau(R)}_p(R/\tau(R), \tau(R)T/R'\tau(R)T) = 0$$

for all $p \geq 1$. Hence we have by [2, VI.4.1.3] that

$$Ext^2_R(\tau(R)T/R'\tau(R)T,\ R'\tau(R)E)$$

$$\cong Ext^2_{R/\tau(R)}(R/\tau(R) \otimes_R \tau(R)T/R'\tau(R)T, R'\tau(R)E) \cong Ext^2_{R/\tau(R)}(0, R'\tau(R)E) = 0.$$

(4) Let D be τ-divisible, and let $\{E_\alpha\}$ be a set of τ-injective modules such that $\bigoplus_\alpha E_\alpha \to D \to 0$. Then

$$R/\tau(R) \bigotimes_R (\bigoplus_\alpha E_\alpha) \to R/\tau(R) \bigotimes_R D \to 0 \text{ is exact;}$$

$$\text{i.e. , } \bigoplus_\alpha E_\alpha/\tau(R)E_\alpha \to D/\tau(R)D$$

is an epimorphism. For a given α, there is a free $R/\tau(R)$-module mapping onto $E_\alpha/\tau(R)E_\alpha$. By (3) $E_\alpha/R'\tau(R)E_\alpha$ is τ-injective for each α; so since $E_\alpha/\tau(R)E_\alpha$ is isomorphic to a direct summand of $E_\alpha/R'\tau(R)E_\alpha$, then $E_\alpha/\tau(R)E_\alpha$ is τ-injective for each α. Thus we have the diagram

$$\begin{array}{ccc} 0 \to \bigoplus_{\beta} R/\tau(R) & \to & \bigoplus_{\beta} Q_\tau \\ \downarrow & \swarrow & \\ E_\alpha/\tau(R)E_\alpha & & \end{array}$$

Hence we can obtain an epimorphism

$$\bigoplus_{\alpha}\left(\bigoplus_{\beta} Q_\tau\right) \to D/\tau(R)D$$

by composing maps. Since $pd_R N \leq 1$, we obtain an exact sequence

$$Ext_R\left(N, \bigoplus_{\alpha}\left(\bigoplus_{\beta} Q_\tau\right)\right) \to Ext_R(N, D/\tau(R)D) \to 0 .$$

Since Q_τ is flat, [2, VI.4.1.3] gives

$$Ext_R\left(N, \bigoplus_{\alpha}\left(\bigoplus_{\beta} Q_\tau\right)\right) \cong Ext_{Q_\tau}\left(Q_\tau \bigotimes N, \bigoplus_{\alpha}\left(\bigoplus_{\beta} Q_\tau\right)\right) .$$

Since $Q_\tau \bigotimes N$ is projective as a Q_τ-module by hypothesis, we must have $Ext_R(N, D/\tau(R)D) = 0$ by exactness.

It now suffices to show that $Ext_R(N, \tau(R)D) = 0$. Consider an epimorphism $\bigoplus_X \tau(R) \to \tau(R)D$. By our hypotheses, we have the exact sequence

$$0 = Ext^1_R(N, \bigoplus_X \tau(R)) \to Ext^1_R(N, \tau(R)D) \to 0 ,$$

so that $Ext_R(N, \tau(R)D) = 0$, as desired.

Theorem 8. Let $\mathcal{L}_\tau$ have a cofinal subset of finitely generated left ideals. Then every τ-torsionfree module is a B^*-module if and only if the following conditions hold:

(1) $\tau(R)$ is a direct summand of ${}_R R$,
(2) τ induces τ_g on $R/\tau(R)$, and
(3) Every τ_g-torsionfree $R/\tau(R)$-module is a B^*-module.

Proof. ($\Rightarrow$) Corollary 2 gives (1), and Lemma 4 yields (2). Thus every τ_g-torsionfree $R/\tau(R)$-module is τ-torsionfree, every $R/\tau(R)$-module of τ_g-bounded order has

τ-bounded order, and every τ_g-divisible $R/\tau(R)$-module is τ-divisible; thus (3) is immediate from the hypothesis.

($\Leftarrow$) Let S be a τ-torsion module with τ-bounded order. Since $Ext^p_R(R/\tau(R), S) = 0$ for all $p \geq 1$ by (1), then [2, VI.4.1.4] yields

$$Ext^1_{R/\tau(R)}(F, Hom_R(R/\tau(R), S)) \cong Ext^1_R(F, S)$$

for all τ-torsionfree ${}_R F$. Thus (2), (3) and Lemma 6 force $Ext_R(F, S) = 0$.

Let N be a τ-torsionfree module. We still need to show that $Ext(N, D) = 0$ for each τ-divisible module D. Note that $\tau(R)N = 0$. By (2), (3) and [15, Theorem 3.1], we must have $pd_{R/\tau(R)} N \leq 1$. By (1), a $R/\tau(R)$-projective resolution is also a R-projective resolution; so $pd_R N \leq 1$.

Since $\mathcal{L}_\tau$ has a cofinal subset of finitely generated left ideals, then by (2) so does the filter for τ_g on $R/\tau(R) - mod$. Hence $R/\tau(R)$ is finite dimensional, so that $Q_{max} = Q_{\tau_g} = Q_\tau$ is a semisimple artinian ring. By (2) τ_g is a perfect torsion theory for $R/\tau(R) - mod$ and hence τ must also be perfect. Thus Q_τ is a flat right R-module. Consequently Lemma 7 gives $Ext(N, D) = 0$ for all τ-divisible D, as desired.

A ring R is a subdirect product of rings R_1, R_2, $\cdots$, R_n if R is a unital subring of $\bigoplus_{i=1}^n R_i$ such that the projection of R to each coordinate of $\bigoplus_{i=1}^n R_i$ is surjective. We say that R is an essential product of rings R_1, R_2, $\cdots$, R_n if R is a subdirect product of $\bigoplus_{i=1}^n R_i$ such that R contains an essential left ideal of $\bigoplus_{i=1}^n R$. (See [6], [11] or [12].)

Theorem 8 states that to determine when every τ-torsionfree module is a B^*-module it is sufficient to consider τ_g over a τ_g-torsionfree ring. Moreover, since $\mathcal{L}_\tau$ has a cofinal subset of finitely generated left ideas, so does our induced $\mathcal{L}_{\tau_g}$. Therefore, we may assume that R is a left nonsingular ring with finite left (Goldie) dimension. We now want to show that R is obtained in a natural way as an essential product of rings with simple artinian left maximal quotient ring, so that we can make the additional assumption that R has a simple left artinian maximal quotient ring.

First, we note that a left finite dimensional left nonsingular ring has a semisimple left maximal quotient ring $\bigoplus_{i=1}^n Q_i$, where each Q_i is a simple artinian ring. Let $p_j : \bigoplus_{i=1}^n Q_i \to Q_j$ be the natural projection. Then R embeds as an essential submodule of $\bigoplus_{i=1}^n p_i(R)$ and each $p_i(R)$ is a left finite dimensional left nonsingular ring.

Let π_j and λ_j be the natural projection and injection associated with the j-th coordinate of $\bigoplus_{i=1}^n R_i$. As usual, we let $H_j = R \cap \lambda_j \pi_j (\bigoplus_{i=1}^n R_i)$ and $E_j = \pi_j H_j$.

Since the singular modules used in the proof of [11, Lemma 11] have bounded order, the proof of [11, Lemma 11] actually establishes the following result.

Lemma 9. Let τ_g have BSP, let H be a two-sided ideal that is essential as a left ideal of the left nonsingular ring R, and let A be a nonsingular left R-module. Then

(1) $Tor_1^R(R/H, A) = 0$.

(2) A/HA is projective left R/H-module.

Consequently, the same proof as [11, Theorem 12] shows mutatis mutandis that the following result holds.

Proposition 10. Let R be an essential product of the left nonsingular rings R_1 and R_2. Then τ_g has BSP for R if and only if

(a) τ_g has BSP for R_1 and R_2 and
(b) if L_i is an essential left R_1-submodule of E_i ($i = 1$ or 2), then E_i/L_i is a direct summand of R_i/L_i (in the above notation).

A result similar to Proposition 10 is easily proved for B^*-modules.

Proposition 11. Let R be an essential product of the left finite dimensional, left nonsingular rings R_1 and R_2. Then every τ_g-torsionfree R-module is a B^*-module if and only if

(a) every τ_g-torsionfree R_i-module is a B^*-module ($i = 1, 2$) and
(b) if L_i is an essential left R_i-submodule of E_i ($i = 1$ or 2), then E_i/L_i is a direct summand of R_i/L_i (in our standard notation).

Proof. Let Q_i be the left maximal quotient ring of R_i; hence $Q \cong Q_1 \bigoplus Q_2$ is the left maximal quotient ring of R.

($\Rightarrow$) Assume that every τ_g-torsionfree R-module is a B^*-module. By Proposition 10 and symmetry, it is sufficient to show that $Ext_{R_1}(N, D) = 0$ for each nonsingular R_1-module N and each τ_g-divisible R_1-module D. Since $\tau_g(R_1) = 0$, there is an epimorphism $\bigoplus Q_1 \to D$, which induces an epimorphism $\bigoplus Q \to D$. By hypothesis, we now obtain $Ext_R(N, D) = 0$. Hence $Ext_{R_1}(N, D) = 0$.

($\Leftarrow$) Assume that (a) and (b) hold. By Proposition 10, we only need to show that $Ext_R(A, C) = 0$ whenever A is a nonsingular left R-module and C is a τ_g-divisible left R-module. Since there is an epimorphism $f : \bigoplus Q \to C$, we consider $C_1 = im(f \mid \bigoplus Q_1)$. Then C_1 is a τ_g-divisible R_1-module. Moreover, $C_2 = C/C_1$ is a homomorphic image of $\bigoplus Q_2$. Hence there is an exact sequence

$$0 \to C_1 \to C \to C_2 \to 0$$

with C_i a τ_g-divisible R_i-module. Thus we can use the proof of [11, Theorem 12] mutatis mutandis to show that $Ext_R(A, C) = 0$ once we make the following observation: there is an exact sequence

$$0 \to \tau_g(C) \to C \to C/\tau_g(C) \to 0$$

with $C/\tau_g(C)$ injective, so that we may assume C to be singular.

As soon as we introduce some notation, we are ready to prove our result that allows us to reduce to the case in which R has a simple artinian left maximal quotient ring.

Let $R_1, R_2, \cdots, R_n$ be rings. For any subset S of $\{1, 2, \cdots, n\}$, let $\pi_S : \bigoplus_{i=1}^n R_i \to \bigoplus_S R_i$ be the natural projection. For $T \subseteq S$, let $\lambda_{S,T} : \bigoplus_T R_i \to \bigoplus_S R_i$ be the natural inclusion map.

Theorem 12. Let R be an essential product of left finite dimensional, left nonsingular rings $R_1, R_2, \cdots, R_n$. For nonempty sets $T \subset S \subseteq \{1, 2, \cdots, n\}$, let $H_{S,T} = \pi_S(R) \cap \lambda_{S,T}(\bigoplus_T R_i)$. Then every nonsingular left R-module is a B^*-module for τ_g if and only if

(i) every nonsingular left R_i-module is a B^*-module for $\tau_g (i = 1, 2, \cdots, n)$, and
(ii) if L is an essential submodule of some $H_{S,T}$, then $H_{S,T}/L$ is a direct summand of $\lambda_{S,T}(\bigoplus_T R_i)/L$.

Proof. ($\Leftarrow$) Assume that (i) and (ii) hold. We use induction on n. By letting $S = \{1,2\}$ and $T = \{1\}$ and $\{2\}$, we obtain Case $n = 2$ from Proposition 11.

For the general case, let R be an essential product of $R_1, R_2, \cdots, R_n$, and let $S = \{1, 2, \cdots, n-1\}$. Then $\pi_S(R)$ is an essential product of $R_1, R_2, \cdots, R_{n-1}$ that satisfies (i) and (ii). By our induction hypothesis, every nonsingular $\pi_S(R)$-module is a B^*-module for τ_g.

Now R is also an essential product of $\pi_S(R)$ and R_n. By hypothesis, R_n satisfies (a) and (b) of Proposition 11. By the preceding paragraph, hypothesis (ii) and the modular law, $\pi_S(R)$ also satisfies (a) and (b) of Proposition 11. Hence every nonsingular R-module is a B^*-module for τ_g by Proposition 11.

($\Rightarrow$) It is straightforward to establish (i) by using the method of Proposition 11, and (ii) can be established by using the proof of [11, Theorem 12] mutatis mutandis.

We have now reduced the problem to characterizing the rings R with simple artinian left maximal quotient ring such that every nonsingular R-module is a B^*-module for τ_g. As in the study of the splitting properties SP and BSP, the results are somewhat different in the case where R has essential socle and the case where $Soc_R R = 0$ (see [5], [6], [10], [11]). We note that since R has a simple artinian maximal quotient ring, then $Soc_R R \neq 0$ implies that $Soc_R R$ is essential in ${}_R R$. For $Soc_R R$ essential in ${}_R R$, BSP for τ_g implies SP; so the study of B^*-modules trivializes, as the rings with SP for τ_g are well understood in this case. For completeness, however, we state a solution to our problem for the case in which $Soc_R R$ is essential in ${}_R R$.

Theorem 13. Let $Soc_R R \neq 0$ and assume that R has a simple artinian left maximal quotient ring. Then every nonsingular left R-module is a B^*-module for τ_g if and only if

$$R \cong \begin{pmatrix} A & B \\ 0 & C \end{pmatrix}$$

where

(a) A is a simple artinian ring,

(b) ${}_A B$ is finitely generated and B_C is faithful, and
(c) C is a semihereditary semiprimary ring possessing a left artinian left classical quotient ring $Q^1 \subseteq T = End_A B$ such that T is a right flat Q^1-module.

Proof. This is easily obtained from [6, Theorem 2.5], [10, Theorem 4.9 and Corollary 5.4], and the fact that R has a simple artinian left maximal quotient ring.

We now turn our attention to the more interesting case in which $Soc_R R = 0$. To obtain our characterization (Theorem 18), we need several lemmas.

Lemma 14. Let R be a left finite dimensional, left nonsingular ring with semisimple maximal quotient ring Q such that $lgldimR \leq 2$ and $pd_R Q \leq 1$. Then $pd_R N \leq 1$ for all nonsingular ${}_R N$.

Proof. Since N is nonsingular and Q is semisimple, then

$$N \subseteq \prod_X Q \subseteq \bigoplus_Y Q$$

for appropriate index sets X and Y. Thus, for any module ${}_R M$, we have an exact sequence

$$0 = Ext_R^2 \left(\bigoplus\nolimits_Y Q, M\right) \to Ext_R^2 (N, M) \to Ext_R^3 \left(\bigoplus\nolimits_Y Q/N, M\right) = 0$$

by our hypotheses. Thus $pd_R N \leq 1$.

Lemma 15. Let

$$R = \begin{pmatrix} A & B \\ 0 & C \end{pmatrix}$$

be a ring where
(a) A is a prime left Goldie ring with quotient ring Q such that $Soc_A A = 0$, τ_g has BSP for A, $lgldimA \leq 2$, and $pd_A Q \leq 1$;
(b) ${}_A B_C$ is a bimodule such that ${}_A B$ is a finite dimensional, nonsingular injective A-module and B_C is a faithful C-module; and
(c) C is a semiprimary ring with two-sided maximal quotient ring $End_A B$.

Consider the extension ring

$$S = \begin{pmatrix} Q & B \\ 0 & C \end{pmatrix}$$

of R. Then S is a left hereditary ring and $R \to S$ is a flat epimorphism (i.e., $S \bigotimes_R S \cong S$ and S_R is flat).

Proof. Since ${}_A B$ is a nonsingular injective module, then B is a Q, C-bimodule; hence S is a ring extension of R. By [6, Theorem 5.1] τ_g has BSP for R. Hence B_C is flat by [6, Lemma 3.7] and C is left hereditary via [10, Lemma 5.2]. Consequently, S is left hereditary by [12, Proposition 4.1].

Now the set

$$\mathcal{L} = \{I \subseteq R \mid I = \begin{pmatrix} E & B \\ 0 & C \end{pmatrix} \ , {}_A E \text{ essential in } {}_A A\}$$

is a filter of left ideals for a torsion theory σ. The localization $Q_\sigma(R)$ is easily seen to be S. Since each essential left ideal E of A contains a regular element d of A, then $\mathcal{L}$ has a cofinal subset of finitely generated projective left ideals of the form

$$\begin{pmatrix} Ad & B \\ 0 & C \end{pmatrix} .$$

Thus σ is a perfect torsion theory; so $R \to Q_\sigma(R) = S$ is a flat epimorphism. (See [13].)

Lemma 16. Let R be a matrix ring that satisfies the hypotheses (a)-(c) of Lemma 15 and the following condition:

(d) For each indecomposable Q-submodule Y of any finite direct sum $\bigoplus_{i=1}^n B$ of copies of ${}_A B$, there exists a nonzero $(c_1, c_2, \cdots, c_n) \in \bigoplus_{i=1}^n C$ and an essential submodule ${}_A B'$ of ${}_A B$ such that $(bc_1, bc_2, \cdots, bc_n) \in Y$ for each $b \in B'$.

Let K be a closed submodule of a free left R-module $F = \bigoplus_x R$, and let

$$G = \bigoplus_x \begin{pmatrix} 0 & B \\ 0 & 0 \end{pmatrix} \subseteq F .$$

Then $K = W \oplus Re_{22}K$, where $W \subseteq Re_{11}F$ and $W \cap G = 0$.

Proof. Note that $(e_{11}R \cap Re_{22})K$ is a nonsingular A-homomorphic image of $\bigoplus_A B$; so $(e_{11}R \cap Re_{22})K$ is an injective A-module. Write $K \cap G = Y^1 \bigoplus (e_{11}R \cap Re_{22})K$ as both R-modules and A-modules. Since K is closed in F, then $K \cap G$ is closed in G. Since ${}_A G$ is an injective A-module, then we can write

$$Re_{11}K = W \bigoplus (K \cap G) = W \bigoplus Y^1 \bigoplus (e_{11}R \cap Re_{22})K$$

as both R-modules and A-modules. It remains to show that $Y^1 = 0$. If not, then Y^1 is a nonzero closed submodule of F (as K is closed). Since $Y^1 \subseteq G$, then Y^1 is a nonsingular injective A-module. Now ${}_A Y^1$ is a direct sum of indecomposable injective A-modules. Thus, since A is left nonsingular, we may assume that F has an indecomposable closed submodule Y such that Y is contained in a finitely generated free submodule $F^1 \cong \bigoplus_{i=1}^n R$ of F and $Y \subseteq G \cap F^1$. But condition (d) gives

$$s = \left(\begin{pmatrix} 0 & 0 \\ 0 & c_1 \end{pmatrix} , \begin{pmatrix} 0 & 0 \\ 0 & c_2 \end{pmatrix} , \cdots , \begin{pmatrix} 0 & 0 \\ 0 & c_n \end{pmatrix} \right) \in F^1$$

such that $s+Y$ is a nonzero element of the τ_g-torsion submodule of F/Y, which contradicts the assumption that Y is closed.

Lemma 17. Let R be a matrix ring that satisfies the hypotheses of Lemma 16. Let K be a closed submodule of a free R-module $F \cong \bigoplus_x R$. Then K is a projective left R-module.

Proof. By Lemma 16, we can write $K = W \bigoplus Re_{22}K$, where $W \cap G = 0$ and

$$G = \bigoplus_x \begin{pmatrix} 0 & B \\ 0 & 0 \end{pmatrix} \subseteq F .$$

Define an A-homomorphism $\theta : W \to \bigoplus_x A$ via the projection

$$w = \left(\begin{pmatrix} a_\alpha & b_\alpha \\ 0 & 0 \end{pmatrix} \right) \mapsto (a_\alpha) .$$

Then $ker\ \theta = W \cap G = 0$, and hence W is isomorphic to a submodule P of $\bigoplus_x A$.

We claim that P is closed in $\bigoplus_x A$. For suppose that $t = (t_\alpha) \in \bigoplus_x A$ has $(P : t)$ essential in ${}_A A$. Since Q is the left classical quotient ring of A, there exists a regular element d of A such that $d \in (P : t)$. Write $d(t_\alpha) = (a_\alpha) \in P$. Since B is a left Q-module, we can consider

$$\left(\begin{pmatrix} t_\alpha & d^{-1}b_\alpha \\ 0 & 0 \end{pmatrix} \right) \in F$$

where the b_α's are determined by the inverse of θ on P. Then

$$\begin{pmatrix} Ad & B \\ 0 & C \end{pmatrix} \left(\begin{pmatrix} t_\alpha & d^{-1}b_\alpha \\ 0 & 0 \end{pmatrix} \right) = \begin{pmatrix} A & 0 \\ 0 & 0 \end{pmatrix} \left(\begin{pmatrix} a_\alpha & b_\alpha \\ 0 & 0 \end{pmatrix} \right) \in W .$$

Since W is closed in F, we must have

$$\left(\begin{pmatrix} t_\alpha & d^{-1}b_\alpha \\ 0 & 0 \end{pmatrix} \right) \in W$$

and $(t_\alpha) \in P$.

But by our hypotheses and Lemma 14, $pd_A\ (\bigoplus_x A/P) \leq 1$, and hence P is projective. Consequently, $W \cong P$ is a projective A-module. But since $Re_{22}W = 0$ and Re_{11} is a direct summand of R, then W is also projective as a left R-module.

Now let $\{k_\alpha\}_{\alpha \in V}$ be a set of generators for the C-module $e_{22}K$. Define a R-homomorphism

$$\psi : \bigoplus_V Re_{22} \to Re_{22}K$$

via the $\alpha - th$ e_{22} maps to $k_\alpha (\alpha \in V)$. Since $\bigoplus_V Re_{22}$ and $Re_{22}K$ are also

S-modules and since $R \to S$ is a flat epimorphism by Lemma 15, then ψ is an S-homomorphism. But $Re_{22}K$ is a projective S-module by Lemma 15 and [2, I.5.4]. Therefore, $Re_{22}K$ is isomorphic to a direct summand of $\bigoplus_V Re_{22}$ as S-modules. This forces $Re_{22}K$ to be a projective left R-module.

Theorem 18. Let R be a ring with $Soc_R R = 0$ and with a simple artinian maximal left quotient ring. Then every τ_g-torsionfree module is a B^*-module if and only if

$$R \cong \begin{pmatrix} A & B \\ 0 & C \end{pmatrix},$$

where

(a) A is a prime left Goldie ring with quotient ring Q such that $Soc_A A = 0$, τ_g has BSP for A, $lgldim A \leq 2$, and $pd_A Q \leq 1$;

(b) ${}_A B_C$ is a bimodule with ${}_A B$ a finite dimensional, nonsingular, injective A-module and B_C a faithful C-module;

(c) C is a semiprimary ring with two-sided maximal quotient ring $T = End_A B$;

(d) For each nonzero indecomposable Q-submodule Y of any finite direct sum $\bigoplus_{i=1}^n B$ of copies of B, there exists a nonzero $(c_1, c_2, \cdots, c_n) \in \bigoplus_{i=1}^n C$ and an essential submodule ${}_A B'$ of ${}_A B$ such that $(bc_1, bc_2, \cdots, bc_n) \in Y$ for each $b \in B'$.

Proof. ($\Rightarrow$) Assume that every τ_g-torsionfree module is a B^*-module. By [6, Theorem 5.1],

$$R \cong \begin{pmatrix} A & B \\ 0 & C \end{pmatrix}$$

where (1) A is a left nonsingular, left finite dimensional semiprime ring such that $Soc_A A = 0$ and τ_g has BSP for A; (2) ${}_A B_C$ is a bimodule such that ${}_A B$ is finite dimensional, injective and nonsingular; and (3) C is a semiprimary ring with two-sided maximal quotient ring $End_A B$. Since R has a simple left maximal quotient ring, then any two uniform left ideals of R must be subisomorphic; hence A is a prime ring and B_C is faithful. Since Re_{11} is a direct summand of ${}_R R$, then any A-projective resolution of a nonsingular left A-module is also a R-projective resolution; hence $pd_A I \leq 1$ for each left ideal I of A and $pd_A Q \leq 1$ via [15, Lemma 2.1] and [2, VI.4.1.3] Therefore, $lgldim A \leq 2$.

For contradiction, suppose that condition (d) fails. Then there exists a nonzero closed submodule Y of a finitely generated free module $F \cong \bigoplus_{i=1}^n R$ with

$$Y \subseteq \oplus \begin{pmatrix} 0 & B \\ 0 & 0 \end{pmatrix}.$$

Since $pd_R F/Y \leq 1$ by [15, Lemma 2.1], then Y is R-projective and hence

A-projective. Since Y and Q are direct sums of (isomorphic) indecomposable injective A-modules, then ${}_A Q$ is projective. Hence $A \cong Q$, which contradicts the hypothesis that $Soc_R\, R = 0$.

($\Leftarrow$) Assume that R is a matrix ring satisfying (a)-(d). Then τ_g has BSP for R by [6, Theorem 5.1]. Since R has a simple artinian left maximal quotient ring, then [15, Theorem 3.1] says that it is sufficient to show that $pd_R\, N \leq 1$ for each nonsingular ${}_R N$. But this is an immediate consequence of Lemma 17.

Remarks.

(1) The structure of the rings A in part (a) of Theorem 18 is not fully understood, but much is known. For example, if A is commutative, then A must be a Dedekind domain. (See [10].)

(2) If A is commutative, then it can be shown that Conditions (a)-(c) of Theorem 18 imply Condition (d).

Example. Let $\mathbf{Q}$ denote the rational numbers, and let F be the quotient field of $\mathbf{Q}[x]$. Then

$$R = \begin{pmatrix} \mathbf{Q}[x] & F & F & F \\ 0 & F & F & F \\ 0 & 0 & \mathbf{Q} & F \\ 0 & 0 & 0 & F \end{pmatrix}$$

is a ring that satisfies the hypotheses of Theorem 18, where $A = \mathbf{Q}[x]$, $B = [F \quad F \quad F]$, and

$$C = \begin{pmatrix} F & F & F \\ 0 & \mathbf{Q} & F \\ 0 & 0 & F \end{pmatrix} .$$

REFERENCES

1. F.W. Call and T.S. Shore, *The splitting of bounded torsion submodules*, Comm. Alg. **9** (1981), 1161-1214.
2. H. Cartan and S. Eilenberg, *Homological algebra*, Princeton, 1956.
3. P.C. Eklof, L. Fuchs and S. Shelah, *Baer modules over domains*, Trans. Amer. Math. Soc. **322** (1990), 547-560.
4. L. Fuchs and G. Viljoen, *A weaker form of Baer's splitting problem for valuation rings*, Quaestiones Math. **14** (1991), 227-236.
5. J.D. Fuelberth and J. Kuzmanovich, *On the structure of splitting rings*, Comm. Alg. **3** (1975), 913-949.
6. J.D. Fuelberth and J. Kuzmanovich, *On the structure of finitely generated splitting rings*, Pacific J. Math. **80** (1979), 389-424.

7. J.D. Fuelberth and M.L. Teply, *The torsion submodule splits off*, Math. Ann. **188** (1970), 270-284.
8. J.D. Fuelberth and M.L. Teply, *The singular submodule of a finitely generated module splits off*, Pacific J. Math. **40** (1972), 73-82.
9. J.S. Golan, *Torsion theories*, Pitman Monographs **29**, Longman Scientific and Technical/John Wiley, New York, 1986.
10. K.R. Goodearl, *Singular torsion and the splitting properties*, Mem. Amer. Math. Soc. **124** (1972).
11. K.R. Goodearl, *Essential products of nonsingular rings*, Pacific J. Math. **45** (1973), 493-505.
12. K.R. Goodearl, *Ring theory: Nonsingular rings and modules*, Pure and Appl. Math. **33**, Marcel Dekker, New York, 1976.
13. B. Stenström, *Rings of quotients*, Die Grund. der Math. Wiss. in Einzeld. **217**, Springer-Verlag, Berlin, 1975.
14. M.L. Teply, *Subidealizer rings and the splitting properties*, Comm. Alg. **13** (1985), 2267-2298.
15. M.L. Teply and B. Torrecillas, *A weaker form of Baer's splitting problem for torsion theories*, Czech. J. Math., to appear.

CHARACTERIZATIONS OF RINGS USING EXTENDING AND LIFTING MODULES

N. VANAJA
Department of Mathematics, University of Bombay
Bombay 400 098, India

H-rings and Co-H-rings have been defined by K. Oshiro in [9]. A ring R is a right H-ring (Co-H-ring) iff every injective (projective) right R-module is lifting (extending). A right H-ring or a right Co-H-ring is a semiprimary QF-3 ring. Co-H-rings and H-rings are closely related to the following conditions on the ring.

$(*)^*$ Every non-cosmall module contains a non-zero projective summand.

$(*)$ Every non-small module contains a non-zero injective summand.

It has been proved by K. Oshiro [9] that i) a ring is a right H-ring iff it satisfies $(*)$ and is right Artinian, and ii) a ring is a right Co-H-ring iff it satisfies $(*)^*$ and has a.c.c. on right annihilator ideals. We consider in this paper the following two more conditions on the ring.

$(*)^*_1$ Every finitely generated non-cosmall module contains a non-zero projective summand.

$(*)_1$ Every finitely embedded non-small module contains a non-zero injective summand.

We first characterize semiprimary QF-3 rings in terms of extending and lifting properties of its modules. Specifically we prove that a ring R is a semiprimary QF-3 ring iff R is right perfect and any (countably generated) projective right R-module whose indecomposable direct summands are non-small is extending, iff R is left perfect and any (countably embedded) injective right R-module whose indecomposable summands are non-cosmall is lifting.

Phan Dan [11] has proved that a right perfect ring R satisfies $(*)^*$ iff R^2 is an extending R-module, iff R^n is an extending module for every $n \in \mathbb{N}$. We generalize this result. We show that a semiperfect ring R satisfies $(*)^*_1$ iff direct sum of any two indecomposable projectives is extending, iff R^n is an extending R-module for every $n \in \mathbb{N}$. We further prove that for a right perfect ring the condition $(*)^*$ is equivalent to the condition $(*)^*_1$. We deduce that a right perfect ring R satisfies $(*)^*$ iff direct sum of any two indecomposable projectives is extending. We also prove that a semiperfect perfect ring R is a Co-H-ring iff $R^{\aleph_0}$ is an extending right R-module.

We next consider rings R which has essential right socle with $J(E(S)) \neq E(S)$ for every simple submodule S of R. Suppose R is also semiperfect. We show that R is a right QF-3 ring iff direct sum of any two indecomposable non-cosmall injective is lifting and that R has a faithful $\sum$-injective right ideal iff any injective R-module with (countably generated) essential socle whose indecomposable direct summands are non-cosmall is lifting. Lastly we prove that a ring R is a right H-ring iff R is right Artinian and any two embedded injective R-module is lifting, iff R has essential right socle with $J(E(S)) \neq E(S)$ for every simple submodule S of R and every countably embedded injective R-module is lifting, iff R has right essential socle and any injective R-module with essential socle is lifting.

Preliminaries

Throughout this paper R denotes a ring with identity. All R-modules are unitary right R-modules unless otherwise stated. We denote by $J(M)$, $Soc(M)$, $E(M)$ and $P(M)$ the Jacobson radical, socle, injective envelope and the projective cover (if it exists) of an R-module M. We denote by J the Jacobson radical of a ring R.

A module M is call *non-small* if M is not a small submodule of $E(M)$. M is called *small* if it is not non-small. A module M is called *non-cosmall* if $M \simeq P/K$, where P is projective and K is not essential in P. M is called *cosmall* if it is not non-cosmall. A module M is called *extending* if for every non-zero submodule N of M there exists a direct summand T of M such that N is essential in T. A module M is called *lifting* if for every proper submodule N of M there exists a direct summand T of M such that $N = T \oplus K$ and K is small in M. It is a well-known fact that a projective module M is lifting iff every factor module of M has a projective cover. A projective lifting module is called *semiperfect module* [8]. A projective module M is called perfect if any direct sum of copies of M is semiperfect. Let M be an R-module. An R-module N is called *M-injective* if every map $L \to N$, where $L \subseteq M$, can be extended to M. A module M is called *quasi-injective* if M is M-injective. Dually *M-projective* and *quasi-projective* modules are defined. A module M is called *local* if it has a unique maximal submodule and is called *completely indecomposable* if its endomorphism ring is a local ring. An R-module M is said to be *co-faithful* if $M \subseteq R^n$.

A ring R is called *right QF-3* if it has minimal faithful module N, minimal in the sense that any faithful module contains a direct summand isomorphic to N. Colby and Rutter [1] has proved that a ring R is semiprimary QF-3 iff R is right perfect and has a faithful $\sum$-injective right ideal, iff R is right perfect and injective envelope of any projective is projective. From this it can be easily seen that a right perfect ring is semiprimary QF-3 iff $E(R)$ is a $\sum$-injective projective module. Colby and Rutter [1] has also proved that a semiprimary ring is right QF-3 iff it is left QF-3. A ring R is called *right 1-Gorenstein* if $E(R)$ is projective.

Let R be a semiperfect ring. Suppose

$$\mathcal{S} = \{S_1, S_2, \cdots, S_n\}$$

is a complete set of nonisomorphic simple R-modules. We denote by

$$\mathcal{S}_1 = \{S \in \mathcal{S} : P(S) \text{ is non-small}\}$$

$$\mathcal{S}_2 = \{S \in \mathcal{S} : E(S) \text{ is non-cosmall}\}$$

$$A = \bigoplus_{S \in \mathcal{S}_1} \quad \text{and} \quad B = \bigoplus_{S \in \mathcal{S}_2} S$$

1. Rings in Terms of Extending Modules

Let R be a right perfect ring. It may be noted that any indecomposable non-small projective R-module is isomorphic to a summand of $P(A)$. We first show that for a right perfect ring R, $\mathcal{S}_1 \neq \emptyset$ and $P(A)$ is co-faithful. We prove that a right perfect ring is 1-Gorenstein iff $P(S \oplus T)$ is extending for every S, T in $\mathcal{S}_1$, iff $P(A^n)$ is extending for $n \in \mathbf{N}$.

Proposition 1.1. If $f : P \to E$ is a projective cover of an injective module E, then any non-zero direct summand of P is non-small. Hence for a right perfect ring $\mathcal{S}_1 \neq \emptyset$.

Proof. Let $P = X \oplus Y$, where X is a non-zero small module. Now $f(X)$ is small and hence small in E (Proposition 1.1 and Lemma 1.1, [4]). This implies $E = f(X) + f(Y) = f(Y)$. As ker f is small in P, it follows that $Y = P$. This contradicts that $X \neq 0$.

□

Proposition 1.2. Let R be a right perfect ring. If M is a small indecomposable projective module, then $M \subseteq \bigoplus_{i=1}^{n} M_i$, where each M_i is a non-small indecomposable projective module. Hence $P(A)$ is co-faithful.

Proof. Let M be a small indecomposable projective R-module. If $f : N \to E(M)$ is a projective cover of $E(M)$, then $N = \oplus e_i R$, where each $e_i R$ is non-small indecomposable projective (Proposition 1.1). As M is projective $\exists\ h : M \to N$ such that $f \circ h$ is the inclusion map $M \to E(M)$. Since h is 1-1 and M is cyclic, we get the proposition.

□

Corollary 1.3. Let R be a right perfect ring. An R-module M is injective iff M is $P(S)$-*injective* $\forall\ S \in \mathcal{S}_1$.

In the following theorem the equivalence of ii) and iii) has been proved by Harada [4]. We include a proof for the sake of completeness.

Theorem 1.4. For a right perfect ring R the following are equivalent.

i) Direct sum of any two non-small indecomposable projective is extending.
ii) Any non-small indecomposable projective is injective.

iii) R is 1-Gorenstein.
iv) Any finite direct sum of non-small indecomposable projective in injective.
v) $P(A^n)$ is extending for all $n \in \mathbf{N}$.
vi) $P(A^2)$ is extending.

Proof. $i) \Rightarrow ii)$: Let S, $T \in \mathcal{S}_1$. Then $P(S)$ and $P(T)$ are completely indecomposable. Also any monomorphism from $P(S)$ to $P(T)$ is onto as $P(S)$ is non-small and $P(T)$ is local. Hence, by Theorem 12 of [6], $P(S)$ is $P(T)$-injective. It follows that $P(A)$ is $P(A)$-injective and hence injective (Corollary 1.3). As any non-small indecomposable projective is isomorphic to a direct summand of $P(A)$, $ii)$ follows.
$ii) \Rightarrow iii)$: By $ii) P(A)$ is injective and $iii)$ follows from Proposition 1.2.
$iii) \Rightarrow ii)$: As $E(R)$ is projective and R is right perfect, $E(R)$ is lifting. If M is non-small indecomposable projective, then $M \subseteq E(R)$ and M is indecomposable implies that M is a direct summand of $E(R)$.
$ii) \Rightarrow iv) \Rightarrow v) \Rightarrow vi) \Rightarrow i)$ are trivial.

□

Theorem 1.5. For a right perfect ring R the following are equivalent.
i) R is a semiprimary QF-3 ring.
ii) Any projective R-module whose indecomposable direct summands are non-small is extending.
iii) $P(A^{|K|})$ is extending for any set K.
iv) Any countably generated projective R-module whose indecomposable direct summands are non-small is extending.
v) $P(A^{\aleph_0})$ is extending.

Proof. $i \Rightarrow ii)$: As $E(R)$ is projective, $P(A)$ is injective (Theorem 1.4). Since $E(R)$ is $\sum$-injective, so also is $P(A)$.
$ii) \Rightarrow iii) \Rightarrow iv) \Rightarrow v)$ are trivial.
$v) \Rightarrow i)$: By Theorem 12 of [6], $P(A^{\aleph_0}) = (P(A))^{\aleph_0}$ is $P(S)$-injective for every $S \in \mathcal{S}_1$. Hence $P(A)^{\aleph_0}$ is injective (Corollary 1.3). This implies that $P(A)$ is $\sum$-injective. As $P(A)$ is faithful, R is a semiprimary QF-3 ring (Theorem 1.3, [1]).

□

Phan Dan [11] has proved that a right perfect ring satisfies $(*)^*$ iff R^2 is an extending right R-module, iff R^n is an extending right R-module for every $n \in \mathbf{N}$. We prove below a similar result for semiperfect rings, which is also a finitely generated version of Theorem 2 of [9] for semiperfect rings. We need some lemmas and the proof of the following lemma is obvious.

Lemma 1.6. P is a projective extending module iff any homomorphic image of P is a direct sum of a singular and a projective module.

Lemma 1.7. Let R be any ring. Suppose P and Q are *R-modules* such that P is a projective extending module, Q is a direct sum of a singular and a projective module, and P is essential in Q. Then Q is projective.

Proof. Let $Q = N \oplus T$, where N is projective and T is singular. Suppose σ is the restriction to P of the projection map from Q onto N. As P is extending $\sigma(P) = X \oplus Y$, where X is projective and Y is singular. If μ is the projection of $\sigma(P)$ onto X, then the map $\mu \circ \sigma$ splits. Then $\ker(\mu \circ \sigma)$ is a projective module contained in the singular module $Y \oplus T$. Hence $\ker(\mu \circ \sigma) = 0$. Thus $T \cap P = 0$. As P is essential in Q, $T = 0$.

□

Corollary 1.8. If every finitely (countably) generated projective module is extending, then any finitely (countably) generated essential extension of a finitely (countably) generated projective is projective.

We prove below that the converse of Corollary 1.8 is true for a semiperfect ring in Theorems 1.10 and 1.13.

Lemma 1.9. Let R be a semiperfect ring and M be a non-cosmall module. Then M contains a cyclic hollow non-cosmall module.

Proof. Consider $f : P \to M$ with P projective and f onto. Now there exists $0 \neq T \subseteq P$ such that $T \cap \ker f = 0$. Let $x \in T$. Then $xR \subseteq Q = P_1 \oplus P_2 \oplus \cdots \oplus P_n$, where the P_i 's are indecomposable projectives and Q is a summand of P. Let $K_i = ker(f \mid P_i)$ and $K = ker(f \mid Q)$. As K is not essential in Q and $K_1 \oplus \cdots \oplus K_n \subseteq K$, K_i is not essential in P_i for at least one i, $1 \leq i \leq n$. Thus $f(P_i)$ is a cyclic hollow non-cosmall submodule of M.

□

Theorem 1.10. Let R be a semiperfect ring. Then the following are equivalent.

i) Direct sum of any two indecomposable projective is extending.
ii) Every indecomposable projective R-module is uniform and any finitely generated essential extension of an indecomposable projective is projective.
iii) R satisfies $(*)_1^*$.
iv) R^n is an extending module $\forall\ n \in \mathbb{N}$.
v) Every finitely generated R-module is a direct sum of a projective module and a singular module.
vi) Every indecomposable projective R-module is uniform and any finitely generated essential extension of a finitely generated projective R-module is projective.

Proof. i) $\Rightarrow$ ii) : Let M be an indecomposable projective module. Then M is uniform as M is extending. Let $M \subseteq K \subseteq E(M)$ be such that K is finitely generated. We can find a generating set $\{x_1, x_2, \cdots, x_n\}$ such that each x_iR is cyclic local and $x_1R = M$. Suppose $L = x_1R + x_2R$. By Lemmas 1.6 and 1.7, L is projective. Since M is uniform and essential in L, L is indecomposable projective. By finite induction we get that K is an indecomposable projective module.

ii) $\Rightarrow$ iii) : Let L be a finitely generated non-cosmall R-module. By Lemma 1.9, L contains a cyclic hollow non-cosmall module N. Since every indecomposable projective is uniform, N is indecomposable projective. The identity map $N \to N$ can be extended to a map $g : L \to E(N)$. As $g(L)$ is finitely generated, it is projective. Hence L contains a projective summand.

iii) $\Rightarrow$ *iv*) : Let M be a non-essential submodule of R^n. As R is semiperfect any projective direct summand of R^n can be expressed as a direct sum of atmost nk indecomposable projective modules, where R is a direct sum of k indecomposable projective modules. We can choose a direct summand P of R^n containing M which has a minimal number of direct summands in its representation as a direct sum of indecomposable projectives. We claim that M is essential in P. If not, then P/M is a finitely generated non-cosmall module and hence has a non-zero projective summand T/M. If $f : P \to P/M$ is the natural map and $g : P/M \to T/M$ is the projection map, then $g \circ f$ splits. This implies that P contains a proper summand containing M, which contradicts the choice of P. Hence R^n is an extending module.

iv) $\Rightarrow$ *i*) and *vi*) $\Rightarrow$ *ii*) are obvious.

iv) is equivalent to *v*) (Lemma 1.6).

iv) $\Rightarrow$ *vi*) (Corollary 1.8).

□

We show next that for perfect ring $(*)^*$ is equivalent to $(*)^*_1$.

Proposition 1.11. Let R be a right perfect ring. The ring R satisfies $(*)^*$ iff it satisfies $(*)^*_1$.

Proof. Suppose R satisfies $(*)^*_1$. Let M be an indecomposable projective module. From Theorem 1.10 every finitely generated submodule of $E(M)$ containing M is indecomposable projective and so has a unique maximal submodule. Hence $E(M)/M$ is a uniserial module. As R is right perfect and there exists only finitely many non-isomorphic indecomposable projectives, it is easily seen that $E(M)/M$ is an Artinian uniserial module. Hence any submodule of $E(M)$ containing M is projective. Let L be a non-cosmall R-module. By Lemma 1.9, L contains a principal cyclic non-cosmall module N. Since every indecomposable projective is uniform, N is indecomposable projective. The identity map $N \to N$ can be extended to a map $g : L \to E(N)$. As $g(L)$ is projective L contains a projective summand.

□

As a corollary we get the following generalization of Theorem 4 of [11].

Corollary 1.12. The following are equivalent for a right perfect ring R.

i) R satisfies $(*)^*$.
ii) R satisfies $(*)^*_1$.
iii) Every finitely generated projective module is extending.
iv) Direct sum of any two indecomposable projective modules is extending.

Next we show that a semiperfect ring R is a right Co-H-ring iff every countably generated projective R-module is extending. This gives us a countably generated version of Theorem II of [9].

Theorem 1.13. For a semiperfect ring R the following are equivalent.

i) R is a right Co-H-ring.
ii) $R^{\aleph_0}$ is extending.
iii) Any countably generated essential extension of a countably generated projective R-module is projective and every indecomposable projective is uniform.
iv) Any countably generated R-module is a direct sum of a projective and a singular module.
v) Every countably generated non-cosmall R-module contains a non-zero projective summand and R satisfies a.c.c. on annihilating right ideals.

Proof. It is obvious that $i) \Rightarrow ii)$, $iii)$, $iv)$ and $v)$.
The equivalence of $ii)$ and $iv)$ follows from Lemma 1.6.

$ii) \Rightarrow ii)$ follows from Corollary 1.8.

$iii) \Rightarrow i)$: Suppose P is a finitely generated projective R-module. If $P \subseteq K \subseteq E(P)$ and K is countably generated, then K is projective. As P is essential in K, K is finitely generated. Hence $E(P)/P$ is a Noetherian module, and every essential extension of P is finitely generated projective. Let $E(R) = e_1 R \oplus \cdots \oplus e_k R$, where each e_i is a primitive idempotent of R. We claim that $E(R)$ is $\sum$-injective. It is enough to prove that for any injective projective indecomposable R-module M, $M^{\aleph_0}$ is injective. Let $N = \bigoplus_{i \in \mathbb{N}} M_i$, where each M_i is isomorphic to M. Suppose $x \in E(N)/N$ and $L = N + xR$. As L is projective $L = L_1 \oplus L_2$, where $x \in L_1$ and L_1 is finitely generated. Now $L_1 + N = L$ implies that the restriction of the projection map of L onto L_2 along L_1 to N is onto L_2, a projective module. Hence $N = (L_1 \cap N) \oplus K$. By Krull-Remak-Schmidt Theorem, $L_1 \cap N$ is isomorphic to a direct sum of copies of M and since L_1 is finite dimensional, $L_1 \cap N$ is injective. Hence $L_1 = (L_1 \cap N) \oplus T$. Then $L = N \oplus T$. As N is essential in L we get $N = L$. Hence N is injective. Therefore, $E(R)$ is $\sum$-injective. It follows that R satisfies a.c.c. on right annihilating ideals. In view of Theorem 3.17 of [9], it is enough to show that R satisfies $(*)^*$. Suppose M is a non-cosmall R-module. By Lemma 1.9, M contains a non-cosmall cyclic hollow submodule N. Since any indecomposable projective R-module is uniform, N is projective. The identity map $f : N \to N$ can be extended to a map $g : M \to E(N)$. Since $g(M)$ is projective, M contains a projective summand.

$v) \Rightarrow i)$: Let M be a indecomposable projective R-module. Any countably generated essential extension of M is projective. As in $iii) \Rightarrow i)$, we get that $E(M)$ is projective and any essential extension of M is projective. Let N be a non-cosmall R-module. By Lemma 1.9, N contains a cyclic non-cosmall hollow module P, which is projective by $v)$. The identity map on P can be extended to a map from N to $E(P)$. Hence N contains a non-zero projective summand. Thus R satisfies $(*)^*$.

□

K. Oshiro has proved that a ring is a generalized uniserial ring iff every quasi-projective is extending (Theorem 2, [10]). For a semiperfect ring we get the following by using Theorem 1 of [13].

Corollary 1.14. Let R be a semiperfect ring. R is a generalized uniserial ring iff every countably generated quasi-projective R-module is extending.

Remark. It is known that a right Co-H-ring, that is every free module is extending, is a right perfect ring. We do not know whether a ring R is semiperfect if $R^{\aleph_0}$ is extending. If it is so, then the assumption that the ring R is semiperfect may be omitted from the hypothesis of Theorem 1.13 and Corollary 1.14.

2. Rings in Terms of Lifting Modules

In this section we prove the dual of the results in Section 1. We consider semiperfect rings R which has essential right socle and $J(E(S)) \neq E(S)$ for every simple submodule S of R. We show that R is a right QF-3 ring iff direct sum of any two indecomposable non-cosmall injective is lifting and that R has a faithful $\sum$-injective right ideal iff any injective R-module with (countably generated) essential socle whose indecomposable direct summands are non-cosmall is lifting. At the end we prove that a ring R is a right H-ring iff $E((R/J)^{\aleph_0})$ is lifting and R is a right or left perfect ring, iff any two-embedded injective R-module is lifting and R is right Artinian, iff R has right essential socle and any injective R-module with essential socle is lifting.

Proposition 2.1. Let R be any ring. If P is a projective R-module, then any non-zero direct summand of $E(P)$ is non-cosmall. For any simple R-module, $E(S)$ is non-cosmall iff S is isomorphic to a submodule of $Soc(R)$.

Proof. We omit the proof as it is easy.

□

Corollary 2.2. If R is a semiperfect ring with essential right socle, then $S_2 = \emptyset$ and $E(B)$ is faithful.

Lemma 2.3. Let R be any ring and M be an R-module such that $M = \bigoplus_{j \in I} M_j$, where each M_j is completely indecomposable, local and any epimorphism M_i to M_j is an isomorphism for i, j in I. If M is a lifting module, then for any i in I, M_i is $\bigoplus_{j \in I'} M_j$-projective, where $I' = I \backslash \{i\}$.

Proof. Let $M' = \bigoplus_{j \in I'} M_j$. Suppose $g : M' \to M'/K$, is the natural map, where $K \subseteq M'$. Let $f : M_i \to M'/K$ be any given map. Consider

$$A = \{x + y \in M : x \in M_i,\ y \in M' \quad \text{and} \quad f(x) = g(y)\}\ .$$

Then A is not small in M. As M is lifting, there exists a direct summand D of M such that $A = D \oplus L$ and L is small in M. Consider the projection $p : M \to M_i$ along M'. Then $M_i = p(A) = p(D) \oplus p(L) = p(D)$. By Krull-Remak-Schmidt Theorem, D is a direct summand of indecomposable modules isomorphic to $M_j's$. There exists a direct summand N of D such that $N \simeq M_k$ for some k in I and

$p(N) = M_i$. By assumption, p is 1-1 on N. Hence $M = N \oplus M'$. Let h be the restriction to M_i of the projection of M to M' along N. Then $g \circ h = f$.

□

Theorem 2.4. Let R be a semiperfect ring with essential right socle for which $J(E(S)) \neq E(S)$ for every $S \in \mathcal{S}_2$. Then the following are equivalent.

i) Direct sum of any two indecomposable non-cosmall injectives is lifting.
ii) Any non-cosmall indecomposable injective is projective.
iii) R is a right QF-3 ring.
iv) $E(B^n)$ is lifting for all $n \in \mathbf{N}$.
v) $E(B^2)$ is lifting.

Proof. *i*) $\Rightarrow$ *ii*) : Let $S, T \in \mathcal{S}_2$. Then $E(S)$ and $E(T)$ are completely indecomposable, local and any epimorphism from $E(S)$ to $E(T)$ is 1-1 for $E(S)$ is uniform and $E(T)$ is non-cosmall. By Lemma 2.3 it follows that $E(S)$ is $E(T)$-projective for every S, T in $\mathcal{S}_2$. Hence $E(B)$ is $E(B)$-projective. Moreover as $JE(S) \neq E(S)$, each $E(S)$ is cyclic. Now $E(B)$ is a faithful finitely generated quasi-projective with a projective cover and hence is projective.

ii) $\Rightarrow$ *iii*) : Now $E(B)$ is a faithful injective projective right ideal of R (Corollary 2.2 and Proposition 2.1). By Theorem 1 of [2], R is a right QF-3 ring.

iii) $\Rightarrow$ *iv*) : As any finitely generated projective module over a semiperfect ring is lifting, *iv*) follows.

iv) $\Rightarrow$ *v*) $\Rightarrow$ *i*) are obvious.

Note: $J(E(S)) \neq E(S)$ for every S in $\mathcal{S}_2$ if either $E(B)$ is a finitely generated or R is right perfect.

Corollary 2.5. Let R be a perfect ring. The $P(A^2)$ is extending iff $E(B^2)$ is lifting and R is finite dimensional. In this case $P(A) \simeq E(B)$.

Proof. Suppose $P(A^2)$ is extending. Then by Proposition 1.2 and Theorem 1.4, R is finite dimensional. Now $E(B) \subseteq E(R)$ and so by Theorem 1.4, $E(B)$ is projective. Now Theorem 2.4 implies that $E(B^2)$ is lifting. Conversely assume that $E(B^2)$ is lifting and R is finite dimensional. Let $P(S) \simeq eR$, where S is simple, be non-small. As Soc (eR) is of finite length, $eR \subseteq \oplus E(S_i)$, where each $S_i \in \mathcal{S}_2$ (Proposition 2.1). Since each $E(S_i)$ is local and eR is non-small, the projection $eR \to E(S_i)$ is onto for at least one i. As $E(S_i)$ is projective, eR is injective. Hence $P(A^2)$ is extending by Theorem 1.5.

□

Theorem 2.6. Let R be a semiperfect ring with essential right socle for which $J(E(S)) \neq E(S)$ for every $S \in \mathcal{S}_2$. Then the following are equivalent.

i) R has a faithful $\sum$-injective right ideal. (In this case R is a right QF-3 ring also.)
ii) $E(B^{|K|})$ is lifting for any set K.

iii) Any injective R-module with essential socle whose indecomposable direct summands are non-cosmall is lifting.
iv) Any countably embedded injective R-module whose indecomposable direct summands are non-cosmall is lifting.
v) $E(B^{\aleph_0})$ is lifting.

Proof. $i) \Rightarrow ii)$: We recall that a direct decomposition $X = \bigoplus_{i \in I} X_i$ of X, where each X_i is completely indecomposable, is called locally semi$-T$ nilpotent if for every sequence $f_n : X_{i_n} \to X_{i_{n-1}}$, $(n \in \mathbf{N})$ of non-isomorphisms, with all i_n distinct, and every $x \in X_{i_1}$, there exists $m \in \mathbf{N}$ with $f_m \circ f_{m-1} \circ \cdots f_1(x) = 0$. Suppose each $X_i = eR$, where e is an indecomposable idempotent of R and I is an infinite set. Any non-isomorphism of $eR \to eR$ is given by a multiplication by an element of eJe on the left. Hence the decomposition of X is locally semi$-T$ nilpotent iff eJ is right T-nilpotent (a subset L of the ring R is called right T-nilpotent if given an infinite set $\{a_n\}$, $n \in N$, of elements in L, there exists an integer m such that a_m, $a_{m-1} \cdots a_1 = 0$).

Let M be a faithful $\sum$-injective right ideal. Then $M = \bigoplus_{i=1}^{k} e_i R$, where each e_i is a primitive idempotent. Let $1 \leq i \leq k$. As $e_i R$ is $\sum$-injective, $N_i = \bigoplus_{j \in K} M_j$, where each $M_j = e_i R$ and K is an infinite set, is locally semi$-T$-nilpotent (cf, [5]). By what we have seen above, we get that $e_i J$ is right T-nilpotent. Hence N_i is a perfect module (Theorem 1, [3]). This implies that M is a perfect module (Theorem 5, [3]). Hence $M^{|K|}$ is lifting module for any set K. It is easy to see $B \simeq$ to a direct summand of M. Hence $ii)$ follows.

$ii) \Rightarrow iii) \Rightarrow iv) \Rightarrow v)$ are obvious.

$v) \Rightarrow i)$: As $E(B^{\aleph_0})$ is lifting, $E(B)$ is $\sum$-injective (Lemma 2.7, [12]). By Theorem 2.4, $E(B)$ is projective and hence $i)$ follows.

□

Any right perfect ring with a faithful $\sum$-injective right ideal is a semiprimary QF-3 ring. Hence combining Theorems 1.6 and 2.6, we get Corollary 2.7.

Corollary 2.7. Let R be a perfect ring. Then the following are equivalent.

i) R is a semiprimary QF-3 ring.
ii) Any projective R-module whose indecomposable direct summands are non-small is extending.
iii) Any countably generated projective R-module whose indecomposable direct summands are non-small is extending.
iv) Any injective R-module whose indecomposable direct summands are non-cosmall is lifting.
v) Any countably embedded injective R-module whose indecomposable direct summands are non-cosmall is lifting.

We next give some characterizations of a ring R in which every finitely embedded injective is lifting.

Proposition 2.8. The following are equivalent for any ring R.

i) R satisfies $(*)_1$.
ii) Every finitely embedded module is a direct sum of an injective module and a small module.
iii) Every finitely embedded injective module is lifting.

Proof. $i) \Rightarrow ii)$: Suppose M is a finitely embedded R-module. If M is non-small, then $M = N \oplus T$, where N is a non-zero injective module. Now $ii)$ follows easily by induction on the length of socle of M.

$ii) \Rightarrow iii)$: Let M be a finitely embedded injective R-module. Suppose N is a submodule of M. Hence by $ii)$ $N = T \oplus L$, where T is injective and L is a small module. Now $M = T \oplus K$, $N = T \oplus (K \cap N)$. As $(K \cap N) \simeq L$, $K \cap N$ is a small module. As M is injective $K \cap N$ is a small submodule of M. Hence $iii)$ follows.

$iii) \Rightarrow i)$ is obvious.

□

The following is dual to Lemma 1.8 and Corollary 1.9, and so its proof is omitted.

Proposition 2.9. Let R be any ring. If M is a small cover of a lifting injective R-module and if M is a direct sum of any injective and a non-small module, then M is injective. Hence if every finitely (countably) embedded injective R-module is lifting, then any finitely (countably) embedded small cover of a finitely (countably) embedded injective R-module is injective and $E(S)$ is a hollow module for every simple R-module S.

We prove below the converse of the last part of the above proposition (finitely embedded case) for a semiperfect ring R with $J(E(S)) \neq E(S)$ for every simple R-module S. This also gives us a finitely generated version of Theorem I in [9].

Theorem 2.10. The following are equivalent for a semiperfect ring R with $J(E(S)) \neq E(S)$.

i) R satisfies $(*)_1$.
ii) Every finitely embedded module is a direct sum of an injective module and a small module.
iii) Every finitely embedded injective module is lifting.
iv) $E((R/J)^n)$ is lifting for every $n \in N$.
v) $E(S) \oplus E(T)$ is lifting for any two simple R-modules S and T.
vi) Every finitely embedded small cover of a finitely embedded injective module is injective and $E(S)$ is hollow for every simple R-module S.

Proof. The equivalence of $i)$, $ii)$, $iii)$ and $iv)$ follows from Proposition 2.8.

$iv) \Rightarrow v)$ is trivial.

$v) \Rightarrow i)$: We first note that for any simple module S, $E(S)$ is a cyclic hollow module. Let M be a finitely embedded non-small module. Then $E(M)$ is finitely generated and $J(E(M)) = E(M)J$ is small in $E(M)$. Hence there exists $x \in M$ such that xR is non-small in $E(M)$. As R is semiperfect, xR contains a non-small cyclic hollow module. Hence it is enough to prove that every cyclic hollow non-small finitely embedded module M is injective. We prove this by induction on $k = \ell(Soc(M))$, the length of Soc (M). By $v)$ the result is true for $k \leq 2$. Assume the result to be true for $k \leq n-1$. Suppose M is a cyclic hollow non-small module with Soc $(M) = \bigoplus_{i=1}^{k} S_i$, where each S_i is simple and $k \leq n$. The map f_i, the restriction of the projection map of $E(M)$ onto $E(S_i)$ to M is onto for at least one i, say for $i = 1$. Then $M \subseteq M_1 + M_2$, where $M_1 = (f_1 + f_2 + \cdots + f_{k-1})(M)$ and $M_2 = f_k(M)$. By induction hypothesis, M_1 is injective. This implies that $k \leq 2$. Hence M is injective.

$iv) \Rightarrow vi)$ follows from Proposition 2.9.

$vi) \Rightarrow i)$: Let M be a finitely embedded non-small R-module, with Soc $(M) = \bigoplus_{i=1}^{k} S_i$, where each S_i is simple. The map f_i, the restriction of the projection map of $E(M)$ onto $E(S_1)$ to M, is onto for at least one i, say for $i = 1$. Let $g : P \to E(S_1)$ be a projective cover. Then there exists a map $h : P \to M$ such that $h \circ f_1 = g$. Now $h(P)$ is a small cover of $E(S_1)$ and hence is injective.

□

Next we give equivalent conditions for every countably embedded injective R-module to be lifting. The proof of Proposition 2.13 is similar to Proposition 2.8 and hence is omitted.

Lemma 2.11. Let R be a ring for which every countably embedded injective R-module is lifting, then R is right Noetherian.

Proof. Let M be a countably generated semisimple R-module. Then $E(M^{\aleph_0})$ is lifting implies that $E(M)$ is $\sum$-injective (Lemma 2.7 [12]). By Theorem 2.4 [7], R is right Noetherian.

□

Lemma 2.12. Let E be an injective module and $\{x_\alpha\}$, $\alpha \in L$ be a generating set for E. If $E^{|L|}$ is lifting, then E contains a cyclic injective submodule.

Proof. Let $M = \bigoplus_{x \in L} E_x$, where $E_x = E$. Consider the map $\phi : M \to E$ where $\phi \mid_{E_x} = 1_E$. Then $\phi\left(\bigoplus_{n \in L} xR\right) = E$. Hence $\left(\bigoplus_{x \in L} xR\right)$ is non-small and so contains a non-zero injective summand K. As K has exchange property, at least one xR contains a cyclic injective submodule isomorphic to a non-zero summand of E.

□

Next we give equivalent conditions for every countably embedded injective R-module to be lifting which can be proved similar to Proposition 2.8

Proposition 2.13. The following are equivalent for a ring R.

i) Every countably embedded injective R-module is lifting.

ii) Every countably embedded R-module is a direct sum of an injective and a small module.

iii) Every countably embedded non-small R-module contains a non-zero injective module and R is right Noetherian.

We give below some equivalent conditions for a right H-ring.

Theorem 2.14. The following are equivalent for a ring R.

i) R is a right H-ring.

ii) R has essential right socle with $J(E(S)) \neq E(S)$, for every simple submodule S of R, and every countably embedded injective R-module M is lifting.

iii) R is right or left perfect and $E((R/J)^{\aleph_0})$ is lifting.

iv) R is right Artinian and any 2-embedded injective R-module is lifting.

v) R has essential right socle and any injective R-module with essential socle is lifting.

Proof. *i*) $\Rightarrow$ *ii*) and *v*) are trivial as right H-ring is right Artinian.

ii) $\Rightarrow$ *iii*) : By Lemma 2.11, $E(R)$ is $\sum$-injective and so it is a finite direct sum of indecomposables. Since R has essential right socle, $E(R) = \bigoplus_{i=1}^{k} E(S_i)$, where each S_i is a simple submodule of R. As $J(E(S_i)) \neq E(S_i)$ and $E(S_i)$ is hollow (being an indecomposable lifting module), $E(S_i)$ is cyclic for each $i = 1, 2, 3, \cdots, k$. Hence $E(R)$ is finitely generated. By Theorem A of [13], R is right Artinian. So *iii*) follows.

iii) $\Rightarrow$ *iv*) : By Lemma 2.11, R is right Noetherian. Hence R is right Artinian.

iv) $\Rightarrow$ *i*) : Let M be a non-small R-module. As R is right perfect, there exists a cyclic non-small submodule N of M. As R is right Artinian, N is finitely embedded and hence has a non-zero injective summand (Theorem 2.10). That is, R satisfies the condition (*). As R is right Noetherian, any injective R-module is lifting.

v) $\Rightarrow$ *i*) : By Lemma 2.11, R is right Noetherian. By Lemma 2.12, $E(S)$ is cyclic for every simple R-module S. Hence $E(R)$ is finitely generated. By Theorem A of [13], R is right Artinian. Thus *i*) follows.

□

REFERENCES

1. R.F. Colby and E.A. Rutter, Jr., *Generalisations of QF-3 algebras*, Trans. Amer. Math. Soc. **153** (1971), 371-386.
2. R.F. Colby and E.A. Rutter, Jr., *QF-3 rings with zero singular ideal*, Pacific J. Math. **28** (1969), 303-308.
3. R.S. Cunnighan and E.A. Rutter, Jr., *Perfect modules*, Math. Zeit. **140** (1974), 105-110.
4. M. Harada, *Non-small and non-cosmall modules*, Ring Theory Conference, University of Antwerpen (1978), 669-690.

5. M. Harada, *On lifting properties of direct sum of hollow modules*, Osaka Jour. Math. **17** (1980), 783-791.
6. M. Harada and K. Oshiro, *On extending property of direct sum of uniform modules*, Osaka Jour. Math. **18** (1981), 767-785.
7. R.P. Kurshan, *Rings whose cyclic modules have finitely generated socle*, Jour. Algebra **15** (1970), 376-386.
8. E.A. Mares, *Semiperfect modules*, Math. Zeit. **2** (1963), 347-360.
9. K. Oshiro, *Lifting modules, extending modules and their applications to QF-rings*, Hokkaida Math. Jour. **13** (1984), 310-338.
10. K. Oshiro, *Lifting modules, extending modules and their applications to generalised uniserial rings*, Hokkaida Math. Jour. **13** (1984) 339-346.
11. Phan Dan, *Right perfect rings with the extending property on finitely generated free modules*, Osaka Jour. Math. **26** (1989), 265-273.
12. N. Vanaja and V.N. Purav, *Characterisation of generalised uniserial rings in terms of factor rings* (to appear in Communications in Algebra).
13. C.I. Visonhaler, *Orders in QF-3 rings*, Jour. Algebra **17** (1971), 149-151.

RINGS ARISING FROM CONDITIONS ON PRERADICALS

ANA M. de VIOLA-PRIOLI and JORGE E. VIOLA-PRIOLI
Departamento de Matemáticas Puras y Aplicadas
Universidad Simón Bolívar
Caracas 89000, Venezuela

ABSTRACT

This article shows how the lattice of kernel functors can be used in order to determine the way properties of a given ring are reflected in the category of modules. Rings whose kernel functors are idempotent, strongly prime rings, linear ordering of the lattice and divisibility of kernel functors are treated. Some open problems are stated.

After the pioneering articles by Gabriel [5] and Goldman [9] and the subsequent papers by the latter ([10] and [11]), it became clear that, at least in the Noetherian case, kernel functors were to play a key role in the description of a non-commutative arithmetic. In the works of Goldman, Teply, Golan and others, we can see the far-reaching consequences of dealing with idempotent kernel functors over Noetherian rings; however, even without the assistance of chain conditions, very interesting results can be obtained when treating kernel functors. The purpose of this expository paper is to show how kernel functors are used both to determine the way properties of a given ring are reflected in the category of modules and to introduce new rings by formulating conditions in terms of kernel functors instead of ideals. To this end, this article presents four problems to which the authors have contributed, gives an up-to-date account of the present state of knowledge, and includes open questions.

Let us recall that a kernel functor (equivalently, a left exact torsion preradical) is a left exact subfunctor of the identity on the category of right R-modules. A kernel functor σ is said to be idempotent if in addition it satisfies $\sigma(M/\sigma(M)) = 0$ for every module M. To every kernel functor there corresponds a unique (Gabriel) filter of right ideals and a unique class of modules, closed under taking submodules, images and arbitrary direct sums. We will refer to these classes as torsion classes, although we warn that this term has been used in a slightly different context elsewhere.

The relevance of the idempotent kernel functors lies in that they correspond to the *hereditary* torsion theories, so establishing the proper setting to build localizing functors and their associated rings and modules of quotients (see [6] and [7]). Particularly chosen idempotent kernel functors (for instance, those corresponding to the Utumi-Lambek dense right ideals or to the Goldie torsion theory) have been extensively researched.

Some (not necessarily idempotent) kernel functors are worth considering in their own right, as is best exemplified by the functors $Z(Z(M)$ being the singular submodule of M) and socle. Why not consider *all* the kernel functors over a given ring and avoid leaving the non-idempotent kernel functors backstage? In so doing, a new approach emerges in which attention is centered in the *set* of kernel functors. In fact, this set $K(R)$ is a complete lattice because there is an ordering defined by setting $\sigma \leq \tau$ if $\sigma(M) \subseteq \tau(M)$ for every R-module M.

The trivial kernel functors, indicated by O and ∞, are defined by setting, respectively, $O(M) = 0$ and $\infty(M) = M$ for every module M. Furthermore, $K(R) - \{\infty\}$ has maximal elements (called coatoms in Golan [8]) and dually $K(R) - \{0\}$ has minimal elements (atoms) (see [21]).

What makes this lattice $K(R)$ the more interesting is the way in which algebraic and ordering notions are intertwined. Giving enough evidence of this is one of the purposes of the rest of this particle.

Absolutely Torsion Free Rings

These rings were introduced by Rubin [20] in the following way: R is a right ATF ring if $\sigma(R) = 0$ for every $\sigma \neq \infty$. Equivalently, a formulation entirely in terms of the lattice $K(R)$ can be given as follows: R is a right ATF ring if there exists a unique idempotent coatom.

As Rubin showed, the ATF rings are prime non-singular rings; they provide a good non-commutative analogue of domains and include, among others, the simple rings and the prime right Goldie rings. This study was extended by Viola-Prioli [24], where an equivalent definition was found and exploited extensively. In fact, the author showed that R is ATF if and only if for every nonzero right ideal I, R embeds into n copies of I (here n is a natural number that depends on I). This, in turn, can be formulated in terms of finitely generated projective modules and their submodules. It is worth noticing that, in their study of a special type of prime rings, Handelman and Lawrence [15] arrived at the ATF rings simultaneously with J. Viola-Prioli. Given the context in which they were working, the rings were called right strongly prime; this and ATF are different denominations for exactly the same class of rings. In [15] an example is presented to show that the property under consideration is not symmetrical.

The approach presented in [24] was successfully extended later by Beachy [1] to obtain the ATF modules (also treated by Wisbauer in [26]). Another way this notion has evolved has been the consideration of strongly prime ideals [12]: P is a *strongly prime ideal* of R whenever R/P is a strongly prime ring. This, in turn, led to the creation of a new radical in the category of rings, the *strongly prime radical*, defined in the obvious way. That this is, in fact, a radical in the sense of Kurosh and Amitsur was shown later by Groenewald and Heyman in [13]. Several authors have studied this radical in specific situations, like the cases of cross products [19] and Ore extensions [4]. Recently, this concept has found a place in the theory of near rings, since Groenewald shows in [14] that the strongly prime radical can be defined and used in the category of near rings as well. In particular, the strongly prime radical contains the Levitzki radical.

Probably the hardest problem to tackle is to determine when a group ring is absolutely torsion free.

Rings for Which Every Kernel Functor is Idempotent

Since the idempotent kernel functors are the tool needed to construct non-commutative rings of quotients, it is natural to ask when these idempotent kernel functors are abundant, a discussion initiated by Viola-Prioli in [25]. The commutative case was treated first, yielding that every kernel functor is idempotent if and only if R is a finite direct product of fields. A surprising consequence of this theorem is that in the commutative case, if every kernel functor is idempotent, $K(R)$ is finite. Whether this last fact remains true in the non-commutative case is still open.

In general, these rings are semiprime non-singular. A curious connection with ATF rings was later established by Handelman [16]: if every kernel functor is idempotent, R is necessarily a finite subdirect product of strongly prime rings. (This condition is not sufficient, however.) Also in the positive direction and generalizing what has been obtained in the commutative case: for polynomial identity rings, if every kernel functor is idempotent, R is semisimple artinian. Again, $K(R)$ turns out to be finite (see [16]).

In the Noetherian case, a contribution by Fenrick [3] must be recalled. In fact, she proved that if R is right Noetherian and every kernel functor is idempotent, then R is a right V-ring (that is, every simple module is injective). Although this result throws some light into these rings, even in the Noetherian case their characterization remains open in spite of what is claimed in [16, p. 122]. Nevertheless, the following three particular instances have been analyzed successfully.

To begin with, say the kernel functor σ *splits* if $\sigma(M)$ is a direct summand of M for every module M. Obviously this includes the case in which $\sigma(M)$ is injective for every M and for every non-trivial kernel functor σ. Clearly, if every σ splits, then every kernel functor is idempotent. In [25] we find the following characterization: $\sigma(M)$ is injective for every M and for every non-trivial kernel functor σ if and only if R is Morita equivalent to a right Noetherian PCI ring. (Recall that R is a PCI ring if every cyclic module non-isomorphic to R is injective.) The second case, also treated in [25], led to the following decomposition theorem: every kernel functor splits if and only if R is Morita equivalent to a finite product of rings, each block being a simple V-domain in which every kernel functor splits. As for the third particular case, we return to Fenrick [3]. A *perfect* kernel functor is what Goldman [9] calls a functor with property T (for tensor); in particular, every perfect kernel functor is idempotent. The characterization due to Fenrick can be summarized as follows: every kernel functor is perfect (both in R-Mod as in Mod-R) if and only if R is a finite product of rings, each block being a simple two-sided Noetherian V-ring with the property that quotients by one-sided essential ideals are semisimple artinian.

Based on all the available examples, it is tempting to conjecture that R is necessarily Noetherian if every kernel functor is idempotent.

$K(R)$ Linearly Ordered

A valuation ring is a commutative ring whose ideals are linearly ordered by inclusion. This notion has been extended to the non-commutative case by several authors who considered rings whose lattice of right ideals are linearly ordered by inclusion, thus creating the so-called generalized valuation rings or right chain rings. In [21] a totally different approach was presented with the introduction of those rings whose lattice of kernel functors are linearly ordered (equivalently, whose Gabriel filters are linearly ordered under inclusion) and evidence was provided that this is an acceptable generalization. Of the many similarities between valuation rings and these rings, only a few samples extracted from that paper will be mentioned here.

These rings include the chain rings, coincide with valuation rings if commutativity is assumed, have their two-sided ideals linearly ordered by inclusion and have a local center. It is, however, unknown whether their center is always a valuation ring.

In [21] it was observed that when $K(R)$ is linearly ordered, there is a unique simple module (up to isomorphisms) and a unique atom, which coincides with socle. There is also a unique coatom, denoted by $\sigma_{\max}$, whose associated filter is the set of weakly essential right ideals that appeared also in [17]. If $\sigma_{\max}$ is idempotent, it follows that R is an ATF ring and the coatom coincides with Z, but it is not known whether this still holds when $\sigma_{\max}$ is not idempotent.

Still assuming that $K(R)$ is linearly ordered, the analysis of the size of $K(R)$ leads to interesting consequences. By Proposition 20 of [21], if the lattice is infinite, either R contains chains of right ideals of any given length or R is non-artinian. On the other hand, if R is right artinian, there is an order-reversing one-to-one correspondence between the kernel functors and the two-sided ideals of R. In this case, if $K(R)$ is linearly ordered, its size coincides with the (finite) number of two-sided ideals of R. It is, therefore, natural to try to decide whether the two-sided ideals are powers of the Jacobson radical J. (Equivalently, is the index of nilpotency of J equal to the cardinality of $K(R)$?) This has been recently answered on the negative by Osofsky and Resco [18]. For finite dimensional F-algebras with linearly ordered two-sided ideals, they proved that the answer is affirmative if and only if F is a perfect field.

There is an abundance of examples of non-commutative rings for which $K(R)$ is linearly ordered both in Mod-R and R-Mod. Despite this fact, the asymmetry of this property was settled in [22], where the authors showed that for every natural number $n \geq 3$ there exists a ring R such that $K(R_R)$ is a linearly ordered lattice of n elements but $K({}_R R)$ is not linearly ordered.

If $K(R)$ is linearly ordered (and R is not simple artinian), it follows that the socle is nilpotent and J contains every proper two-sided ideal of R. (Note how this compares with the valuation ring situation.) Therefore, if R is regular, it is simple. It could be interesting to analyze the maximal ring of quotients of R in the light of the classification of regular rings due to Goodearl.

Every valuation ring has an associated valuation semigroup, whose construction in terms of principal ideals is well known. Let us see how that could be recreated in our general context. For any given ring R, $K(R)$ admits a semigroup structure in the following way: given σ and λ in $K(R)$, let $\mathcal{A}$ and $\mathcal{B}$ be their corresponding torsion classes. Consider the collection $\mathcal{C} = \{M_R\,;$ there exists an exact sequence $(0) \to A \to M \to B \to (0)$ with $A \in \mathcal{A}$ and $B \in \mathcal{B}\}$. This is a torsion class and we let denote by $\sigma * \lambda$ the corresponding kernel functor. This operation is associative and $(K(R), *)$ is a semigroup. When $K(R)$ is linearly ordered and every kernel functor is idempotent, $(K(R), *)$ is a commutative semigroup. Whether this could be the valuation semigroup of a valuation ring was answered in [21]: if $K(R)$ is linearly ordered and every kernel functor is idempotent, then $(K(R), *)$ is the valuation semigroup of a valuation ring if and only if R is simple artinian. Naturally, this should be followed by the analysis of situations (other than requiring that every kernel functor be idempotent) that lead to the commutativity of the binary operation $*$ and, in turn, to a valuation semigroup. On the other hand, taking a sub-semigroup of $K(R)$ could well be worth considering.

Divisibility of Kernel Functors

There is a certain type of non-commutative valuation ring, known as generalized discrete valuation ring, that appears rather frequently in Ring Theory, particularly in the form of localization of iterated skew polynomial rings.

In the terminology of Brungs [2], R is called a (right) *generalized discrete valuation ring* whenever every non-empty collection of right ideals has a unique maximal element. In other words, the right ideals of R are inverse well-ordered by inclusion, which, in turn, means that the lattice of right ideals is a chain, where every element is indicated by an ordinal number belonging to a certain segment $[0, \tau]$.

From the intrinsic point of view Brungs' analysis was complete and left no important feature of these rings unaccounted for. For this reason, in [23] these rings are studied from a different perspective, focusing on the behavior of their torsion classes. Since we are dealing with chain rings, we know that $K(R)$ is linearly ordered; what remains to be seen is how the discreteness is reflected in the lattice $K(R)$. The answer, however, does not lie on the natural ordering of $K(R)$ but on the semigroup $(K(R), *)$, and we proceed to explain why this is so.

Let us extract from [2] two properties that will be used in this section: if R is a generalized discrete valuation ring, it is a right principal ideal ring and every right ideal is two-sided (that is, R is *right invariant*). In addition, since there is a valuation onto $[0, \tau]$, the right ideals are linearly ordered under divisibility. This is the fact that one would like to see reproduced in $K(R)$.

We will say $K(R)$ *is linearly ordered under divisibility*, if, given σ and λ in $K(R)$, there exists $\mu \in K(R)$ such that either $\sigma = \mu * \lambda$ or $\lambda = \mu * \sigma$, that is either λ divides σ or σ divides λ. It is, therefore, natural to ask whether $K(R)$ is linearly ordered under divisibility when R is a generalized discrete valuation ring. An affirmative answer was provided in [23]. Here some comments on the proof will be inserted.

a) In terms of their corresponding filters, the product of two given kernel functors amounts to $\mathcal{F} \cdot \mathcal{G} = \{I_R ; \exists\, G \in \mathcal{G}$ such that $G \supseteq I$ and $\forall\, g \in G, (I : g) \in \mathcal{F}\}$, a fact that was used in [23] for computations.
b) Since a generalized discrete valuation ring is a right chain ring, the filters are easily surveyed (see [21]).
c) Since a generalized discrete valuation ring is a right invariant ring, the product of filters can be computed in manageable terms.
d) To see that $\mathcal{G}$ divides $\mathcal{F}$, a filter $\mathcal{X}$ must be found that satisfies $\mathcal{F} = \mathcal{X} \cdot \mathcal{G}$. This is done by considering the many cases that may occur, with the help of the expansion of every ordinal number as a sum of powers of ω. The fact that R is Noetherian is crucial and ensures that a limit process needed along the way works.

Also in [23] it is shown rather easily that the condition that $K(R)$ be linearly ordered under divisibility is not symmetrical. The article concludes with the construction of a valuation ring whose kernel functors are not linearly ordered under divisibility, on either side. Given that $\sigma * \lambda \geq \sigma$ always holds, that example shows that, of the two orderings mentioned here, one is strictly stronger than the other.

It should be remarked that there exists a non-division ring that satisfies the four properties considered in this article. In fact, the ring of differential polynomials over a Kolchin universal field is such an example.

In closing, the lattice $K(R)$ holds significant information about a given ring and is the source of many interesting problems. This makes of $K(R)$ an object that deserves further research.

REFERENCES

1. J. Beachy and W.D. Blair, *Rings whose faithful left ideas are cofaithful*, Pac. J. Math. **58** (1975), 1-13.
2. H. Brungs, *Generalized discrete valuation rings*, Can. J. Math. **21** (1969), 1404-1408.
3. M. Fenrick, *Conditions under which all preradical classes are perfect hereditary torsion classes*, Comm. Algebra **2**(4) (1974).
4. M. Ferrero, *The strongly prime radical of an Ore extension*, Comm. Algebra, **17**(2) (1989), 351-376.
5. P. Gabriel, *Des catégories abéliennes*, Bull. Soc. Math. France **90** (1962), 323-448.

6. J. Golan, *Localization of Noncommutative Rings*, Marcel Dekker, New York (1975).
7. J. Golan, *Torsion Theories*, Longman Scientific and Technical, Harlow (1986).
8. J. Golan, *Linear Topologies on a Ring: An Overview*, Longman Scientific and Technical, Harlow (1987).
9. O. Goldman, *Rings and modules of quotients*, J. Algebra **13** (1969), 10-47.
10. O. Goldman, *A Wedderburn-Artin-Jacobson structure theorem*, J. Algebra **34** (1974).
11. O. Goldman, *Elements of noncommutative arithmetic I*, J. Algebra **35** (1975), 308-341.
12. K. Goodearl, D. Handelman and J. Lawrence, *Strongly Prime and Completely Torsion-Free Rings*, Monograph, Carleton University Press (1974).
13. N.J. Groenewald and G.A. Heyman, *Certain classes of ideals in group rings II*, Comm. Algebra **9**(2) (1981), 137-148.
14. N.J. Groenewald, *Strongly prime near-rings*, Proc. Edinburgh Math. Soc. **31**(2) (1988), No. 3, 337-343.
15. D. Handelman and J. Lawrence, *Strongly prime rings*, Trans. Amer. Math. Soc. **211** (1975), 209-223.
16. D. Handelman, *Strongly semiprime rings*, Pacific J. Math. **60**(1) (1975), 115-122.
17. W.K. Nicholson & B. Sarath, *Rings with a largest linear topology*, Comm. Algebra **13** (1985), 769-780.
18. B. Osofsky & R. Resco, *Artinian Rings with Linearly Ordered Two-Sided Ideals*, Proceedings of the Colorado Springs Methods in Module Theory Conference, to appear.
19. M.M. Parmenter, D.S. Passman and P.N. Stewart, *The strongly prime radical of crossed products*, Comm. Algebra **12**(9) (1984), 1099-1113.
20. R.A. Rubin, *Absolutely torsion-free rings*, Pacific J. Math. **46** (1973), 503-514.
21. A.M. de Viola-Prioli & J. Viola-Prioli, *Rings whose kernel functors are linearly ordered*, Pacific J. Math. **132** (1988), 21-34.
22. A.M. de Viola-Prioli & J. Viola-Prioli, *Asymmetry in the lattice of kernel functors*, Glasgow Math. Journal **33** (1991), 95-97.
23. A.M. de Viola-Prioli & J. Viola-Prioli, *Divisibility of kernel functors*, to appear.
24. J. Viola-Prioli, *On absolutely torsion-free rings*, Pacific J. Math. **56** (1975), 275-283.
25. J. Viola-Prioli, *When is every kernel functor idempotent?*, Canad. J. Math. **27** (1975), 545-554.
26. R. Wisbauer, *On prime modules and rings*, Comm. Algebra **11**(20) (1983), 2249-2265.

RIGHT LOCALLY DISTRIBUTIVE RINGS

MARY H. WRIGHT
Department of Mathematics
Southern Illinois University
Carbondale, IL 62901 - 4408, USA

ABSTRACT

A right locally distributive ring is a semiperfect ring each of whose local direct summands (as a right module) has a distributive lattice of submodules. It is shown that a number of results known for right serial rings hold in this more general setting, particularly with respect to prime ideals and right nonsingular rings. Also Jacobson's conjecture holds in two-sided Noetherian locally distributive rings.

0. Introduction

All rings are associative with identity. Modules are generally right modules, unless otherwise indicated. $J(R)$ (or just J) denotes the Jacobson radical of the ring R. A module is *local* provided that it has a unique maximal proper submodule. An element x in a module M is *local* if xR is a local module. When the ring R is semiperfect, for each local element x in a module M, there exists a local idempotent $e \in R$ such that $x = xe$.

Definition 0.1. A right module M over a ring R is *distributive* if its lattice of submodules is distributive: $\forall A_R, B_R, C_R \leq M_R$, $A \cap (B + C) = (A \cap B) + (A \cap C)$. Equivalently, $\forall A_R, B_R, C_R \leq M_R$, $A + (B \cap C) = (A + B) \cap (A + C)$.

Distributive modules have been studied by Stephenson [7] and Camillo [1]. Camillo characterized distributive modules by the property that each factor module contains at most one copy of each simple module. The following results due to Stephenson will be used frequently in this article:

Proposition 0.2. (Stephenson [7, Prop. 1.1]). A right module M_R over any ring R is distributive iff for all submodules A, $B \leq M$, $Hom_R(A/A \cap B, B/A \cap B) = 0$.

Proposition 0.3. (Stephenson [7, Prop. 1.2, 1.3]). Given a family $\{A_i | i \in I\}$ of right modules over a ring R, the following are equivalent:

(i) $\bigoplus_{i \in I} A_i$ is distributive.

(ii) $\forall X_R < \bigoplus_{i \in I} A_i$, $X = \bigoplus_{i \in I} (X \cap A_i)$.

(iii) $\forall i, j \in I$, $\forall B_i < A_i$, $\forall B_j < A_j$, $Hom(A_i, A_j/B_j) = 0$.

Proposition 0.4. (Stephenson [7, Thm. 1.6, Cor. 2]). Any finite direct sum of cyclic submodules of a distributive module is cyclic.

Definition 0.5. A ring R is *right locally distributive*(denoted $r.LD$) if R is semiperfect and for each local idempotent $e \in R$, eR is distributive.

A locally distributive (LD), (resp. Noetherian, etc.) ring is understood to be both left and right locally distributive (resp. Noetherian, etc.). Recall that a module is *uniserial* if its submodules are linearly ordered by inclusion and a ring is *right serial* if it is a direct sum of uniserial right ideals. It was observed by Fuller in [4] that a ring R is $r.LD$ iff 1 is a sum of local orthogonal idempotents, $1 = \sum\{e|e \in F\}$ (for some set F), and for all e, $f \in F$, eRf is a uniserial right fRf module. For this reason, we may view $r.LD$ rings as generalizations of right serial rings. From this point of view, it is natural to ask how much of the known theory of right serial rings generalizes to $r.LD$ rings.

1. The Unique Maximal Essential Direct Sum of Uniform Submodules

The following provides a useful tool throughout.

Proposition 1.1. If X is a distributive module over a $r.LD$ ring R, then X contains a unique maximal essential direct sum of uniform submodules. The number of summands is less than or equal to the cardinality of a complete set of local idempotents in R.

Proof. By considering projective covers and applying Prop. 0.3(iii), if 1 is a sum of n local orthogonal idempotents in R, then the number of summands in any direct sum of (nonzero) submodules of X is bounded above by n. Let $A = A_1 \oplus A_2 \oplus \ldots \oplus A_t$ be a direct sum of nonzero submodules of X with t maximal. Then $t \leq n$, A is essential in X and A_i is uniform for $i = 1, \ldots, t$. If $B = B_1 \oplus B_2 \oplus \ldots \oplus B_s$ is another essential direct sum of (nonzero) uniform submodules of X, then $A_1 \cap B \neq 0$. Distributivity implies $A_1 \cap B = (A_1 \cap B_1) \oplus \ldots \oplus (A_1 \cap B_s)$. Since A_1 is uniform, re-indexing if necessary, we may assume $A_1 \cap B_1 \neq 0$ and $A_1 \cap B_j = 0$ for $j = 2, \ldots, s$. Also, $B_1 \cap A_i = 0$ for $i = 2, \ldots, t$. It follows that $s = t$, and (re-indexing if necessary), $A_i \cap B_i \neq 0$ and $A_i \cap B_j = 0$ for $i \neq j$.

We will show that A_1 is an essential submodule of $A_1 + B_1$. Indeed, given $0 \neq a + b \in A_1 + B_1$ (with $a \in A_1$, $b \in B_1$), there exists $r \in R$ such that $0 \neq (a+b)r \in A_1 \oplus A_2 \oplus \ldots \oplus A_t$. Then $br \in (A_1 \oplus A_2 \oplus \ldots \oplus A_t) \cap B_1 = A_1 \cap B_1$. Hence $0 \neq (a+b)r \in A_1$. Moreover, the sum $\sum_{i=1}^{n} (A_i + B_i)$ is direct. It follows easily that X contains a unique maximal essential direct sum of uniform submodules.

Following Warfield [9], given a finitely generated module M over a semiperfect ring R, $Gen(M)$ is defined to be the number of simple summands in a decomposition of M/MJ as a direct sum of simple R-modules.

Corollary 1.2. Let R be $r.LD$ with a complete set F of n local orthogonal idempotents. Suppose that M is a module with $Gen(M) = m$. Then for any $f.g.$ submodule $N \leq M$, $Gen(N) \leq nm$.

Proof. By induction on m. The case $m = 1$ is covered by Prop. 1.1. If $Gen(M) = m > 1$, select a local cyclic submodule bR, where $b = bg \in M - MJ$ for some local idempotent g. Consider the exact sequence $0 \to bR \cap N \to N \to N/bR \cap N \to 0$. Since R/J is semisimple and bR is distributive, $bR \cap N/(bR \cap N)J$ is a direct sum of k simple modules, for some $k \in \{0, \ldots n\}$. Hence $GenN \leq n + Gen(N/bR \cap N) = n + Gen(bR + N/N) \leq n + n(Gen(M/bR)) \leq n + n(m-1) = nm$.

Corollary 1.3. Let R be $r.LD$ and Y a $f.g.$ module. Then Y contains an essential finite direct sum of local cyclic uniform submodules.

Proof. Cor. 1.2. implies that the number of (nonzero) summands in a direct sum of submodules of Y is bounded above by $nGen(Y)$. The argument of the first part of Prop. 1.1 completes the proof.

2. Non–singular Right Locally Distributive Rings

Throughout this section, R denotes a $r.LD$ ring, Z the right singular ideal of R. For a module M, $Z(M)$ denotes the singular submodule of M. In this section, we show that right locally distributive rings share a nice non–singularity property with their right serial cousins (see Thm. 2.1; cf. Upham [8], Prop. 1.1). We also give a characterization of nonsingular right locally distributive rings in the spirit of Warfield's characterization of nonsingular right serial rings.

Theorem 2.1. Let R be a $r.LD$ ring. Then for any module M, $M/Z(M)$ is a nonsingular R/Z-module.

Proof. Let $(\,)'$ denote classes modulo $Z(M)$ (resp. Z). Suppose $0 \neq m' \in Z(M'_{R'})$. Without loss of generality, we may assume $m = me$ for some local idempotent $e \in R$. There is an essential right ideal I' in R' such that $m'I' = 0'$. In view of Prop. 0.4 and 1.1, we may assume $I' = (eyR + (1-e)R + Z)/Z$ for some $y = ey = \sum_{i=1}^{t} yf_i$, where the f_i are distinct local orthogonal idempotents for $i = 1, \ldots, t$. Since my $\in Z(M)$, there is an essential right ideal $A \leq R$ such that $myA = 0$. But since $m \notin Z(M)$, $eyA \oplus (1-e)R$ is not essential in R. Again by Prop. 1.1 and 0.4, there exists $x \in eR$ such that $yA \oplus xR$ is essential in eR and $x = \sum_{j=1}^{s} xe_j$ for some subset $\{e_j\}$ of a complete set of local orthogonal idempotents. As $m \notin Z(M)$, we may assume without loss of generality that $mxe_1 \neq 0$. If $mxe_1 \in Z(M)$, set $y^{(1)} = 1$. If not, by iterating the above argument, there exists $y^{(1)} = e_1y^{(1)}$ such that $mxy^{(1)} \in Z(M)$ and then there is an essential right ideal B in R such that $mxy^{(1)}B = 0$. As above, $yA \oplus xy^{(1)}B$ is not essential in eR so there exists $x^{(1)}$ such that $yA \oplus xy^{(1)}B \oplus x^{(1)}R$ is essential in eR. Since the number of nonzero summands in a direct sum of submodules of eR is bounded above, this process must terminate with the conclusion that $m \in Z(M)$.

Corollary 2.2. If R is a $r.LD$ ring, then R/Z is a right nonsingular $r.LD$ ring.

Corollary 2.3. If R is a $r.LD$ ring, then R/Z has *acc* and *dcc* on right annihilators.

Proof. This follows from Chatters and Hajarnavis [3], Lemma 6.11.

Recall that if R is a right serial ring, then R is right nonsingular if and only if R is right semihereditary (Warfield [9], Thm. 4.1). We now develop a parallel result for $r.LD$ rings (Thm. 2.6). Let F denote a complete set of local orthogonal idempotents for R. Given $e \in F$, $\tilde{U}(e)$ denotes the maximal essential direct sum of uniform submodules of eR, and its summands are denoted by $U_i(e)$ for $i = 1, \ldots, t(e)$. Recall that a submodule N in a module M is *closed in* M provided N has no proper essential extension in M.

Proposition 2.4. With notation as above, suppose X is a nonsingular and distributive module with maximal essential direct sum of uniform submodules $\tilde{V} = V_1 \oplus \ldots \oplus V_s$. Then $\forall k = 1, \ldots, s, V_1 \oplus \ldots \oplus V_k$ is closed in X.

Proof. By induction on k. If $k = 1$ and A is a proper essential extension of V_1 in X, then by maximality of $\tilde{V}$, $A \cap (V_2 \oplus \ldots \oplus V_s) \neq 0$. Since V_1 is essential in A, we have $(A \cap V_1) \cap (V_2 \oplus \ldots \oplus V_s) \neq 0$, contradiction.

Assume that $V_1 \oplus \ldots \oplus V_k$ is closed in X and suppose A is a proper essential extension of $V_1 \oplus \ldots \oplus V_{k+1}$ in X. Select an injective hull W of $V_1 \oplus \ldots \oplus V_{k+1}$ in $E(X)$ such that $W \supseteq A$. Then $W = W_1 \oplus \ldots \oplus W_{k+1}$, where each W_i is an injective hull of V_i. Let $a \in A - (V_1 \oplus \ldots \oplus V_{k+1})$. Without loss of generality, we may assume $a = af$ for some $f \in F$. Then $a = w_1 + \cdots + w_{k+1}$ for some $w_j = w_j f \in W_j$, $j = 1, \ldots, k+1$. If $w_{k+1} = 0$, then $aR + (V_1 \oplus \ldots \oplus V_k) = (w_1 + \cdots + w_k)R + (V_1 \oplus \ldots \oplus V_k)$ is a proper essential extension of $V_1 \oplus \ldots \oplus V_k$ in X, contradicting the induction hypothesis.

For each $j = 1, \ldots, k+1$, the obvious map $\varphi_j : fR \to w_j R$ (sending f to $w_j f$) has non-essential kernel, since X is non-singular. Hence for each $j = 1, \ldots, k+1$, there is at least one summand $U_{i(j)}(f)$ of $\tilde{U}(f)$ such that $\varphi_j|_{U_{i(j)}(f)}$ is injective. Then for $m \neq i(j)$ in $\{1, \ldots, t(f)\}$, $U_m(f) \subseteq \ker \varphi_j$. Otherwise $w_j U_{i(j)}(f) \cap w_j U_m(f) \neq 0$, which yields local elements $x \in U_{i(j)}(f)$ and $y \in U_m(f)$ with the same projective cover, contradicting Prop. 0.3(iii). Hence for each $j = 1, \ldots, k+1$, there is a unique $i = i(j) \in \{1, \ldots, t(f)\}$ such that $\varphi_j|_{U_{i(j)}(f)}$ is injective and $\bigoplus_{m \neq i(j)} U_m(f) \subseteq \ker \varphi_j$.

Now suppose that for $h \neq j \in \{1, \ldots, k+1\}$ we have $i(j) = i(h) = i$. Then $0 \neq w_j U_i(f) \cap V_j$ and $0 \neq w_h U_i(f) \cap V_h$. Say $0 \neq v_j = w_j x \in V_j$ and $0 \neq w_h y \in V_h$, with $x = fx$ and $y = fy \in U_i(f)$. Since $U_i(f)$ is uniform, $\exists 0 \neq z \in xR \cap yR$, and we may assume $z = zg$ for some $g \in F$. Since φ_j and φ_h are injective on $U_i(f)$, we have $0 \neq w_j z \in V_j$ and $0 \neq w_h z \in V_h$, which again contradicts Prop. 0.3(iii).

Thus, after re-indexing if necessary, we may assume that for $j = 1, \ldots, k+1$, $w_j U_j(f) \neq 0$, and $w_j U_h(f) = 0$ whenever $h \neq j$. We than have $aR \cap V_{k+1} \supseteq aU_{k+1}(f) \cap V_{k+1} = w_{k+1} U_{k+1}(f) \cap V_{k+1} \neq 0$. Hence $V_{k+1} + A$ is a proper essential extension of V_{k+1} in X, contradiction.

Corollary 2.5. With the same hypotheses as Prop. 2.4, $X = \tilde{V}$.

Corollary 2.6. If R is $r.LD$ and right nonsingular, then for each local idempotent $e \in R$, eR is uniform.

Proof. Prop. 2.4 implies that $\tilde{U}(e) = eR$, from which it follows that $t(e) = 1$.

Theorem 2.7. For a $r.LD$ ring with a complete set F of local orthogonal idempotents, $t.f.a.e$:

(i) R is right nonsingular.

(ii) Every nonzero right ideal of R contains an essential finitely generated projective submodule.

(iii) Every nonzero submodule of a $f.g.$ projective right module contains an essential $f.g.$ projective submodule.

Proof. (i) $\Rightarrow$ (ii) Let X be any nonzero right ideal of R. By Cor. 1.3, X contains an essential finite direct sum of local uniform cyclic submodules. Since R is right nonsingular, for each $e \in F$, eR is uniform and for any $0 \neq a = ae$ in X, the obvious map $eR \to aR$ is an isomorphism. Hence X contains an essential $f.g.$ projective submodule.

(ii) $\Rightarrow$ (iii) If K is a submodule of a $f.g.$ projective module, we may assume K is a submodule of $\bigoplus_{i=1}^{k} Q_i$ for some local projective modules Q_i and some $k \in \mathbb{N}$. Then for at least one canonical projection $\pi : \oplus Q_i \to Q_h$, $\pi(K) \neq 0$. By (ii), $\pi(K)$ contains an essential $f.g.$ projective submodule P. Consider

$$\begin{array}{ccccc} \pi^{-1}(P) & \hookrightarrow & K & \hookrightarrow & \bigoplus_{i=1}^{k} Q_i \\ \downarrow & & \downarrow & & \downarrow \\ P & \hookrightarrow & \pi(K) & \hookrightarrow & Q_h \end{array}$$

Then there exists $\mu : P \to \pi^{-1}(P)$ such that $\pi^{-1}(P) = \mu(P) + (1-\mu\pi)(\pi^{-1}P)$. Now $(1-\mu\pi)(\pi^{-1}P) \subseteq \ker \pi \subseteq \bigoplus_{i \neq h} Q_i$, so inductively, we may assume $(1-\mu\pi)(\pi^{-1}P)$ contains a $f.g.$ essential projective module Q. Then $\mu(P) + Q$ is essential, $f.g.$ and projective in $\pi^{-1}(P)$, which in turn is essential in K.

(iii) $\Rightarrow$ (i) Given a local idempotent $e \in F$, if $e \neq 0$, then $eZ \cap U_i(e) \neq 0$ for some uniform direct summand $U_i(e)$ of $\tilde{U}(e)$. By (iii), $eZ \cap U_i(e)$ contains an essential, *f.g.* projective submodule P, which is necessarily isomorphic to fR for some $f \in F$. Say $P = xR$ with $x = xf$. Then the kernel of the obvious homomorphism $fR \to P$ is both essential in fR and a direct summand of fR, contradiction.

Applying well-known results (see for example Stenstrom [6], sections 19–22) we have for a right non-singular, *r.LD* ring R, the maximal ring of quotients, Q, is isomorphic to $End(E(R))$; Q is semisimple and R has *acc* on right annihilators of subsets of $E(R)$. If in addition R is semiprime, Q coincides with the classical ring of right quotients of R.

3. Prime Ideals in Right Locally Distributive Rings

In this section we investigate properties of prime ideals in *r.LD* rings. We show that the local direct summands of a semiprime *r.LD* ring are uniform. Generalizing results known for serial rings, we show that incomparable prime ideals of a *r.LD* ring are comaximal (cf. Mueller and Singh [5]). Throughout this section, R denotes a *r.LD* ring with a complete set of local orthogonal idempotents F.

Proposition 3.1. If R is a prime *r.LD* ring, then for each local idempotent $e \in F$, eR is uniform.

Proof. By Prop. 1.1, given $e \in F$, eR contains an essential direct sum of local cyclic uniform modules $y_1R \oplus \ldots \oplus y_tR$ for some $t \geq 1$. By Prop. 0.3(iii) for $i = 2, \ldots, t$, $y_1Ry_i = 0$. Since R is prime and $y_1 \neq 0$, it follows that $t = 1$.

Proposition 3.2. Incomparable prime ideals of a *r.LD* ring R are comaximal.

Proof. For any local idempotent $e \in F$, eRe is uniserial, hence for prime ideals P and Q, either $ePe \subseteq eQe$ or vice versa. If P and Q are incomparable, it follows that for all $e \in F$, either $e \in P$ or $e \in Q$. Hence $1 \in P + Q$.

Corollary 3.3. (cf. Mueller and Singh [5]). Suppose P and Q are prime ideals of a *r.LD* ring R.

(i) If M is a local cyclic module such that $MP \subset M$ and $MQ \subset M$, then P and Q are comparable.

(ii) If $\forall e \in F$, $e \in P \Rightarrow e \in Q$, then P and Q are comparable.

(iii) If in addition to (ii) $\exists f \in Q - P$, then $P \subset Q$.

Following Mueller and Singh, given a subset $T \subseteq F$, define $\mathcal{P}(T)$ to be $\{P \in \operatorname{Spec} R | P \cap F = T\}$ and let $\mathcal{T} = \{T \subseteq F | \mathcal{P}(T) \neq \emptyset\}$. The proofs of the following are essentially unchanged from the corresponding results on right serial rings (cf. [5]):

Proposition 3.4. Let R be a *r.LD* ring with complete set of local orthogonal idempotents F.

(i) For any subset $T \subseteq F$, $\mathcal{P}(T)$ is a chain (possibly empty).

(ii) Given $T, T' \subseteq F$, then $T \subset T' \iff \forall P \in \mathcal{P}(T), \forall P' \in \mathcal{P}(T'), P \subset P'$.

(iii) If P is a minimal prime, then $P \cap F$ is minimal in $\mathcal{T}$. Conversely, if T is minimal in $\mathcal{T}$, then $\cap\mathcal{P}(T)$ is a minimal prime ideal of R.

(iv) Every semiprime ideal S in R is the intersection of a finite set of primes minimal over S.

Corollary 3.5. A semiprime $r.LD$ ring is a direct product of prime $r.LD$ rings. In this case, each local direct summand of R is uniform.

4. Noetherian LD Rings

In this section, we extend some results on Noetherian serial rings to Noetherian LD rings. In particular we show that Jacobson's conjecture holds for Noetherian LD rings.

Lemma 4.1. (cf [5, Lemma 1.6]). Let M be a distributive module over a right LD ring R and $\{I_s | s \in S\}$ a set of ideals of R. Then for all local elements $x \in M$, $x(\bigcap_{s\in S} I_s) = \bigcap_{s\in S} xI_s$.

Proof. Without loss of generality, $x = xe$ for some local idempotent e. Trivially $x(\bigcap_{s\in S} I_s) \subseteq \bigcap_{s\in S} xI_s$. Suppose $y \in \bigcap_{s\in S} xI_s$. Without loss of generality, $y = yf$ for some local idempotent $f \in R$. Fix some $s_0 \in S$. Then $\forall s \in S$, $\exists u_s \in eI_sf$ such that $y = xu_s$. Hence, $\forall s \in S$, $u_s - u_{s_0} \in r_{eR}(x)$ and so $u_{s_0} \in r_{eR}(x) + u_sR$. It follows that $\forall s \in S$, $u_{s_0}R = u_{s_0}R \cap (r_{eR}(x) + u_sR) = (u_{s_0}R \cap r_{eR}(x)) + (u_{s_0}R \cap u_sR)$. Therefore, $\forall s \in S$, $\exists x_s$, $t_s \in fRf$ such that $u_{s_0} = u_{s_0}t_s + u_sx_s$ with $xu_{s_0}t_s = 0$. If for some $s \in S$, t_s is a unit of fRf, then $y = xu_{s_0} = 0 \in x(\bigcap_{s\in S} I_s)$, and we are done. If this is not the case, then $\forall s \in S$, $u_{s_0}(f - t_s) = u_sx_s$ with $f - t_s$ a unit of fRf; hence $u_{s_0}R \subseteq u_sR \, \forall s \in S$ and $y \in x(\bigcap_{s\in S} I_s)$.

Corollary 4.2. (Corollary to Proof). Let R be a (right and left) locally distributive ring, e, e' and e'' local idempotents of R, $b = e'be''$ an element of R such that $e'be''Re'' = e'Re'be''$ and $a = eae' \in R$. Then for any set $\{I_s | s \in S\}$ of ideals of R, $\bigcap_{s\in S} aI_sb = a(\bigcap_{s\in S} I_s)b$.

Proof. Given $w \in \bigcap_{s\in S} aI_sb$, $\forall s \in S$, $\exists u_s \in e'I_se'$ such that $w = au_sb$. Fix $s_0 \in S$. The proof of Prop. 4.1 essentially shows that either $w = 0$ or $\forall s \in S$, $\exists w_s \in e''Re''$ such that $u_{s_0}b = u_sbw_s$. The hypothesis gives that $\forall s \in S$, $\exists z_s \in e'Re'$ such that $bw_s = z_sb$. Hence $u_{s_0}b = u_sz_sb \in \bigcap_{s\in S} I_sb$. Hence $w = au_sb \in a(\bigcap_{s\in S} I_sb) = a(\bigcap_{s\in S} I_s)b$.

Proposition 4.3. Let R be a Noetherian LD ring in which 1 is a sum of n local orthogonal idempotents. Then $\bigcap_{i=1}^{\infty} J^i = 0$.

Proof. First observe that if T is a semiperfect and right Noetherian ring, then for all right ideals $I \leq T$, for all local idempotents e, $f \in T$, eIf is a $f.g.$ right fTf-module. Using the fact that eRf is uniserial both as a left $eRe-$ and a right fRf-module, for each $k \in \mathbb{N}$, there are local idempotents $e_1(k), \ldots, e_k(k)$ such that $eJ^{k+1}f = eJe_1(k)Je_2(k)\ldots Je_k(k)Jf$. When $k = mn+1$, at least one $e_i(k)$ occurs at least m times. Then, in the sequence $e_1(n+1)$, $\ldots$, $e_{n+1}(n+1)$, $e_1(2n+1)$, $\ldots$, $e_{2n+1}(2n+1)$ $\ldots$, $e_{mn+1}(mn+1)$, $\ldots$, at least one local idempotent, denoted e', occurs infinitely many times. Then $e(\bigcap_{i\in\mathbf{N}} J^i)f \subseteq \bigcap_{i\in\mathbf{N}} eR(e'Je')^iRf$. Since R is Noetherian, $eRe' = ae'Re'$ for some $a = ea \in R$. Since both $e'Je'Rf$ and $e'RfJf$ are properly contained in $e'Rf$, given $b \in e'Rf - (e'Je'Rf \cup e'RfJf)$, we have $e'Rf = e'Re'bf = e'bfRf$. By Cor. 4.2, $e(\bigcap_{i\in\mathbf{N}} J^n) \subseteq a(\bigcap_{i\in\mathbf{N}}(e'Je')^1)b$. Since $e'Re'$ is a Noetherian serial ring, $\bigcap_{i\in\mathbf{N}} (e'Je')^i = 0$.

5. REFERENCES

1. V. Camillo, Distributive modules, *J. Alg.* **36** (1975), 16–25.

2. A. W. Chatters, Serial rings with Krull dimension, *Glasgow Math J.* **32** (1990), 71–78.

3. A. W. Chatters and C.R. Hajarnavis, *Rings with Chain Conditions*, Pitman, London, 1980.

4. K. R. Fuller, Rings of left invariant module type, *Comm. in Algebra* **6** (1978), 153–167.

5. B. J. Mueller and S. Singh, Uniform modules over serial rings, McMaster Math. Reports, Preprint Series No. **6**, 1989-90.

6. B. Stenstrom, *Rings and Modules of Quotients*, Lecture Notes in Mathematics, **237**, Springer-Verlag, New York, 1971.

7. W. Stephenson, Modules whose lattice of submodules if distributive, *Proc. London Math. Soc* (3) **28** (1974), 291–310.

8. M. H. Upham, Serial rings with Krull dimension one, *J. Algebra* **109** (1987), 319–333.

9. R. B. Warfield, Jr., Serial rings and finitely presented modules, *J. Algebra* **37** (1975), 187–222.

ON WEAKLY SEMIPRIMITIVE RINGS

JULIUS M. ZELMANOWITZ
Department of Mathematics, University of California,
Santa Barbara, CA 93106, USA

ABSTRACT

The objective of this article is to show that weak semiprimitivity is a property of rings which is inherited by matrix rings and polynomial rings.

§1. Introduction.

Weakly semiprimitive rings were introduced in [5] as a simultaneous generalization of semiprimitive rings (that is, rings with zero Jacobson radical) and weakly primitive rings (see [3]). As defined in [5], a weakly semiprimitive ring is a ring with a faithful module which is polyform, locally finite dimensional and weakly compressible. A module with the latter three properties is also called a *weakly semisimple module* for consistency with classical terminology. (The reader should be aware that Jain, López-Permouth and Singh use the term "weakly semisimple ring" for a different concept in [2].)

To recall the relevant definitions: a *polyform module* is one in which partial endomorphisms have essentially closed kernels; a *locally finite dimensional* module is one in which finitely generated submodules have finite uniform dimension (denoted dim); and a module M is *weakly compressible* if for every submodule $N \neq 0$ there exists a homomorphism from M into N whose square is not zero.

Weakly semisimple modules should be thought of as simultaneous generalizations of semisimple modules and of monoform compressible modules. The relationship with semisimple modules is perhaps best illustrated by the following observation: a ring R with identity element is semisimple artinian if and only if ${}_RR$ is a semisimple module; whereas a ring R with regular elements is a left order in a semisimple artinian ring if and only if ${}_RR$ is a weakly semisimple module [5, Theorem 3.1].

In [5, Theorem 2.2], weakly semiprimitive rings were shown to have a density property which generalizes both the Chevalley-Jacobson density theorem for semiprimitive rings [1, page 127] and the density theorem for weakly primitive rings [3, Theorem 2.2]. It states, in brief, that a ring R is a "local order" in a direct product of full linear rings if and only if R has a faithful weakly semisimple module.

In the next sections, it is shown that weak semiprimitivity is a Morita invariant of rings and survives passage to polynomial rings in any number of variables.

1991 *Mathematics Subject Classification.* 16N60, 16D60.

This is accomplished by demonstrating that a weakly semisimple module induces a corresponding weakly semisimple module over the Morita-equivalent ring or the polynomial ring.

Throughout this paper, it is assumed that every R-module M satisfies the property that $Rm \neq 0$ whenever $0 \neq m \in M$. In particular, the right annihilator of R is zero. R^1 will denote the ring obtained from R by adjoining an identity element in the customary manner. By an R-module, we will always mean a left module; homomorphisms will always be placed on the side opposite to the scalars. The notation $N < M$ will indicate that N is a submodule of M.

§2. Passage through Morita Contexts.

The objective in this section is to show that weak semiprimitivity is a Morita invariant.

Let $(R, {}_RP_S, {}_SQ_R, S)$ be a Morita context with bilinear mappings $(\, , \,) : P \times Q \to R$ and $[\, , \,] : Q \times P \to S$. For each $p \in P$ define $p^* \in \mathrm{Hom}_{\mathbf{Z}}(Q \otimes_R M, M)$ on generators of $Q \otimes_R M$ by $p^*(q \otimes m) = (p, q)m$. Set $K = \cap_{p \in P} \ker p^*$, a left S-submodule of $Q \otimes_R M$. We may also regard p^* as a homomorphism from $Q \otimes_R M/K \to M$, and we set $M^* = Q \otimes_R M/K$. For $L \subset M^*$, set $P^*(L) = \sum_{p \in P} p^*(L)$. Observe that $P^*(x) \neq 0$ for any $0 \neq x \in M^*$, and that for any $p \in P$, $q \in Q$ and $x \in M^*$, $q \otimes p^*(x) + K = [q, p]x$.

Proposition 2.1. *Assume that (R, P, Q, S) satisfies $(P, Q)m \neq 0$ for any $m \neq 0$ in M. Then the following statements are true.*

1. *For any S-submodule $L < M^*$, $\dim_S L = \dim_R P^*(L)$.*
2. *M^* is a polyform S-module when M is a polyform R-module.*
3. *M^* is a (locally) finite dimensional S-module when M is a (locally) finite dimensional R-module.*
4. *M^* is a (weakly) compressible S-module when M is a (weakly) compressible R-module.*

Proof. (2) and (3) were proved in Proposition 5.2 of [4], where the inequality $\dim_S L \leq \dim_R P^*(L)$ was also proved. The inequality $\dim_S L \geq \dim_R P^*(L)$ can be demonstrated by a similar argument. So it remains to verify statement (4).

It is quite straightforward to verify the following assertions. (i) A (non-zero) partial R-endomorphism $f : N \to N'$ of M induces a (non-zero) partial S-endomorphism $f^* : (Q \otimes N + K)/K \to (Q \otimes N' + K)/K$ of M^* defined by $\sum q_i \otimes n_i + K \to \sum q_i \otimes n_i f + K$ for $q_i \in Q$ and $n_i \in N$. (ii) A partial S-endomorphism $g : L \to L'$ of M^* induces a partial R-endomorphism $g^* : P^*(L) \to P^*(L')$ of M defined by $\sum p_i{}^*(x_i) \to \sum p_i{}^*(x_i f)$ for $p_i \in P$ and $x_i \in L$. (iii) With f as in (i), $f^{**} = f$ on $(P, Q)N$ and, when $N' < N$, $(f^*)^2 = (f^2)^*$. The proofs are left to the reader.

Now suppose that ${}_RM$ is weakly compressible and that L is a non-zero S-submodule of M^*. Then $P^*(L)$ is a non-zero R-submodule of M and so there exists an R-homomorphism $f : M \to P^*(L)$ with $f^2 \neq 0$. Then $f^* : M^* \to$

$(Q \otimes P^*(L) + K)/K = [Q, P]L < L$. Since $f^2 \neq 0$, $(f^*)^2 = (f^2)^* \neq 0$, and so M^* is weakly compressible.

Note also that $P^*(\ker f^*) \subset \ker f^{**} \subset \ker f$. From this it follows that f^* is a monomorphism if f is. Thus ${}_SM^*$ is compressible when ${}_RM$ is.

Theorem 2.2. *Let M be a weakly semisimple R-module and let (R, P, Q, S) be a Morita context satisfying $(P, Q)m \neq 0$ for any $m \neq 0$ in M. Then M^* is a weakly semisimple S-module with $\dim_S M^* = \dim_R M$. Furthermore, M^* is a faithful S-module if, in addition, $(Ps, Q) \neq 0$ whenever $s \neq 0$ in S.*

Proof. Everything except the last sentence follows from the preceding Proposition. For the final assertion, observe that $\{s \in S \mid sM^* = 0\} = \{s \in S \mid (Ps, Q)M = 0\}$.

Corollary 2.3. *Weak semiprimitivity is a Morita invariant of rings with identity.*

Corollary 2.4. *If R is a weakly semiprimitive ring then so is $M_n(R)$, the ring of n-by-n matrices over R.*

§3. Semigroup Rings and Polynomial Rings.

Polynomial rings can be thought of as semigroup rings in the following way. A semigroup G is called a UP-semigroup (or "unique product semigroup") if given finite subsets $\{a_1, \ldots, a_m\}$ and $\{b_1, \ldots, b_n\}$, at least one of the $a_i b_j$ has a unique representation as such a product. A polynomial ring $R[x_i | i \in I]$ can be thought of as a semigroup ring RG where G is the UP-semigroup generated by $\{x_i | i \in I\} \cup \{1\}$. UP-semigroups are a generalization of totally ordered semigroups.

For M a left R-module we set $MG = R^1 G \otimes_R M$. We may identify the elements of MG as unique sums $x = \sum_{g \in G} m_g g$, where $m_g = 0$ for almost all $g \in G$. Given such an element $x = \sum_{g \in G} m_g g$, we set $\|x\| = |\{g | m_g \neq 0\}|$. Observe that, when G is a UP-semigroup, if $gx = 0$ with $g \in G$ and $x \in MG$ then $x = 0$.

Now suppose that G is a UP-semigroup, that L is a left RG-submodule of MG, and that $f \in \mathrm{Hom}_{RG}(L, MG)$ is a non-zero partial endomorphism of MG. Set $v_f = \min\{\|xf\| \mid x \in L \text{ and } xf \neq 0\}$, $\mu_f = \min\{\|x\| \mid x \in L \text{ and } \|xf\| = v_f\}$ and $L_f = \{x \in L | \ \|x\| \leq \mu_f \text{ and } \|xf\| \leq v_f\}$. Fix an element $x \in L_f$ with $\|x\| = \mu_f$ and $\|xf\| = v_f$. Then $f|_{Rx}$ may be regarded as a non-zero R-homomorphism from Rx onto $R(xf)$.

Lemma 3.1. *If $0 \neq f \in \mathrm{Hom}_{RG}(L, MG)$ is given with kernel f an essential RG-submodule of L, then there exists $x \in L$ such that $f|_{Rx} \neq 0$ and kernel $f|_{Rx}$ is an essential R-submodule of Rx.*

Proof. Choose $x \in L_f$ as above with $\|x\| = \mu_f$ and $\|xf\| = v_f$. Let $r \in R$ be arbitrary. If $(rx)f = 0$, then $rx \in \ker f$, so we may assume that $(rx)f \neq 0$. Since $\|(rx)f\| = \|r(xf)\| = v_f$ we may write $(rx)f = m_1 g_1 + \cdots + m_v g_v$ where each $0 \neq m_i \in M$, each $g_i \in G$, and $v = v_f$. By hypothesis, there exists $y \in RG$ with $0 \neq yrx \in \ker f$. We proceed by induction on $\|y\|$ to prove that there exists $s \in R$ with $0 \neq s(rx) \in \ker f|_{Rx}$.

If $\|y\| = 1$, write $y = sh$ for some $s \in R$ and $h \in G$. Then $0 = (yrx)f = y((rx)f) = \sum_{1\leq i\leq v} sm_i hg_i$. From the UP property, $sm_{i'} = 0$ for some i'. Hence $\|s((rx)f)\| < v$, and it then follows from the choice of v that $s((rx)f) = 0$. So $s(rx) \in \ker f|_{Rx}$ and $s(rx) \neq 0$ because $y(rx) \neq 0$.

Now suppose that $\|y\| = k > 1$ and write $y = s_1h_1 + \cdots + s_kh_k$ where each $0 \neq s_j \in R$ and each $h_j \in G$. Then $0 = (yrx)f = y((rx)f) = \sum_{1\leq j\leq k}\sum_{1\leq i\leq v} s_j m_i h_j g_i$. From the UP property, $s_{j'}m_{i'} = 0$ for some choice of j' and i'. Hence $\|(s_{j'}(rx))f\| = \|s_{j'}((rx)f)\| < v$ and so $s_{j'}(rx) \in \ker f|_{Rx}$. If $s_{j'}(rx) \neq 0$ we are done, so we may assume that $s_{j'}(rx) = 0$. Set $y_1 = y - s_{j'}h_{j'}$. Then $y_1(rx) = y(rx) - s_{j'}h_{j'}(rx) = y(rx) \neq 0$, so $0 \neq y_1(rx) \in \ker f$ with $\|y_1\| = k - 1$. The induction hypothesis now completes the proof.

Proposition 3.2. *Suppose that G is a UP-semigroup. Then the following statements are true.*

(1) *MG is a faithful RG-module when M is a faithful R-module.*
(2) *MG is a polyform (respectively, monoform) RG-module when M is a polyform (respectively, monoform) R-module.*
(3) *MG is a (weakly) compressible RG-module when M is a (weakly) compressible R-module.*
(4) *Assume, in addition, that for every uniform R-submodule N of M, NG is a finite dimensional RG-submodule of MG. Then MG is a (locally) finite dimensional RG-module when M is a (locally) finite dimensional polyform R-module.*

Proof. The verification of (1) is straightforward.

(2) Suppose that M is a polyform left R-module and let $f \in \mathrm{Hom}_{RG}(L, MG)$ be given with L an RG-submodule of MG and with kernel f an essential RG-submodule of L. If $f \neq 0$ then, from the preceding Lemma, we know that there exists $x \in L$ with $f|_{Rx} \neq 0$ and with kernel $f|_{Rx}$ an essential R-submodule of Rx. Since $Rx \subset M^{(\mu)}$ and $R(xf) \subset M^{(v)}$ for some positive integers μ and v, $f|_{Rx}$ can be regarded as a partial endomorphism of $M^{(\lambda)}$ with $\lambda = \max\{\mu, v\}$. From Proposition 3.3 of [4], it follows that kernel $f|_{Rx}$ is a closed submodule of Rx. Hence $f|_{Rx} = 0$, a contradiction from which we conclude that $f = 0$. This establishes that MG is polyform as an RG-module.

(3) Suppose that RM is weakly compressible (respectively, compressible) and let L be a non-zero RG-submodule of MG. Choose $x \in L$ with $\|x\|$ minimal among all $x \neq 0$ in L. Write $x = \sum_{1\leq i\leq k} m_i g_i$ with each $m_i \in M$ and each $g_i \in G$. The choice of x implies that for $r \in R$, $rm_i = 0$ for some $1 \leq i \leq k$ if and only if $rx = 0$ if and only if $rm_i = 0$ for all $1 \leq i \leq k$.

Fix $g_o \in G$, and use the UP property to choose $1 \leq i', j' \leq k$ such that $g_o g_{i'} g_{j'}$ is uniquely represented in the set $\{g_o g_i g_j | 1 \leq i, j \leq k\}$. By renumbering if necessary, we may assume that $i' = k$.

By hypothesis, there exists an R-homomorphism (respectively, R-monomorphism) $\alpha : M \to Rm_k$ with $\alpha^2 \neq 0$. α extends in the natural way to an RG-homomorphism from MG into Rm_kG, which we also denote by α. Next, define $\beta : Rm_kG \to L$ by $\sum_{g\in G} r_g m_k g \to \sum_{g\in G} r_g g x$ ($r_g = 0$ for almost all $g \in G$). Using the fact that G is a UP-semigroup, it is straightforward to check that β is an RG-monomorphism. Set $\theta = \alpha\beta : MG \to L$. (Observe that when ${}_RM$ is compressible, θ is an RG-monomorphism, thus proving that ${}_{RG}MG$ is compressible.) It remains only to show that $\theta^2 \neq 0$.

Choose $m \in M$ with $m\alpha^2 \neq 0$, and write $m\alpha = rm_k$ and $m_i\alpha = r_i m_k$ for $1 \leq i \leq k$ and some r, $r_i \in R$. Then $(mg_o)\theta^2 = (mg_o)\alpha\beta\theta = ((m\alpha)g_o)\beta\theta = (rm_kg_o)\beta\theta = (rg_ox)\theta = (\sum_{1\leq i\leq k} rm_ig_og_i)\alpha\beta = \sum_{1\leq i\leq k}((rm_i)\alpha g_og_i)\beta = \sum_{1\leq i\leq k}(rr_im_kg_og_i)\beta = \sum_{1\leq i\leq k} rr_ig_og_ix = \sum_{1\leq i,j\leq k} rr_im_jg_og_ig_j$. Suppose that $(mg)\theta^2 = 0$. From the choice of $g_og_{i'}g_{j'}$, we would then have that $rr_{i'}m_{j'} = 0$ from which it follows that $rr_kx = rr_{i'}x = 0$, contradicting $m\alpha^2 \neq 0$. Hence $\theta^2 \neq 0$.

(4) This proof is accomplished in a series of steps.

First, we note that if N is an essential R-submodule of M then NG is an essential RG-submodule of MG. For, given $0 \neq x \in MG$, write $x = \sum_{1\leq i\leq k} m_ig_i$ with each $m_i \neq 0$ in M and each $g_i \in G$. Fix $g \in G$ and suppose that $gg_{i'}$ is uniquely represented in the set $\{gg_i | 1 \leq i \leq k\}$. Since ${}_RM$ is polyform, there exists $r \in R$ such that $rm_i \in N$ for each $1 \leq i \leq k$ and $rm_{i'} \neq 0$. Hence $(rg)x = \sum_{1\leq i\leq k} rm_igg_i \neq 0$ in NG.

Second, it is easy to check that if $\sum_{i\in I} \oplus M_i \subset M$ is a direct sum of R-submodules of M, then $\sum_{i\in I} M_iG$ is a direct sum of RG-submodules of MG.

Suppose that ${}_RN \subset M$ is finite dimensional. There exists a direct sum $\sum_{1\leq i\leq k} \oplus M_i$ of uniform R-submodules of M which is essential in N. From the hypothesis and the previous paragraphs, we know that $\sum_{1\leq i\leq k} M_iG$ is a direct sum of finite dimensional RG-submodules of MG which is essential in NG. Hence NG is a finite dimensional RG-module. Taking $N = M$ establishes that ${}_{RG}MG$ is finite dimensional when ${}_RM$ is.

Now suppose that ${}_RM$ is locally finite dimensional. Given an element $x = \sum_{1\leq i\leq k} m_ig_i$ in MG, $RGx \subset \sum_{1\leq i\leq k} Rm_iG$. Each Rm_i is finite dimensional and so, by the previous paragraph, each Rm_iG is a finite dimensional RG-submodule of MG. From (2), MG is a polyform RG-module, so $\sum_{1\leq i\leq k} Rm_iG$ is also a finite dimensional RG-module ([4], page 1152). Hence RGx is finite dimensional. Since $x \in RG$ was arbitrary, this proves that ${}_{RG}MG$ is locally finite dimensional.

We now turn to polynomial rings. For M an R-module, and X any set of indeterminates over R, set $M[X] = R^1[X] \otimes_R M$; $M[X]$ has a natural structure of $R[X]$-module.

Theorem 3.3. *If R is a weakly semiprimitive ring then so is $R[X]$.*

Proof. Let M be a faithful weakly semisimple R-module. We already know from the previous Proposition that $M[X]$ is a faithful polyform and weakly compressible

$R[X]$-module. So it remains only to show that $M[X]$ is locally finite dimensional. This follows from (3) of the previous Proposition and the well known fact that $M[X]$ is a uniform $R[X]$-module when M is a uniform R-module.

References

1. N. Jacobson, *Structure of Rings*, Amer. Math. Soc. Colloq. Publ. **37** (1973), Providence.
2. S. K. Jain, S. R. López-Permouth, and S. Singh, *On a class of QI-rings*, Glasgow Math. J. **34** (1992), 75-81.
3. J. M. Zelmanowitz, *Weakly primitive rings*, Comm. in Algebra **9** (1981), 23-45.
4. J. M. Zelmanowitz, *Representations of rings with faithful monoform modules*, J. London Math. Soc. **29** (1984), 237-248.
5. J. M. Zelmanowitz, *Weakly semisimple modules and density theory*, Comm. in Algebra, in press.

A CONVENIENT SOURCE OF HOMOLOGICAL EXAMPLES OVER ARTINIAN RINGS

B. ZIMMERMANN HUISGEN
Department of Mathematics, University of California, Santa Barbara, CA 93106, USA

ABSTRACT

We introduce a simplified quiver for artinian rings R with $J(R)^2 = 0$ which allows finding various homological dimensions without computational effort. For these rings, we prove that the categories of finitely generated modules of finite projective dimension are contravariantly finite in R-mod, resp., mod-R, and we show how to determine the projective dimensions of the corresponding minimal approximations of the simple modules. For arbitrary artinian rings, we present a description of those projective modules which appear as direct summands of higher syzygies.

1. Introduction and notation

At the 1991 summer workshop on homological aspects of finite dimensional algebras, M. Auslander posed the question whether, given an Artin algebra R, the left and right finitistic dimensions of R coincide provided that both the categories $\mathcal{P}^\infty({}_RR)$ and $\mathcal{P}^\infty(R_R)$ of finitely generated left and right R-modules of finite projective dimension are contravariantly finite in R-mod and mod-R, respectively. Recall that $\mathcal{P}^\infty({}_RR)$ is contravariantly finite in R-mod, in the sense of Auslander-Smalø [2] and Auslander-Reiten [1], in case each finitely generated left R-module X has a right $\mathcal{P}^\infty({}_RR)$-approximation, i.e., a homomorphism $f : A \to X$ with A in $\mathcal{P}^\infty({}_RR)$ such that each map from an object of $\mathcal{P}^\infty({}_RR)$ to X factors through f. One of the purposes of this note is to describe a graphically intuitive road to a source of homological examples that can be evaluated without any computational effort. In particular, this class contains specimens answering the above question in the negative.

We start by exhibiting a speedy optical method for determining various homological dimensions, including the finitistic dimensions, for left artinian rings with zero radical square. This is done by replacing the quiver $\Gamma(R)$ of such a ring R by a simplified quiver $\widetilde{\Gamma}(R)$ which, e.g., reduces the computation of the projective dimension of a minimal right $\mathcal{P}^\infty({}_RR)$-approximation of a simple left R-module to a count of the lengths of paths leaving the corresponding vertex in $\widetilde{\Gamma}(R)$. This last application

1991 *Mathematics Subject Classification.* 16E10,16P10,16P20.

This research was partially supported by a National Science Foundation grant.

of $\widetilde{\Gamma}(R)$ is relevant inasmuch as we prove that right $\mathcal{P}^\infty({}_RR)$-approximations always exist for this class of rings. In other words, the $\mathcal{P}^\infty$-categories are contravariantly finite whenever R is artinian with vanishing radical square.

In the third section we upgrade a criterion of Bass [3] and Auslander-Reiten [1] for the vanishing of the *left* finitistic dimension of a *right* artinian ring R (here no restriction is placed on the radical). More specifically, we characterize the projective summands of higher syzygies as follows: an indecomposable projective left module Re is isomorphic to a direct summand of some syzygy $\Omega^d(M)$, where $d \geq 1$ and M is an R-module, if and only if the simple right module eR/eJ does not occur in the right socle of R; here e is a primitive idempotent and J the Jacobson radical of R. For the case of a path algebra modulo relations, we then give a graph-theoretic interpretation of this criterion which will permit us to recognize projective summands of syzygies at a glance.

Throughout R will be an (at least one-sided) artinian ring with Jacobson radical J, and $e_1, \ldots, e_n$ will be a complete set of primitive idempotents. Recall that the left quiver $\Gamma(R)$ of R has vertices $e_1, \ldots, e_n$ and m arrows from e_i to e_j if m is the multiplicity of the simple left R-module $S_j = Re_j/Je_j$ in Je_i/J^2e_i. Whenever we refer to a path in a quiver, we mean a *directed* path, and given paths p, q such that the endpoint of p coincides with the starting point of q, the composite path is denoted qp.

The full subcategory of R-mod having as objects the modules of finite projective dimension will be abbreviated by $\mathcal{P}^\infty(R) = \mathcal{P}^\infty({}_RR)$; the definition of $\mathcal{P}^\infty(R_R)$ is analogous. The left (little) finitistic dimension of R, finally, is

$$\operatorname{l\,fin\,dim} R = \sup\{\operatorname{p\,dim} M \mid M \text{ is an object of } \mathcal{P}^\infty({}_RR)\},$$

where $\operatorname{p\,dim} M$ denotes the projective dimension of M. Observe that, when $J^2 = 0$, we have

$$\operatorname{l\,fin\,dim} R = 1 + \sup\{\operatorname{p\,dim} S \mid S \subseteq J \text{ is a simple module in } \mathcal{P}^\infty({}_RR)\}.$$

2. The homology of left artinian rings with $J^2 = 0$

The following lemma will make it a cinch to find $\operatorname{l\,fin\,dim} R$ on the basis of $\Gamma(R)$.

Lemma 1. *Let R be left artinian with $J^2 = 0$, and let $i \in \{1, \ldots, n\}$. Then the projective dimension of S_i is finite if and only if the quiver $\Gamma(R)$ contains no (directed) path of length ≥ 0 starting in the vertex e_i and ending on an oriented cycle.*

In case this condition is satisfied, $\operatorname{p\,dim} S_i$ equals the length of the longest path of $\Gamma(R)$ starting in e_i.

Proof. Suppose first that $\Gamma(R)$ contains a path starting in e_i and ending in a vertex that belongs to an oriented cycle. Let p be a path with this property having minimal length $\ell \geq 0$. We show $\operatorname{p\,dim} S_i = \infty$ by induction on ℓ. If $\ell = 0$, then the vertex e_i sits on an oriented cycle, of length $r \geq 1$ say. Since clearly S_j is isomorphic to a direct summand of $\Omega^1(S_i)$ if and only if there exists an arrow $e_i \to e_j$, we see that S_i is isomorphic to a direct summand of $\Omega^r(S_i)$, and as $r \geq 1$ we conclude that indeed $\operatorname{p\,dim} S_i = \infty$. Now let $\ell \geq 1$ and write $p = p'\alpha$, where α is an arrow and $\operatorname{length}(p') \geq 0$. If the endpoint of α is e_k, we know that S_k is a direct summand of $\Omega^1(S_i)$. Since $\operatorname{length}(p') = \ell - 1$, the induction hypothesis yields $\operatorname{p\,dim} S_k = \infty$ and thus also $\operatorname{p\,dim} S_i = \infty$.

Now suppose that $\Gamma(R)$ contains no path starting in e_i and ending on an oriented cycle. We prove $\operatorname{p\,dim} S_i = d$ by induction on the number

$$d = \max\{\operatorname{length}(q) \mid q \text{ is a path in } \Gamma(R) \text{ starting in } e_i\}.$$

If $d = 0$, our claim is trivial, so suppose that $d \geq 1$, and let $i(1), \ldots, i(m)$ be the endpoints of arrows starting in e_i. Then $\Omega^1(S_i)$ is a direct sum of copies of the simple modules $S_{i(1)}, \ldots, S_{i(m)}$ and, by the induction hypothesis, $\operatorname{p\,dim}(\bigoplus_{1 \leq j \leq m} S_{i(j)}) = d - 1$. This completes the proof. □

For less confusing optics, we consider the following simplified (left) quiver $\widetilde{\Gamma}(R)$ of R. To obtain $\widetilde{\Gamma}(R)$ from $\Gamma(R)$, delete all those arrows of $\Gamma(R)$ which can be extended to a path terminating on an oriented cycle; in particular, this requires deletion of all arrows that themselves belong to an oriented cycle.

Proposition 2. *If R is left artinian with $J^2 = 0$, then* $\operatorname{l\,fin\,dim} R$ *equals the length of the longest path of the simplified quiver* $\widetilde{\Gamma}(R)$.

Proof. In view of the obvious fact that

$$\operatorname{l\,fin\,dim} R = 1 + \sup\{\operatorname{p\,dim} S \mid S \subseteq J \text{ is a simple left } R\text{-module with } \operatorname{p\,dim} S < \infty\},$$

the claim follows from Lemma 1. □

Our first example illustrates the merits of considering the simplified quiver.

Example 3. Let k be a field and R the path algebra $k\Gamma/I$, where Γ is the quiver

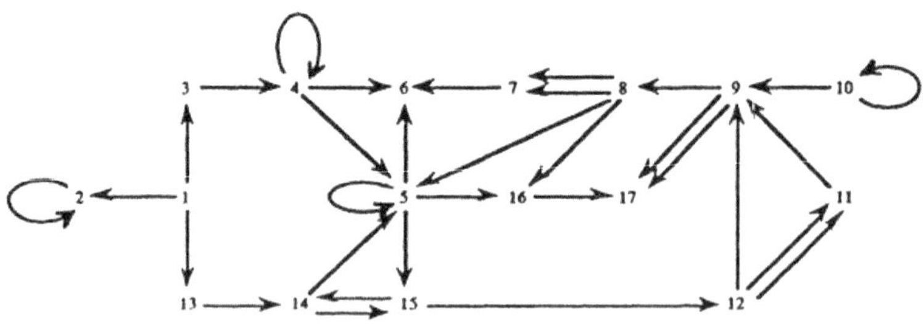

and I the ideal of $k\Gamma$ generated by all paths of length 2. Then the simplified quiver $\widetilde{\Gamma}(R)$ is

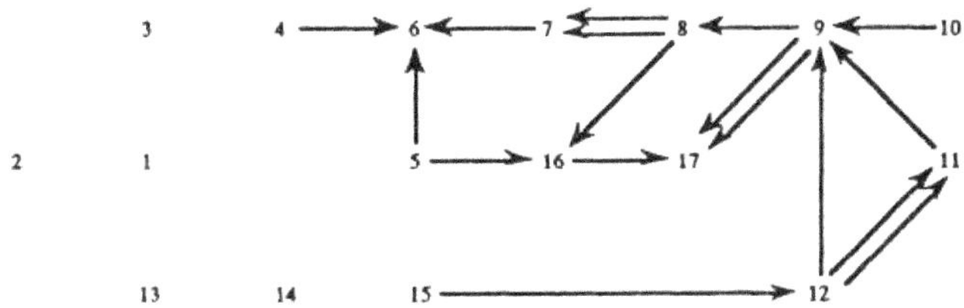

and $\operatorname{l fin dim} R = 6$.

If we reverse the arrow $4 \to 6$ in Γ and let R' be the corresponding algebra with vanishing radical square, then the simplified quiver $\widetilde{\Gamma}(R')$ is

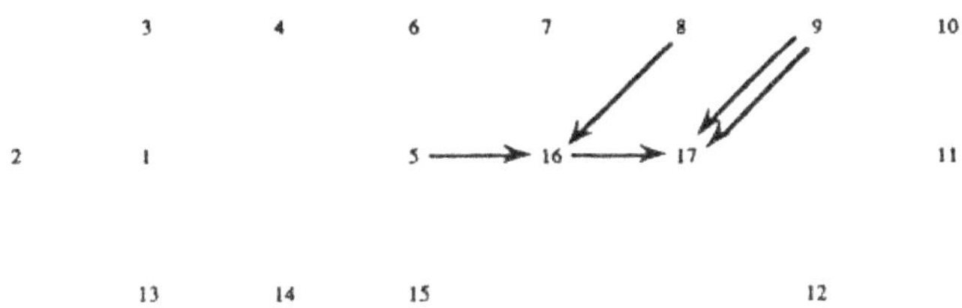

and $\operatorname{l fin dim} R' = 2$. □

Suppose that R is left artinian and X a finitely generated left R-module. Recall the definition of a *right* $\mathcal{P}^\infty(R)$*-approximation* of X from the introduction. A $\mathcal{P}^\infty(R)$-approximation $f : A \to X$ is *minimal* provided A has minimal composition length with this property; in case of existence, minimal $\mathcal{P}^\infty(R)$-approximations are unique up to isomorphism [**1**]. If right $\mathcal{P}^\infty(R)$-approximations exist globally, i.e. for all objects in R-mod, one says that $\mathcal{P}^\infty(R)$ is *contravariantly finite* in R-mod.

In view of an example of Igusa, Smalø, Todorov [**6**], contravariant finiteness of $\mathcal{P}^\infty(R)$ in R-mod may fail, even for finite dimensional algebras with $J^3 = 0$. We include this example to illustrate its simplicity.

Example 4. (See [**6**]) Let Γ be the quiver

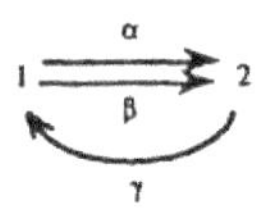

k any field and $R = k\Gamma/\langle\gamma\beta, \alpha\gamma, \beta\gamma\rangle$, i.e., the graphs of the indecomposable projective left R-modules (in the sense of [**4,5**]) are:

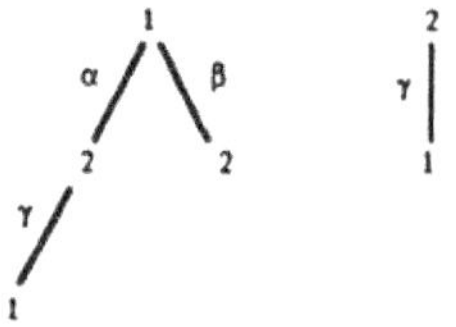

Then Re_1/Je_1 does not have a right $\mathcal{P}^\infty(R)$-approximation by [**6**], and hence $\mathcal{P}^\infty(R)$ is not contravariantly finite in R-mod. Moreover, using [**4**, Section 5] we immediately compute that $\operatorname{l\,fin\,dim} R = 1$. On the other hand, it is clear from the graphs above that

$$\operatorname{Soc}({}_RR) = (Re_1/Je_1)^2 \bigoplus (Re_2/Je_2),$$

whence $\operatorname{r\,fin\,dim} R = 0$ (see Section 3). In particular, $\mathcal{P}^\infty(R_R)$ is contravariantly finite in mod-R. □

By contrast, contravariant finiteness of $\mathcal{P}^\infty(R)$ is guaranteed if $J^2 = 0$. In fact, assuming $J^2 = 0$, we can explicitly describe the minimal right $\mathcal{P}^\infty(R)$-approximation of an arbitrary finitely generated left R-module X.

Start with the obvious fact that any semisimple left R-module Y has a unique decomposition $Y = Y_{\text{fin}} \bigoplus Y_{\text{inf}}$, where Y_{fin} is the sum of the simple submodules of Y having finite projective dimension and Y_{inf} is the sum of the simple submodules of infinite projective dimension.

Theorem 5. *If R is left artinian with $J^2 = 0$, then $\mathcal{P}^\infty(R)$ is contravariantly finite in R-mod. The minimal right $\mathcal{P}^\infty(R)$-approximation of a finitely generated left R-module X is the canonical map*

$$\overline{f} : P/(\Omega^1(X))_{\text{fin}} \to X,$$

where $0 \to \Omega^1(X) \to P \xrightarrow{f} X \to 0$ is a projective cover of X.

In particular, the canonical maps $A_i = Re_i/(Je_i)_{\text{fin}} \to Re_i/Je_i = S_i$ are the minimal right $\mathcal{P}^\infty(R)$-approximations of the simple modules S_i. An alternate description of A_i is as follows:

$$A_i = Re_i/\Big(\bigoplus_{1 \le j \le m(i)} R\alpha_{ij}\Big),$$

where $\alpha_{i1}, \ldots, \alpha_{i,m(i)}$ are precisely the arrows departing from e_i in $\widetilde{\Gamma}(R)$, and $\operatorname{p\,dim} A_i$ is the length of the longest path with starting point e_i in $\widetilde{\Gamma}(R)$.

Proof. To back up the first assertion, let $g \in \operatorname{Hom}(B, X)$, where B is any finitely generated module of finite projective dimension. For a projective cover

$$0 \to \Omega^1(B) \to Q \xrightarrow{h} B \to 0$$

this means that $\Omega^1(B) = \mathrm{Ker}(h) \subseteq (JQ)_{\mathrm{fin}}$. In view of the projectivity of Q we moreover obtain a commutative diagram

$$\begin{array}{ccccc}
Q & \xrightarrow{\ell} & P & \xrightarrow{f} & X \\
\| & & \downarrow{\scriptstyle \mathrm{canon}} & & \| \\
Q & \xrightarrow{\overline{\ell}} & P/(\Omega^1(X))_{\mathrm{fin}} & \xrightarrow{\overline{f}} & X \\
\| & & & & \| \\
Q & \xrightarrow{h} & B & \xrightarrow{g} & X
\end{array}$$

Now $\ell(\Omega^1(B)) \subseteq \ell((JQ)_{\mathrm{fin}}) \subseteq (JP)_{\mathrm{fin}}$, and $\ell(\Omega^1(B)) \subseteq \mathrm{Ker}(f) = \Omega^1(X)$, since $f\ell(\Omega^1(B)) = gh(\Omega^1(B)) = 0$. We conclude that $\ell(\Omega^1(B)) \subseteq (JP)_{\mathrm{fin}} \cap \Omega^1(X) = (\Omega^1(X))_{\mathrm{fin}}$. Consequently, $\overline{\ell}(\Omega^1(B)) = 0$, where $\overline{\ell}$ is as indicated in the above commutative diagram, and so $\overline{\ell} = g'h$ for a suitable homomorphism $g' : B \to P/(\Omega^1(X))_{\mathrm{fin}}$. Since $gh = \overline{f}\overline{\ell} = \overline{f}g'h$ and h is onto, it follows that $g = \overline{f}g'$. Thus the canonical map $\overline{f} : P/(\Omega^1(X))_{\mathrm{fin}} \to X$ is indeed a right $\mathcal{P}^\infty(R)$-approximation of X.

To see that the composition length of $P/(\Omega^1(X))_{\mathrm{fin}}$ is minimal among the lengths of the right $\mathcal{P}^\infty(R)$-approximations of X, let $\phi : A \to X$ be any right $\mathcal{P}^\infty(R)$-approximation. On one hand, the projective cover $f : P \to X$ then factors through ϕ, and hence ϕ is onto. On the other hand, ϕ factors through $\overline{f} : P/(\Omega^1(X))_{\mathrm{fin}} \to X$, i.e., there exists $g \in \mathrm{Hom}_R(A, P/(\Omega^1(X))_{\mathrm{fin}})$ such that $\overline{f}g = \phi$. But $\overline{f}$ having superfluous kernel, this shows us that g is in turn onto, and hence the length of $P/(\Omega^1(X))_{\mathrm{fin}}$ cannot exceed that of A. This proves our first claim.

The supplementary statements about A_i are now immediate consequences of Lemma 1. □

Let us compute the projective dimensions of the minimal right $\mathcal{P}^\infty(R)$- approximations $A_i \to S_i$ of the simple left modules S_i in Example 3. In view of the simplified quiver exhibited there, Theorem 5 immediately yields that $\mathrm{p\,dim}\, A_i = 0$ for $i \in \{1, 2, 3, 6, 13, 14, 17\}$, while $\mathrm{p\,dim}\, A_4 = 1$, $\mathrm{p\,dim}\, A_5 = 2$, $\mathrm{p\,dim}\, A_7 = 1$, $\mathrm{p\,dim}\, A_8 = 2$, $\mathrm{p\,dim}\, A_9 = 3$, $\mathrm{p\,dim}\, A_{10} = 4$, $\mathrm{p\,dim}\, A_{11} = 4$, $\mathrm{p\,dim}\, A_{12} = 5$, $\mathrm{p\,dim}\, A_{15} = 6$, $\mathrm{p\,dim}\, A_{16} = 1$.

Example 6. For any positive integer n, we now present a finite dimensional algebra R such that $\mathcal{P}^\infty(R) \subseteq R$-mod and $\mathcal{P}^\infty(R_R) \subseteq$ mod-R are contravariantly finite, while $\mathrm{l\,fin\,dim}\, R - \mathrm{r\,fin\,dim}\, R = n$. Let k be a field and $R = k\Gamma/I$, where Γ is the quiver

and I the ideal of the path algebra $k\Gamma$ generated by all paths of length 2. Then both contravariant finiteness conditions are satisfied by Theorem 5. Moreover, the simplified left quiver $\widetilde{\Gamma}(R)$ is

$$1 \longrightarrow 2 \longrightarrow \cdots \longrightarrow n \longrightarrow n+1,$$

whence $\operatorname{l\,fin\,dim} R = n$ by Lemma 3, and the simplified right quiver of R is

$$1 \quad 2 \quad \cdots \quad n \quad n+1,$$

whence $\operatorname{r\,fin\,dim} R = 0$. $\square$

This graphical method can be regarded as a supplement to the algorithms presented in [**4**] and [**5**].

3. Projective syzygies

Definition. Call a left R-module P a *projective syzygy* if P is projective and isomorphic to a direct summand of a syzygy $\Omega^k(M)$ for some $k \geq 1$ and R-module M.

Note that, given any non-projective module M of finite projective dimension k, $\Omega^k(M)$ is a projective syzygy. The following characterization of projective syzygies was inspired by Bass [**3**, Theorem 6.3].

Proposition 7. *For a* <u>*right*</u> *artinian ring R and a primitive idempotent $e \in R$ the following conditions are equivalent:*

(1) *Re is a projective syzygy.*

(2) *Re embeds into a finite direct sum $J \bigoplus \cdots \bigoplus J$.*

(3) *The simple module eR/eJ does not occur in the right socle, $\operatorname{Soc}(R_R)$, of R.*

Proof. Without loss of generality, R is basic and $e = e_1$.

'(1)$\Longleftrightarrow$(2)' is obvious.

'(2)$\Longrightarrow$(3)'. By (2), there exist a positive integer t and an element $a = (a_1, \ldots, a_t)$ in $J^{(t)}$ with $e_1 a = a$ and $\operatorname{l.ann}_R(a) = \operatorname{l.ann}_R(e_1) = \bigoplus_{2 \leq j \leq n} Re_j$. Observe that the right ideal $A = \sum_{2 \leq j \leq n} e_j R + \sum_{1 \leq i \leq t} a_i R$ of R is contained in the maximal right ideal $B = \bigoplus_{2 \leq j \leq n} e_j R \bigoplus e_1 J$ because $a_i \in e_1 J$ for $1 \leq i \leq t$. Next note that occurrence of a copy of $e_1 R/e_1 J$ in $\operatorname{Soc}(R_R)$ is tantamount to $\operatorname{l.ann}_R B \neq 0$; so, to prove non-occurrence of $e_1 R/e_1 J$, it suffices to show that $\operatorname{l.ann}_R A = 0$. But for any $r \in \operatorname{l.ann}_R A$ we have $r(a_1, \ldots, a_t) = 0$ and hence $re_1 = 0$, which, in view of $re_2 = \cdots = re_n = 0$, implies $r = 0$.

'(3)$\Longrightarrow$(2)'. If B is as in the preceding paragraph, then (3) says $\operatorname{l.ann}_R B = 0$. This clearly implies that $\operatorname{l.ann}_{Re_1}(e_1 J) = 0$. Since R is right artinian, $e_1 J = \sum_{j=1}^{s} c_j R$ for suitable elements $c_j \in e_1 J$. Then $c = (c_1, \ldots, c_s) \in J^{(s)}$ satisfies

$e_1c = c$ and $\operatorname{l.ann}_{Re_1} c = 0$. But this means that the map $Re_1 \to Rc \subseteq J^{(s)}$ which sends r to rc is an embedding. □

Since, clearly, $\operatorname{l fin dim} R = 0$ is equivalent to the non-existence of nonzero projective left syzygies, the following result of Bass [3] and Auslander-Reiten [1] is an immediate consequence.

Corollary 8. *If R is right artinian, the following statements are equivalent:*
(a) $\operatorname{l fin dim} R = 0$.
(b) *Each simple right R-module occurs in* $\operatorname{Soc}(R_R)$. □

For quick detection of projective syzygies over path algebras modulo relations, the following reformulation of Proposition 7 is helpful.

Observation 9. *Let $R = k\Gamma/I$ be a finite dimensional path algebra modulo relations over a field k and $e_i \in R$ a primitive idempotent which we identify with a vertex of Γ. Then the following are equivalent:*
(1) *Re_i is a projective syzygy.*
(2) *Whenever x is a linear combination in $k\Gamma$ of paths starting in e_i such that $x\alpha \in I$ for all arrows α, then $x \in I$.* □

We use this observation to compute the projective syzygies in a number of examples.

Examples 10. Let Γ be the quiver

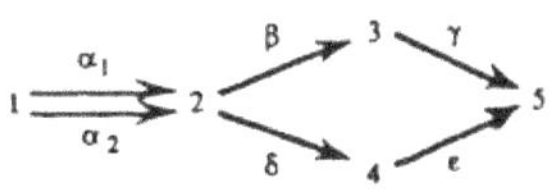

and $R_j = k\Gamma/I_j$ for $j = 1, \ldots, 5$, where

$$
\begin{aligned}
I_1 &= \langle \gamma\beta\alpha_1 - \epsilon\delta\alpha_1 \rangle \\
I_2 &= \langle \gamma\beta\alpha_i - \epsilon\delta\alpha_i \mid i = 1, 2 \rangle \\
I_3 &= \langle \gamma\beta - \epsilon\delta \rangle \\
I_4 &= \langle \gamma\beta\alpha_1,\ \gamma\beta\alpha_2 \rangle \\
I_5 &= \langle \gamma\beta\alpha_1 \rangle .
\end{aligned}
$$

For all of the rings $R = R_j$, $j = 1, \ldots, 5$, the left modules Re_3, Re_4, Re_5 are projective syzygies by Observation 9, while Re_1 is never a projective syzygy. Using that same observation, we see that the module Re_2 is a projective syzygy if R is R_1, R_3, or R_5, but not if R is R_2 or R_4. □

6. References

1. M. Auslander and I. Reiten, Applications of contravariantly finite subcategories, *Advances in Math.* **86** (1991) 111-152.

2. M. Auslander and S. O. Smalø, Preprojective modules over Artin algebras, *J. Algebra* **66** (1980) 61-122.

3. H. Bass, Finitistic dimensions and a homological generalization of semi-primary rings, *Trans. Amer. Math. Soc.* **95** (1960) 466-488.

4. B. Zimmermann Huisgen, Predicting syzygies over monomial relation algebras, *manuscripta math.* **70** (1992) 157-182.

5. ________, Homological domino effects and the first finitistic dimension conjecture, *Invent. Math.* **108** (1992) 369-383.

6. K. Igusa, S. O. Smalø, and G. Todorov, Finite projectivity and contravariant finiteness, *Proc. Amer. Math. Soc.* **109** (1990) 937-941.

OPEN PROBLEMS

G. Azumaya

Problem 1. Characterize those rings over which every projective left module is a direct sum of finitely generated (projective) submodules. (Since every projective module is a direct sum of countably generated submodules by Kaplansky's theorem, we may restrict ourselves to countably generated projective left modules; moreover, every ring over which every finitely generated submodule of a countably generated free left module has a projective cover, and so in particular every semi-perfect ring and every von Neumann regular ring give examples of our rings due to Oberst-Schneider theorem.)

Problem 2. Characterize those rings over which every left module is a direct sum of countably generated submodules. (It is known that such a ring is left Artinian.)

J. Bergen

Problem 1. Let L be a Lie algebra over a field K of characteristic 0. If L has no nonzero finite dimensional ideals, must $U(L)$ be equal to its symmetric Martindale quotient ring?

Problem 2. Let L be a Lie algebra over a field K of characteristic 0 such that L is the join of its finite dimensional ideals. Does every nonzero ideal of $U(L)$ contain a semi-invariant?

Problem 3. Let H be a Hopf algebra. If H is Artinian as an algebra, must H be finite dimensional?

G.F. Birkenmeier, H.E. Heatherly and E.K. Lee

Problem 1. Is the direct product of an arbitrary family of 2-primal rings a 2-primal ring?

Problem 2. Is the ring of formal power series over a 2-primal ring also a 2-primal ring?

Problem 3. If the prime radical of a ring R contains all nilpotent elements of index two, is R a 2-primal ring? This is related to a question of C. Faith [*Strongly regular extensions of rings*, Nagoya Math. J. **20** (1962), 169-183] in which he asks if the center of a ring contains all nilpotent elements of index two, does the center contains all nilpotent elements?

Problem 4. Determine "nice" conditions on a 2-primal ring which will imply that every prime ideal is completely prime.

Problem 5. Is there a "condition" on ideals, weaker than the 2-primal condition, such that an ideal is completely prime if and only if it is prime and satisfies the "condition"?

K.R. Goodearl

Let $A = A_1(\mathrm{C}) = \mathrm{C}[x][y; d/dx]$ be the first Weyl algebra over the complex field C, and let $B = \mathrm{C}(x)[y; d/dx]$. Block [1] classified the simple left A-modules as follows. First note that any simple A-module is either torsion or torsionfree as a $\mathrm{C}[x]$-module. The $\mathrm{C}[x]$-torsion simple A-modules are precisely the modules $A/A(x-\alpha)$ for $\alpha \in \mathrm{C}$. There are inverse bijections between the collections of isomorphism classes of $\mathrm{C}[x]$-torsionfree simple A-modules and simple left B-modules, given by $M \mapsto B \otimes_A M$ and $N \mapsto soc_A(N)$[1, Theorem 4.4]. Since B is a PID, the simple left B modules are precisely the modules B/Bb where b is an irreducible element of B, and simple modules B/Bb and B/Bb' are isomorphic if and only if b and b' are *similar* in a suitable sense.

Problem 1. Find a way to list and/or generate the irreducible elements of B.

Problem 2. Find an algorithm to decide whether a given element of B is irreducible.

Problem 3. Find an algorithm to decide whether two given elements of B are similar.

For any collection S of (isomorphism types of) simple left A-modules, let T_S be the torsion theory generated by the modules in S, and let A_S denote the corresponding localization of A. Localization induces a bijection between the collections of isomorphism classes of those simple left A-modules not in S and the simple left A_S-modules [2, Theorem 3]. Moreover, since A is an HNP ring, it follows from results of Kuzmanovich [4] and Sandomierski [7] that A_S is an HNP ring (see [2, Proposition 2]). It is not hard to see that A_S is an Ore localization of A in case S consists of all but one simple module [3, Theorem 3.2]–in which case A_S has a unique simple left module (up to isomorphism), or in case S contains all the $\mathrm{C}[x]$-torsion simple A-modules [5, Lemma 10]. In general, A_S will be an Ore localization provided $A_{\{M\}}$ is an Ore localization for each simple module $M \in S$. Müller and

Zhang have shown that $A_{(M)}$ is an Ore localization for any simple left A-module M such that $\dim_{C(x)}(B \otimes_Z M) \leq 1$[8; 5, Theorem 7], and have considered the question for other simple modules [5, 6, 8].

Problem 4. If M is an arbitrary simple left A-module, is the ring $A_{(M)}$ an Ore localization of A?

Problem 5. Suppose that M is a simple left $A_n(\mathbf{C})$-module for some $n > 1$. Does there exist an Ore set C in $A_n(\mathbf{C})$ such that $C^{-1}M$ is the unique simple left $C^{-1}A_n(\mathbf{C})$-module (up to isomorphism)?

References

[1] R.E. Block, *The irreducible representations of the Lie algebra sl(2) and of the Weyl algebra*, Advances in Math. **39** (1981), 69-110.

[2] K.R. Goodearl, *Localization and splitting in hereditary noetherian prime rings*, Pacific J. Math. **53** (1974), 137-151.

[3] ―――――, *Linked injectives and Ore localizations*, J. London Math. Soc. **37**(2) (1988), 404-420.

[4] J.J. Kuzmanovich, *Localizations of Dedekind prime rings*, J. Algebra **21** (1972), 378-393.

[5] B.J. Müller and Y.-L. Zhang, *Ore localization in the first Weyl algebra*, in Noncommutative Ring Theory (Athens, Ohio, 1989), Lecture Notes in Math. **1448** (Springer-Verlag, Berlin, 1990), pp. 148-154.

[6] ―――――, *The irreducible representations of the first Weyl algebra*, Comm. in Algebra **20** (1992), 1493-1509.

[7] F.L. Sandomierski, *Nonsingular rings*, Proc. Amer. Math. Soc. **19** (1968), 225-230.

[8] Y.-L. Zhang, Doctoral dissertation, McMaster University, 1990.

B.J. Müller

Problem 1. Let R be a left and right self-injective ring (so $J(R) = Z(_R R) = Z(R_R)$). Show $Z_2(_R R) = Z_2(R_R)$, or provide a counter example.

Problem 2. Let B be a left and right self-injective regular ring. Does there exist a bimodule $_A X_B$ (A any ring) which, on both sides, is injective, balanced (i.e., $A \cong End(X_B)$, $B \cong End(_A X)$ naturally), and singular (i.e., $X = Z(_A X) = Z(X_B)$)? [This question is closely related to Problem 1.]

B. Osofsky

Problem 1. For what rings R is every R-projective module projective? (Note: R-injective modules are always injective). Conjecture: One of the following holds. Either (a) R-projectives are projective $\Rightarrow$ R semiperfect (or R has finite Goldie dim r?) or (b) The answer is dependent on the set theory you are working in.

Problem 2. For what rings R does every nonzero R-module contain a nonzero uniform submodule? This holds for noetherian R, uniserial R, semi-artinian R.

Problem 3. Does there exist a two-sided artinian ring R which has a Morita duality on one side but not on the other? The 'standard' way of getting a two-sided artinian ring R without a Morita duality on one side is $\begin{pmatrix} E & D \\ 0 & D \end{pmatrix}$ $E \subset D$ division rings (cf. Rosenberg and Zelinsky) $[D:E]_R = \infty$, $[D:E]_L < \infty$ (right comp. length $= 3$; left comp. length $= 2 + [D:E]_L$). Then on the right $\begin{pmatrix} 0 & 0 \\ 0 & D \end{pmatrix}$ looks like $D_E \oplus D_D$ which does not have finite length. On the left $\begin{pmatrix} E & 0 \\ 0 & 0 \end{pmatrix}$ looks like $E \oplus_D Hom({}_E D_D, {}_E E)$. For the examples in P.M. Cohn's Skew Ring Constructions, ${}_D Hom_E(D, E)$ is infinite dimensional for essentially the same reason that $[D:E]_R = \infty$ and there does not appear to be an easy way to modify the example to avoid this. Perhaps the examples in Schofield representations of skew fields might work.

B. Zimmermann Huisgen

Problem 1. Characterize the finite dimensional monomial relation algebras A (in terms of quivers and relations) which have the property that the full subcategory P-infinity of A-mod, having as objects all finitely generated modules of finite projective dimension, is contravariantly finite in A-mod. This latter condition means that each finitely generated left A-module has a right approximation by a module of finite projective dimension in the sense of [Auslander and Reiten, *Applications of contravariantly finite subcategories*, Advances in Math. **86** (1991), 111-152]. In the positive cases, determine the structure of the minimal right 'P-infinity approximations' of the simple left A-modules. See [Auslander and Reiten, loc. cit.] to understand how the minimal approximations of the simples provide a strong clue towards an understanding of arbitrary objects in P-infinity.

[Recall that a monomial relation algebra over a field k is an algebra of the form kQ/I, where kQ is the path algebra of a quiver Q and I is an ideal in kQ which is generated by paths of length > 1.]

Problem 2. Does contravariant finiteness of P-infinity in A-mod imply the equality of the left little and big finitistic dimensions of A? [Equality of these dimensions means that each number arising as the projective dimension of some left A-module occurs as the projective dimension of some finitely generated left A-module.]

In general, these finitistic dimensions need not coincide, not even for finite dimensional monomial relation algebras [Z. Huisgen, *Homological domino effects and the first finitistic dimension conjecture*, Invent. Math., to appear]. I see this problem as a first test question in trying to determine how contravariant finiteness of P-infinity in A-mod affects the full module category A-Mod.

Problem 3. The following problem is being independently explored by Kirkman/ Kuzmanovich and Goodearl/Huisgen/Odenthal: Given a DVR D with quotient field K and a classical order Lambda in the algebra of $n \times n$ matrices over K, is the big finitistic dimension of Lambda finite, in other words, is there an upper bound on the natural numbers occurring as projective dimensions of left Lambda-modules. Find bounds in terms of n whenever possible.

This problem is motivated by the (refuted) Tarsy-Conjecture concerning the global dimension of classical orders; see [Fujita, *Tiled orders of finite global dimensions*, Trans. Amer. Math. Soc. **322** (1990), 329-341] and [Kirkman and Kuzmanovich, *Global dimensions of a class of tiled orders*, J. Algebra **127** (1989), 57-72] for more detail.

www.ingramcontent.com/pod-product-compliance
Ingram Content Group UK Ltd.
Pitfield, Milton Keynes, MK11 3LW, UK
UKHW020401150926
8624IPUK00004B/278

9 789810 213855